4·21·98

D0934624

ENVIRONMENTAL, SAFETY, AND HEALTH ENGINEERING

ENVIRONMENTAL, SAFETY, AND HEALTH ENGINEERING

GAYLE WOODSIDE
DIANNA KOCUREK

JOHN WILEY & SONS, INC.

New York • Chichester • Weinheim • Brisbane • Singapore • Toronto

This text is printed on acid-free paper.

Copyright © 1997 by John Wiley & Sons, Inc.

All rights reserved. Published simultaneously in Canada.

Reproduction or translation of any part of this work beyond
that permitted by Section 107 or 108 of the 1976 United
States Copyright Act without the permission of the copyright
owner is unlawful. Requests for permission or further
information should be addressed to the Permissions Department,
John Wiley & Sons, Inc., 605 Third Avenue, New York, NY
10158-0012.

This publication is designed to provide accurate and
authoritative information in regard to the subject
matter covered. It is sold with the understanding that
the publisher is not engaged in rendering legal, accounting,
or other professional services. If legal advice or other
expert assistance is required, the services of a competent
professional person should be sought.

Library of Congress Cataloging in Publication Data:
Woodside, Gayle, 1951–
    Environmental, safety, and health engineering / Gayle Woodside,
Dianna Kocurek.
        p.   cm.
    Includes bibliographical references.
    ISBN 0-471-10932-0 (cloth : alk. paper)
    1. Environmental engineering.   2. Industrial safety.
    3. Industrial hygiene.   I. Kocurek, Dianna S.   II. Title.
TD145.W67   1997
628—dc20   96-44717

Printed in the United States of America

10 9 8 7 6 5 4 3 2 1

This book is dedicated to our parents,
John and Donna Woodside, and Adolph and Ethel Stalinsky

CONTENTS

**PART IV   PRINCIPLES OF INDUSTRIAL HYGIENE AND
OCCUPATIONAL HEALTH ENGINEERING**

PREFACE

Over the last decade, the environmental, safety, and health (ESH) fields have dissolved many of their traditional boundaries and have moved toward one, integrated professional field. Thus, it is appropriate for ESH engineers and other professionals to concern themselves with all three disciplines concurrently. This book, *Environmental, Safety, and Health Engineering*, embodies the new approach that professionals are taking in respect to the ESH disciplines—that is, it takes down the "walls" and merges the three disciplines. We feel that although this book is a "first" of its kind, it is certainly one that is long overdue. The industrial professional has needed to be conversant in all aspects of the ESH field for many years now. We hope that this book gives the professional a basis for putting together an integrated ESH program that is efficient as well as comprehensive.

Traditional topics associated with the ESH field are covered. In addition, new topics of current interest in the field have been added where pertinent. Part I presents an overview of both EPA and OSHA regulations; Part II delves into environmental topics such as hazardous waste treatment, air emissions abatement and management, wastewater discharges and management, and waste storage and containment technologies. Less traditional, but important, environmental topics such as environmental statistics and calculations for local publicly owned treatment works (wastewater) limits are included. Part III covers safety topics, such as equipment safety, fire and life safety, confined space, and safety management, including a discussion of OSHA's voluntary protection programs. Also found in this section are chapters on process safety and reliability and construction safety. Part IV addresses industrial hygiene and occupational health topics such as chemical exposure assessment, personal protective equipment, industrial ventilation, radiation safety, and noise assessment and management. A chapter related to ergonomics is also included. Finally, the book includes several appendices; one is a useful safety checklist that can be customized for use in an industrial safety program.

Environmental, Safety, and Health Engineering is created for the use of professionals who are working in the ESH field, as well as process managers and engineers, facilities managers and engineers, and students and professors who are interested in ESH topics. The book provides numerous solved problems that apply to practical situations. Figure, tables, and checklists are included for ease of use. We tried to incorporate information that we often use ourselves, to enhance its technical appeal. We hope that you, as reader, professional, or student, will find the ESH topics included in this book interesting and pertinent.

ACKNOWLEDGMENTS

We would like to thank the following people for their help with this book: John Prusak, John Woodside, Bruce Almy, Lial Tischler, Kathy Bonner, and IBM-Austin's Environmental Engineering Department—Bonnie Blam, Chris Bauer, Barbara Casey, David Dalke, Jeff Erb, Stuart Hurwitz, Rich Reich, Ken Takvam, Bob Tassan, Chau Vo, and Kelvin Langlois, our photographer. Also, thanks to the editorial staff at John Wiley & Sons—Dan Sayre, Lis Cobas, and Millie Torres.

ENVIRONMENTAL, SAFETY, AND HEALTH ENGINEERING

PART I

REGULATORY OVERVIEW

1

ENVIRONMENAL PROTECTION AGENCY

The Environmental Protection Agency (EPA) was established December 2, 1970, by executive order of President Nixon, as an independent agency in the executive branch of U.S. government. The agency regulates numerous and varied aspects of environmental protection including air quality, water quality, solid waste, the Superfund, toxic substances, pesticides, radon, and pollution prevention.

Congress has passed environmental legislation since the creation of the EPA in the form of several key Acts, which typically have empowered EPA to create regulations to accomplish the goals set forth in the Acts. EPA also has the responsibility for permitting and enforcement. When possible, EPA delegates permitting and enforcement responsibilities to a state, if the state has demonstrated that it has effective plans for ensuring proper environmental protection. A state also has the power to create its own regulations, as long as these regulations are at least as stringent as EPA's.

KEY ENVIRONMENTAL REGULATIONS

National Environmental Policy Act of 1969

The National Environmental Policy Act (NEPA) was the first Act that made environmental protection a national policy. The Act established the Council on Environmental Quality and set lofty national goals. Included in the Act is a charge to the federal government to:

- Fulfill the responsibilities of each generation as trustee of the environment for succeeding generations

3

- Assure for all Americans safe, healthful, productive, and aesthetically and culturally pleasing surroundings
- Attain the widest range of beneficial uses of the environment without degradation, risk to health or safety, or other undesirable and unintended consequences

These goals have helped shape environmental legislation as we know it today. In addition to setting policy, NEPA mandated that federal agencies submit environmental impact statements for all proposed actions. The environmental impact statement has now become common for almost all major projects that have the potential to affect the environment, including the siting of power plants, industrial facilities, municipal water and wastewater treatment facilities, and other such facilities.

Clean Air Act Amendments of 1970, 1977, and 1990

The Clean Air Act (CAA) was originally passed in 1963; however, most of the elements of the Act as we know it today have been provided by the CAA Amendments of 1970, 1977, and 1990. The regulations are defined in 40 CFR Parts 1–99.

In 1970, the CAA Amendments required EPA to set National Ambient Air Quality Standards (NAAQS) for six air pollutants—particulates, sulfur oxides, carbon monoxide, nitrogen dioxide, ozone, and lead—that posed a significant threat to human health. The CAA Amendments of 1977 required installation of Best Available Control Technology (BACT) on new processes or sources in order to further progress toward meeting NAAQS. In addition, sources in areas that had not attained the NAAQS standards (termed nonattainment areas) required additional controls, and proportional decreases (offsets) for emissions were required for each emission increase. National standards were also set for specific pollutant sources, such as dry cleaners and plastics manufacturers, and were termed New Source Performance Standards (NSPS). All new sources constructed or modified after the promulgation of NSPS had to meet the new, more stringent standards.

Because almost all major cities and many medium-sized cities were unable to attain NAAQS during the 1980s, the CAA Amendments of 1990 altered significantly the national air pollution control strategy. A summary of the key topics covered by in the Act is presented in Table 1.1.

All new sources are still required to have at least BACT under the 1990 CAA amendments; however, new sources in nonattainment areas have more stringent requirements. These sources must meet the Lowest Achievable Emission Rate (LAER) for pollutants that exceed NAAQS. Existing sources in nonattainment areas must retrofit emission sources with Reasonably Available Control Technology (RACT). In addition, decreases in emissions at ratios greater than one to one (1:1) must be demonstrated before a new source is permitted in a nonattainment area.

TABLE 1.1 Key Topics Covered by the CAA Amendments of 1990

*Title I—Provisions for Attainment and Maintenance of
National Ambient Air Quality Standards*

- General planning requirements and general provisions for nonattainment areas
- Additional provisions for ozone nonattainment areas
- Additional provisions for carbon monoxide nonattainment areas
- Additional provisions for particulate matter (PM_{10}) nonattainment areas
- Additional provisions for areas designated nonattainment for sulfur oxides, nitrogen dioxide, or lead
- Miscellaneous and other provisions

Title II—Provisions Relating to Mobile Source

- Control of vehicle refueling emissions
- Emission standards for conventional motor vehicles
- Mobile source related air toxics
- Motor vehicle testing and certification and emission control diagnostic systems
- Enforcement
- Reformulated and oxygenated gasoline
- Other fuel provisions
- Urban buses
- Clean-fuel vehicles
- Numerous other mobile source provisions

Title III—Hazardous Air Pollutants

- Hazardous air pollutants, including listed hazardous air pollutants, list of source categories, emission standards and schedule, modifications, state programs, prevention of accidental releases, and numerous other provisions
- Risk assessment and management commission
- Chemical process safety management
- Solid waste combustion
- Ash management and disposal

Title IV—Acid Deposition Control

- Acid deposition control, fossil fuel use, acid deposition standards, acid deposition research, and other provisions
- Sulfur dioxide allowance program, Phase I and Phase II requirements
- Repowered sources
- Excess emission penalty
- Nitrogen oxides emission reduction program
- Clean coal technology regulatory incentives
- Monitoring, reporting, and recordkeeping requirements
- General compliance, enforcement, and other provisions

(Continued)

TABLE 1.1 *Continued*

Title V—Permits

- Permit programs
- Permit applications
- Permit requirements and conditions
- Small business stationary source technical and environmental compliance assistance program
- Other provisions

Title VI—Stratospheric Ozone Protection

- Listed substances and phaseout of Class I and Class II substances
- Monitoring, reporting, and labeling
- National recycling and emission reduction program
- Safe alternatives
- Miscellaneous provisions

Other Titles

- Title VII—Provisions relating to enforcement
- Title VIII—Miscellaneous provisions
- Title IX—Clean air research
- Title X—Disadvantaged business concerns
- Title XI—Clean air employment transition assistance

The regulation of hazardous air pollutants was also changed by the CAA Amendments of 1990. Rather than regulating these pollutants on a specific basis, EPA has established source categories, such as refineries and ethylene manufacturers, and hazardous air pollutants are regulated within the category. Maximum Achievable Control Technology (MACT) is required for these sources, and these sources must be retrofit to meet MACT within three years.

Hazardous air pollutants from other industries can still be regulated if such an industry is considered a major source of any of the listed hazardous air pollutants. A major source is one that emits 10 tons per year or more of a listed hazardous chemical or that emits an aggregate of 25 tons per year of all hazardous air pollutants.

In addition, mobile sources are more stringently regulated under the 1990 amendments, especially in areas that have not reached attainment for ozone. Fuel alternatives, enhanced vehicle inspection and maintenance, ride sharing and other transportation control measures, and other requirements are currently being implemented in nonattainment areas, with more stringent requirements being imposed on cities and areas with the worst air pollution.

CAA Amendments of 1990 require each state to establish an operating permit program that meets EPA requirements. Moreover, each state must prepare a plan sufficient to show that the state will attain ambient standards by certain deadlines and will maintain compliance thereafter. A typical State Implementation Plan (SIP) prescribes emission limits for sources, stringent enough to meet attainment on a timely basis.

Clean Water Act, Formerly the Federal Water Pollution Control Act of 1972, Amended 1977, 1987 (Reauthorization Pending)

The basic statutory structure for water quality management was set in place in 1972 with the Federal Water Pollution Control Act. This Act set up the framework for the establishment of minimum acceptable requirements for water quality and wastewater management. This Act and the Safe Drinking Water Act of 1974 provide regulation for essentially all aspects of water quality management.

In 1977, the Federal Water Pollution Control Act was renamed the Clean Water Act (CWA) and was amended to provide increased controls on toxic pollutants. In 1987 the Water Quality Act set forth further enhancements to the control of toxic pollutants, and now nearly 200 pollutants are regulated, including heavy metals, toxics, and other chemicals deemed harmful to human health and aquatic life. Point source discharges regulated by CWA include wastewater and storm water discharges from industrial and commercial facilities, municipalities, publicly owned treatment works (POTWs), private treatment plants, and other sources. Regulations governing these discharges are found in several places in the *Code of Federal Regulations*. EPA has defined general pretreatment standards in 40 CFR Part 403; effluent guidelines are found in 40 CFR Parts 401–471; and regulations pertaining to permitting are located in 40 CFR Parts 122–125. Water quality standards have been outlined in 40 CFR Part 131.

General pretreatment standards regulate nondomestic discharges to POTWs. These standards, including the development of local limits, are covered in detail in Chapter 6.

The effluent guidelines for wastewater discharges cover a limited, but diverse, number of specific industrial and commercial categories such as hospitals, food processors, metal finishers, semiconductor manufacturers, and chemical manufacturers. Both facilities that discharge directly to water bodies and those that discharge to POTWs are covered by effluent guidelines.

The National Pollution Discharge Elimination System (NPDES) process administers the permitting of wastewater and storm water discharges. Wastewater discharges that require permitting include those covered by effluent guidelines, and wastewater discharges from POTWs, wastewater discharges from industrial and other privately owned treatment facilities. Storm water discharges that require permitting include storm water discharges from urban municipalities, construction sites, and certain industries.

Water quality standards define the quality goals of a water body. They designate the use or uses for the body of water, such as for the protection and propagation of fish or for recreational purposes. Water quality standards also set criteria necessary to protect these uses. Once established, water quality standards become the basis for the establishment of water-quality based treatment controls and for long-term strategies for water quality protection, such as zero discharge and other strategies.

Safe Drinking Water Act of 1974, Amended in 1986 (Reauthorization Pending)

The Safe Drinking Water Act (SDWA) granted EPA authority to protect public drinking water by establishing standards and regulations for maximum contaminant levels for those substances determined by the agency to be harmful to human health. Primary and secondary drinking water standards are delineated under 40 CFR 141 Subpart B and 40 CFR 143.3, respectively. Under the Act, states are to enforce primary drinking water standards; however, EPA can assume responsibility if a state fails to enforce the standards.

Another part of the Act, added in the 1986 amendments, governs underground injection of waste materials. Requirements for geologic assessments and other measures are outlined to assure that underground drinking water sources are not contaminated by the use of underground injection wells. States must have programs designed to control underground injections of waste materials, and some states have opted to ban underground injection altogether. Protection of sole source aquifers, or critical aquifer protection areas, are also addressed in the Act.

The 1986 SDWA amendments also required EPA to set national standards limiting contaminants in water supplies and to set best available technology for meeting these standards. EPA was required to set standards limiting more than 80 contaminants in drinking water. Technical assistance programs for small water systems were established to help meet the new requirements. In addition, drinking water coolers with lead-lined tanks were addressed in the 1986 amendments. These tanks were required to be replaced by lead-free tanks by 1988.

Toxic Substance Control Act of 1976, Amended in 1988

The Toxic Substance Control Act (TSCA) was enacted to ensure that chemicals manufactured in or imported into the United States do not present an unreasonable risk to health or the environment. Thus, all chemicals manufactured or imported must be registered and listed in the TSCA registry. Regulations pertaining to this Act are outlined in 40 CFR Parts 700–799.

In order to receive authorization to manufacture or import a new chemical for commercial purposes, the manufacturer or importer must file a premanufacture notice (PMN) with EPA. Research and development facilities, however, are generally exempt from filing a PMN if certain requirements are met. As a part of the registration process, environmental fate and health effects studies must be developed and submitted. Other aspects of the regulation require reporting of production/importation quantities and record keeping and reporting of allegations of health/environmental effects and associated investigations.

Facilities that use chemicals in their processes must ensure that appropriate chemical information accompanies the chemical and that the chemical is registered under TSCA before it is received. These facilities are also subject to record keeping and reporting requirements related to health/environmental allegations.

Polychlorinated biphenyls (PCBs) and asbestos are also regulated under TSCA. Requirements for PCBs are outlined in 40 CFR Part 761. Asbestos requirements are found in 40 CFR Part 763. In addition to EPA requirements, asbestos is regulated by OSHA under three separate standards—general industry standards, construction standards, and standards pertaining to shipbuilding and repair.

The TSCA amendments of 1988 authorized federal aid to help states develop programs for mitigating the effects of radon gas in homes, schools, and other buildings. As provided by the amendments, EPA set up an information clearinghouse of radon information and provides technical assistance to states for the implementation of radon surveys and mitigation techniques.

Resource Conservation and Recovery Act of 1976 and Amendments (Reauthorization Pending)

EPA promulgated regulations for managing solid waste under the Resource Conservation and Recovery Act (RCRA), and these rules are outlined in 40 CFR Parts 190–259 for nonhazardous solid waste and in 40 CFR Parts 260–270 for hazardous waste. Technical standards and corrective action requirements for owners and operators of underground storage tanks are outlined in 40 CFR Part 280.

Wastes considered solid wastes by EPA include waste solids, sludges, liquids, and containerized gases. Wastes considered to be hazardous by EPA can be those with a hazard characteristic, such as ignitability or corrosivity, or one of several hundred waste chemicals listed by the EPA that can become hazardous if they are discarded or spilled. In addition, some wastes from specific or nonspecific sources are defined as hazardous, such as wastewater treatment sludges generated from a specific process or certain spent halogenated and nonhalogenated solvents. All hazardous wastes must be managed from generation through final disposal. The regulations set forth standards for hazardous waste generators, transporters, and owners and operators of treatment, storage, and disposal facilities. Included in the standards are facility permitting requirements, transporter identification and tracking requirements, record keeping and inspection requirements, requirements for financial bonding, and other requirements. Regulatory citations for the major elements covered by RCRA for hazardous waste include the following:

- Part 260—Hazardous waste management system: General
- Part 261—Identification and listing of hazardous waste
- Part 262—Standards applicable to generators of hazardous waste
- Part 263—Standards applicable to transporters of hazardous waste
- Part 264—Standards for owners and operators of hazardous waste treatment, storage, and disposal facilities
- Part 265—Interim status standards for owners and operators of hazardous waste treatment, storage, and disposal facilities
- Part 266—Standards for the management of specific hazardous wastes and specific types of hazardous waste management facilities
- Part 268—Land disposal restrictions
- Part 270—EPA-administered permit programs: The Hazardous Waste Permit Program
- Part 271—Requirements for authorization of state hazardous waste programs
- Part 272—Approved state hazardous waste management programs
- Part 279—Standards for the management of used oil
- Part 280—Technical standards and corrective action requirements for owners and operators of underground storage tanks (UST)
- Part 281—Approval of state underground storage tank programs

Hazardous and Solid Waste Amendments (HSWA) of 1984. HSWA was enacted by Congress to strengthen RCRA. Regulations pursuant to these amendments provide technical standards for landfill disposal, leak detection systems, and underground storage of petroleum products and hazardous substances. The regulations also prohibit specified hazardous wastes from land disposal unless certain treatment standards are met. HSWA also regulates continuing releases from solid waste management units.

Medical Waste Tracking Act of 1988. This Act required EPA to set up a demonstration program for characterizing and tracking medical waste and for evaluating treatment techniques for this waste. The requirements of the demonstration program and other medical waste management requirements are delineated in 40 CFR Part 259.

Pollution Prevention Act of 1990. This Act focuses on prevention of waste—through recycling, source reduction, elimination of toxic materials, and other methods—rather than management of hazardous waste. The Act required

EPA to develop programs to minimize pollution and to develop a list of priority chemicals that will be the targets of minimization (and in some cases elimination) programs.

Comprehensive Environmental Response, Compensation, and Liability Act of 1980 and Amendments

The Comprehensive Environmental Response, Compensation, and Liability Act—termed CERCLA or Superfund, interchangeably—and several other Acts passed under the CERCLA/Superfund umbrella provided for liability, compensation, cleanup, and emergency response for hazardous substances released into the environment. Under the Act, a National Priority List of hazardous waste sites to be remediated was established and government monies were allocated for cleanup. This list of Superfund sites, located in 40 CFR Part 300, includes commercial and federal sites. Inasmuch as the cleanup costs of these sites far exceeded the allocations, parties that were responsible (or potentially responsible) for the damage, as well as landowners, were assessed cleanup costs.

In addition to cleanup of hazardous waste sites, the Act also addresses spill reporting. The National Response Center (NRC) must be notified immediately if a material is released to the environment, without a permit, in amounts greater than the reporting quantity. The report to the NRC should include information such as name and telephone number of reporter, name and address of facility, chemical/waste released, quantity released, possible hazards to human health or the environment, and the extent of injuries. Once reported, incidents become part of the public record.

Emergency Planning and Community Right-to-Know Act of 1986. The Emergency Planning and Community Right-to-Know Act (EPCRA) is often referred to as SARA Title III, because the provisions of the Act are incorporated in Title III of the Superfund Amendments and Reauthorization Act (SARA) of 1986. Regulations pertaining to EPCRA are defined in 40 CFR Parts 355, 370, and 372. Basic elements of the regulations include the following:

- Part 355—This part requires the establishment of local emergency planning and release notification. Each state was required to set up a Local Emergency Planning Committee (LEPC), which was to develop an emergency response plan by October 1988. Off-site release notification is also defined, and any off-site release of a listed chemical above a specified quantity is required to be reported to the LEPC and the State Emergency Response Commission (SERC).
- Part 370—This part requires facilities to provide information to local communities about chemicals stored at the facilities. Such information is gathered through a report called the Tier I/Tier II report.

- Part 372—This part requires facilities to report releases of chemicals to the environment. A toxic release inventory (TRI) report is required for facilities that manufacture or process certain chemicals in quantities of more than 25,000 pounds or that use these chemicals in quantities of more than 10,000 pounds. TRI data includes estimates or actual data pertaining to on-site waste treatment and treatment efficiencies, releases to wastewater and air, on-site and off-site disposal of certain chemicals, on-site and off-site waste recycling, energy recovery from disposal of materials, and source reduction activities used at the facility. The report has been used by EPA to set pollution prevention goals, and it has been employed by the public to put pressure on industry to reduce chemical waste.

Radon Gas and Indoor Air Quality Research Act of 1986. Incorporated into the SARA amendments is the provision under Title IV for setting up a research program for radon gas and indoor air quality. The program is intended to identify, characterize, and monitor the sources and levels of indoor air pollution, particularly radon. Control technologies and other mitigation measures to prevent or abate radon gas and other indoor air pollution are to be researched and developed.

Federal Insecticide, Fungicide, and Rodenticide Act of 1972, Amended in 1988

The Federal Insecticide, Fungicide, and Rodenticide Act (FIFRA) was originally passed in 1947, but amendments that strengthened the Act and that provide most of the elements in today's regulations were passed in 1972 and 1988. Regulations are outlined in 40 CFR Parts 150–189. In essence, regulations pertaining to this Act require pesticides distributed or sold in the United States, with certain exceptions, to be registered with the EPA. There are approximately 400 pesticides that are regulated under FIFRA. Steps necessary for registration include submission of a statement to EPA including such information such as the name of the applicant, the name of the pesticide, the label developed for the pesticide, including precautions and directions for its use, the formula of the pesticide, and the use for which it is to be classified.

All new pesticides must undergo extensive testing before they are approved and allowed on the market. In addition, pesticides already being distributed when the FIFRA amendments of 1988 were enacted must be reregistered—with testing data included in the registration package—according to a specified timetable. EPA has set maximum safe levels of pesticide residues in human and animal food. Further, EPA can suspend the registration of a pesticide if that material is an imminent hazard to human health, although the suspension typically includes extensive testing. Other aspects of pesticide management governed by FIFRA are the registration of pesticide producers, a national plan for monitoring pesticides, container rinsate and disposal management (cross referenced to RCRA), and others.

Noise Control Act of 1972, Amended as the Quiet Communities Act of 1978

The Noise Control Act requires EPA to set noise emission standards for commercial products. The Act also addresses the need to abate aircraft noise and sonic boom and requires railroad noise and motor carrier noise emission standards. These standards are in conjunction and coordination with other agencies such as the Department of Transportation. In addition, the Act, as amended in the Quiet Communities Act of 1978, directed EPA to disseminate information and educational materials on the public health and other effects of noise and on the most effective means for noise control, as well as to conduct or finance research on the effects, measurement, and control of noise.

Marine Protection, Research and Sanctuaries Act of 1972

The Marine Protection, Research and Sanctuaries Act, also termed the Ocean Dumping Act, outlawed dumping of waste into oceans without an EPA permit. The Act also requires that EPA designate sites to be used by permit holders. The secretary of the U.S. Army Corps of Engineers is given authority to allow dumping of dredged material into the ocean, but only after public notice and public review and only if the act of dumping will not unreasonably degrade or endanger human health, welfare, or amenities, or the marine environment, ecological systems, or economic potentialities. The EPA administrator must be notified of the U.S. Army Corps of Engineers' intent to dump and can disapprove the action.

EPA LIST OF LISTS

EPA has provided numerous lists of regulated hazardous chemicals and wastes, as well as other important elements that are regulated. Citations of key lists are presented in Table 1.2.

REFERENCES

Code of Federal Regulations, 40 CFR Parts 1–99, U.S. Environmental Protection Agency, Washington, DC.

Code of Federal Regulations, 40 CFR Parts 122–125, 131, and 401–471, U.S. Environmental Protection Agency, Washington, DC.

Code of Federal Regulations, 40 CFR Parts 140–150, U.S. Environmental Protection Agency, Washington, DC.

Code of Federal Regulations, 40 CFR Parts 150–189, U.S. Environmental Protection Agency, Washington, DC.

Code of Federal Regulations, 40 CFR Parts 190–280, U.S. Environmental Protection Agency, Washington, DC.

Code of Federal Regulations, 40 CFR Parts 300, 355, 370, and 372, U.S. Environmental Protection Agency, Washington, DC.

7 U.S.C. §§136–136y, Federal Insecticide, Fungicide, and Rodenticide Act.

15 U.S.C. §§2601–2671, Toxic Substances Control Act.

33 U.S.C. §1401 et seq., Marine Protection, Research, and Sanctuaries Act, as amended.

33 U.S.C. §§1251–1387, Federal Water Pollution Control Act (Clean Water Act).

42 U.S.C. §§300f–300j-26, Safe Drinking Water Act.

42 U.S.C. §4901 et seq., Noise Control Act.

42 U.S.C. §§6901–6992K, Resource Conservation and Recovery Act.

42 U.S.C. §§7401–7626, Clean Air Act, as amended.

42 U.S.C. §§9601–9675, Comprehensive Environmental Response, Compensation, and Liability Act, as amended.

42 U.S.C. §§11001–11050, Emergency Planning and Community Right-to-Know Act (Title III of the Superfund Amendments and Reauthorization Act).

TABLE 1.2   Citations of Key EPA Lists Concerning Hazardous Substances and Other Pollutants

Clean Air Act

- 40 CFR 50.4—National primary ambient air quality standards for sulfur oxides (sulfur dioxide)
- 40 CFR 50.5—National secondary ambient air quality standards for sulfur oxides (sulfur dioxide)
- 40 CFR 50.6—National primary and secondary ambient air quality standards for particulate matter
- 40 CFR 50.8—National primary ambient air quality standards for carbon monoxide
- 40 CFR 50.9—National primary and secondary ambient air quality standards for ozone
- 40 CFR 50.11—National primary and secondary ambient air quality standards for nitrogen oxide
- 40 CFR 50.12—National primary and secondary ambient air quality standards for lead
- 40 CFR 60.01—List of hazardous air pollutants
- 40 CFR 60.617, 60.667, and 60.707—Regulated chemical emissions from the synthetic organic chemical manufacturing industry (SOCMI)
- 40 CFR 61, Subparts B–FF—List of NESHAP chemicals
- 40 CFR 63.74—List of high-risk pollutants for early reduction program
- 40 CFR 68.130, Table 1—Regulated toxic substances and threshold quantities for prevention of accidental releases
- 40 CFR 68.130, Table 3—Regulated flammable substances and threshold quantities for prevention of accidental releases
- 40 CFR 82 Subpart A, Appendices A and B—Class I and Class II controlled substances

TABLE 1.2 *Continued*

- 42 USC 7412; Clean Air Act, Title I, Part A, §112 (as amended, 1990)—List of hazardous air pollutants

Clean Water Act

- 40 CFR 116.4, Table 116.4A—List of hazardous substances by name
- 40 CFR 116.4, Table 116.4B—List of hazardous substances by CAS number
- 40 CFR 117.3—Reportable quantities for hazardous substances
- 40 CFR 122, Appendix D, Table II—Organic pollutants of concern in NPDES discharges
- 40 CFR 122, Appendix D, Table III—Listed metals, cyanides, and phenols of concern in NPDES discharges
- 40 CFR 129—Toxic pollutant effluent standards
- 40 CFR 401.15—List of toxic pollutants
- 40 CFR 401.16—List of conventional pollutants
- 40 CFR Part 423, Appendix A—Priority pollutants
- 57 FR 41331, September 9, 1992—Section 313 water priority chemicals

Safe Drinking Water Act

- 40 CFR 141.11–141.12—Maximum primary contaminant levels for inorganic and organic chemicals
- 40 CFR 141.13—Maximum contaminant levels for turbidity
- 40 CFR 141.15–141.16—Maximum contaminant levels for radioactivity
- 40 CFR 141.50–141.51—Maximum contaminant level goals for organic and inorganic contaminants
- 40 CFR 141.61–141.62—Revised (and expanded) maximum primary contaminant levels for organic and inorganic contaminants
- 40 CFR 143.3—Maximum secondary contaminant levels
- 40 CFR 148.10–148.148.17—Lists of waste prohibited from underground injection
- 47 FR 9352—Contaminants required to be regulated under the SDWA Amendments of 1986
- 55 FR 1470, January 14, 1991—Priority list of drinking water contaminants

Resource Conservation and Recovery Act

- 40 CFR 261.24—List of maximum concentrations of contaminants for the toxicity characteristic leaching procedure (TCLP) limits
- 40 CFR 261.31—List of hazardous wastes from nonspecific sources
- 40 CFR 261.32—List of hazardous wastes from specific sources
- 40 CFR 261.33(e)—List of acutely hazardous wastes
- 40 CFR 261.33(f)—List of toxic wastes

(Continued)

TABLE 1.2 *Continued*

- 40 CFR Part 261, Appendix VII—Basis for listing wastes from specific and nonspecific sources
- 40 CFR Part 261, Appendix VIII—List of hazardous constituents
- 40 CFR Part 264, Appendix IX—Practical quantitation limits for groundwater constituents
- 40 CFR 268.30–268.32—Prohibitions for solvent wastes and dioxin-containing wastes
- 40 CFR 268.32—California list of wastes with specific prohibitions
- 40 CFR 268.33–268.35—Lists of land restricted wastes
- 40 CFR 268.40—Treatment standards for land restricted wastes
- 40 CFR 268.45—Treatment standards for hazardous debris
- 40 CFR 268.48—Universal treatment standards
- 40 CFR Part 268, Appendix III—List of halogenated organic compounds regulated under 268.32

Comprehensive Environmental Response, Compensation, and Liability Act

- 40 CFR 302.4, Table 302.4—List of hazardous substances and reportable quantities
- 40 CFR 302.4, Appendix A—Sequential CAS registry number list of CERCLA hazardous substances
- 40 CFR 302.4, Appendix B—List of radionuclides

Emergency Planning and Community Response Act

- 40 CFR 355, Appendix A—List of extremely hazardous substances for emergency planning and notification
- 40 CFR 372.65—List of toxic chemicals for toxic release inventory reporting

Toxic Substance Control Act

- 40 CFR Part 716, Subpart B—List of chemical substances
- 40 CFR 700 Index (Finders' Aid)—Toxic substances CAS number/chemical

Federal Insecticide, Fungicide, and Rodenticide Act

- 40 CFR Part 180—List of pesticides

BIBLIOGRAPHY

Bregman, Jacob I. and Kenneth M. Mackenthun (1992), *Environmental Impact Statements*, CRC Press, Boca Raton, FL.

Brownell, F. William et al. (1993), *The Clean Air Handbook*, Government Institutes, Rockville, MD.

Egan, James T. (1991), *Regulatory Management*, CRC Press, Boca Raton, FL.

Fire, Frank L., Nancy K. Grant, and David H. Hoover (1990), *SARA Title III*, Van Nostrand Reinhold, New York.

Government Institutes (1994), *Book of Lists for Regulated Hazardous Substances*, Government Institutes, Rockville, MD.

_____ (1994), *Environmental Statutes*, Government Institutes, Rockville, MD.

Mackenthun, Kenneth M., and Jacob I. Bregman (1991), *Environmental Regulations*, CRC Press, Boca Raton, FL.

McGregor, Gregor I. (1994), *Environmental Law and Enforcement*, CRC Press, Boca Raton, FL.

Stensvaag, John-Mark (1991), *Clean Air Act Amendments: Law and Practice*, John Wiley & Sons, New York.

_____ (1989), *Hazardous Waste Law and Practice*, John Wiley & Sons, New York.

2

OCCUPATIONAL SAFETY AND HEALTH ADMINISTRATION

The Occupational Safety and Health Administration (OSHA) was established by the Occupational Safety and Health (OSH) Act of 1970 as an agency within the Department of Labor. The main purpose and duty of OSHA are to develop and enforce mandatory job safety and health standards. OSHA also provides research in occupational safety and health to develop innovative ways of handling occupational safety and health problems, training programs to increase the number and competence of occupational safety and health personnel, and a reporting and record keeping system to monitor job-related injuries and illnesses. This chapter discusses the key elements of the OSH Act and details information about OSHA such as consultation services for the employer, OSHA inspections, and employee workplace rights.

OCCUPATIONAL SAFETY AND HEALTH ACT OF 1970

The OSH Act of 1970 was enacted to assure safe and healthful working conditions for working persons in the United States and its territories. Numerous workplace safety and health standards have been promulgated by OSHA since 1970. OSHA regulations are enumerated under 29 CFR 1900–1999 and include occupational safety and health standards for general industry, the ⋯ ⋯ and repair industry, and the construction industry.

Health Standards

⋯ tandards for general industry are detailed under 29 CFR ⋯d by these standards are presented in Table 2.1. As can be seen in the table, the standards provide a comprehensive set of regulations for virtually all aspects of workplace safety and health.

General Duty Clause

In addition to complying with safety and health standards promulgated under the OSH Act, the employer also has a general duty to furnish each employee with employment and places of employment that are free from recognized hazards causing or likely to cause death or serious physical harm. This duty is commonly referred to as the "general duty clause," and it is invoked when there are no specific standards applicable to a particular recognized hazard.

Recent Enhancements to OSHA Standards

There have been several rules promulgated or updated over the last decade that have significantly enhanced OSHA's standards for general industry. The following paragraphs detail these new or updated rules.

Hazard Communication Standard (Effective 1986). The Hazard Communication Standard affects facilities that manage hazardous materials. Regulations pertaining to this standard are delineated under 29 CFR 1910.1200. Included are requirements for training employees about the hazards associated with chemicals used in the workplace. The training program and other elements of hazard communication must be established in writing and kept at the facility. In addition, material safety data sheets (MSDSs) must be available at the workplace and accessible to employees at all times, and chemical containers—including stationary containers—must be labeled as to the hazards associated with the chemicals they contain.

Hazardous Waste Operations and Emergency Response (Effective 1990). This rule was issued under the Superfund Amendments and Reauthorization Act of 1986 and is administered by OSHA. The rule is outlined under 29 CFR 1910.120, and it details requirements for hazardous waste operations at Superfund and other cleanup sites. Other requirements pertain to workers involved in hazardous material incident responses. Included are requirements for worker training, a safety and health program, an emergency response plan, personal protective equipment, medical surveillance, and other requirements.

Control of Hazardous Energy—Lockout/Tagout (Effective 1990). This standard helps safeguard employees from unexpected start-up of machinery or equipment or the release of hazardous energy while they are performing servicing or maintenance. The rule requires that, in general, before service or maintenance is performed on machines or equipment, the machines or equipment must be turned off and disconnected from the energy source, and the energy-isolating device must be either locked out or tagged out. Written procedures, training, and record keeping are also required.

TABLE 2.1 Topics Addressed in 29 CFR 1910 Subparts A–Z

Subpart	Topics Included
Subpart A	General—definitions; petitions for the issuance, amendment, or repeal of a standard; applicability of standards, incorporation by reference; and definition and requirements for a nationally recognized testing laboratory
Subpart B	Adoption and extension of established federal standards
Subpart C	General safety and health provisions—access to employee exposure and medical records
Subpart D	Working-walking surfaces—stairs, ladders, and guarding of floor and wall openings
Subpart E	Means of egress
Subpart F	Powered platforms, manlifts, and vehicle-mounted work platforms
Subpart G	Occupational health and environmental control—ventilation, occupational noise exposure, and ionizing and nonionizing radiation
Subpart H	Hazardous materials—compressed gases, flammable and combustible materials, explosives and blasting agents, storage and handling of liquified petroleum gases, storage and handling of anhydrous ammonia, process safety management of highly hazardous chemicals, and hazardous waste operation and emergency response
Subpart I	Personal protective equipment—eye and face protection, respiratory protection, head and foot protection, electrical protective equipment, and hand protection
Subpart J	General environmental controls—sanitation, safety color code for marking physical hazards, specifications for accident prevention signs and tags, permit-required confined spaces, control of hazardous energy (lockout/tagout), and other topics
Subpart K	Medical and first aid
Subpart L	Fire protection—portable fire suppression equipment, fixed fire suppression equipment, and other fire protection systems
Subpart M	Compressed gas and compressed air equipment
Subpart N	Materials handling and storage—servicing multipiece and single piece rim wheels, powered industrial trucks, overhead gantry cranes, crawler locomotive and truck cranes, derricks, slings, and other topics
Subpart O	Machinery and machine guarding—requirements for all machinery, woodworking machinery, cooperage machinery, abrasive wheel machinery, mills and calenders, mechanical power presses, forging machines, and mechanical power-transmission apparatus
Subpart P	Hand and portable powered tools and other hand-held equipment—guarding of portable powered tools, general requirements, and other topics

20

TABLE 2.1 *Continued*

Subpart	Topics Included
Subpart Q	Welding, cutting, and brazing—oxygen-fuel gas welding and cutting, arc welding and cutting, and resistance welding
Subpart R	Special industries—pulp, paper, and paperboard mills, textiles, bakery equipment, laundry machinery and operations, saw mills, pulpwood logging, agricultural operations, telecommunications, electric power generation, transmission, and distribution, and grain handling facilities
Subpart S	Electrical—design safety standards for electrical systems and safety-related work practices
Subpart T	Commercial diving operations—personnel requirements, general operations procedures, specific operations procedures, equipment procedures and requirements, and recordkeeping
Subparts U–Y	Reserved
Subpart Z	Toxic and hazardous substances—asbestos, vinyl chloride, inorganic arsenic, lead, cadmium, benzene, coke oven emissions, bloodborne pathogens, cotton dust, acrylonitrile, formaldehyde, hazard communication, and requirements for numerous other substances

Standard for Process Safety Management of Highly Hazardous Chemicals (Effective 1992). This rule, published under 29 CFR 1910.119, details requirements for employers that manufacture, store, or use any of more than 130 specific toxic and reactive chemicals in quantities that exceed a certain threshold. If covered under the standard, an employer must conduct a process hazard analysis, provide employee training on process hazards, and meet numerous other requirements.

Standard for Occupational Exposure to Bloodborne Pathogens (Effective 1992). OSHA's bloodborne pathogens standard applies to every employer with one or more employees who can reasonably be expected to come into contact with blood or other specified body fluids while carrying out or performing their duties. The rule, outlined under 29 CFR 1910.1030, requires employers to establish a written exposure control plan, make available free of charge and at a reasonable time and place the hepatitis B vaccine and vaccination series, and provide training to employees who have occupational exposure. Recordkeeping, labeling, and other requirements are included.

Standard for Permit-Required Spaces (Effective 1993). This rule, found in 29 CFR 1910.146, enhances requirements for practices and procedures to protect employees in general industry from the hazards of entry into permit-required confined spaces. Included in the standards are requirements for extensive testing of the permit space, a permit system, training, and other requirements.

Asbestos Standard (Effective 1994). OSHA regulates asbestos under three separate standards—29 CFR 1910.1001 for general industry, 29 CFR 1926.1001 for construction; and 29 CFR 1915.1001 for shipyards. The updated standard focuses on practical ways to allow the removal of in-place asbestos-containing materials while controlling worker exposure to asbestos fibers. The standard specifies a permissible exposure limit for asbestos as an 8-hour time-weighted average of 0.1 foot per cubic centimeter (f/cc), and an excursion limit of 1 f/cc averaged over 30 minutes. Training, medical surveillance, and specific work practices are also required.

INCORPORATION OF MATERIAL BY REFERENCE

OSHA has incorporated material from agencies or standards organizations into some parts of its regulations by reference. These agencies and organizations are listed in Table 2.2.

OSHA-APPROVED PROGRAMS

The OSH Act encourages individual states (and U.S. territories) to develop and operate, under OSHA guidance and approval, state occupational safety and health programs. Once a state plan is approved, OSHA funds up to 50% of the program's operating costs.

Approved state plans are required to provide standards and enforcement programs, as well as voluntary compliance activities, that must be at least as effective as federal standards. There are currently 25 states and territories that have OSHA-approved plans—23 that cover the private and public sectors and two that cover the public sector only. Approved states include Alaska, Arizona, California, Connecticut (public sector only), Hawaii, Indiana, Iowa, Kentucky, Maryland, Michigan, Minnesota, Nevada, New Mexico, New York (public sector only), North Carolina, Oregon, Puerto Rico, South Carolina, Tennessee, Utah, Vermont, Virgin Islands, Virginia, Washington, and Wyoming.

CONSULTATION SERVICES

Employers who want help in recognizing and correcting safety and health hazards and in improving their safety and health programs can get that help from a free consultation service, which is largely funded by OSHA. The service is delivered by state governments, using well-trained professional staff. The

TABLE 2.2   Agencies or Standards Organizations That Have Material Approved for Incorporation by Reference in 29 CFR 1900–1910

Agencies

- Department of Commerce
- Department of Health and Human Services
- General Services Administration
- National Institute for Occupational Safety and Health
- Public Health Service

Standards Organizations

- American National Standards Institute
- American Conference of Governmental Industrial Hygienists
- American Society of Agricultural Engineers
- Agriculture Ammonia Institute—Rubber Manufacturers Association
- American Society of Mechanical Engineers
- American Society for Testing and Materials
- American Welding Society
- Compressed Gas Association
- Crane Manufacturers Association of America
- Institute of Makers of Explosives
- National Electrical Manufacturers Association
- National Fire Protection Association
- National Food Plant Institute
- Society of Automotive Engineers
- The Fertilizer Institute
- Underwriters Laboratories

consultation program addresses immediate problems and offers advice and help in maintaining continued effective protection. On-site consultants offer the following services:

- Help recognize hazards in the workplace
- Suggest approaches or options for solving a safety or health problem
- Identify sources of help available if further assistance is needed
- Provide a written report that summarizes findings
- Assist in developing or maintaining an effective safety and health program
- Offer training and education for employees
- As applicable, recommend recognition of a site for a one-year exclusion from general schedule enforcement inspections

The service is confidential, and the firm's name and any information about unsafe or unhealthful work practices is not routinely reported to the OSHA inspection staff. The employer, however, is obligated to correct any imminent dangers and other serious job safety and health hazards in a timely manner.

OSHA INSPECTIONS

OSHA has the right to inspect facilities covered by the OSH Act, although technically a facility can refuse admission of the compliance safety and health officer (CSHO) unless a warrant for admission has been obtained first. OSHA inspection priorities are as follows:

- Investigation of imminent dangers
- Investigations of employee complaints
- Programmed high-hazard inspections
- Reinspections

If a CSHO finds a violation, it is placed in one of four categories: imminent danger, which is any existing condition or practice that could reasonably be expected to cause death or serious physical harm immediately or before the imminence of such danger can be eliminated through the enforcement procedures; serious violation, which involves hazardous conditions that could cause death or serious physical harm and that the employer knew or should have known existed; other-than-serious violation, which is cited in situations where an accident or illness that would most likely result from a hazardous condition would probably not cause death or serious physical harm, but would have a direct and immediate relationship to the safety and health of employees; and willful violation, which is either an intentional violation or indifference to OSHA requirements.

BIBLIOGRAPHY

NSC(1992), *Accident Prevention Manual for Business & Industry: Administration & Programs*, 10th ed., National Safety Coucil, Chicago.

OSHA (1994), OSHA Field Inspection Reference Manual, OSHA Instruction CPL 2.103, U.S. Department of Labor, Occupational Safety and Health Administration, Washington, DC.

_____ (1990), *Consultation Services for the Employer*, OSHA 3047, U.S. Department of Labor, Occupational Safety and Health Administration, Washington, DC.

U.S. DOL (latest ed.), *Code of Federal Regulations*, 29 *CFR* 1910, U.S. Department of Labor, Washington, DC.

PART II

PRINCIPLES OF
ENVIRONMENTAL ENGINEERING

3

POLLUTION PREVENTION

Pollution prevention, or P2 for short, has become an integral part of environmental regulations. Throughout the 1980s interest in pollution prevention grew steadily as a result of increasing environmental awareness and concern, tightening waste regulations, and liabilities incurred for contaminated sites. In 1990 pollution prevention became national policy when Congress passed the Pollution Prevention Act.

The ultimate goal of pollution prevention is to reduce or eliminate pollution at its source. Where source reduction is not feasible, then recycling should be considered. Where recycling is not feasible, then proper treatment of a waste is required. Any remaining residuals must be properly disposed of in an environmentally safe manner.

This chapter discusses various pollution prevention programs and information sources, pollution reporting through the Toxic Release Inventory program, and techniques for implementing pollution prevention.

P2 PROGRAMS AND INFORMATION SOURCES

There are a great number of programs and information sources available on the subject of pollution prevention, many of which are government based or supported. Brief descriptions of some of these resources (presented alphabetically) are described in this section.

American Institute for Pollution Prevention (AIPP)

The American Institute for Pollution Prevention (AIPP) is a nonprofit organization of individuals representing industry trade associations and professional societies. AIPP is devoted to fostering and enhancing sustainable pollution prevention solutions to environmental problems. Sponsor organizations are the EPA and the Department of Energy (DOE). AIPP's goals are as follows:

- To serve as a bridge for communication by providing information on pollution prevention resources available from its members and others;
- To promote policies and provide for the transfer of information to encourage the culture shifts necessary for implementation of pollution prevention technologies throughout society; and
- To proactively set future directions for pollution prevention through cooperative and collaborative efforts among industry, government, and the public.

Enviroene

Enviroene, funded by EPA and the Strategic Environmental Research and Development Program, maintains a wide collection of pollution prevention inforformation accessible through the World Wide Web (http://www.envirosense.com) and an EPA bulletin board. It contains or is linked to pollution prevention resources and programs, including several of the resources described elsewhere in this section. Enviroene has a great deal of information on pollution prevention, including technical information on grants, international resources, training opportunities, information notebooks and case studies for specific industries, and solvent substitution (known as the Solvent Umbrella). Abstracts and information texts can be searched by keywords.

Pollution prevention notebooks are available for the 18 industrial sectors shown in Table 3.1. Each notebook contains a comprehensive environmental profile, industrial process information, pollution prevention techniques, pollutant release data, regulatory requirements, compliance and enforcement history, innovative programs, and names of contacts for further information.

"Guided tours" of the web database are also being developed for industrial sectors to help users find information. The tour or content guides include Frequently Asked Questions (FAQs), visual content maps, subject indices, and core document lists.

National Pollution Prevention Center for Higher Education (NPPC)

In 1991, the EPA created the National Pollution Prevention Center for Higher Education (NPPC) to collect, develop, and disseminate educational materials on pollution prevention. The NPPC, located at the University of Michigan, is a collaborative effort among business and industry, government, nonprofit organizations, and academia. The NPPC develops pollution prevention educational materials for college faculty. These materials, designed so that the principles of pollution prevention may be incorporated into existing or new courses, contain resources for professors as well as assignments for students.

TABLE 3.1 Industrial Sectors with Pollution Prevention Notebooks Available through EnviroSenSe

Industrial Sector	Special Keywords for Sector
Dry Cleaning	Perchloroethylene, tetrachloroethylene, dry cleaner, NESHAP, MACT, HAP, wet cleaning, small business, solvent, PCE
Electronics and Computers	Semiconductor, printed circuit board, printed wiring board, cathode ray tube, solder, electroplating, etching, computer, photoetch, photoresist
Fabricated Metal Products	Electroplating, anodizing, conversion coatings, organic coatings, metal forming, chromium, cadmium, nickel, zinc, cyanide, VOCs
Inorganic Chemicals	Mercury cell, diaphragm cell, chlorine electrolysis, salt, caustic soda, brine, hydrogen gas
Iron and Steel	Steelmaking, minimills, integrated mills, coke making, iron making, particulate matter emissions, common sense initiative
Lumber and Wood Products	Forest products, plywood, veneer, panel, particleboard, MDF, OSB, preserving, treatment, composite wood, sawmills, creosote
Metal Mining	Mines, metallic, extraction, leach, pit, acid drainage, rock, slag, gold, silver, iron, copper, beneficiation
Motor Vehicle Assembly	Automobile manufacturing, automotive, truck, automotive parts, transportation equipment, car
Nonferrous Metals	Aluminum, lead, copper, zinc, smelting, refining, slag, sulfur recovery, particulates
Non-Fuel, Non-Metal Mining	Nonmetallic, mineral, mines, extraction, pit, rock, aggregates, limestone, clay, phosphate
Organic Chemicals	Ethylene, propylene, benzene, vinyl chloride, xylene, toluene, butadiene, methane, butylene, flow diagrams, building blocks
Petroleum Refining	Refinery, oil, crude, distillation
Printing	Flexography, lithography, screen printing, gravure, graphic arts, rotogravure, letterpress, imaging, platemaking, post-press, substrate, ink, photograph, pre-press, press, finishing
Pulp and Paper	Kraft, sulfite, secondary fiber, chlorine, BOD, TSS, VOC, particulate matter, dioxin, delignification, effluent, MACT

(Continued)

TABLE 3.1 *Continued*

Industrial Sector	Special Keywords for Sector
Rubber and Plastic	Rubber products, tire manufacturing, plastics products, VOC, styrene, scrap tires, 33/50 program
Stone, Clay, Glass, and Concrete	Cement, glassware, pottery, fiberglass, brick, gypsum, plaster, particulates
Transportation Equipment Cleaning	Washing, fueling, loading and unloading, transport
Wood Furniture and Fixtures	Surface coating, consumer durables, household furniture, office products

Educational materials are available in a variety of disciplines and offer background material, case studies, problem sets, and multimedia resources to use in the classroom. The NPPC's research program develops new frameworks, principles, and tools for achieving sustainable development through industrial ecology and life-cycle assessment and design.

The NPPC offers an internship program, professional education and training, and conferences. Professional education courses are offered through the University of Michigan College of Engineering Continuing Education program and special courses for industry and government. The internship program promotes hands-on pollution prevention experience for undergraduate and graduate students in engineering, business, natural resources, and other fields of study. Students work with faculty mentors to develop educational materials based on their internships. Conferences and workshops organized by the NPPC are designed to identify and discuss critical issues in pollution prevention education and research.

National Pollution Prevention Roundtable (NPPR)

The National Pollution Prevention Roundtable (NPPR), located in Washington, D.C., is a membership organization that provides a national forum for pollution prevention. Voting membership in the Roundtable includes state, local, and tribal government pollution prevention programs. Affiliate members include representatives from federal agencies, nonprofit groups, and private industry. The Roundtable serves its members by:

- Providing access to the latest information on legislative and regulatory pollution prevention developments;
- Preparing and distributing information on pollution prevention technologies and technical assistance practices;
- Serving as a national pollution prevention information clearinghouse, containing publications of state, local, and other related programs; and
- Sponsoring semiannual national meetings that attract the largest gathering of pollution prevention experts from across the country.

The NPPR publishes a national directory, called *P2 Yellow*, of state, regional, and local pollution prevention programs, which includes relevant information, areas of expertise, contact names, and a description of each program.

P2Info

The DOE provides a clearinghouse of information on pollution prevention technologies and vendors, called P2Info. The clearinghouse is operated by Pacific Northwest Laboratory (see the following related resource).

Pacific Northwest Pollution Prevention Research Center (PPRC)

The Pacific Northwest Pollution Prevention Research Center (PPRC) is a nonprofit organization serving EPA Region 10 (Alaska, Idaho, Oregon, and Washington) and British Columbia. PPRC publishes a bimonthly newsletter, *Pollution Prevention Northwest*, which covers topics such as industry-specific research and state pollution prevention programs. Industry-specific information is available for the fields of electronics, forestry, sulfite pulp processing, fish processing, wood products, and oil and gas. PPRC also maintains an online Pollution Prevention Research Projects Database.

Pollution Prevention Information Clearinghouse (PPIC)

EPA's Pollution Prevention Information Clearinghouse (PPIC) offers EPA documents and fact sheets about pollution prevention, including case studies and the *Pollution Prevention Directory*. PPIC also provides a referral service for technical questions.

Waste Reduction Innovative Technology Evaluation (WRITE)

EPA's Waste Reduction Innovative Technology Evaluation (WRITE) program focuses on pilot projects in pollution prevention performed in cooperation with state and local governments. Program priorities are new source reduction and recycling technologies. The goals of the program include identifying and evaluating new or improved economically favorable technologies for pollution prevention in industries presently experiencing waste problems. The program is designed to assist federal, state, and local governments, as well as small and midsized industries by providing performance and cost information on pollution prevention technologies.

Waste Reduction Resources Center (WRRC)

The Waste Reduction Resources Center (WRRC) is a clearinghouse for pollution prevention information, focusing on industries in EPA Regions 3 and 4 (Alabama, Delaware, the District of Columbia, Florida, Georgia, Kentucky,

Maryland, Mississippi, North Carolina, Pennsylvania, South Carolina, Tennessee, Virginia, and West Virginia). WRRC maintains articles, case studies, and technical reports covering issues such as economic analyses, process descriptions, pollution prevention techniques, and implementation strategies. Upon request by an industry within its area, WRRC will prepare reports and information packets for specific facility/waste problems. These industries can also obtain on-site technical training assistance and pollution prevention audits. For facilities outside its area, WRRC provides case summaries, contacts and referrals, and vendor information.

Toxic Release Inventory (TRI)

The Toxic Release Inventory (TRI) was established under the Emergency Planning and Community Right-to-Know Act (EPCRA) of 1986. TRI reports must be submitted by manufacturing facilities to EPA, which compiles these reports into a publicly available database. A TRI report includes the following information on each chemical meeting certain threshold criteria at a facility: onsite releases into air, water, land, and underground injection wells; offsite transfers to publicly owned treatment works (POTWs), treatment, disposal, recycle, and energy recovery; and source reduction and recycling activities. The original TRI list of chemicals contained well over 300 compounds. In 1993, EPA added 31 chemicals and 2 chemical categories to the list, and later, in 1994, EPA added another 286 chemicals. As a result of the TRI data—which focused public scrutiny on the chemical releases that were reported—EPA started the 33/50 Program in which companies were asked to voluntarily reduce releases of 17 of the most prevalent and toxic chemicals (see Table 3.2)

TABLE 3.2   Toxic Chemicals in the 33/50 Program

Benzene
Cadmium
Carbon tetrachloride
Chloroform
Chromium
Cyanide
Dichloromethane
Lead
Mercury
Methyl ethyl ketone
Methyl isobutyl ketone
Nickel
Tetrachloroethylene
Toluene
1,1,1-Trichloroethane
Trichloroethylene
Xylene

by 33% from 1988 to 1992 and by 50% from 1988 to 1995. The TRI and its 33/50 program have been a major factor in stimulating pollution prevention projects within industry, because both EPA and the public regularly use and publish the rankings of companies in respect to chemical releases.

IMPLEMENTING POLLUTION PREVENTION

Beyond certain fairly simple and common pollution prevention techniques, most other techniques are process specific and require a great deal of study and planning. For this reason, this chapter provides only a general outline for pollution prevention approaches, which includes establishing a commitment for effective implementation, prioritizing projects, tracking progress, identifying chemicals of concern, life cycle analysis, process modifications, good housekeeping and preventative maintenance, recycling, and reducing household hazardous waste. Some general guidelines developed by the AIPP to encourage pollution prevention in the workplace are summarized in Table 3.3.

Commitment

Pollution prevention has obvious rewards: potential cost savings, reduction in waste quantities or toxicities, and improved company image. However, other aspects of pollution prevention hinder many projects: reluctance to change from tried-and-true methods or processes (particularly when such changes can impact product quality); time, expense, and difficulty in studying alternatives; initial capital expense; and structural or space constraints at existing locations. Consequently, implementation of significant pollution prevention projects usually requires a commitment from those persons with the authority to make things happen—that is, management. However, employee commitment is important too. Therefore, it is not surprising that the most successful pollution prevention projects are those that have the commitment of both management and employees.

Management can start by developing a pollution prevention policy or mission statement that gives a clear idea of the company's goals, objectives, and approach. As such, the policy or statement can be used as a general gauge to assess progress. ISO 14000, an international standard for environmental management, requires top management usually the chairman or equivalent in most companies—to set policy. Employees set goals and objectives along with management. The firm's pollution prevention policy should be communicated to all employees so that they understand the company's commitment and how they, as employees, play a part in it.

To show commitment, it is important to set clear, measurable pollution prevention goals. These goals should be set beyond mere compliance with environmental regulations; however, they must be realistic and achievable,

TABLE 3.3   Actions to Encourage Pollution Prevention in the Workplace

Recommended by the American Institute for Pollution Prevention

General

1. Enlist support for pollution prevention from top management. Have management demonstrate that support by providing a written policy statement.
2. Appoint an unsinkable champion and give that champion the power to implement.
3. Keep hammering away at the nonbelievers—be patient and persistent.
4. Believe in yourself.
5. Identify and publicize low-tech or retro-technology options in place of high impact processes.
6. Seek a fundamental understanding of the sources of waste.
7. Focus on optimizing the use of resources consumed in the process.
8. Use pollution prevention as a competitive strategy in public relations and to attract a better caliber workforce.
9. Devote adequate resources (people, money, and energy) to pollution prevention.
10. Set up a structure for recycling that complements source reduction.

Culture

11. Establish a clear pollution prevention goal that everyone in the organization feels empowered to put into practice.
12. Reestablish the culture of not wasting.
13. Instill a philosophy of continuous improvement.
14. Link zero discharge, total quality management, and pollution prevention into a working program.
15. Publicize pollution prevention accomplishments.
16. Convince all personnel that they have a role to play in pollution prevention; no one is exempt from pollution prevention, no matter what the job.
17. Share money saved through pollution prevention with the originator(s) of the idea.
18. Incorporate pollution prevention into performance evaluations for middle management.
19. Establish teams to promote pollution prevention, receive ideas and solicit suggestions, evaluate projects and champion implementation.
20. Have top management personally hand out all pollution prevention awards; participate in state/regional awards programs.

Workplace Education

21. Produce innovative and exciting training for teaching pollution prevention concepts.
22. Have primary contractors train and assist subcontractors to practice pollution prevention.
23. Educate each individual on what pollution prevention means.

TABLE 3.3 *Continued*

Decision Making

24. Use expert systems and process simulation modeling to develop effective pollution prevention strategies.

25. Use and develop life cycle studies for processes and products.

26. Develop and implement methods for measuring progress to support continued pollution prevention investments.

27. Develop performance-based specifications rather than prescriptive/design specifications.

28. Set a goal: Think of a bubble around your facility—nothing comes out but finished product.

29. Plan your waste reduction work and work your waste reduction plan.

30. Use expert systems to assist the design of new processes that avoid pollution in the first place.

31. Design products with zero ultimate waste potential, with cradle-to-grave functionality.

32. Incorporate pollution prevention into the development of new products and processes.

33. Assign life cycle responsibility to production management, linked to cost and liability.

34. Charge the true cost of the waste created to the operating unit and make operating units responsible for liabilities, management, and costs of waste streams.

35. Develop analytical tools for accountants and financial managers to recognize the full environmental costs of unwanted environmental programs.

Outreach

36. Establish an industry roundtable to communicate and share methods for compliance and low-impact processes.

37. Develop corporate-public partnerships to protect local resources and to explore solutions to difficult problems.

38. Develop pollution prevention in all locations of the company on an equal basis. Treat foreign facilities with the high level of expectation applied to domestic facilities.

39. Establish a process (newsletter, etc.) for communicating progress.

Communication

40. Develop an internal information system to exchange good ideas and technical knowledge within your business unit, division, or company.

41. Shift from paper systems to paperless communication (e.g., e-mail, electronic bulletin boards).

42. Create effective, friendly, human information networks for effective pollution prevention.

43. Encourage policies that allow plant engineers and managers to spend ample time on the factory floor.

(Continued)

TABLE 3.3   *Continued*

Recycling and Resource Conservation Examples

44. Xeriscape—use low water consumption, drought resistant native plants and low-impact irrigation in corporate landscaping.
45. Make recycled writing tablets out of paper used on one side.
46. Establish a chemical inventory program to reduce redundant purchases, track chemical use more efficiently, and avoid disposal of unused material.
47. Change computer printer to print only "flag" sheet plus manuscript—not additional sheets.
48. Use two-sided copying.

Source: AIPP (http://www.envirosense.com/aipp/48_pg.html).

inasmuch as a company stands to lose credibility if it cannot meet its stated goals. Management should be prepared to provide the necessary resources—personnel, information, and money—to achieve its pollution prevention goals. Setting goals without making it possible to achieve them is only giving lip service.

Management can show its commitment to its employees by valuing employee input and rewarding employees who help achieve pollution prevention goals. Environmental staff, who are usually on the front line of pollution prevention because of compliance pressures, should be able to communicate directly to management. Environmental managers should have a presence in upper management.

Evaluating Performance

Tracking progress in pollution prevention allows a company to see how its stated goals are being met, to document its commitment to its employees and the public, to provide feedback to ongoing and future projects, and to justify cost or environmental benefits. Because many larger companies are required to report chemical releases annually under the TRI, they can use these data to track and document reductions. Other regulatory reports can provide waste data for tracking, such as RCRA hazardous waste elimination and biennial reports, state or local waste reports, waste manifests, wastewater discharge monitoring reports, and air emission inventory reports. Other sources of data are environmental audit reports, waste analyses and waste profiles, material inventory and usage records, continuous emission monitoring, production records, and industrial sector waste surveys. Results of pollution prevention efforts should be reported to management to provide feedback on what does and does not work and where the greatest potential benefits are.

Identifying Chemicals of Concern

In seeking to reduce the quantity of hazardous waste or at least reduce the degree of hazard associated with such wastes, wastes should be characterized by hazard and hazardous constituents. There are many regulations that focus on different types of hazardous chemicals, materials, and wastes, and these can be reviewed for those chemicals that might be of concern and would be candidates for pollution prevention efforts. For example, EPA has challenged industry to reduce wastes containing any of 17 specific toxic chemicals under the TRI 33/50 program. A "list of lists" can be developed from the various regulations as a database for pollution prevention projects. All personnel involved with pollution prevention should be made aware of these lists so that they can focus on important chemicals or characteristics. Such information is particularly important to those persons in research, development, and manufacturing, because they are the ones most likely to have detailed information on products and waste streams, as well as the expertise to suggest modifications and alternatives. Examples of regulated chemicals are found in the RCRA hazardous waste regulations, the TRI reporting regulations, CERCLA lists of hazardous and extremely hazardous substances, CAA lists of hazardous and highly hazardous chemicals, and the CAA list of ozone-depleting substances. A more comprehensive list of regulations and the materials or chemicals they cover are identified in Table 1.2 of Chapter 1.

Process Modifications

Process modifications, naturally, are easiest to make during the process development and design phase. Retrofitting and modifications of an existing process are hampered by ongoing operations, available space, and existing layout. Understandably, they can be disruptive and expensive—two big drawbacks that most people like to avoid. Even so, many companies have made such modifications and have been rewarded not only with pollution reductions, but with tangible and immediate cost savings. Process modifications, both in the design/development stage and in existing operations, are normally quite specific to a particular process. This section discusses process modifications in a broad and general sense.

Material substitution or improvements should be considered for feedstock and secondary materials used in a process. Such process changes can include switching to a less toxic or less corrosive material. Using a material of higher purity or developing a more selective catalyst may reduce by-product formation, improve product yield, and reduce the transfer of unreacted materials into the waste stream.

Other substitutions include using less toxic and volatile solvents and cleaners or those that are not ozone depleting. For example, alternative cleaners and solvents include aqueous and emulsion cleaners, detergent cleaners, and terpene-based chemicals. EPA has developed or is working on various guides to alternative solvents. One of these, called SAGE, is for solvents in the TRI 33/50 program. Other guides being developed are CAGE for coating processes, and AGE for adhesives. EPA maintains a solvent substitution database on the Internet (see "Enviroene" earlier in this chapter), including the following data systems:

- Solvent Alternatives Guide (SAGE)—a logic tree system that can be used to evaluate a current operating scenario and then identify possible surface cleaning alternative chemistries and processes that best suit the defined operating and material requirements.

- Hazardous Solvent Substitution Data System (HSSDS)—an online information system on alternatives to hazardous solvents and related subjects. HSSDS contains product information, material safety data sheets, and other related information.

- Department of Defense (DOD) Pollution Prevention (P2) Technical Library—an electronic resource maintained by the Naval Facilities Engineering Service Center (NFESC). To foster technology transfer with other organizations, the NFESC has made this library available by direct access or by search facilities.

- DOD Ozone Depleting Chemical/Substance Information—the U.S. Air Force Ozone Depleting Chemicals (ODC) Information Exchange provides an information exchange for the Air Force weapons system community to aid in complying with federally mandated ODC reduction goals, courtesy of the Brooks Air Force Base Human Systems Center. The DOD ODS MILSPEC database contains a listing of military and federal specifications that may require the use of Class I ozone depleting substances (ODSs). This database contains information on each document, including identity of the ODS, how it is used, whether non-ODS alternatives are specified, potential substitutes for the ODS called out in the documents, and modification/cancellation information. The U.S. Navy CFC Halon Clearinghouse provides users of ODSs with a central point of contact for information, data, and expertise on Navy ODS policy.

- Solvent Handbook Database System (SHDS)—a database providing access to environmental and safety information on solvents used in maintenance facilities and paint strippers. The database also contains empirical data from laboratory testing.

- National Center for Manufacturing Sciences (NCMS) Solvent Alternatives Database and NCMS Materials Compatibility Database—contributed by the NCMS, a manufacturing consortium of 175 members that includes some of the largest manufacturers in the United States.

When designing chemical reactors with pollution prevention in mind, it is important to ensure proper mixing to maximize product yield and minimize the formation of by-products. Unwanted by-products can result from poor mixing, fluctuating residence times, impurities in process materials and feed-stocks, hot spots on catalysts, and catalytic effects on construction materials. Reactors should be designed to drain completely and clean easily.

Piping designed for pollution prevention should have a minimal run and be able to drain completely. It is particularly important to consider the collection of individual streams (both process materials and wastes) in segregated piping, because the trend in environmental regulations is moving from management of commingled materials (end-of-pipe) further into the process area before streams are combined. The number of valves and fittings should be minimized. Any drain lines, sample lines, and vents should be routed to recycle, treatment, or disposal, as appropriate. Piping should be avoided that drains into the process pad (later collected in process washdown) or into open trenches or ditches.

Process control can be designed and optimized to maximize product yield and minimize waste streams. Process controls can include instrumentation, remote monitoring and data collection, statistical analysis, and inventory management. Methods of analysis, such as statistical process control (SPC), which are commonly used in total quality management programs to assess product quality, can be used in innovative ways to control process chemicals. SPC in conjunction with an automated control system allows the process to be optimized using real-time data, which can restrict chemical usage to much tighter tolerances. Material inventory management can be optimized to reduce the wastage of unused or expired chemicals.

Good Housekeeping and Preventative Maintenance

Good housekeeping and preventative maintenance can not only help to reduce waste generation, but also help to provide a safe and clean working environment. Simple housekeeping tasks should include inspection of material storage areas, spill prevention and control, segregation of incompatible or inappropriate materials, and container labeling. Preventative maintenance tasks should include regularly scheduled checks or maintenance, recordkeeping of inspections and repairs, testing of overflow alarms and release detection equipment, replacing damaged or leaking containers, and maintaining up-to-date equipment manuals.

Implementing good housekeeping practices and effective preventative maintenance requires not only employee training, but cooperation as well. Employees must understand and value the importance of a clean and well-maintained workplace and how this avoids waste generation. Management provides the training and workplace environment; however, that in itself is not enough—management must lead by example and enforcement of company policies and practices. Those employees directly involved in housekeeping and maintenance should to be asked for their input on how to make a task easier while getting the job done.

Reuse, Reclamation, and Recycling

Most people are familiar with recycling in their everyday lives. Household and office waste that is commonly collected for recycling includes aluminum cans, white ledger paper, newspaper, cardboard, glass, cans of metals other than aluminum, and plastic. Recycling is also common at commercial and industrial facilities. Many processes incorporate recycling because it conserves raw materials and reduces waste generation. Some of the most common recycling processes are solvent recovery and wastewater recycling.

Common processes used in solvent recovery include distillation, steam stripping, carbon adsorption, and membrane separation (described in Chapter 5, "Waste Treatment and Disposal Technologies"). Solvent recovery may be closed-loop and integral to the manufacturing process. Alternatively, spent or contaminated solvents may be removed from the process for recovery onsite or offsite. In some cases, a solvent that is lightly contaminated from use in one process can be reused in another process where a lesser-quality solvent is acceptable.

In a similar fashion, process water that is slightly contaminated, such as high-purity rinse waters, may be reused where process requirements are less stringent. Effluent from a facility's wastewater treatment plant may be recycled back to a process area or to a cooling tower for makeup water, thus minimizing water use as well as wastewater discharges into surface waters. Treated sanitary (nonindustrial) wastewater may be reused as drinking water by using it to recharge ground water supplies (health concerns and public perceptions about wastewater have inhibited direct reuse).

Oil recovery is common at many industrial facilities, particularly petroleum refineries and chemical manufacturers, because of the types of materials that are handled and because high concentrations of oil are unacceptable to biological wastewater treatment systems. The oil in wastewaters is usually removed in some type of oil/water separator and/or dissolved air floatation unit. Recovered oil may be recycled back to the process or burned for its fuel value onsite. It may also be sent offsite for material or energy recovery. Waste oil from the maintenance and cleaning of equipment and vehicles can also be recovered.

Wastewaters containing metals may be treated to recover the metals, depending on both the value and the concentration of the particular metal. For example, the silver in photographic wastewaters can be recovered. Metals in wastewaters may be recovered by chemical precipitation, ion exchange, membrane separation, or electrowinning.

Solvents, metals, oil, and water are some of the most common recovered or recycled materials. There are many others, of course, some that are specific to a particular process and others that have a more universal application. Some of these applications use wastes that otherwise have few options other than disposal. For example, soil contaminated with petroleum products or metals may be converted into asphalt-type and paving materials, inorganic residues from supercritical oxidation may be incorporated into bricks, and foundry sand may be used as daily cover for landfills or in the construction of dikes, roads, and parking lots. Ideas for recycling materials are constantly evolving.

Prioritizing Projects

The most effective pollution prevention projects are those that receive input from people who are knowledgeable about the process or will be impacted by it. Usually, representatives from various departments at the site form a pollution prevention team that develops and evaluates the pollution prevention projects. Members of the team may include, as appropriate, those from the following areas: environmental, health, and safety; production; engineering; research and development; maintenance; purchasing and accounting; legal; and information systems. Outside consultants may be added to the team to provide special expertise or to represent an unbiased (noncompany) view.

When a facility is assessing possible pollution prevention projects for more than one process or several alternatives for a single process, it helps to have a way to prioritize projects. There are many different ways to approach the problem; however, most follow a typical sequence: characterizing material streams, identifying pollution prevention alternatives, ranking the alternatives, and, finally, choosing an alternative.

The first step in characterizing material streams associated with a process is to identify them. The types of information that might be included in the characterization are listed in Table 3.4. The information that is collected for each stream should be summarized in a standard form to make sure that characterizations are complete and to make comparisons among streams easier.

Next, pollution prevention alternatives should be listed for each material stream. Source reduction techniques should be given special consideration in keeping with the goal of pollution "prevention." Recycling, treatment, and disposal alternatives should also be listed. Some judgment will be needed to screen out those alternatives that are obviously unsuitable for one reason or

TABLE 3.4 **Information for Material Stream Assessments**

Source (from which part of process)
State or federal waste identification code
Physical characteristics (solid, liquid, gas, semisolid, etc.)
Generation rate
Hazardous properties (toxic, flammable, reactive, corrosive, etc.)
Hazardous constituents and quantities released (i.e., TRI reports)
Other health, safety, or environmental issues
Applicable regulations (existing, pending, or potential)
Management costs (treatment, disposal, transport, analysis, etc.)

another (project scope, technical feasibility, budget constraints, high cost with little benefit). Care must be taken not to strike down alternatives too quickly, lest a truly promising alternative be overlooked. Alternatives that make the final list will then undergo a more intensive review.

Criteria that may be used to rank alternatives are listed in Table 3.5. Each criterion may be assigned a weighting factor to represent its importance relative to the others. Pollution prevention alternatives can then be assigned scores for each criterion. Using the weighted sum method, the score for each criterion is multiplied by its weighting factor. The weighted scores are summed for each alternative, the result representing its overall score.

Another method of comparing alternatives is to use leverage, which is defined as a function of priority and feasibility:

$$\text{leverage} = \text{priority} \times \text{feasibility}$$

Feasibility can be viewed as the possible percentage reduction or the probability of minimizing the stream. Priority is a function of potential environmental impact and can be further defined as:

$$\text{priority} = \text{environmental units} \times \text{fate factor}$$

where environmental units are assigned to a stream as a function of its mass and its toxicity:

$$\text{environmental units} = \text{mass} \times \text{toxicity}$$

TABLE 3.5   Criteria for Ranking Pollution Prevention Alternatives

Reduction in quantity generated
Reduction in hazards
Persistence of hazardous constituents and assimilative capacity of environmental media
Effect on product quality
Regulatory compliance
Technical feasibility
Performance track record
Ease of implementation
Time to implement
Effect on existing operations
Capital cost
Operating cost
Management cost
Process material cost
Liability and insurance cost
Training requirements

Toxicity is determined by health data such as an LD_{50}, permissible exposure limit, leachability such as the TCLP value, or other data uniformly applied to all the streams being ranked. Other factors can be included in the toxicity value, such as hazardous constituents on the CERCLA, SARA, or other list, and contribution to the greenhouse effect or ozone depletion.

The fate factor should account for the potential impact of the stream on health and the environment, based on how it is managed. Streams that are recycled or that are incinerated will be given a lower fate factor than those that are released to the air or to surface water. The hierarchy of fate factors is source reduction, recycling, incineration, treatment or hazard reduction, discharge to air, water, and land.

Leverage can also be defined in terms of cost, where management cost is substituted for priority in the preceding equations.

Life Cycle Assessment

Life cycle assessment or analysis (LCA), as applied to pollution prevention, is a broader approach, which looks at a product or service from its inception to final disposition. In LCA, a product or service may begin with the acquisition of raw materials, followed by manufacturing or development of the product, then recycling or maintenance of the product, and, finally, waste management of the product or its residuals. LCA is useful because, in addition to considering impacts on the ecosystem and human health, it also considers impacts on resources such as raw materials and energy.

LCA is generally conducted in three phases: inventory, impact assessment, and improvement assessment. In the inventory phase, the pollution prevention team first defines its goals and how broad the "life cycle" of the product will be. The life cycle stages that may be included are marketing, research and development, raw material acquisition, material manufacture, production, packaging, distribution, consumer use, and disposal. After the scope of the life cycle is defined, the usage and generation rates of the different material and energy streams are estimated.

The second phase of LCA, impact assessment, itself consists of three phases: classification, characterization, and valuation. In the classification phase, inventoried items are assigned to impact categories of environmental stressors. In the characterization phase, the values assigned to inventoried items are converted into impact descriptors, such as direct measure and equivalency factors. In the third phase, valuation, impacts are compared by methods that generally incorporate some type of weighting factors applied to dissimilar impacts to produce a single value.

Following impact assessment is the improvement assessment phase of LCA. In this phase options are identified that will improve negative impacts associated with a product. These options are then evaluated and compared.

Household Hazardous Waste

Many private citizens are interested in practicing pollution prevention in their own lives, particularly with materials that are hazardous. Cities and communities can help by distributing information to let people know how they can reduce the use of hazardous materials that they may end up discarding. For those materials that are discarded, it helps to know how to dispose of them properly. Communities can also arrange for special hazardous waste collection days (annually or more frequently) to give their citizens an incentive to get rid of accumulated materials as well as a means of disposing of them safely. Ideas for reducing household hazardous waste are summarized in the following lists:

General Recommendations

- If available, use an alternative product that is less toxic or less hazardous.
- Before buying a product, be certain that it can do the job so that it will not be thrown out partially or completely unused.
- Buy only the amount of product that is actually needed or will be used within a reasonable time period so that extra or expired quantities do not have to be thrown out.
- Use the product correctly so that it has the best chance of doing its job properly.

Specific Recommendations

- Household cleaners: Use less hazardous cleaners such as baking soda, vinegar, borax, detergents, and lemon juice.
- Paints: Use latex or water-based paints, recycled paints made from mixing discarded paints (prepared during waste collection drives), natural earth pigment finishes, limestone-based whitewash, and casein-based paints.
- Pesticides: Reduce the need for fungicides by keeping areas clean, dry, and not overwatered. Use naturally derived pesticides such as pyrethrum, rotenone, sabadilla, nicotine, and insecticidal soap.

BIBLIOGRAPHY

Chemical Manufacturers Association (1991), *CMA Pollution Prevention Resource Manual*, Washington, DC.

Hanus, Denise (1995), "Assessments Aid Waste Minimization Efforts," *Pollution Engineering*, Vol. 27/No. 8, Cahners Publishing Company, Highlands Ranch, CO.

Luper, Deborah (1995), "Process, Design Considerations Minimize Waste," *Pollution Engineering*, Vol. 27/No. 6, Cahners Publishing Company, Highlands Ranch, CO.

Mooney, Gregory A. (1992), "Pollution Prevention: Shrinking the Waste Stream," *Pollution Engineering*, Vol. 24/No. 3, Cahners Publishing Company, Highlands Ranch, CO.

Quinn, Barbara (1995), "Beyond the Big Stick: EPA as Business Aide," *Pollution Engineering*, Vol. 27/No. 8, Cahners Publishing Company, Highlands Ranch, CO.

_____ (1995), "Finding the Right Recipes for Pollution Prevention," *Pollution Engineering*, Vol. 27/No. 6, Cahners Publishing Company, Highlands Ranch, CO.

_____ (1995), "Info Sources Help Develop the Substance in the Middle," *Pollution Engineering*, Vol. 27/No. 12, Cahners Publishing Company, Highlands Ranch, CO.

_____ (1995), "Panning for Gold," *Pollution Engineering*, Vol. 27/No. 5, Cahners Publishing Company, Highlands Ranch, CO.

Texas Natural Resource Conservation Commission (1996), *Household Hazardous Wastes: Alternatives and General Storage Directions*, (brochure), Austin, TX.

Water Environment Federation (undated), *Household Hazardous Waste: What You Should and Shouldn't Do*, (brochure), Alexandria, VA.

Woodside, Gayle (1993), *Hazardous Materials and Hazardous Waste Management: A Technical Guide*, John Wiley & Sons, Inc., New York.

Woodside, Gayle and Dianna S. Kocurek (1994), *Resources and References: Hazardous Waste and Hazardous Materials Management*, Noyes Publications, Park Ridge, NJ.

4

AIR POLLUTION ENGINEERING

The Clean Air Act (CAA) Amendments of 1990 touch on virtually every aspect of air pollution law. Mobile and stationary sources, large and small businesses, routine and toxic emissions, and consumer products are all regulated under the CAA. In addition to regulating air pollution sources, it is the intent of the Act to focus on areas of poor air quality, particularly large cities that have serious air pollution problems of many years. A description of the CAA and a list of its elements is presented in Chapter 1. This chapter focuses on air emissions modeling, sampling methods, and air pollution control technologies.

AIR EMISSIONS MODELING

Air emissions models are used to assess industrial point source air quality impacts, be they from routine stack emissions or from unplanned releases, tank ruptures, or other catastrophic events. Typical data that are required for most air emissions models include:

- Meteorological data—such as wind velocity, temperature, relative humidity, turbulence, and net radiation flux
- Site information—such as topography, equipment operating parameters, and property boundaries
- Source data—such as physical characteristics, chemical characteristics, geometry of source, and release rates
- Receptor information—such as receptor location and distance between receptors.

A sample plot of receptors used in a typical model is presented in Figure 4.1.

AIR SAMPLING METHODS

EPA has published guidance on sampling methods acceptable for determining the amount of toxic organic compounds in ambient air (EPA 1988). Test methods listed in the guidance document and applicabilities of the test methods are presented in Table 4.1.

Other analytical methods for analyzing compounds in ambient air are as follows:

- Ultraphotometric ozone analyzer—Used for continuous analysis of ozone
- Nondispersive infrared spectrometry—Used for continuous analysis of carbon monoxide

Figure 4.1 Sample plot of receptors used in air modeling.

TABLE 4.1   EPA Test Methods and Applicabilities for Sampling Toxic Organic Compounds in Ambient Air

---

Method T01

Description—Tenax gas chromatograph (GC) adsorption and gas chromatograph/ mass spectrometer (GC/MS) analysis

Types of Compounds Determined—Volatile, nonpolar organics having boiling points in the range of 80°C to 200°C

Examples of Specific Compounds Determined—Benzene, carbon tetrachloride, chlorobenzene, chloroform, chloroprene, 1,4-dichlorobenzene, ethylene dichloride, methyl chloroform, nitrobenzene, perchlorethylene, toluene, trichloroethylene, o,m,p-xylene

Method T02

Description—Carbon molecular sieve adsorption and GC/MS analysis

Types of Compounds Determined—Highly volatile, nonpolar organics having boiling points in the range of −15°C to 120°C

Examples of Specific Compounds Determined—Acrylonitrile, allyl chloride, benzene, carbon tetrachloride, chloroform, ethylene dichloride, methyl chloroform, methylene chloride, toluene, trichloroethylene, vinyl chloride, vinylidene chloride

Method T03

Description—Cryogenic trapping and GC/FID or ECD analysis

Types of Compounds Determined—Volatile, nonpolar organics having boiling points in the range of −10°C to 200°C

Examples of Specific Compounds Determined—Acrylonitrile, allyl chloride, benzene, carbon tetrachloride, chlorobenzene, chloroform, 1,4-dichlorobenzene, ethylene dichloride, methyl chloroform, methylene chloride, nitrobenzene, perchlorethylene, toluene, trichloroethylene, vinyl chloride, vinylidene chloride, o,m,p-xylene

Method T04

Description—High volume PUF sampling and GC/ECD analysis

Types of Compounds Determined—Organochlorine pesticides and PCBs

Examples of Specific Compounds Determined—4,4′-DDE, 4,4′-DDT

Method T05

Description—Dinitrophenylhydrazine liquid impinger sampling and HPLC/UV analysis

Types of Compounds Determined—Aldehydes and ketones

Examples of Specific Compounds Determined—Acetaldehyde, acrolein, benzaldehyde, formaldehyde

TABLE 4.1 *Continued*

Method T06

Description—HPLC analysis
Compound Determined—Phosgene

Method T07

Description—Thermosorb/N adsorption
Compound Determined—N-nitrosodimethylamine

Method T08

Description—Sodium hydroxide liquid impinger with high performance liquid chromatography
Compounds Determined—Cresol, phenol

Method T09

Description—High-volume PUF sampling with HRGC/HRMS analysis
Compound Determined—Dioxin

Method T10

Description—Low-volume PUF sampling with GC/EDC analysis
Types of Compounds Determined—Pesticides
Examples of Specific Compounds Determined—Alochlor 1242, 1254, and 1260, captan, chlorothalonil, chlorpyrifos, dichlorovos, dicofol, dieldrin, endrin, endrin aldehyde, folpet, heptachlor, heptachlor epoxide, hexachlorobenzene, α-hexachlorocyclohexane, hexachlorocyclopentadiene, lindane, methoxychlor, mexacarbate, mirex, trans-nonachlor, oxychlordane, pentachlorobenzene, pentachlorphenol, p,p'-DDE, p,p'-DDT, ronnel, 1,2,3,4-tetrachlorobenzene, 1,2,3-trichlorobenzene, 2,4,5-trichlorophenol

Method T011

Description—Adsorbent cartridge followed by HPLC detection
Types of Compounds Determined—Aldehydes
Examples of Specific Compounds Determined—Acetaldehyde, acrolein, butyraldehye, crotonaldehyde, 2,5-dimethylbenzaldehyde, formaldehyde, hexanaldehyde, isovaleraldehyde, propionaldehyde, o-toluene, m-toluene, p-toluene, valeraldehyde

Method T012

Description—Cyrogenic PDFID
Compounds Determined—Non-methane organic compounds

(Continued)

TABLE 4.1 *Continued*

Method T013

Description—PUF/XAD-2 adsorption with GC and HPLC detection

Types of Compounds Determined—Polynuclear aromatic hydrocarbons

Examples of Specific Compounds Determined—Acenaphthene, acenaphthylene, anthracene, benzo(a)anthracene, benzo(a)pyrene, benzo(b)fluoranthene, benzo(e)pyrene, benzo(g,h,i,)perylene, benzo(k)fluoranthene, chrysene, dibenzo(a,h)anthracene, fluoranthene, fluorene, indeno(1,2,3-cd)pyrene, naphthalene, phenanthrene, pyrene

Method T014

Description—SUMMAR® passivated canister sampling with gas chromatography

Types of Compounds Determined—Semivolatile and volatile organic compounds

Examples of Specific Compounds Determined—Acenaphthene, acenaphthylene, benzene, benzyl chloride, carbon tetrachloride, chlorobenzene, chloroform, 1,2-dibromomethane, 1,2-dichlorobenzene, 1,3-dichlorobenzene, 1,4-dichlorobenzene, 1,1-dichloroethane, 1,2-dichloroethylene, 1,2-dichloropropane, 1,3-dichloropropane, ethyl benzene, ethyl chloride, ethylene dichloride, 4-ethyl toluene, freon 11, freon 12, freon 113, freon 114, methyl benzene, methyl chloride, methyl chloroform, perchlorethylene, 1,1,2,2-tetrachlorobenzene, toluene, 1,2,3-trichlorobenzene, 1,2,4- trichlorobenzene, 1,1,2-trichlorobenzene, trichloroethylene, 1,2,4-trimethylbenzene, 1,3,5-trimethylbenzene, vinyl benzene, vinyl chloride, vinylidene chloride, o,m,p-xylene

---

Source: Information from EPA (1988).

Note: Acronyms used are as follows:
ECD—electron capture detection
GC—gas chromatography
GC/MS—gas chromatography/mass spectrometry
GC/FID—gas chromatography/flame ionization detection
HPLC—high performance liquid chromatography
HRGC/HRMS—high-resolution gas chromatography/ high-resolution mass spectrometry
PDIFD—Preconcentration and direct flame ionization detection
PUF—polyurethane foam
UV—ultra violet
XAD-2—type of resin adsorbent

- Visible absorption spectrometry—Used for analyzing metals such as hexavalent chromium that are collected on a polyvinyl chloride membrane or equivalent filter
- Inductively coupled plasma-atomic emission spectrometry—Used for analyzing metals that are collected on a mixed cellulose ester filter or equivalent
- Constant volume sampling analyzer—Used to analyze carbon monoxide, carbon dioxide, and nitrogen oxides, generally from automobile emissions

- Superfluid extraction/gas chromatograhy—Used to analyze volatile and semivolatile compounds collected on a sorbent bed
- Colorimetric analysis—Used to analyze mercaptans collected with a midget bubbler with coarse porosity frit
- Gravitation measurements—Used to analyze for dusts and other heavy particulates that are collected on filters
- Optical sizing—Used for quantifying small particulates that are collected on filters

Methods of point source emissions monitoring, also termed "stack sampling," are regulated under 40 CFR Part 60, Appendix A. Selected examples of these EPA-approved monitoring methods are presented in Table 4.2.

AIR POLLUTION ABATEMENT EQUIPMENT

Air pollution abatement equipment is used for sources that are regulated or that have the potential to emit significant amounts of toxic or other compounds into the atmosphere. Examples of types of air pollution abatement

TABLE 4.2   Selected EPA-Approved Monitoring Methods for Point Source Emissions

Method	Description/Applications
Method 2	Method uses a Type S pitot tube to determine stack velocity and volumetric flow rate.
Method 4	Method uses sampling train with probe heater, condenser, and vacuum system to determine moisture content of stack gases.
Method 5	Method uses sampling train with a glass fiber filter and impingers to collect particulates for analysis.
Method 6C and 7E	Methods allow for use of a gas analyzer to measure concentrations of sulfur dioxide and nitrogen oxides emissions, respectively, from stationary sources.
Method 12	Method allows for use of inorganic lead sample train to collect samples for analysis by atomic absorption spectrophotometer.
Method 17	Method allows for in-stack filtration (collection) of particulates.
Method 21	Method allows for determination of volatile organic compound leaks
Method 25A	Method allows for determination of total gaseous organic concentration using a flame ionization analyzer.

Source: Information from 40 CFR Part 60, Appendix A.

equipment include gravitational and inertial separation, filtration, gas-solid adsorption, electrostatic precipitators, liquid scrubbers, combustion units, and combination systems (Clayton and Clayton 1991).

Gravitational and Inertial Separation

Gravitational and inertial separators are typically used for particulate removal. Also known as "mechanical collectors," these abatement units use gravity or inertia to remove relatively large particles (i.e., 50 μm or greater) from suspension in a moving gas stream. Abatement unit types include settling chambers, inertial separators, dynamic separators, and cyclones.

Settling Chambers. This type of air abatement equipment represents the oldest type of air pollution control device. Essentially, a settling chamber uses the forces of gravity to separate large particles, such as dust and mist, from a gas stream. The device is typically used as a precleaning device before the gas stream is sent to a more energy-intensive and sophisticated air pollution control device. An example of this type of emission control is presented in Figure 4.2 (from Clayton and Clayton, 1991).

Inertial Separators. Inertial separators are simple and cost effective air pollution control devices; however, they are not particularly efficient and, consequently, are used mainly as precleaning devices. Inertial separators include dry-type collectors that utilize the inertia of particles to provide for particulate-gas separation.

Dynamic Separators. Dynamic separators are moving devices. Also termed rotary centrifugal separators, these devices use centrifugal force to separate particulates from the airstream. The particulates are concentrated on impellers, and the filter cake is dropped into a hopper.

Cyclones. Cyclones are simple and economical mechanical collectors. There are three major elements of a cyclone: a gas inlet that produces the vortex; an outlet for discharged, cleaned gas; and a discharge for particulates (dust). An example of a cyclone that has a tangential inlet and an axial discharge is presented in Figure 4.3 (from Clayton and Clayton, 1991).

Wet Cyclones. Although cyclones typically are thought of as dry mechanical devices, there are also wet cyclones that have an inlet for water or some other fluid to be impacted with the incoming particles. The fluid rinses the particles from the airstream. Although it costs more to operate this type of cyclone than to operate a dry cyclone, the efficiencies are typically greater.

Multiple Cyclones. Multiple cyclones consist of several cyclones—usually in parallel, and infrequently in series. Actually, this configuration is not typical,

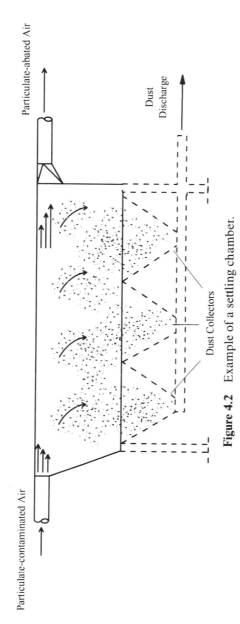

Figure 4.2 Example of a settling chamber.

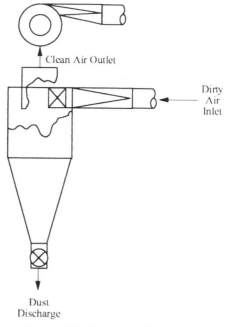

Figure 4.3   Example of a cyclone.

but might be used in cases where there is heavy dust load and a single cyclone would prove inefficient.

Filtration

Filtration uses a filtering medium for removal of particulate matter from gas streams by retention of particles in and around the pores of the filter. Air emissions filtration devices include woven fibrous mats and aggregate beds, paper filters, and fabric filters. Filtration units are typically considered reliable and inexpensive to operate.

Fibrous Mats and Aggregate Beds—General. These types of filtration devices have high porosity, the predominant forces for filtration being impaction, impingement, and surface attraction. Mats may be made of glass fibers, stainless steel, brass, or aluminum. Efficiencies can be very high, even with low dust loadings in the input airstream.

Designs for aggregate beds are numerous. Sand is often used, although glass fiber beds have been demonstrated. In addition, many designs are available for "self-cleaning" the beds or filters. An example of a self-cleaning mat that uses a water spray is shown in Figure 4.4 (from Clayton and Clayton, 1991). This type of filter, commonly termed a "wet filter," is sprayed continuously while the airstream flows through the filter. The particles are dislodged by the water and flow with it into a sump.

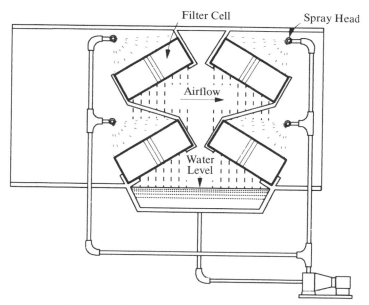

Figure 4.4 Example of a self-cleaning fiber mat.

Paper Filters. These filters are typically used for applications that require ultrahigh efficiencies, such as clean rooms in semiconductor manufacturing or other equivalent applications in the aerospace industry, hospitals, and data processing centers. Although termed "paper filters," these filters can be made of glass microfibers, minerals, or other materials. Sizes and shapes vary, depending on the application. Diffusion is the common capture mechanism for this type of filter, and efficiencies are typically as high as 99.97%.

Fabric Filters. Fabric filters, also termed "baghouses," are suitable for removing particles as small as 0.5 µm. The filters are typically made of woven or felted synthetic fabric. The particle-ladened gas passes through the filter, and the particles are collected through several mechanisms. These include interception and impaction of the particles on the fabric filters, diffusion, electrostatic attraction, and gravitational setting within the filter pores. Once a cake is built up, there is a mechanism, such as a shaker or air pressure, to remove the cake, which falls into a hopper or other containment. An example of a shaker-type baghouse is shown in Figure 4.5 (from Clayton and Clayton, 1991).

Gas-Solid Adsorption

Adsorption is a diffusion process whereby certain gases are retained selectively on the surface or in the pores of a selected solid. Typical materials used in this method of air pollution abatement include activated carbon, alumina, bauxite, and silica gel; activated carbon is the most popular type because of its

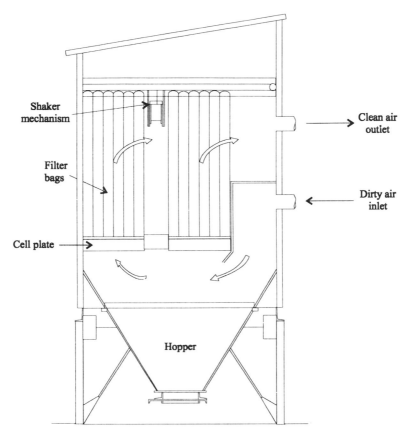

Figure 4.5   Example of a fabric filter.

low cost and its ability to be regenerated. In addition, the micropore structure of the carbon allows for a large surface area for a given weight of carbon. Types of gas-solid adsorption include physical adsorption, polar adsorption, and chemical adsorption.

Physical Adsorption.   This process involves a stream of gas passing through the adsorption material, with the result that adsorbent molecules in the gas stream are "driven" from the relatively homogeneous gas phase into the pores of the material. At the surface of the adsorbing solid, molecules are attracted and deposited at a rate that depends on concentration, ease of saturation of the material, and temperature.

Once the solvent adsorbed on the sorbent material reaches a saturation point, "break through" occurs. Usually, physical adsorption units are operated in series and in round-robin fashion, so that the unit that is saturated can be taken off-line and regenerated. A pictorial example of this type of operation is presented in Figure 4.6.

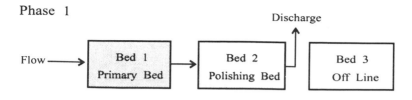

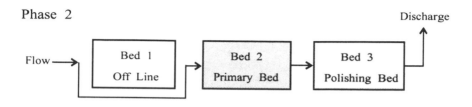

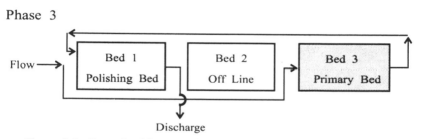

Figure 4.6 Round-robin operation of a three-bed carbon adsorption unit.

Polar Adsorption. Polar adsorption is a process whereby molecules of a given adsorbate are attracted to and deposited on the surface of an adsorbing solid by virtue of the polarity of the adsorbate. The medium is typically siliceous adsorbents, which provide for selective removal of gas molecules from an airstream. Some synthetic materials of uniform pore sizes have the ability to segregate molecules based on their shape.

Chemical Adsorption. Chemical adsorption takes place when a chemical reaction occurs between the adsorbed molecule and the solid adsorbent, resulting in the formation of a chemical compound. Typically, in this type of adsorption the forces holding the adsorbate to the solid are so strong that the reaction is irreversible, so that onsite regeneration is usually not an option.

Electrostatic Precipitation

During the electrostatic precipitation process, particulates are separated from a gaseous stream through the use of electrostatic forces. As with several other types of air pollution control devices, the particulates are deposited on solid surfaces for subsequent removal. Removal of particles 20 μm or less (with removals to the submicron level, depending on the device) are common. Airstreams that have particles greater than 20 μm must first pass through a mechanical precleaning device, such as a settling chamber or cyclone, before being sent to the electrostatic precipitation unit. The devices are often operated in sequential stages to afford very high efficiencies.

The mechanisms involved in electrostatic precipitation include:

- Electrically charging entrained particulate matter in the first stage of the unit
- Providing a voltage gradient to propel the charged particulate matter onto a grounded surface, on which it is collected
- Removal of the particulate matter from the collection surface, generally through "rapping" of the surface at selected points

Plate-Type Precipitators. This type of electrostatic precipitator uses a series of parallel, vertical grounded steel plates between which the gas flows. Spacing, size, and shape of the plates are dependent on application.

Pipe-Type Precipitators. This type of precipitator is typically configured with high-voltage electrodes positioned at the center line of a grounded pipe. Dust is deposited on the inner walls of the grounded pipe. Sizing of the unit is based on airstream parameters, such as particle size and air velocity, as well as dust resistivity and total collection area.

Liquid Scrubbing

During liquid scrubbing, a suitable liquid is used to remove contaminants— such as particulates, chlorides, sulfur compounds, and other compounds— from a gas stream. Removals can be in the submicron range. Scrubbing geometry, media, and scrubbing liquor are the design variables that must be optimized for optimal efficiency of the unit. The spent scrubbing liquor is typically sent for treatment before discharge.

Spray Chambers. Spray-type scrubbers are useful in collecting particulate matter. The device uses spray nozzles to atomize the liquid. Particles collide with the droplets and are entrained. Because of the simple design of the system, the scrubbing liquid can be recycled and still be effective even with relatively high concentrations of suspended solids. A mist eliminator is necessary for this type of device. An example of a vertical spray tower is shown in Figure 4.7.

Packed Towers. Packed towers are essentially spray chambers that contain packing material, as shown in Figure 4.8. The packing material provides a

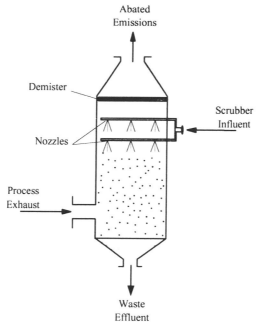

Figure 4.7 Example of a vertical spray tower.

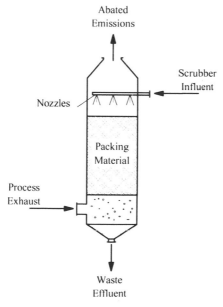

Figure 4.8 Example of a packed tower.

large surface area for optimizing mass transfer of the particulates to the liquid. Packed towers are typically used to remove gaseous contaminants from airstreams. The packing may serve as a mist eliminator, so a special device for this purpose may not be needed with some configurations.

Orifice Scrubbers. Orifice scrubbers use the carrier gas velocity and turbulence to promote dispersal of the scrubbing liquid and contact of the contaminated gas stream with the dispersed liquid. These devices make use of internal geometric designs in an attempt to supply scrubbing liquid uniformly across the cross section of the gas flowing through the unit. In the contact zone, the intensely turbulent motions provide violent contact between particulate matter and the larger scrubbing liquid droplets, which have high inertia.

Although there are numerous geometries and internal designs for orifice scrubbers, the venturi scrubber—which operates at very high velocities—is a common design. This technology uses a venturi oriface (an inlet with a constricted throat to create pressure differntial) and a cyclone separator.

Combustion

Combustion, also known as incineration, converts matter entrained in exhaust gases to a less toxic form—principally carbon dioxide and water. Combustion equipment typically uses either direct flame or catalytic combustion.

Direct Flame Applications

Flares. Flares are used for the oxidation of combustible gases, typically in emergency situations. Processes that utilize flares include petroleum refining processes and research/pilot plants for the petroleum refining industry. Design features include gas stream parameters and velocity, addition of auxiliary fuel and/or air (optional), fuel-air mixing, ratio of hydrogen to carbon in the materials being combusted, flare height, heat and light emitted from the flare, and heat recovery.

Direct Flame Combustion. Direct flame combustors (incinerators) may or may not use auxiliary fuel and heat recovery to ensure efficient combustion, in order to minimize operation costs. The key factors for the success of this type of combustor include sufficient residence time of the gas in the chamber, sufficient fuel, and sufficient mixing. Generally, a heat exchanger is used to preheat incoming gases to a temperature high enough to allow for as complete combustion as possible. Applications include removal of combustible particulate matter, particularly of particulate matter of submicron size.

Catalytic Combustion. Catalytic combustors use a catalyst composed of platinum, palladium, or a similar metal to accelerate the rate of combustion and minimize the requirements for auxiliary fuel. A typical system includes a preheat burner, an exhaust fan, a catalyst, a heat exchanger, and safety mechanisms. Catalytic combustors are used where concentrations of hydrocarbons are relatively low. In some cases, the airstream is incompatible with this method of treatment, because selected constituents can "poison" the catalyst and render it ineffective. Further, this system is incompatible with airstreams containing particulates unless there is a precleaning device incorporated into the system.

REFERENCES

Clayton, George D., and Florence E. Clayton, eds., *Patty's Industrial Hygiene and Toxicology*, vol. 1 Part A, 4th ed., John Wiley & Sons, New York, 1991.

U.S. Environmental Protection Agency, *Compendium of Methods for the Determination of Toxic Organic Compounds in Ambient Air*, prepared by Engineering-Science for the U.S. Environmental Protection Agency, Research Triangle Park, NC, 1988.

BIBLIOGRAPHY

Godish, Thad (1990), *Air Quality*, 2d ed., Lewis Publishers, Boca Raton, FL.

Stensvaag, John-Mark (1991), *The Clean Air Act of 1990*, John Wiley & Sons, New York.

Wuebbles, Donald J. (1991), *Primer on Greenhouse Gases*, Lewis Publishers, Boca Raton, FL.

5

WASTE TREATMENT AND DISPOSAL TECHNOLOGIES

Waste treatment and disposal systems can incorporate many different unit processes. For example, a wastewater treatment system may include oil/water separation, neutralization, biological treatment, sedimentation, and sludge dewatering. There are many well known and widely used technologies such as activated sludge, incineration, carbon adsorption, air stripping, metal precipitation, ion exchange, reverse osmosis, distillation, landfilling, land treatment, and deep well injection. However, more stringent environmental regulation and an increasing knowledge of chemical hazards continually challenge the capabilities of waste treatment and monetary resources. These challenges have given rise to a keen interest in the development of new, innovative technologies. Many of these new technologies use familiar unit processes, but in innovative ways. Recycling technology, an important part of pollution prevention, incorporates both old and new unit processes.

In this chapter, the discussion of waste treatment and disposal technologies is divided into five major groups reflecting either the type of technology or major environmental area: physical/chemical processes, biological processes, thermal processes, soil and ground water treatment, and land-based systems.

PHYSICAL/CHEMICAL

Air Stripping

Compounds that have relatively high volatilities and low water solubilities can be transferred (stripped) from aqueous streams into the air or an airstream. Examples of compounds that can be easily air stripped include gasoline and jet fuels (with benzene, toluene, ethylbenzene, and xylenes (BTEX) as primary components of interest), solvents such as trichloroethene and tetrachloroethene, and ammonia. The strippability of a compound is related to its Henry's law constant, which is the ratio of the vapor phase concentration of the chemical in equilibrium to its concentration in water. In general, the higher the Henry's law constant, the more strippable is the compound. Similarly, inasmuch as Henry's law constant increases with temperature, higher temperatures generally improve air-stripping efficiencies.

Types of air strippers include packed towers (most common), tray towers, and spray towers. Packed towers are packed or filled with small forms made of polyethylene, stainless steel, polyvinyl chloride (PVC), or ceramic that provide large surface area-to-volume ratios that increase transfer rates into the airstream. Packed towers operate in countercurrent mode; that is, the aqueous stream enters at the top of the tower while air is blown in from the bottom, as the example schematic in Figure 5.1 shows. Channeling or short circuiting of the aqueous stream is minimized by distribution plates placed at intervals throughout the column. Design criteria for packed towers include surface area provided by the packing media, column height and diameter, and air-to-water flow rates. Tray towers also operate countercurrently; however, the aqueous stream contacts air as the liquid cascades from tray to tray to the bottom of the tower. Spray towers operate by spraying the aqueous stream into the top of the tower while air flows countercurrent to the liquid. No packing media are used in a spray tower.

The airstream exiting a stripper generally requires some type of emissions control, depending on local and regulatory requirements. Carbon adsorption is often used; catalytic oxidation is another option.

Carbon Adsorption

Carbon adsorption is a well established and widely used technology for the removal of organics from wastewaters and gaseous streams. Carbon adsorption is identified as a "best available technology" for meeting maximum contaminant levels for organic contaminants in drinking water at 40 CFR §141.61(b). Thus, as a proven technology for drinking water, carbon adsorption can reduce concentrations to very low or nondetectable levels.

In carbon adsorption, contaminants are physically attracted or adsorbed on the surface of the carbon. The adsorption capacities of carbon are high because its porosity provides a large surface area relative to its volume. "Activated" carbon is prepared from lignite, bituminous coal, coke, wood, or other organic materials such as coconut shells.

Carbon adsorption is most effective at removing organic compounds that have low polarity, high molecular weight, low water solubility, and a high boil-

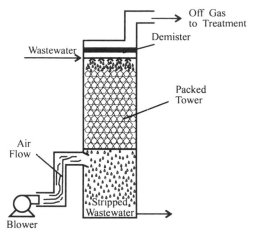

Figure 5.1   Example of a countercurrent packed tower.

ing point. The following are some general guidelines for assessing carbon adsorption efficiency (WEF 1990):

- Branch-chain compounds are more sorbable than straight-chain compounds.
- Compounds low in polarity and water solubility (less than 0.1 mg/L) are more sorbable.
- For compounds that have similar chemical characteristics, those of larger molecular size are more sorbable.
- There is a wide range in adsorption efficiency for inorganics.
- A low pH promotes adsorption of organic acids.
- A high pH promotes adsorption of organic bases.

Common examples of compounds that are amenable to carbon adsorption include aromatics such as benzene and toluene and chlorinated organics such as trichloroethene, trichloroethane, tetrachloroethene, polychlorinated biphenyls (PCBs), DDT, and pentachlorophenol. Compounds that are not adsorbed effectively by carbon include ethanol, diethylene glycol, and numerous amines (butylamine, triethanolamine, cyclohexylamine, hexamethylenediamine) (WEF 1990). In general, to be suitable for carbon adsorption, the organic concentration of a wastewater should be less than 5,000 mg/L (total organic carbon).

Most carbon adsorption units use granular activated carbon (GAC), which has an average particle diameter of about 1,500 microns. The powdered form of activated carbon (PAC) typically is less than 100 microns in diameter. PAC is used in the treatment of drinking water and in wastewater treatment (see "Activated Sludge" in the "Biological" section of this chapter). It may also be used to reduce dioxins in incinerator emissions (Roeck and Sigg 1996).

GAC may be used in fixed or "moving" beds and in the downflow or upflow mode. Fixed beds are operated in the downflow mode and, as such, provide some amount of solids filtration; however, influent solids concentration must be kept low (less than 5 mg/L suspended solids) to prevent rapid plugging of the bed. Filtered solids are periodically removed by backwashing. Upflow beds are more tolerant of solids, because they are fluidized and expanded by the wastewater entering at the bottom. In moving beds, the flow is counter-current and makeup fresh carbon is added continuously at the top of the unit while an equal amount of spent carbon is removed from the bottom.

When the adsorption capacity of a carbon unit is exceeded, there is a breakthrough of the contaminant in the treated stream. Fixed beds may be operated in series to allow continuous treatment while spent or exhausted units are recharged with fresh carbon. In series operation, there are two or more units. The majority of the contaminant is removed by the first unit in the series, with the downstream units acting as polishing units. When breakthrough occurs in the first or primary unit, it is recharged with fresh carbon and becomes a polishing unit while the next unit in the series, takes over and becomes the primary treatment unit.

Spent carbon is usually regenerated, but if the quantity is small, it may instead be disposed of by incineration or landfilling. Regeneration can be done by several different methods and may actually be part of the treatment process. Regeneration technologies include steam or thermal desorption, incineration, acid and base washing, and solvent extraction. Organic-laden streams from the regeneration process are treated, recovered, or incinerated. Regeneration by incineration results in some loss of carbon and deterioration and loss of active surface area.

Design criteria for carbon adsorption include type and concentration of contaminant, hydraulic loading, bed depth, and contact time. Typical ranges are 2 to 10 gpm/ft^2 for hydraulic loading, 5 to 30 feet for bed depth, and 10 to 50 minutes for contact time (WEF 1990). The adsorption capacity for a particular compound or mixed waste stream can be determined as an adsorption iso-therm and tested in a pilot unit. The adsorption isotherm relates the observed effluent concentration to the amount of material adsorbed per mass of carbon.

New areas in adsorption technology include carbonaceous and polymeric resins (Musterman and Boero 1995). Based on synthetic organic polymer materials, these resins may find special uses where compound selectivity is important, low effluent concentrations are required, carbon regeneration is impractical, or the waste to be treated contains high levels of inorganic dis-solved solids.

Dissolved Air Flotation

Dissolved air flotation (DAF) is used to separate suspended solids and oil and grease from aqueous streams and to concentrate or thicken sludges. It is especially effective for emulsions after they have been chemically broken. DAF is capable of producing an effluent with low oil concentration. DAF is used in many wastewater treatment systems, but it is perhaps best known in respect to hazardous waste in its association with the listed waste, K048, DAF float solids from petroleum refining wastewaters. Of course, the process itself is not what is hazardous, but the materials it helps to remove from refining wastewaters.

With DAF, air bubbles carry or float the solids and oils to the surface, from which they can be removed. The air bubbles are formed by pressurizing either the influent wastewater or a portion of the effluent in the presence of air. When the pressurized stream enters the flotation tank, which is at atmospheric pressure, the dissolved air comes out of solution as tiny, microscopic bubbles. Coagulant and flocculant chemicals are often needed to improve process removal efficiencies, as well as the dewatering characteristics of the DAF float.

Distillation

Distillation separates volatile components from a waste stream by taking advantage of differences in vapor pressures or boiling points among volatile fractions and water. There are two general types of distillation, batch or differential distillation and continuous fractional or multistage distillation.

Batch distillation is typically used for small amounts of solvent wastes that are concentrated and consist of very volatile components that are easily separated from the nonvolatile fraction. Batch distillation, which is amenable treatment to small quantities of spent solvents, allows these wastes to be recovered onsite. With batch distillation, the waste is placed in the unit and volatile components are vaporized by applying heat through a steam jacket or boiler. The vapor stream is collected overhead, cooled, and condensed. As the waste's more volatile, high-vapor-pressure components are driven off, the boiling point temperature of the remaining material increases. Less volatile components will begin to vaporize, and once their concentration in the overhead vapors becomes excessive, the batch process is terminated. Alternatively, the process can be terminated when the boiling point temperature reaches a certain level. The residual materials that are not vaporized are called still bottoms. A schematic of an example batch distillation unit is shown in Figure 5.2.

If a waste contains a mixture of volatile components that have similar vapor pressures, it is more difficult to separate these components and continuous fractional distillation is required. In this type of distillation unit, a packed tower or tray column is used. An example of such a unit is shown in Figure 5.3.

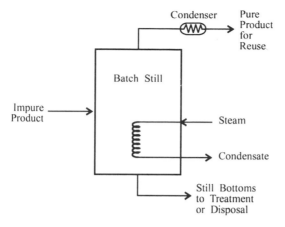

Source: Adapted from EPA (1987).

Figure 5.2 Example of a batch distillation unit.

Steam is introduced at the bottom of the column while the waste stream is introduced above and flows downward, countercurrent to the steam. As the steam vaporizes the volatile components and rises, it passes through a rectification section above the waste feed. In this section, vapors that have been condensed in the process are refluxed to the column, contacting the rising vapors and enriching them with the more volatile components. The vapors are then collected and condensed. Organics in the condensate may be gravity-separated from the aqueous stream, after which the aqueous stream can be recycled to the column.

Factors affecting distillation include the vapor pressures of the volatile components; column temperature, pressure, and internal structure (packing or trays); and concentrations of oil and grease, total dissolved solids, and total dissolved volatile solids. The higher the vapor pressure of a volatile component, the more easily it can be distilled or stripped from solution. Column pressure influences the boiling point of the liquid stream. Column temperature can be reduced by operating the column under a vacuum. Trays used in columns (bubble cap, sieve, valve, and turbo-grid) do not plug as easily with solids as do packed media and can be used with a wider range of liquid and vapor flow rates. Packed columns have lower pressure drops per stage, are more corrosion resistant, and reduce foam by distributing the liquid flow more uniformly. Oil and grease, total dissolved solids, and total dissolved volatile solids affect the partial pressures and solubilities of volatile components.

Ion Exchange

Ion exchange is an adsorption process in which ionic species are adsorbed from solution by changing places with similarly charged ions on the exchange

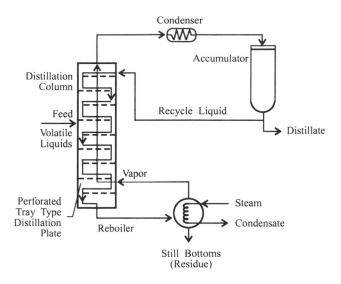

Source: EPA (1987).

Figure 5.3   Example of a continuous fractional distillation system.

media. Ion exchange is used primarily to remove metals, although non-metallic inorganic and organic ions can also be removed. Metals that can be removed by ion exchange include barium, cadmium, hexavalent chromium, copper, lead, mercury, nickel, selenium, silver, uranium, and zinc (Krishnan et al. 1993). Nonmetallic ions that may be removed by ion exchange include nitrate, sulfate, cyanide (when complexed with iron), and some organic acids, phenols, and amines under special conditions.

The adsorptive materials used for ion exchange are called zeolites. Naturally occurring zeolites belong to a group of hydrous aluminum silicate minerals. Synthetic zeolites or resins are based on organic polymers and are more commonly used today. Ion exchange resins that adsorb positively charged ions are cationic, and those that adsorb negatively charged ions are anionic. These resins can be designed to remove certain ions with greater specificity than other ions; however, ions with the same charge (positive, negative) can still compete for exchange sites and adversely affect the removal efficiency of the targeted ion if they are present in high enough concentrations.

Ion exchange units can be operated in a parallel or series and in downflow or upflow (fluidized) mode. A unit is regenerated when the target ion appears in the effluent (breakthrough). Ion exchange resins are regenerated by washing with a concentrated solution containing the original exchangeable ion. This results in the desorption of the target ion from the resin and replacement by the original ion. Hydrogen-based cationic resins are regenerated with acid solutions such as sulfuric, nitric, or hydrochloric acid. Anionic resins that are hydroxide based are regenerated with caustic solutions such as sodium hydroxide. Sodium chloride solution is used to regenerate sodium-based cationic and chloride-based anionic resins. Concurrent regeneration occurs when the regenerant solution flows through the unit in the same direction as the wastewater; countercurrent regeneration flows in the opposite direction and requires less solution.

Pretreatment of aqueous streams may be required prior to using ion exchange. Suspended solids that can plug an ion exchange unit should be reduced to the 10 micron level. Organics that can foul resins can be removed by carbon adsorption. Iron and manganese, commonly present in ground waters, should be removed because they precipitate on the resin.

Membrane Filtration

Membrane filtration describes a number of well-known processes, including reverse osmosis, ultrafiltration, nanofiltration, microfiltration, and electrodialysis. The basic principle behind this technology is the use of a driving force (electricity or pressure) to filter particles, ions, and organic molecules through a membrane, producing a "clean" stream on one side and a concentrated stream on the other. Membrane filtration, discussed in this section in relation to waste treatment, also has many industrial applications, such as water desalination and softening, product concentration and purification, and metal recovery. Applications in waste treatment include metals removal/recovery, oil/water separation, and removal of toxic organic compounds.

The individual membrane filtration processes are defined chiefly by pore size, although there is some overlap. The smallest membrane pore size is used in reverse osmosis (0.0005 to 0.002 microns), followed by nanofiltration (0.001 to 0.01 microns), ultrafiltration (0.002 to 0.1 microns), and microfiltration (0.1 to 1.0 microns). Electrodialysis uses electric current to transport ionic species across a membrane. Micro- and ultrafiltration rely on pore size for material separation, reverse osmosis on pore size and diffusion, and electrodialysis on diffusion. Separation efficiency does not reach 100% for any of these membrane processes. For example, when used to desalinate/soften water for industrial processes, the concentrated salt stream (reject) from reverse osmosis can be 20% of the total flow. These concentrated, yet still dilute streams, may require additional treatment or special disposal methods.

Reverse osmosis and electrodialysis are used to separate metals from electroplating bath rinse waters in the metal finishing industry. Ultrafiltration is used for oil/solids/water separation, metals recovery, and paint recovery from anodic paint processes. It has also been tested as an innovative technology for removal of solids and residual organics following biological treatment of contaminated ground water (EPA 1995e). Microfiltration is used for oil/solids/water separation and metals removal/recovery. Although nanofiltration has been used successfully for water treatment (removal of hardness, organic compounds, and viruses), it has not been used much for waste treatment. Testing has also been conducted with another type of membrane filtration process, pervaporation, which uses a vacuum to induce transfer of volatile organics from a liquid across the membrane (EPA 1994d). Membrane filtration of gaseous streams, which is used on industrial process streams to recover volatile organics, has been tested as posttreatment for air stripping and soil-venting gases.

Membranes are subject to fouling, which can be caused by metal oxides, precipitating salts, colloids, and biological growth. Cleaning agents include acids, alkalines, oxidizers, detergents, and organic solvents. Pretreatment prior to membrane filtration is required to reduce heavy solids and remove free oil.

Neutralization

Neutralization is the pH adjustment of an acidic or caustic waste to a more neutral range. Neutralization is a very common treatment step for wastewaters and gases; it is used less frequently with solid wastes because the pH is less of a problem with these wastes. The most commonly used neutralizing agents are lime or calcium hydroxide for acidic wastewaters and sulfuric acid for caustic wastewaters. Other chemicals used for wastewater neutralization include sodium hydroxide and magnesium hydroxide for acidic streams and hydrochloric acid and carbon dioxide for caustic streams. At facilities where both acidic and caustic streams are generated, commingling these streams helps reduce the amount of neutralizing chemicals required. These acid/caustic waste streams, however, can be commingled only where they are compatible, that is, where the mixture will not generate unwanted precipitants or chemical reactions. Neutralization of acid gases with liquid caustic solutions (wet scrubbing) is very common. For example, the flue gas from an incinerator may be scrubbed with a soda ash (sodium carbonate) solution to remove acid gases. Wet scrubbing is normally done in a packed tower; the packing referred to usually consists of small plastic forms that provide a high degree of surface area per unit volume in order to maximize the contact of the gas with the scrubbing solution. There are also dry scrubbing systems for acid gases whereby the neutralizing agent—for example, lime—is sprayed into the gas stream.

Oil/Water Separation

Gravity separation of oil and water is a very simple process, often used as a pretreatment step in an overall wastewater treatment system. Under quiescent conditions, free oil floats to the surface where it can be skimmed off. Oil/water emulsions may be broken for separation by using coagulants, acids, pH adjustment, heat, centrifuging, or high-potential alternating current (Noyes 1994). The API separator and corrugated plate interceptor (CPI) are common types of oil/water separators. API separator sludge from petroleum refining wastewaters is listed as a hazardous waste, K051. It is not the API separator itself that makes K051 hazardous, but, as in the DAF process discussed earlier in this section, the hazardous materials the waste contains.

Oxidation and Reduction

Oxidation and reduction (redox) reactions are used for both partial and complete degradation of many organic and inorganic compounds. A substance is oxidized when its oxidation state is increased; likewise, it is reduced when its oxidation state is reduced. For inorganic compounds, oxidation involves the loss of electrons and reduction means a gain in electrons. Redox reactions for organics, however, are more complicated and may include the transfer of electrons, a hydrogen atom, an oxygen atom, a hydroxyl radical, a chlorine atom, a chlorinium ion (Cl^+), or similar species (Weber 1972). Examples of compounds that are treated by redox processes include alcohols, phenols, and cyanide. Common chemical oxidants for waste treatment include chlorine, ozone, and hydrogen peroxide.

Oxidizers do not discriminate between compounds and are capable of reacting with any oxidizable compounds in a waste stream. Oxidation is used either to completely degrade a compound or, quite often, to partially degrade a compound to a less toxic form or intermediate that can be discharged or, if necessary, treated further by another process.

Factors affecting oxidation processes include pH, the type and quantity of oxidizable compounds in the waste, and metals that can react with the oxidizing agent. The pH affects the oxidation rate by changing the free energy of the overall reaction, changing the reactivity of the reactants, and/or affecting specific OH^- ion or H_3O^+ ion catalysis (Weber 1972). The presence of oxidizable compounds in addition to the target compound increases the amount of oxidant required. Metal salts, especially those of lead and silver, can also react with the oxidizer and increase the required dosage or interfere with the treatment process (Noyes 1994).

Chlorine. Chlorine is a well-known disinfectant for water and wastewater treatment; however, it can react with organics to form toxic chlorinated compounds such as the trihalomethanes bromodichloromethane, dibromochloromethane, and chloroform. Chlorine dioxide may be used instead, inasmuch as it does not produce the troublesome chlorinated by-products generated by chlorine. In addition, by-products formed by chlorine dioxide oxidation tend to be more readily biodegradable than those produced by chlorine; however, chlorine dioxide is not suitable for waste streams containing cyanide.

Cyanide destruction by alkaline chlorination is a widely used process. In alkaline chlorination, cyanide is first converted to cyanate with hypochlorite at a pH greater than 10. A high pH is required to prevent the formation of cyanogen chloride, which is toxic and may evolve in gaseous form at a lower pH. With additional hypochlorite, cyanate is then oxidized to bicarbonate, nitrogen gas, and chloride. The pH for this second stage is 7 to 9.5 (Weber 1972).

Ozone. A primary advantage of oxidation with ozone is the avoidance of chlorinated by-products of the reaction. Excess ozone that is not consumed in the reaction decomposes to oxygen. The main disadvantage is the high electrical cost of producing ozone, which is generated onsite from dry air or oxygen by high-voltage electric discharge.

Ozone can be used to completely oxidize low concentrations of organics in aqueous streams or to partially degrade compounds that are refractory or difficult to treat by other methods. Compounds that can be treated with ozone include alkanes, alcohols, ketones, aldehydes, phenols, benzene and its derivatives, and cyanide. Ozone readily oxidizes cyanide to cyanate, however, further oxidation of the cyanate by ozone proceeds rather slowly and may require other oxidation treatment, such as alkaline chlorination, to complete the degradation process.

Ozone is only slightly soluble in water. Thus, factors that affect the mass transfer between the gas and liquid phases are important; these include temperature, pressure, contact time, contact surface area (bubble size), and pH.

Ozonation can be enhanced by the addition of ultraviolet (UV) radiation. This combination can be effective in degrading chlorinated organic compounds and pesticides. In addition, metal ions such as iron, nickel, chromium, and titanium can act as catalysts, as can ultrasonic mixing.

Hydrogen Peroxide. Hydrogen peroxide is typically used as an oxidizer in combination with UV light, ozone, and/or metal catalysts. Hydrogen peroxide with iron as a catalyst is known as Fenton's reagent. Hydrogen peroxide/UV light has been shown to be effective in oxidizing benzene, chlorobenzene, chloroform, chlorophenol, 1,1-dichloroethane, dichloroethene, phenol, tetrachloroethene, 1,1,1-trichloroethane, trichloroethene, toluene, xylenes (EPA 1993a, 1993c, 1993d, 1994e), and many other organic compounds.

Precipitation

Precipitation processes have been used for many years to remove metals from aqueous streams. Metals precipitation is accomplished by pH adjustment and the addition of a chemical reagent which forms a precipitant with the metal that can be settled out and separated from the aqueous stream. An example of a chemical precipitation system is shown in Figure 5.4. Chemical reagents such as calcium hydroxide (lime), sodium hydroxide (caustic), and magnesium oxide/hydroxide can be used to precipitate arsenic, cadmium, trivalent chromium, copper, iron, lead, manganese, nickel, and zinc. Hydroxide precipitation, particularly with lime, is the most common precipitation process, because the chemical reagents are readily available, relatively inexpen-

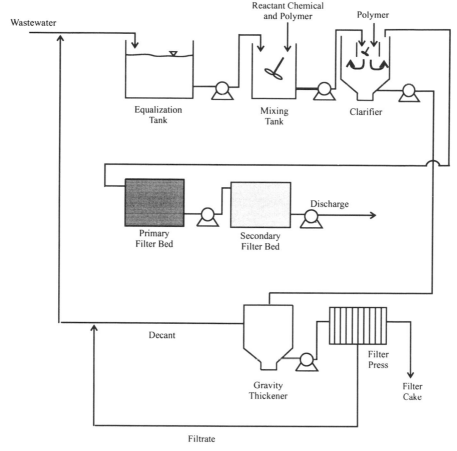

Figure 5.4 Example of a chemical precipitation system.

sive, easy to handle, and form sludges that dewater easily. Sulfide precipitation with reagents such as sodium sulfide and ferrous sulfide can be used with cadmium, cobalt, copper, iron, mercury, manganese, nickel, silver, tin, and zinc. Carbonate precipitation with reagents such as sodium carbonate (soda ash) and calcium carbonate can be used with cadmium, lead, nickel, and zinc. Because the sulfide forms of metals are much less soluble than their hydroxide counterparts, sulfide precipitation can treat metals to lower concentrations. Although very insoluble, metal sulfides form very fine, hard-to-settle particles and generate large quantities of sludge.

The typical precipitation process, which takes place in a series of tanks, begins with chemical addition, followed by flocculation and coagulation of precipitated solids, sludge thickening, and finally, sludge dewatering. Sand filtration of the wastewater effluent to remove residual solids may be used as a final polishing step. Chemicals may be added to aid flocculation and coagulation and to produce a denser and more easily dewatered sludge. Typical chemical aids are alum (aluminum sulfate), ferric chloride, and proprietary polyelectrolytes. Precipitation of arsenic first requires oxidation of the arsenite form to arsenate. Hexavalent chromium requires reduction to the trivalent form.

Precipitation is affected by pH, solubility product of the precipitant, ionic strength and temperature of the aqueous stream, and the presence of metal complexes. For each metal precipitant, there is an optimum pH at which its solubility is lowest and, hence, the highest removal rate may be achieved. When an aqueous stream contains various metals, the precipitation process cannot be optimized for each metal, sometimes making it difficult to achieve effluent targets for each. Solubility products depend on the form of the metal compound and are lowest for metal sulfides, reflecting the relative insolubility of these compounds. For example, the solubility product for lead sulfide is on the order of 10^{-45}, as compared with 10^{-13} for lead carbonate. Metal sulfides and hydroxides are more soluble at higher temperatures, whereas carbonates are less soluble. When metals are complexed or chelated with other ions or molecules (for example, cyanide and ethylenediaminetetraacetic acid (EDTA)), they are more soluble, may not precipitate easily, and require additional or alternate forms of treatment.

Solvent Extraction

With solvent extraction, organics are separated from a waste by mixing the waste with a solvent capable of dissolving or extracting the organics. Solvent extraction may also be referred to as liquid extraction or liquid-liquid extraction when used to treat liquids or wastewaters. The extraction process can have multiple stages and can be operated where the solvent and waste pass concurrently or countercurrently.

After mixing, the solvent and waste are separated. The solvent with dissolved organics is called the extract. The waste remaining after extraction is called the raffinate. The extract may be sent to a distillation or steam stripping unit to separate the dissolved organics from the solvent, and the solvent can be recycled back to the extraction process. The raffinate may require additional treatment or may be disposed of or incinerated.

Factors affecting solvent extraction include type of solvent and the relative solubilities of the organic components in the solvent, selectivity value (the ratio of the equilibrium concentrations of the organic in the solvent and in the waste), temperature, pH, mixing, and settling time for separation. Temperature and pH affect equilibrium conditions. The type and quantity of mixing affects the degree of contact and transfer efficiency between solvent and waste. Sufficient time must be allowed for settling to separate the extract and raffinate (EPA 1988).

Stabilization/Solidification

Stabilization and solidification are technologies that are often used together to improve or strengthen the physical nature of a waste and to bind, immobilize, or otherwise prevent the migration of toxic constituents contained in the waste. Stabilization generally refers to processes that reduce the toxicity, leaching, or mobility of toxic constituents, perhaps through chemical reactions, but the physical form of the waste is not necessarily changed or improved. Solidification generally refers to processes that change the physical nature of the waste to make it more manageable, increase its structural strength and load-bearing capacity, or reduce its permeability, but not necessarily by chemical reaction. Many treatment technologies involve characteristics of both stabilization and solidification, which makes distinctions between the two processes unclear at times, so that the two terms are often used together. Applying the correct definition is not as important as understanding the changes in the waste characteristics: Are toxic constituents rendered less toxic or mobile, and can the waste be handled easily and withstand the physical stresses during and after disposal?

Some of the most common stabilization/solidification processes are those using cement, lime, and pozzolanic materials. These materials are popular because they are very effective, plentiful, and relatively inexpensive. Other stabilization/solidification technologies include thermoplastics, thermosetting reactive polymers, polymerization, and vitrification. Vitrification is discussed in the "Soil and Ground Water" section of this chapter; the other stabilization/solidification processes are discussed below in the following paragraphs.

Cement, Lime, and Pozzolanic Materials. Cement is the chemical binder used in making concrete, a chemically cured mixture of cement, water, and aggregate (usually sand and gravel). With sufficient water for hydration, cement forms a calcium alumino-silicate crystalline structure that can physically and/or chemically bind with toxic constituents such as metal hydroxides and carbonates. Use of lime (calcium hydroxide) involves similar reactions where alumino-silicates are supplied by the waste or treatment additives. Pozzolans are siliceous materials that can combine with lime to form cementitious materials.[1] Common pozzolanic materials are fly ash, blast furnace slags, and cement kiln dust. The high pH of cementitious binders protects against acid conditions that can cause metal leaching. Organics interfere with hydration and the formation of the crystalline structure binding the waste. This interference can be minimized by treatment additives such as clays (natural and organically modified), vermiculite, or organically modified silicates. Other waste constituents that can reduce the effectiveness of cementitious processes include sulfates, chlorides, borates, and silt.

[1] Pozzolans derive their name from Pozzuoli, Italy (near Mount Vesuvius), where a siliceous volcanic ash is found.

Thermoplastics. Wastes containing heavy metals and/or radioactive materials may be mixed with thermoplastics. In this process the waste is mixed with a molten thermoplastic material such as bitumen, asphalt, polyethylene, or polypropylene. Wastes are normally dried, heated, mixed with the thermoplastic, cooled, and finally placed in containers for disposal. The process is not suitable for organic wastes unless air emission controls are installed, because the high temperatures will drive off volatile components. It is also not suitable for hygroscopic wastes, which absorb water, expanding and cracking the thermoplastic. Strongly oxidizing constituents, anhydrous inorganic salts, and aluminum salts, likewise, are unsuitable for treatment with thermoplastics.

Thermosetting Reactive Polymers. Materials used as thermosetting polymers include reactive monomers urea-formaldehyde, phenolics, polyesters, epoxides, and vinyls, that form a polymerized material when mixed with a catalyst. The treated waste forms a sponge-like material that traps the solid particles, but not the liquid fraction; the waste must usually be dried and placed in containers for disposal. Because the urea-formaldehyde catalysts are strongly acidic, urea-based materials are not generally suitable for metals that can leach into the untrapped liquid fractions. Thermosetting processes have good utility for radioactive materials and acid wastes.

Polymerization. Spills of chemicals that are monomers or low-order polymers can be polymerized by adding a catalyst. Compounds that may be treated by polymerization include aromatics, aliphatics, and other oxygenated monomers such as vinyl chloride and acrylonitrile.

Steam Stripping

Steam stripping is used to separate volatile components from wastewater. This technique is very similar to distillation, which is discussed earlier in this chapter. Steam stripping differs from distillation, however, in that there is no rectifying section and the waste stream is introduced at the top of the column (packed or tray). Steam is introduced at the bottom, and as it rises countercurrent to the downflowing wastewater, it vaporizes the volatile components. The vapor stream is collected overhead and condensed. Steam stripping is amenable to treating volatile components that phase-separate from water. The organic and aqueous phases are separated after condensation; the organics may be recovered or disposed of, and the aqueous stream can be recycled to the stripper. A column with rectification (i.e., distillation) is required if the volatile components are not phase-separable, in order to obtain a high concentration of volatiles in the overhead stream (EPA 1995f).

Supercritical Fluid Extraction

Supercritical fluid extraction takes advantage of the enhanced solvent power of compounds at supercritical temperature and pressure conditions to extract organics from oily and/or organic liquids and sludges. Carbon dioxide is a particularly attractive solvent at supercritical conditions (greater than 73 atm and 31°C), because it is nontoxic and leaves no residue when it is reverted to gaseous form during the recovery step. Carbon dioxide is not suitable for wastes with high alkalinity, however, because excessive amounts of the supercritical solvent are consumed in the formation of carbonates and bicarbonates (Noyes 1994). Applications using carbon dioxide have been developed to replace toxic organic solvents in the cleaning of products and equipment.

Supercritical Water Oxidation

Supercritical water oxidation is like wet air oxidation (discussed in the following paragraph) in that it uses high temperature and high pressure to oxidize organics. Temperature and pressure conditions in this case, however, are higher (greater than 218 atm and 374°C), so that water has supercritical properties, between those of a liquid and a gas. Organics are easily solubilized in supercritical water, and, once solubilized, they are completely oxidized under the high temperature and pressure conditions. Reaction times are very short (ranging from a few seconds to less than a minute) and treatment efficiencies very high (essentially 100%). Supercritical water oxidation has the capability of treating many different kinds of organics including polycyclic aromatic hydrocarbons, chlorinated hydrocarbons, PCBs, paint, oil, dyes, pulp and paper wastes, chemical warfare agents, and missile propellants (Jensen 1994). Wastes must be either aqueous or in slurry form.

Reaction vessels for supercritical water oxidation must be highly corrosion resistant because of the aggressive nature of supercritical water and oxidation reaction products at extreme temperatures and pressures. Supercritical oxidation of PCBs and some chlorinated hydrocarbons can be difficult, because acids are formed that are highly corrosive to almost any materials used for reactor components (Jensen 1994). Inorganic salts can be a problem, because they are sparingly soluble under these conditions and will precipitate on the vessel walls and components.

Wet Air Oxidation

With wet air oxidation, increased temperature and pressure are used to oxidize dilute concentrations of organics and some inorganics, such as cyanide, in aqueous wastes that contain too much water to be incinerated but are too toxic to be treated biologically. In general, wet air oxidation provides primary treatment for wastewaters that are subsequently treated by conventional methods. This technology can be used with wastes that are pumpable, such as slurries and liquids.

Waste streams that are treated by wet air oxidation generally are those having dissolved or suspended organic concentrations from 500 to 50,000 mg/L. Below 500 mg/L, oxidation rates are too slow, and above 50,000 mg/L, incineration may be more feasible (EPA 1991).

Operating parameters include temperature, pressure, oxygen concentration, and residence time. Materials of construction include stainless steel, nickel, and titanium alloys (the latter for extremely corrosive wastes containing heavy metals). Vented gases from the process may require scrubbing or other emission controls.

Because of the high energy input required, wet air oxidation is relatively expensive. To reduce energy demand, the incoming wastewater can be preheated by heat exchange with the treated effluent. Operating temperatures may be lowered by using a catalyst. Typically, removal of the catalyst from the effluent is necessary to avoid contamination and to minimize operating costs.

BIOLOGICAL

Biological processes are effective in treating a wide variety of organic and inorganic compounds. Most biological processes are aerobic and use either the oxygen available in supplied air or pure oxygen. The chief products of aerobic biodegradation of carbonaceous and nitrogenous compounds are carbon dioxide, ammonia, water, and biomass (cell growth of the microorganisms). The ease with which a compound is degraded is related to its chemical structure. Some compounds are not degraded completely, but are instead broken down into simpler compounds.

The biodegradability of hazardous compounds according to general structure has been summarized by Pitter and Chuboda (1990) from a number of studies. They have found that the biodegradability of hydrocarbons generally decreases in the following order: alkanes and alkenes with straight alkyl chains, mono- and dicycloalkanes, and aromatic hydrocarbons, including those that are polycyclic. Mono- and dicyclic aromatic hydrocarbons are usually more biodegradable than cycloalkanes. Chain branching in alkanes and alkenes decreases biodegradability, although the biodegradability of alkanes and alkenes is essentially the same if they have the identical chain configurations. Monocycloalkanes are usually readily biodegradable, although exceptions have been noted with specific compounds. Polycycloalkanes are less biodegradable, especially tetra- and higher polycycloalkanes. Naphthalene, lower alkylbenzenes, and phenanthrene are aromatic hydrocarbons that are biodegraded relatively easily. The biodegradability of alkylbenzenes depends on the number and position of the alkyl groups. Polyalkylation of aromatic hydrocarbons usually decreases biodegradability. With phenylalkanes, biodegradability depends on the branching of long alkyl chains, and those phenylalkanes with a quaternary carbon atom, particularly in the terminal position, are quite resistant to biodegradation. Polycyclic aromatic hydrocarbons (except for naphthalene) are relatively slow to biodegrade, particularly tetra- and higher aromatics.

General guidelines for biodegradability have been developed by others. Eckenfelder and Musterman (1995) give the following general guidelines for the biological treatment of industrial wastewaters:

- Nontoxic aliphatic compounds containing carboxyl, ester, or hydroxyl groups are readily biodegradable. Those with dicarboxylic groups require longer acclimation times than those with a single carboxyl group.
- Compounds with carbonyl groups or double bonds are moderately degradable and slow to acclimate.
- Compounds with amino or hydroxyl groups decrease in biodegradability relative to the degree of saturation, in the following order: primary, secondary, then tertiary carbon atom of attachment.
- Biodegradability decreases with increasing degree of halogenation.

Noyes (1994) presents a simpler and more general order of decreasing biodegradability: straight-chain compounds, aromatic compounds, chlorinated straight-chain compounds, and chlorinated aromatic compounds.

A variety of different factors influence the performance of biological treatment systems. Although each specific biological process has special requirements, factors common to biological processes, besides biodegradability of the waste constituents, include organic concentration, temperature, pH, nutrients, and oxygen (aerobic or anaerobic).

Activated Sludge

Activated sludge is widely used to treat both municipal and industrial wastewater. Used in a variety of process configurations and operating modes, the activated sludge process contains three basic elements—an aeration basin, clarification, and sludge recycle—as shown in Figure 5.5. The aeration basin contains a suspension of *activated* microorganisms, primarily bacteria and protozoa, that biodegrade the wastewater constituents under aerobic conditions. Aerators supply oxygen for the microorganisms and mixing energy to keep them in suspension and in contact with the waste. The microorganisms are separated from the treated wastewater by clarification, and a portion of this biological sludge is recycled back to the aeration basin to maintain the microbial population and begin a new treatment cycle.

Although the activated sludge process is relatively robust and able to handle variable wasteloads and operating conditions, it must be protected from shock loadings that are toxic or excessively high in concentration. Influent variability, both in flow and quality, can be reduced through equalization prior to the aeration basin. Other common types of pretreatment that may be necessary to protect the aeration basin are neutralization, oil/water separation, and grit or solids removal.

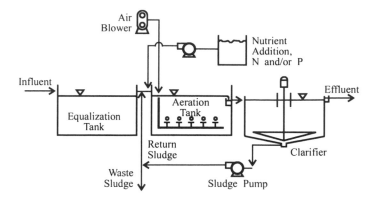

Figure 5.5   Example of an activated sludge process (*Source:* EPA 1987).

The microorganisms grow in response to the food source supplied in the wastewater and produce more biological sludge than is needed to maintain the process. This excess sludge must be wasted from the process and is usually treated by dewatering and aerobic or anaerobic digestion. The digested sludge may be disposed of by incineration or landfilling or may be used beneficially in land application for its nutrient (nitrogen, phosphorus) and organic value.

Aeration basins can be constructed as concrete or steel tanks or earthen impoundments, although tanks are currently more common because of ground water problems caused by leakage from impoundments and the stringent regulation of impoundments for the treatment of hazardous waste.

Process Variations. There are many variations of the activated sludge process that have been developed in response to different wastewater characteristics and process requirements. The mixing regime of the process may be plug flow, complete mixing, or, more often, something between the two. In a plug-flow system, influent wastewater and return activated sludge are mixed at the head of the aeration basin and travel as a *plug* along the length of the basin. Therefore, the organic concentration is greatest at the head of the basin and diminishes as the wastewater is treated traveling toward the basin outlet. In a completely mixed activated sludge system, the influent wastewater and return activated sludge are uniformly or completely mixed in the aeration basin.

Mixing the influent wastewater with the larger volume of the aeration basin allows high organic loads and toxic or inhibitory compounds to be diluted, so as to protect the system from shock, an advantage over plug flow systems. In general, complete mix systems have higher mixed liquor concentrations than plug-flow systems and, consequently, can treat higher organic loads. However, complete mix systems tend to produce a sludge that is less dense and may be more difficult to settle. In reality, pure plug flow and complete mixing do not exist, as the process operates somewhere in between. Some of the most common process variations of activated sludge systems are described in the following paragraphs.

Tapered Aeration. In this process, aeration is provided in a tapered or stepped fashion to match oxygen requirements in a plug flow system. Oxygen requirements are highest at the head of the aeration basin, where the organics-laden influent enters. Aeration is decreased along the basin as the organics are degraded. Toward the end of the basin, it may be necessary to maintain a certain level of aeration in excess of oxygen requirements in order to provide enough mixing energy to keep the microorganisms in suspension.

Contact Stabilization. This process takes advantage of the ability of the activated sludge to quickly adsorb organics. Contact stabilization is operated either as a two-unit system or in two zones within a single aeration basin. In the two-unit system, the influent wastewater is first mixed with return activated sludge in a contact basin. Contact time is short, 30 to 90 minutes, but sufficient for the sludge to adsorb the organics. The mixture then flows to a clarifier, where the sludge and adsorbed material are settled out. The treated effluent is discharged from the clarifier. A portion of the settled sludge is wasted as excess biomass, and the rest enters the stabilization basin, where the adsorbed organics are biodegraded over a period of 3 to 6 hours, thereby reactivating the sludge. In the two-zone system, the aeration basin is operated as a plug-flow system. The contact zone for the influent wastewater is near the end of the aeration basin, where organic concentrations are lowest and the sludge is ready to adsorb additional organics. The stabilization zone begins at the head of the aeration basin, where the settled sludge with its adsorbed organic load enters. Contact stabilization requires about 50% less aeration basin, capacity than conventional systems and produces less biomass, which settles better. On the downside, effluents are poorly nitrified and the process is better suited to wastewaters that contain high proportions of biochemical oxygen demand (BOD) as colloidal or suspended matter.

Extended Aeration. As its name implies, extended aeration operates with a long aeration period. With a long sludge age and low food-to-microorganism ratio, sludge production is low and the sludge settles well. Because aeration and basin volume requirements are high, extended aeration treatment of domestic sewage is normally used for small flows and small communities. With industrial wastewaters, extended aeration is common when there are a high concentration of organics and low biodegradation rates of some of the organic constituents. The oxidation ditch is a common form of the extended aeration process. Viewed from above, an oxidation ditch looks like a racetrack around which the wastewater flows. Aeration is provided by rotors or brushes, which also provide the energy to keep the water flowing around the track.

Influent wastewater usually enters upstream of the aerator. Anoxic zones can be set up in oxidation ditches, allowing nitrogen removal through denitrification. In addition to the racetrack design, oxidation ditches can be of the carousel type, which resembles a long racetrack folded in half. The carousel has two aerators that allow larger flows to be treated.

Pure Oxygen. Activated sludge systems can be designed to use pure oxygen instead of air to increase oxygen transfer rates. Advantages of pure oxygen systems are increased BOD removal rates, smaller aeration basin volumes, lower sludge production, and improved sludge settling. The aeration basin is covered and divided into a series of stages. Influent wastewater, return activated sludge, and oxygen are introduced into the first stage. In each progressive stage, both the BOD and oxygen content decrease, acting like a tapered aeration plug-flow system. Operating parameters for pure oxygen systems (mixed liquor suspended solids (MLSS), mixed liquor oxygen levels, food-to-microorganism (F/M) loading rate) are higher than for conventional activated sludge processes. Disadvantages of pure oxygen systems include the high cost of oxygen production and the need for safety precautions and monitoring of combustible hydrocarbons in high-oxygen environments.

PACT® System. Powdered activated carbon treatment, or PACT®, is the simple addition of powdered activated carbon to an activated sludge system. This process is used to prevent inhibitory effects of toxic compounds, remove refractory organics or color, or remove effluent toxicity. Other benefits of the process are reduced emissions of volatile compounds from wastewaters and improved sludge settling. Carbon dosage rates depend on the given situation but usually range from 50 to 3,000 mg/L (Eckenfelder and Musterman 1995).

Process Control. There are many different process control variables that determine the efficiency of the activated sludge process. Effluent quality depends primarily on wastewater characteristics, plant loading (organics and flow), residence time in the system, temperature, type and concentration of microorganisms degrading the waste, oxygen supply, and degree of mixing. Major process control variables are described in the following paragraphs.

Hydraulic Retention Time. The hydraulic retention time (HRT) is a measure of the average time the wastewater spends in the aeration basin. It is calculated by dividing the volume of the aeration basin by the influent wastewater flow rate. Because the flow rate of recycle sludge is not included in the calculation, the actual retention time would be less. The HRT depends on the type of activated sludge process being used. Conventional systems for domestic sewage and some readily degradable industrial wastewater have HRTs of about 4 to 8 hours. Extended aeration systems can have much longer HRTs, ranging from about 18 hours to as long as 72 hours.

MLSS/MLVSS. The concentration of microorganisms in the aeration basin determines the organic loading and the efficiency of the process. The contents of the aeration basin, the mixture of wastewater and microorganisms, is referred to as mixed liquor. The mixed liquor suspended solids (MLSS) and mixed liquor volatile suspended solids (MLVSS) are both used as measures of the microorganism concentration in the aeration basin. MLVSS more closely approximates the biologically active portion of the solids in the mixed liquor, because microbial cellular material is organic and volatilizes or burns at 500°C (the temperature for volatile solids analysis). MLSS includes both the volatile and inert solids in the mixed liquor. It is often used, or at least monitored more frequently, than MLVSS, because the analytical test for suspended solids is quicker than the test for volatile suspended solids. MLSS concentration for conventional systems ranges from 1,500 to 3,500 mg/L. The volatile fraction representing the MLVSS varies, but a typical value is 0.8. MLSS concentrations are higher in other process configurations. For example, pure oxygen systems may have MLSS concentrations from 5,000 to 8,000 mg/L. Specially acclimated systems may have MLSS concentrations of 10,000 mg/L and higher. MLSS concentrations are limited by the availability of oxygen in the aeration basin and by recycle sludge flow rates.

F/M Ratio. The food-to-microorganism ratio (F/M) is a measure of the organic loading on the biological population. The F/M ratio is calculated as the mass of BOD per day divided by the mass of MLSS or MLVSS in the aeration basin. Thus, the F/M ratio has units of day^{-1}. The F/M ratio can be controlled by adjusting both the flow rate and concentration of the sludge recycled to the aeration basin. When based on MLVSS, typical values for the F/M ratio for conventional systems range from 0.2 to 0.4 day^{-1}. The F/M ratio is higher for contact stabilization (0.2 to 0.6 day^{-1}) and pure oxygen systems (0.4 to 1.0 day^{-1}), and lower for extended aeration systems (0.03 to 0.15 day^{-1}).

Mean Cell Residence Time. The mean cell residence time (MCRT) is a measure of the average length of time (in days) spent in the system by the microorganisms and relates to the overall growth phase of the microbial population. It is calculated as the mass of microorganisms in the aeration basin, as represented by MLSS or MLVSS, divided by the mass of microorganisms removed from the system through sludge wastage and in the effluent discharge. The MCRT may also be referred to as the solids retention time, sludge residence time (SRT), or sludge age. The MCRT in conventional systems typically ranges from 3 to 5 days. Pure oxygen systems have higher MCRTs (8 to 20 days) as do extended aeration systems (6 to 40 days). Systems used for nitrification operate at MCRTs greater than 10 days.

Sludge Recirculation. The concentration of microorganisms in the aeration basin is achieved by recirculating settled sludge from the clarifier. Both the flow rate and the concentration of the return activated sludge determine the solids in the aeration basin. The recirculation rate is calculated as the flow rate of the return activated sludge divided by the flow rate of the influent wastewater. Typical values for conventional systems are 0.25 to 0.5. Pure oxygen systems have similar rates, whereas the recirculation rates are higher for both contact stabilization (0.25 to 1.0) and extended aeration (0.75 to 1.5).

Aeration. Aeration not only delivers oxygen to the activated sludge system, but provides mixing energy to keep the microorganisms in suspension. Aeration is provided by mechanical aeration using surface aerators, or by diffusers on the bottom of the aeration basin. Both types of aeration equipment are common in conventional and pure oxygen systems, while horizontal surface aerator brushes are used in oxidation ditches. In activated sludge systems using air, oxygen concentrations in the aeration basin are usually maintained between 1.5 and 2.0 mg/L. In pure oxygen systems, the concentrations are much higher (4 to 8 mg/L).

Aerated Lagoons and Polishing Ponds

Surface impoundments (ponds) can be used as activated sludge basins. Ponds are also used in biological wastewater treatment systems as aerated lagoons and polishing ponds. As in an activated sludge system, surface aerators in an aerated lagoon provide oxygen and mix the wastewater and microorganisms together. A settling pond is used after the aerated lagoon to separate the microorganisms from the treated wastewater. Alternatively, a section of the aerated lagoon downstream of the aerators can be used for settling under quiscient conditions. In either case, the settled sludge is not recirculated back to the aeration basin, which distinguishes the aerated lagoon from the activated sludge system. A polishing pond, following an activated sludge system or aerated lagoon, has no mechanical aeration and is used to "polish" the wastewater by providing additional biodegradation and solids settling.

Rotating Biological Contactors

A rotating biological contactor (RBC) is physically quite different from an activated sludge system. With an RBC, the biological solids are attached to a solid surface as a "fixed film," rather than being suspended in the wastewater as in an activated sludge system. An RBC is a series of large closely spaced plastic disks on a rotating horizontal shaft placed in a rectangular basin containing the wastewater. The disks, which are several feet in diameter, are partially submerged in the wastewater. The shaft rotates the disks slowly through the wastewater and out again, contacting the biological film growing on the disks with the wastewater and allowing reaeration. The biological film continuously grows and sloughs off the disks into the wastewater. Downstream of the RBC, the treated wastewater and biological solids are separated in a clarifier. RBCs require less space and energy than an activated sludge plant; however, they are used much less frequently. They are not well suited for high-organic-strength industrial wastewaters because of insufficient oxygenation and the plugging of the space between disks with excessive biomass that does not slough off.

Trickling Filters

Trickling filters are similar to RBCs in that the microorganisms grow as a film on support media. The support media in a trickling filter can be rock or other inert natural material or a synthetic plastic. The support media is usually contained within a circular basin. A central shaft rotates a set of distributor arms that "trickle" the wastewater over the media. The wastewater is continuously aerated, because air is able to pass freely through the void spaces in the bed. Excess biological growth sloughs off and is separated from the treated wastewater in a clarifier. Trickling filters can be operated as low- or high-rate systems. Low-rate systems are highly dependable, perform consistently, and nitrify (remove ammonia) well; however, odor and filter flies can be nuisance problems. High-rate systems eliminate the odor and fly problems by recirculating part of the clarified wastewater back to the filter. The primary reason for recirculation, however, is to increase treatment efficiency so that higher organic loadings can be achieved.

Sludge Digestion

Organic sludges from biological wastewater treatment systems are normally treated by aerobic or anaerobic digestion to reduce the organic content and thereby stabilize the material before land application or landfilling. Waste activated sludge and primary sludge (if a solids clarifier precedes the aeration basin) are combined for digestion. Anaerobic digestors that are not mixed form an aqueous layer (supernatant) above the digesting solids. Because it is rich in soluble biodegradation by-products, the supernatant is returned to the wastewater treatment system. The methane generated by anaerobic digestion can be used onsite to heat both the sludge digestor and facility buildings. Advantages of anaerobic digestion are that it involves no oxygen requirements, which makes treatment of high-strength wastes economical, low residual sludge quantities, methane production, and easily dewatered sludges. Disadvantages of anaerobic digestion are the larger digestor volumes required to maintain a longer MCRT, heating requirements, and sensitivity to upsets and environmental changes. Aerobic digestors are mixed to distribute the supplied oxygen and are usually unheated. Digested solids are separated from the supernatant in a sludge thickener tank.

THERMAL

Thermal treatment is used to destroy, break down, or aid in the desorption of contaminants in gases, vapors, liquids, sludges, and solids. There are a variety of thermal processes that destroy contaminants, most of which are classified as incineration. To incinerate literally means "to become ash."[2] In respect to the incineration of hazardous wastes regulated under the Resource Conservation and Recovery Act (RCRA), however, there is a strict legal definition of what constitutes an incinerator.[3] A more simplified definition describes it as a unit that combusts materials in the presence of oxygen at temperatures normally ranging from 800°C to 1650°C. Common types of incineration units discussed in this section are liquid injection, rotary kiln, multiple hearth, fluidized beds, and catalytic oxidation. Thermal desorption is also discussed in this section. Prior to descriptions of these individual technologies, an overview of the main factors affecting incinerator performance is given in the following paragraphs.

Factors Affecting Performance

There are many factors that affect both the choice of a particular thermal treatment and its performance. Chief among these are waste characteristics, temperature, residence time, mixing or turbulence, and air supply.

Waste Characteristics. The physical form of the waste restricts the applicability of some thermal technologies. A convenient summary of physical characteristics and thermal processes is presented by Cheremisinoff (1994). Those processes that can treat the widest range in physical form are incineration, pyrolysis, calcination, and microwave discharge. These processes are capable of treating wastes in essentially any form: containerized, solid, sludge, slurry, liquid, or fume. Molten salt reactors and plasma technology can be used with any of these types of wastes, except fumes. Evaporative processes are amenable to treatment of solids, sludges, slurries, and liquids. Wet air oxidation, sometimes considered thermal treatment but also described as a physical/chemical process, can be used for slurries and liquids. Distillation (with and without steam) and steam stripping, also considered physical/chemical treatment, are limited to liquid treatment. Catalytic incineration is limited to fume treatment.

Other waste characteristics that affect the choice of thermal treatment and its operation include heating value, contaminant boiling points, metal content, heterogeneity, moisture content, decomposition products and by-products, and the presence of explosive constituents. The heating value and moisture content of the waste determine whether combustion is self-supporting or if auxiliary fuel will be required. Elemental constituents in the waste may be carried in the exhaust gases and must be removed. Problem metals are antimony, arsenic, cadmium, lead, and mercury. Acid gases con-

[2]From Medieval Latin, *incinerare*, in or into ashes.
[3]The definition of *incinerator* in the federal hazardous waste regulations at 40 CFR §260.10 is "any enclosed device that: (1) uses controlled flame combustion and neither meets the criteria for classification as a boiler, sludge dryer, or carbon regeneration unit, nor is listed as an industrial furnace; or (2) meets the definition of infrared incinerator or plasma arc incinerator."

taining chlorine, sulfur, fluorine, or bromine that are generated must be neutralized and scrubbed from the flue gases. The heterogeneity of the waste is determined by the distribution and concentration of combustible organics; localized concentrated materials can produce hot spots in the combustion chamber. Heterogeneity is also defined by the type and amount of debris that may be included in excavated wastes. These materials can interfere with the process by shielding the waste, fouling equipment, or slagging in the reactor. Undesirable decomposition products and by-products include nitrogen oxides and dioxins.

Temperature. The temperature for combustion processes must be balanced between the minimum temperature required to combust the original contaminants completely and any intermediate by-products and the maximum temperature at which the ash becomes molten. Typical operating temperatures for thermal processes are the following: incineration (750°C–1650°C), catalytic incineration (315°C–550°C), pyrolysis (475°C–815°C), and wet air oxidation (150°C–260°C at 1500 psig) (Cheremisinoff 1994). Pyrolysis is thermal decomposition in the absence of oxygen or with less than the stoichiometric amount of oxygen required. Because exhaust gases from pyrolytic operations are somewhat "dirty," with particulate matter and organics, pyrolysis is not often used for hazardous wastes.

Residence Time. For cost-efficiency, residence time in the reactor should be minimized but must be long enough to achieve complete combustion. Typical residence times for various thermal processes are as follows: incineration (0.1 second to 1.5 hours), catalytic incineration (1 second), pyrolysis (12 to 15 minutes), and wet air oxidation (10 to 30 minutes) (Cheremisinoff 1994).

Turbulence. Turbulence is important to achieve efficient mixing of the waste, oxygen, and heat. Effective turbulence is achieved by liquid atomization (in liquid injection incinerators), solids agitation, gas velocity, physical configuration of the reactor interior (baffles, mixing chambers), and cyclonic flow (by design and location of waste and fuel burners) (Noyes 1994; WEF 1990).

Air Supply. Oxygen in excess of stoichiometric requirements for complete combustion is needed, because incineration processes are not 100% efficient and excess air is needed to absorb a portion of the combustion heat to control the operating temperature. In general, units that have higher degrees of turbulence, such as liquid injection incinerators, require less excess air (20% to 60%), whereas units with less mixing, such as hearth incinerators, require more (30% to 100%).

Catalytic Oxidation

Catalytic oxidation is used only for gaseous streams, because combustion reactions take place on the surface of the catalyst which precludes being applicable to a liquid or solid material. Common catalysts are palladium and platinum. Because of the catalytic boost, operating temperatures and residence times are much lower, which reduces operating costs. Catalysts in any treatment system are susceptible to poisoning (masking of or interference with the active sites). Catalysts can be poisoned or deactivated by sulfur, bismuth, phosphorus, arsenic, antimony, mercury, lead, zinc, tin, or halogens (notably chlorine); platinum catalysts can tolerate sulfur compounds, but can be poisoned by chlorine.

Fluidized Bed

In a fluidized bed incinerator, the waste and an inert bed material (sand, normally, or alumina) are fluidized by blowing heated air through a distributor plate at the bottom of the bed. Fluidization results in good mixing and uniform distribution of materials within the bed, which allows lower operating temperatures (450°C to 710°C) and lower excess air requirements (Noyes 1994). Fluidized beds can be used to incinerate gases, liquids, or solids, although the each type of waste is introduced into the bed differently. Fluidized beds are susceptible to seizing or binding up or agglomeration on individual bed particles, which can lead to seizing up. Bed seizure can occur with wastes containing clays, inorganics (salts), or high concentrations of lime (Noyes 1994).

Liquid Injection

Liquid injection units are currently the most common type of incinerator for the destruction of liquid hazardous wastes such as solvents. Atomizers break the liquid into fine droplets (100 to 150 microns) (LaGrega et al. 1994), which allows the residence time to be extremely short (0.5 to 2.5 seconds). The viscosity of the waste is very important; the waste must be both pumpable and capable of being atomized into fine droplets. Both gases and liquids can be incinerated in liquid injection units. Such gases include organic streams from process vents and those from other thermal processes; in the latter case, the liquid injection incinerator operates as an afterburner (WEF 1990). Aqueous wastes containing less than 75% water can also be incinerated in liquid injection units (LaGrega et al. 1994).

Multiple Hearth

Although well suited for combustion of wet sludges, particularly municipal biological wastewater sludges, multiple hearth incinerators are not often used for hazardous wastes because operating temperatures (800°C to 1000°C) of this type of incinerator are too low for many hazardous compounds (WEF 1990). The multiple hearths are stacked vertically, and sludges are introduced into the top hearth. A series of "rabble" arms extending from a rotating center shaft mix and move the waste along the circular hearth until the waste falls through holes into the hearth below. For incineration, a minimum of six hearths is usually required; more hearths are required for pyrolysis (Noyes 1994). Liquid wastes may be incinerated along with sludges when injected through burner nozzles.

Rotary Kiln

A rotary kiln is a long, cylindrical incinerator that is sloped a few degrees from the horizontal. Waste is introduced at the upper end. The gentle slope and slow rotation of the kiln continually mix and reexpose the waste to the hot refractory walls, moving it toward the exit point. Rotary kilns operate in continuous or batch mode, although continuous operation is more cost-efficient and less wearing on the refractory material. Operating temperatures range from about 800°C to 1650°C (WEF 1990). They can be operated in either an ashing or slagging mode; the former is common in the United States. Rotary kilns are relatively large incinerators, but because of their versatility in treating a wide range of wastes and their treatment efficiency, they are often used for incinerating hazardous wastes. Rotary kilns can be used to incinerate gases, liquids, sludges, and solids. However, highly aqueous organic sludges tend to form a ring on the walls of the kiln that prevents the discharge of ash, and they can also form sludge balls that are not completely incinerated (WEF 1990).

Thermal Desorption

Thermal desorption is an innovative treatment that has been applied primarily to soils (see the following section "Soil and Ground Water"). Wastes are heated to temperatures of 200°C to 600°C to increase the volatilization of organic contaminants. Volatilized organics in the gas stream are removed by a variety of methods, including incineration, carbon adsorption, and chemical reduction.

SOIL AND GROUND WATER

Contaminated soil and ground water present special challenges, because the contamination is diffuse and difficult to access. In situ processes, those that leave both soil and ground water in place during treatment, are often favored because they are less disruptive, they take advantage of natural, existing conditions, and they preserve the existing subsurface environment. Ex situ and pump-and-treat processes remove soil and ground water for treatment. They may be favored over in situ treatment where they will reduce cleanup times, their operation and capabilities are considered more reliable or better understood, or they can achieve lower cleanup levels. Both in situ and ex situ treatment for soil and ground water rely on a combination of unit processes, which often include biological degradation of organics.

Nonaqueous phase liquids (NAPLs) present special problems for soil and ground water cleanup. Contaminant transport through ground water depends in part on the water solubility of the compound. Because NAPLs cling to subsurface particles and are slow to dissolve in ground water, they hinder cleanups and prolong cleanup times. Dense nonaqueous phase liquids (DNAPLs) migrate downward in the aquifer and can collect in pools or pockets of the substructure. Examples of DNAPLs are the common solvents tetrachloroethylene (also known as perchloroethylene or PCE) and trichloroethylene (TCE), which were used extensively at many facilities before the extent of subsurface contamination problems was realized.

Today, strict regulation of hazardous waste and material management promises to minimize future soil and ground water contamination; however, past practices have left us with a large legacy of contaminated sites that are often difficult and costly to remediate. This is the driving force behind the development of innovative technologies for soil and ground water cleanup. EPA strongly supports the development of innovative technologies and maintains the Vendor Information System for Innovative Treatment Technologies (VISITT) database to provide information on availability, performance, and cost of these technologies.

In this section, soil and ground water cleanup technologies are divided into five subsections: plume containment, bioremediation, physical/chemical in situ treatment, pump-and-treat technology for ground water, and ex situ soil treatment. Many of the unit processes for soil and ground water treatment are general waste treatment technologies, but are discussed here in the context of soil and ground water cleanup. A more general discussion of these technologies is found elsewhere in this chapter.

Plume Containment

Wells can be placed at a contaminated site to prevent the contamination from spreading further or migrating offsite. In the past, containment efforts often relied on physical methods such as bentonite slurry trenches, grout curtains, sheet pilings, well points, and fixative injections (Bouwer and Cobb 1987). Containment by judiciously placed wells generally costs less than these techniques, takes less time to install, and is more flexible because pumping rates and locations can be varied (Sims et al. 1992).

Four typical well patterns for contaminant plume containment are described by Sims and associates (1992). The first is a pair of injection-production wells. The second is a line of downgradient pumping wells. The third is a pattern of injection-production wells around the boundary of a plume. The fourth, the *double-cell* system, uses an inner cell and outer recirculation cell, with four cells along a line bisecting the plume in the direction of flow. Two other methods of plume containment are biofilters and a funnel-and-gate system, which are described in a subsequent subsection under "In situ bioremediation."

Bioremediation

Bioremediation has great appeal. It is a natural process that degrades hazardous organic chemicals into innocuous carbon dioxide and water or nonhazardous by-products, and it is often less expensive and more effective than pump-and-treat methods. Articles on bioremediation appear regularly in environmental journals, and EPA has its own regular series of reports on current activities called *Bioremediation in the Field.*

On the downside, there is much that is not known or completely understood about the process, which makes it difficult to design and prove up in field applications. It is a relatively slow process, and complete remediation of a site can take several years. The efficiency of the bioremediation cannot always meet target cleanup levels. Bioremediation has been very successful in cleaning up sites contaminated with gasoline and other fuels, because the primary contaminants—benzene, toluene, ethylbenzene, and xylenes (BTEX)—are relatively easy to extract and degrade. Although other compounds have been biodegraded successfully, there are many that are still subjects of bioremediation research. Nevertheless, what is known and has been done so far with bioremediation is quite encouraging and points to the great potential of bioremediation in the future of contaminated ground water and soil remediation.

Biodegradation Principles. Microorganisms degrade organic compounds to synthesize cellular materials and to obtain energy to grow and reproduce. Anabolism results in the synthesis of new cellular material, and catabolism (degradation) produces chemicals with lower free energy and, in the case of aerobic biodegradation, often results in complete decomposition of a hydrocarbon to carbon dioxide and water. The principal microorganisms involved in the degradation of organic chemicals in water and soil are bacteria and fungi.

Microorganisms can degrade compounds under aerobic or anaerobic conditions, although the types of microorganisms and compounds and the degradation process and products may differ. In aerobic environments, free oxygen is present. In anaerobic conditions, although free oxygen is not present, oxygen may be present in the form of compounds such as nitrates, sulfates, and iron oxide.

Respiration, or biological oxidation, is the use of oxygen as an electron receptor in the catabolic degradation of an organic and can occur either aerobically or anaerobically. Aerobic respiration uses free oxygen as an electron receptor, whereas anaerobic respiration uses inorganic oxygen. In both cases, however, water and carbon dioxide are the principal end products, as in any oxidation reaction.

Fermentation is an anaerobic catabolic process that uses organics as electron receptors. Because fermentation produces organic products that have lower free energy than their precursors, it is useful in remediation. The lowest free energy form of carbon produced is methane.

Although both fermentation and respiration degrade organics, microorganisms that use respiration to meet their energy requirements for synthesis and reproduction generally grow much more quickly. This, and the fact that respiration completely oxidizes organics to carbon dioxide and water, usually make aerobic biodegradation the treatment of choice. However, because certain synthetic organics do not or are difficult to degrade aerobically, there is much interest in anaerobic processes.

Cometabolism refers to situations in which a compound cannot be biodegraded effectively unless another food source is available. The recalcitrant compound, such as TCE, does not provide the energy to allow the microorganisms to grow and thrive. When another food source is available, such as methane, energy is produced for growth and enzymes are produced to metabolize the energy-providing food source and cometabolize the recalcitrant compounds.

Types of Treatment. Bioremediation can be conducted either ex situ or in situ. An ex situ process is one in which ground water is pumped to the surface, or soil is excavated, for treatment. Pump and treat refers to the ex situ removal and treatment of ground water. After treatment, the ground water can either be reinjected or discharged; treated soils can be redeposited, landfilled, or recycled. In situ bioremediation takes place in the ground; neither ground water or soil is extracted, although ground water may be pumped to the surface for the addition of oxygen and nutrients. In situ bioremediation has become very popular because the process takes full advantage of natural conditions (microorganisms, soil, water) as a bioreactor, little site disturbance is necessary, and it is less equipment intensive than ex situ treatment.

Ex situ bioremediation may use various biological wastewater treatment processes, soil piles, or land application. With in situ bioremediation, the basic process is the same—microbes, soil, and water working together as a bioreactor. Where the in situ techniques differ are in how contaminants and microbes are brought in contact and how oxygen, nutrients, and other chemical supplements are distributed in the soil/water/air matrix. Typical in situ bioremediation techniques include natural or intrinsic attenuation, air sparging, and bioventing. Treatment processes for both ex situ and in situ bioremediation are discussed in the following sections.

In Situ Bioremediation. In situ bioremediation can be an aerobic or anaerobic process, or a combination of the two. In designing an in situ bioremediation system, one should consider the types of microorganisms available (naturally in place or added), the structural and chemical makeup of the soil matrix, types of contaminants, oxygen and nutrient addition and distribution, and temperature. These factors are discussed in the following paragraphs, prior to the introduction of the individual techniques for in situ bioremediation.

Aerobic, Anaerobic, and Combined Systems. The vast majority of in situ bio-remediations are conducted under aerobic conditions because most organics can be degraded aerobically and more rapidly than under anaerobic conditions. Some synthetic chemicals are highly resistant to aerobic biodegradation, such as highly oxidized chlorinated hydrocarbons and polynuclear aromatic hydrocarbons (PAHs). Examples of such compounds are PCE, TCE, benzo(a)pyrene, PCBs, and pesticides.

Besides being slower, anaerobic treatment is more difficult to manage and can generate by-products that are more mobile or toxic than the original compound; for example, the daughter products of TCE—dichloroethenes and vinyl chloride. It also requires a longer acclimation period, which means slower start-ups in the field. The microbial processes for anaerobic treatment are less well understood and, hence, less controllable than those for aerobic systems.

Nevertheless, an anaerobic system may be the method of choice under certain conditions: (1) contamination with compounds that degrade only or better under anaerobic conditions, (2) low-yield aquifers that make pump-and-treat methods or oxygen and nutrient distribution impractical, (3) mixed waste contamination, in which oxidizable compounds drive reductive dehalogenation of chlorinated compounds, or (4) deep aquifers that make oxygen and nutrient distribution more difficult and costly (Sewell et al. 1990).

Anaerobic respiration can degrade organics by using nitrate or sulfate as oxygen sources. Compounds that have been shown to be biodegraded by nitrate-respiring microorganisms are methanes, carbon tetrachloride, m-xylene, and some phenols, cresols, and PAHs. Chloroform and stable chlorinated ethanes and ethenes are not degraded by nitrate respiration. Reductive dehalogenation, not degradation, has been observed of naturally occurring aromatics where sulfate respiration was occurring. Sulfate respiration can be inhibited by the accumulation of its by-product, hydrogen sulfide (King et al. 1992).

Combined aerobic/anaerobic systems use sequenced aerobic and anaerobic conditions to degrade compounds that are resistant to aerobic biodegradation. Ground water passes through the anaerobic zone first, which partially degrades the resistant compounds. These degradation products are then further degraded in an aerobic zone. A two-zone plume interception approach was tested under EPA's Superfund Innovative Technology Evaluation (SITE) Emerging Technologies Program to degrade chlorinated compounds (Hasbach 1993). The first zone was anaerobic, where methanogenic (methane-producing) microorganisms were expected to promote reductive dechlorination of chlorinated solvents. The second zone was aerobic, where the partially dechlorinated products were to be biodegraded. Combined systems do not necessarily have to comprise separate zones in the aquifer. Aerobic and anaerobic conditions can be alternated by eliminating the oxygen source at timed intervals.

In one study a coarse-grained sand aquifer was injected with methane and oxygen to stimulate the production of methane monooxygenase (MMO) enzyme, which is capable of degrading TCE (Semprini et al. 1987). TCE, added at 60 to 100 μg/L, was degraded by 20% to 0%. Injected concentrations of methane and oxygen were approximately 20 mg/L and 32 mg/L, respectively.

Design Considerations. The effectiveness of in situ bioremediation is influenced by many factors, including microorganisms, soils, oxygen, pH, temperature, type and quantity of contaminants, and nutrients. These factors are discussed in the following paragraphs.

MICROORGANISMS. A large number of naturally occurring bacteria and fungi can use hydrocarbons for growth. The most commonly found genera of bacteria and fungi that are capable of hydrocarbon biodegradation are (1) bacteria—pseudomonas, arthrobacter, alcaligenes, corynebacterium, flavobacterium, achromobacter, micrococcus, nocardia, and mycobacterium—and (2) fungi—thricoderma, penicillium, aspergillus, and mortierella (Miller 1990).

Soils contain a diverse culture of microorganisms, with the number and type present being a function of local environmental conditions. Usually, the limiting factors on the microbial population at an uncontaminated site are the amount and type of carbonaceous material available for biodegradation (Miller 1990). However, there are almost always some microorganisums present in every soil environment that will take advantage of a new or different carbon source when it becomes available and will rapidly increase their populations to metabolize it. The addition of more or specially cultured microbes has been studied and applied in the field; however, some people believe that all that is needed in most cases is to simply optimize environmental conditions for the naturally occurring microbes. Yet for recalcitrant compounds or for more rapid biodegradation, specialized microbes may have potential. For example, EPA has agreed to license specialized microbes that degrade chlorinated aromatic and chlorinated aliphatic compounds such as TCE under aerobic conditions, which avoids the generation of vinyl chloride as a by-product (EPA 1993b).

SOILS. Soils with hydraulic conductivities greater than 10^{-4} centimeters per second (cm/s) are good candidates for in situ bioremediation because they are permeable enough to allow the transport of oxygen and nutrients through the aquifer (Sims et al. 1992). The degree of homogeneity of the soil is important to predicting ground water flow patterns for plume containment and distribution of oxygen and nutrients.

CONTAMINANTS. The type and concentration of contaminants in an aquifer dictate what type of in situ bioremediation system—aerobic, anaerobic, or combination—will be most effective.

OXYGEN. Many microorganisms require either free oxygen or inorganic oxygen for growth. Some bacteria can use either free oxygen or inorganic oxygen for respiration; thus, these are called facultative bacteria. Many bacteria, though, require free oxygen for growth.

A rule of thumb used in wastewater treatment is that at least 1.5 mg/L of dissolved oxygen is needed to prevent oxygen from being a growth limiting factor. Kemblowski et al. (1987) documented a threshold dissolved oxygen concentration of 0.5 mg/L in ground water. Below 0.5 mg/L, it was found that the biodegradation rates of benzene, toluene, and xylene were retarded. It was also shown that biodegradation rates increased until the ground water dissolved oxygen concentration reached 2 mg/L, at which point no further increase was observed. Thus, it appears that a target of 1.5 to 2.0 mg/L dissolved oxygen will maximize biodegradation rates and put an upper limit on the cost to supply the oxygen.

Field data collected from ground water in spill areas have shown that dissolved oxygen concentrations are typically reduced below 1 mg/L as a result of biodegradation. For example, at the site of a gasoline spill where gasoline-degrading bacteria had been identified, dissolved oxygen in the contaminated ground water ranged from 0 to 0.5 mg/L, whereas in uncontaminated areas levels were ranged from 2.3 to 4 mg/L (Raymond et al. 1976). In another study of an aquifer contaminated with chlorinated solvents, dissolved oxygen was always less than 0.2 mg/L (Semprini et al. 1987). These studies demonstrate that biological growth and biodegradation of hydrocarbons is oxygen-limited, a fact that stimulates a continuing interest in developing effective oxygen sources and the means of distributing oxygen in ground water and soil.

Hydrogen Peroxide. Hydrogen peroxide is an effective means of getting more oxygen into ground water for bioremediation and has many advantages over air or pure oxygen. Hydrogen peroxide has been demonstrated not only to boost dissolved oxygen levels, but also to stimulate microbial activity. It has the further advantage of chemically degrading contaminants partially or fully. Hydrogen peroxide mixes intimately with ground water for better distribution in the aquifer and can prevent injection well plugging caused by bacterial growth.

On the downside, various studies (Chambers et al., 1991; Huling et al., 1990; Pheiffer, 1990a) have shown that hydrogen peroxide decomposes rapidly after soil contact, it is cytotoxic at a 3% solution, and, unless stabilized, oxygen bubbles can escape prematurely through the unsaturated zone before they have a chance to disperse well in the ground water. Catalase enzymes generated by aerobic bacteria near the injection wells appear to be the major cause of hydrogen peroxide decomposition and oxygen off-gassing. Others have noted that hydrogen peroxide will also react with dissolved iron, manganese, and other inorganics to produce oxides that can plug wells and formations (D. Russell, 1992; Ware, 1993). Precipitation of inorganic compounds during bioremediation has been studied in more detail by Morgan and Watkinson

(1992). Concentrated solutions of hydrogen peroxide (30%) used in bioremediation must be handled carefully to avoid skin burns.

Reports of ground water remediation studies using hydrogen peroxide are found frequently in the literature. A field-scale study of an area contaminated by an aviation gasoline fuel spill showed that hydrogen peroxide did increase the concentration of oxygen downgradient of the injection well (Huling et al., 1990). Hydrogen peroxide was used to remediate ground water after a jet fuel leak at Eglin Air Force Base in Florida (Pheiffer 1990a). Field applications using hydrogen peroxide for ground water cleanup are continuously reported by EPA in its *Bioremediation in the Field* series. Treatment conditions common to the EPA studies are the use of hydrogen peroxide and nutrient addition to stimulate biodegradation by indigenous microorganisms.

Most reports citing hydrogen peroxide do not specify dosage rates. Concentrations as high as 500 mg/L have been shown to be nontoxic to most microorganisms (D. Russell 1992). Other studies report dosage rates of 100 to 500 mg/L, and unpublished reports of rates as high as 1,000 to 10,000 mg/L are not uncommon (Huling et al. 1990). Research studies have shown that a 0.05% solution (500 mg/L) was the maximum tolerated by a mixed culture of microorganisms capable of degrading gasoline. Concentrations of up to 2,000 mg/L could be tolerated if the dosage was stepped up incrementally; however, most toxicity studies showing no toxicity at relatively high concentrations are usually based on bacteria counts. Other studies have shown that concentrations greater than 100 mg/L decrease the *oxygen utilization rate* by microorganisms, which is counterproductive to biodegradation and, therefore, increases the time needed for remediation.

Premature decomposition of hydrogen peroxide can be inhibited by adding mono-basic potassium phosphate as a stabilizer (Huling et al. 1990). The phosphorus in the solution has the added benefit of being a nutrient for the microorganisms. A concentration of 190 mg/L potassium phosphate was used in Huling's field study of ground water contaminated with aviation gasoline. The phosphate is added ahead of the hydrogen peroxide until breakthrough occurs in the downgradient wells. When it is added first, the phosphate binds with the inorganics in the soils and allows the hydrogen peroxide to be distributed further. Other chemicals that have been used or studied to increase the stability of hydrogen peroxide are polyphosphates, stannate or phosphate, sodium pyrophosphate, citrate, and sodium silicate (Elizardo 1993).

Hydrogen peroxide is also used as a treatment to clean wells of biological growth. One method is to pour a 0.5% solution into a well and allow it to sit for four hours, after which pumping is resumed (King et al. 1992). Some treatments use solution concentrations as high as 10% to remove biofouling from wells (Sims et al. 1992).

Pure Oxygen. By increasing the oxygen level in soils and ground water, pure oxygen or oxygen-enriched streams stimulate bacterial activity (Chambers et al. 1991). Through the use of bottled oxygen, 35 to 40 mg/L of dissolved oxygen can be supplied at most field temperatures (King et al. 1992).

Being less stable than hydrogen peroxide, oxygen has the disadvantage that it may not distribute as well in the aquifer, although surfactants can promote highly stable microbubbles that are better able to travel throughout the ground water. Unlike hydrogen peroxide, oxygen does not aid in keeping an injection well free from biological growth that can plug the well.

Nitrates and Sulfates. There is not much information on actual applications of nitrates or sulfates as oxygen sources. The advantages of nitrate are that it is more soluble than oxygen, less costly, and less toxic than hydrogen peroxide (Hutchins et al. 1991). Regulatory agencies are less inclined to approve the use of nitrates in ground water cleanup as an oxygen or nutrient supplement, because it can potentially contaminate drinking water supplies and plug the aquifer with biological growth. If a nitrate is used, dosages must be carefully calculated and controlled to avoid these problems. In Hutchins and associates' study (1991), nitrate was used as the oxygen source for microbial respiration in the cleanup of a drinking water aquifer contaminated with jet fuel and concentrations of BTEX were treated below 5 µg/L. There is less literature on sulfate as an oxygen source than on nitrate. Most references merely state that sulfate can be used as an oxygen source for anaerobic respiration.

Solid Phase Oxygen. Solid phase oxygen in the form of peroxides (calcium peroxide, magnesium peroxide) has been investigated as a slow, continuous release system that avoids problems associated with the transient nature of molecular oxygen and hydrogen peroxide (Brubaker 1995, Fagan 1994). Calcium peroxide releases oxygen into the system by first breaking down into hydrogen peroxide and calcium hydroxide; the hydrogen peroxide then decomposes to oxygen and water. Cages or socks of the material are suspended directly in the ground water well, and the material can be blended into the soil for ex situ treatment and landfarming. For ground water treatment, this form of oxygen has the most potential for containment or remediation of low concentrations of dissolved organics.

Iron Oxides. Chelated iron oxide, using nitrilotriacetic acid and EDTA, has been studied as an alternative oxygen source (WEF 1994). Iron oxide, which is often difficult for microbes to access, is made more available by chelating agents.

NUTRIENTS. In addition to carbon and oxygen, other nutrients are needed for microbial growth. Nitrogen, phosphorus, potassium, calcium, magnesium, sodium, sulfur, chlorine, iron, manganese, cobalt, copper, boron, zinc, molybdenum, and aluminum are all present in the microbial cellular material (Clifton 1957). With the possible exceptions of nitrogen and phosphorus, which are present in the cell in relatively large amounts, these nutrients are needed at only trace levels and there is almost always enough in the natural environment to prevent growth-limitation.

In the presence of large amounts of degradable carbon, the naturally available nitrogen and phosphorus could potentially limit growth and thus biodegradation rates. Based on a literature review conducted by Miller (1990), typical aerobic soil microorganisms contain 5 to 15 parts of carbon per part of nitrogen, 10 parts being an acceptable average. This is about twice the 5:1 ratio of carbon to nitrogen usually assumed to represent the composition of the biomass in wastewater treatment processes (Eckenfelder 1966). If this difference exists, it may be due to the specific environmental conditions in a particular location, inasmuch as such conditions are known to influence the amount of nitrogen accumulated with cellular material (Eckenfelder 1966). The phosphorus requirement is about one fifth that of nitrogen (Eckenfelder 1966, Miller 1990).

These estimates of nitrogen and phosphorus do not consider the amounts that are reintroduced by biological recycling. Bacterial growth and death continually recycle some fraction of the synthesized nitrogen and phosphorus. Miller cites an optimum carbon-to-nitrogen-to-phosphorus ratio of 250:10:3 for biodegradation of carbon compounds in soil, which accounts for the carbon oxidized to carbon dioxide and that assimilated into cellular material. Assuming that one part of carbon is incorporated into cellular material for every two parts of carbon that are used for energy and are converted to carbon dioxide, the ratio of carbon to nitrogen in the cell is about 8.3 to 1, a reasonably conservative estimate. Miller states that a C:N:P ratio of 300:10:1 should be adequate for biodegradation, assuming no recycle.

Nutrients are usually added at concentrations ranging from 0.005% to 0.02% by weight (Sims et al. 1992). In a field application using hydrogen peroxide, nutrients were added to the injected water at the following concentrations: 380 mg/L ammonium chloride, 190 mg/L disodium phosphate, and 190 mg/L potassium phosphate, the latter used primarily to complex with iron in the formation to prevent decomposition of hydrogen peroxide (Huling et al. 1990).

TEMPERATURE. Ambient temperature is a key factor in successful biodegradation. In the range of 15°C to 35°C, bacterial growth rates double for every 10°C rise in temperature. For most soil microorganisms, the optimal temperature for maximum growth is 30°C to 35°C (Miller 1990). Inasmuch as vadose zone and ground water temperatures are typically in the range of 10° to 25°C (Hart and Couvillion 1986), microbial growth will not be prevented but will be at less than the maximum possible rates.

pH. A pH between 6 and 8 should be maintained for in situ bioremediation. Because most soil pHs are within this range, pH should not be a controlling factor for biodegradation of hydrocarbons unless the soil has also been contaminated with acidic or caustic substances.

Air Sparging. With this technique, air is injected in the saturated zone to volatilize organics and to deliver oxygen for biodegradation. Soil vapor extraction may be used in conjunction with air sparging to capture the volatilized organics. Air sparging is used frequently at sites that have been contaminated with volatile compounds such as BTEX. However, performance data are lacking and difficult ro measure (EPA 1995b; API 1995).

Biosparging. Biosparging is a form of air sparging, but the difference is that the primary purpose of biosparging is to deliver just enough air to meet oxygen requirements for bioremediation. Volatilization of organics may be an added benefit, but it is secondary to oxygen delivery. An enhancement is to sparge ozone in place of air (Nelson 1995). Besides providing oxygen for biodegradation, ozone aids in breaking down recalcitrant compounds such as chlorinated ethenes, polycyclic aromatic hydrocarbons, and pentachlorophenol.

Biofilters. Biofilters, also known as bio-barriers or microbial fences, are used to hinder migration of a contaminant plume. A biofilter is essentially a zone of biological activity that treats the contaminant as the ground water flows through the area. The biofilter zone can be established by installing a line of air-sparging wells perpendicular to the direction of ground water flow.

Bioventing. Bioventing is soil venting that enhances biodegradation while extracting volatile compounds from the unsaturated zone.

Funnel-and-Gate. This in situ bioremediation method is so named because ground water is funneled through openings or gates in an impermeable sheet piling or slurry wall. Zones of biological activity are created at the gates through air sparging. Contaminants are biodegraded as ground water is forced to pass through the gates. The funnel-and-gate method (also known as flume-and-gate) is used where ground water gradients are small. Because smaller treatment zones are created, operating expenses can be reduced and there is better process control (Brubaker 1995).

Phytoremediation. Phytoremediation is a developing technology that uses plants, trees, and grasses to biodegrade, extract, or stabilize organic and metal contaminants in soil and ground water. Engineered wetlands is a well-known form of phytoremediation. Phytoremediation is likely to be used in conjunction with other treatment processes, primarily as a polishing step (Matso 1995). As reported by Matso, this method has potential in the cleanup of residual contamination in soil micropores, hydraulic control (poplars and weeping willows pump 50 to 350 gallons per day per tree), alternative landfill caps for leachate control, leachate treatment, and buffer zones for plume containment.

Ex Situ Bioremediation. Ground water can be pumped and soils can be excavated for biological treatment ex situ. Ex situ methods include biological wastewater treatment for ground water and slurry-phase treatment and "pile" type treatment for excavated soils.

Biological Wastewater Treatment. Ground water and leachate can be treated in a biological wastewater treatment system using rotating biological contactors (RBCs), trickling filters, sequencing batch reactors, fluidized bed reactors, activated sludge, or aerated impoundments. For biological treatment to be feasible, the ground water or leachate must contain sufficient organics to support microbial growth; otherwise, dilute streams should be treated by physical/chemical means or combined with higher-strength wastewaters for biological treatment in a larger wastewater treatment system (for example, in an onsite wastewater plant treating process wastewaters at a manufacturing facility). A sequencing batch reactor is a form of activated sludge process, by which treatment takes place in one tank in batch mode, starting with biodegradation with mixing, followed by quiescent settling and sludge withdrawal. A fluidized bed reactor uses granular activated carbon or sand as the support media for microbial growth. The ground water to be treated is introduced through the bottom of the reactor, which serves to fluidize or suspend the support media. Pretreatment to remove solids is required to avoid plugging the reactor.

Slurry-Phase Soil Treatment. Contaminated soil can be treated biologically in slurry form, in a process not unlike the biological treatment of wastewater. However, one important difference in treating contaminated soil is that the contaminants preferentially adsorb into the soil particles and must be desorbed before the microorganisms can effectively degrade the compounds. Contaminated soil may be pretreated prior to the treatment in the bioreactor, for example, by soil washing to separate the fine particles that normally contain most of the contamination. Contaminated soil can be slurried by mixing with water, wastewater, or contaminated ground water. The solids concentration of the slurry can be fairly high, up to 50%, which makes mixing to keep the soil in suspension an important design parameter.

Soil Heaps, Piles, Beds, and Windrows. Contaminated soil may be excavated and placed in heaps, piles, beds, or windrows. Other organic or bulking agents, such as wood chips or straw, may be added to aid in composting, mixing, and aerating the soil. In a windrow system, the soil is placed in long rows and turned periodically to mix and aerate it. Soil piles and heaps differ in size; heaps are larger and used where large volumes of contaminated soil must be treated. Piles and heaps may be covered or uncovered, depending on the volatility of the contaminants and air emissions requirements and whether a vacuum is drawn through or air is blown into the soil. With treatment beds, contaminated soil is placed on top of a liner system consisting of a synthetic liner, sand, and leachate drainage pipe. The bed may be covered or enclosed. In any of these treatment systems, volatilized contaminants that are removed by vacuum may be treated by incineration, carbon adsorption, or vapor-phase biodegradation. Leachate may be treated biologically in a wastewater treatment system.

Physical/Chemical/Thermal In Situ Treatment

Electroosmosis. Electroosmosis is the basis of an innovative technology dubbed the "lasagna" process, which was created to treat difficult wastes in low permeability, silt- and clay-laden soils (EPA 1994a). The lasagna process is so named because it consists of a number of layered subsurface electrodes and treatment zones. These layers can be constructed either horizontally, whereby contaminants are forced to move upward, or vertically where lateral contaminant movement is desired.

A low-voltage electric current is applied with subsurface electrodes, which forces the migration of contaminants from low-permeable soils into high-permeable treatment zones. These treatment zones may be created by hydro-fracturing (injecting a slurry to create porous pancake-shaped zones), directional drilling, or sheet piling. Treatment in these zones may include physical, chemical, or biological processes, such as adsorption by activated carbon, which would then serve as a biodegradation matrix.

In Situ Air Stripping. An innovation to conventional pump-and-treat air stripping is in situ air stripping (Hazen 1995). Two horizontal wells are installed, one below the water table and one in the vadose zone. Air is injected in the lower well while contaminated soil vapor is extracted by vacuum through the upper well.

Soil Flushing. Soil flushing is similar to soil washing of excavated soils, except that flushing is done in situ below the ground surface. With soil flushing, the contaminated area is flooded with water. Surfactants or detergents may be added to the water to enhance the removal of organics.

Soil Vapor Extraction. Volatile compounds can be extracted from subsurface soils by applying a vacuum. An added benefit is that the vacuum pulls in air from the ground surface, which stimulates biodegradation through increased oxygen transport. The ground water below the vadose zone can also be aerated to volatilize contaminants in the ground water (air sparging), which then move into the vadose zone, from which they are removed along with the soil vapors.

An EPA study (Corbin et al. 1994) showed that soil vapor extraction (SVE) is an effective treatment for removing volatile contaminants from the vadose zone. Sandy soils are more effectively treated than clay or soils with higher organic content, because higher air flows are possible in sand and clays/organic soils tend to adsorb or retain more contaminants. Removal of volatiles is rapid in the initial phase of treatment and decreases rapidly thereafter—an important consideration in the design of air emissions control over the life of the project.

An innovative companion technology to SVE is radio frequency heating of the soil (EPA 1995c, EPA 1995d). Heating the soil increases the volatilization of the contaminants that are removed by SVE. Antennae are installed near the center of the contaminated area; the radio frequency energy is applied through the antennae to heat the soil to target levels of 100°C to 150°C.

Vitrification. Vitrification is an innovative treatment that turns contaminated soils into a glasslike monolithic mass. Heat is applied through electrodes placed in the ground, the soil reaching temperatures of 2900°F to 3600°F (1600°C to 2000°C). A layer of graphite and glass frit is first placed on the surface of the ground between the electrodes to act as the initial conductive starter path. The conductive layer and adjacent soils become a molten mass that acts as the primary electrical conductor and heat transfer medium. As heat continues to be supplied by the electrodes, the molten mass moves both outward and downward. As organic contaminants in the soil are heated, they begin to vaporize and eventually pyrolize with the rising temperature. Inorganic contaminants are immobilized in the molten material. Off-gases, which will include vaporized organics and the by-products of organic and inorganic pyrolysis, are captured above the site and treated to meet air emissions standards. Once cooled, the vitrified mass is very stable with low leaching potential.

Vitrification is effective at destroying and immobilizing hazardous materials, but it is highly energy intensive and thus expensive. Consequently, it is used primarily where wastes are difficult to treat or destruction/immobilization of contaminants is very important, such as with radionuclides.

Pump-and-Treat

In the early years of ground water and soil remediation, pump-and-treat was the conventional technology. Contaminated ground water is pumped to the surface, where it is treated and reinjected or discharged to surface waters or to wastewater treatment plants. Reinjection may be used to stimulate in situ bioremediation, wherein the treated ground water is enriched with oxygen and nutrients prior to reinjection. Many years of experience with pump-and-treat methods have brought to light certain limitations, as well as improvements and enhancements. Therefore, pump and treat technology must be reviewed for its applicability to a particular site.

Pump-and-treat technology is inherently slow, because it depends on ground water for transport of the contaminant to the extraction well. This characteristic is particularly troublesome when the contaminant is only slightly water soluble, adheres to the soil, or collects in pools within the aquifer.

There are many cases of contamination by DNAPLs that have frustrated pump-and-treat efforts. The general consensus is that pump-and-treat can reduce contamination or keep it from spreading, but it has failed in many cases to remediate aquifers to stringent cleanup goals.

Common technologies for surface treatment of contaminated ground water are air stripping, carbon adsorption, and biodegradation. If cleanup goals are in the μg/L range, then typically pump-and-treat options are limited to air stripping and carbon adsorption (D. Russell 1992). Conventional biological treatment is usually not suitable for the relatively low levels of contaminants found in ground water (Reidy et al. 1990). Biological treatment of extracted ground water is discussed earlier in this chapter under "Ex Situ Bioremediation." Air stripping, carbon adsorption, and other pump-and-treat technologies are discussed in the following paragraphs. In addition, because the treatment technologies used for pump-and-treat operations are common technologies also used for many other types of wastes, further discussion may be found under "Physical/Chemical," a section near the beginning of this chapter.

Extraction Techniques. Ground water can be extracted through vertical or horizontal wells; vertical wells are the more common type. Well systems are designed to extract ground water for treatment and may also be designed for containment of the contaminant plume and reinjection of treated ground water. Although more costly and difficult to install, horizontal wells may be the method of choice where buildings or other surface features, such as roads or streams, prevent direct access to the contamination from the surface. In addition, the greater lengths of horizontal wells translate into higher ground water extraction rates.

Two-Phase Vacuum Extraction (TPVE) is a way of pumping both contaminated air and ground water through a single well (Costa 1995). A vacuum is applied through a well that is screened through the vadose and ground water zones. Both air and ground water are pumped to the surface, where they are separated for treatment. The advantages of this system are that it does not require separate wells for air extraction and ground water pumping and that water pumping rates can be greatly increased. A variation of this technique is "bioslurping," used to extract light nonaqueous phase liquids (LNAPLs) that float on the surface of the ground water and are retained in the vadose zone (Baker 1996). Bioslurping uses a slurp tube within a well to vacuum-extract LNAPLs, ground water, and contaminated vapors. The tube can be raised or lowered to extract these phases separately or as mixtures.

Air Stripping. Volatile contaminants can be removed from ground water by an air stripper. Air, blown into a column or tank where ground water is introduced, strips out the volatile compounds through volatilization. Air strippers are in widespread use in ground water treatment and in other environmental areas (see "Air Stripping" in the section "Physical/Chemical") because the technology is well understood, design is rather straightforward, and the technology is proven and cost-effective; however, they are not as effective in removing contaminants as other treatment methods, such as carbon adsorption (Reidy et al. 1990).

There are four basic types of air strippers used in ground water treatment: packed towers, diffused aeration, spray aeration, and tray aeration (Reidy et al. 1990). Packed towers are the most effective and widely used, packed with material that has a high surface-to-volume ratio and rising to more than 30 meters in height. Contaminated ground water is fed through the top of the column while air is blown in through the bottom. Packed towers can remove 90% to 99% of volatile contaminants. In diffused aeration, air is blown through diffuser pipes in the bottom of a tank or pond through which the ground water passes. Removal rates for diffusers are 70% to 90% for many volatile chemicals. In spray aeration, ground water is dispersed through nozzles over a pond or basin to strip volatiles into the ambient air. The disadvantage to this system is that it is difficult to contain and treat the contaminated air stream if required by local conditions or regulations. Spray aeration, however, can be used to recharge the aquifer if it is operated over the recharge area. Tray aeration, which is not as efficient as other air stripping methods, is sometimes used for pretreatment of ground water. In tray aeration, ground water is fed through the top of a column filled with slat trays while air is blown through the bottom of the column.

Factors affecting the design of an air stripping system include contaminant vapor pressure, water solubility, and concentration; ground water temperature; metal precipitates; solids; cleanup levels; and emission requirements (Reidy et al. 1990). Air strippers are effective for contaminants that have high vapor pressures and low water solubilities. They are most effective with concentrations above 100 mg/L and become much less effective when influent concentrations drop to less than 1 mg/L. Vapor pressure increases with temperature, so ground water temperature is an important factor. High concentrations of iron and manganese in the ground water can precipitate on the packing material and retard contaminant removal. Likewise, suspended solids are a problem in packed towers because they can cause plugging or short-circuiting. Cleanup levels determine the type and size of air stripper required. The most stringent cleanup levels require packed towers. Treatment of the air stream leaving the stripper may be required, depending on regulatory and local conditions. Control devices include demisters, fume or vapor incinerators, carbon columns, and vapor condensers.

Air strippers are capable of treating ground waters to low μg/L levels. If designed properly, air strippers can remove petroleum-type contaminants to less than 1 μg/L, meeting most regulatory criteria (D. Russell 1992). At Wurtsmith Air Force Base in Michigan, TCE concentrations were reduced from 6,000 μg/L to less than 1.5 μg/L (H. Russell 1992). At another site reported by H. Russell, total chlorinated hydrocarbons were reduced from 120,000 μg/L to less than a detection limit of 1 μg/L. At a TCE contamination site in Des Moines, optimum stripper performance was 98% for influent concentrations averaging 900 μg/L (Pheiffer 1990b). Removals of trans-1,2-dichloroethene were more variable than for TCE and ranged from 85% to 96%. Vinyl chloride influent concentrations ranged from 1 to 38 μg/L and were nondetectable in the effluent stream. Air stripping at a municipal well field site demonstrated removals below 1 µg/L with influent concentrations of less than 3 µg/L for 1,1-dichloroethene; 2 µg/L for 1,1-dichloroethane; 15 µg/L for 1,1,1-trichloroethane; and less than 12 µg/L for TCE (Pheiffer 1990c).

Pretreatment of the ground water prior to air stripping may be required to equalize flows and concentrations; remove high concentrations of suspended solids; remove immiscible liquids such as NAPLs; remove iron, manganese, or water hardness; remove dissolved metals; or adjust the pH (EPA 1995a).

Carbon Adsorption. Carbon adsorption is widely used in ground water remediation where very high-quality effluent is desired, for example, at or beyond drinking water standards. Carbon adsorption is used on ground waters that have only low levels of contaminants, making pretreatment of some ground waters necessary. Carbon is often used to treat the off-gas, or contaminated air stream, from air strippers.

Contaminated ground water is fed through the unit, where contaminants are adsorbed onto the surface of the carbon. When the adsorptive capacity of the carbon is exceeded, the spent carbon is either regenerated or disposed of. GAC, rather than PAC, is commonly used in ground water treatment. A more detailed discussion of general carbon adsorption technology is given in this chapter in the section "Physical/Chemical."

Factors affecting the design of carbon adsorption units are contaminant or total organic concentration, contaminant molecular weight and water solubility, metal precipitates, and solids (Reidy et al. 1990). Ground water with high contaminant levels will quickly overload a carbon unit and should be pretreated. Total organic carbon concentrations of less than 1 mg/L are best for carbon units; those greater than 5 mg/L are considered poor candidates for carbon adsorption. Higher molecular weight compounds and those with low water solubilities are easily adsorbed on carbon. Dissolved concentrations of iron and manganese greater than 5 mg/L can interfere with adsorption when they precipitate on the carbon surface; concentrations below 0.2 mg/L are desirable. Suspended solids concentrations in the feed to a carbon unit are usually kept below 5 mg/L so as not to plug or short-circuit the unit; concentrations above 20 mg/L are not acceptable.

Other Technologies. Chemical oxidation is often used in ground water treatment as a pretreatment step; for example, increasing biodegradability can be increased by partial oxidation of refractory organics or by oxidizing arsenite to arsenate. Chemical oxidants commonly used are hydrogen peroxide, ozone, and chlorine. Ultraviolet radiation may be used in combination with hydrogen peroxide and/or ozone.

Ion exchange is most often used to remove metals from ground water. It can be used to remove sulfates, nitrates, and radionuclides (EPA 1995a). Pretreatment of the ground water may be required; for example, 10-µm filtration to remove solids; carbon adsorption to remove organics that foul strong base resins; dechlorination to neutralize chlorine; and aeration, precipitation, or filtration to remove iron and manganese.

Chemical precipitation may be used to remove metals such as arsenic, cadmium, chromium, lead, mercury, selenium, and silver from ground water.

In photocatalytic oxidation, platinized titanium dioxide is used as a catalyst and ultraviolet radiation (UV) is supplied by UV lamps or the sun to produce hydroxyl radicals that oxidize organic compounds in contaminated ground water. Pretreatment of the ground water may be necessary to remove solids that interfere with light transmission or clog the reactor bed and to remove ionic species that can foul the catalyst.

Ex Situ Soil Nonbiological Treatment

Soil Leaching. Soil leaching or acid extraction uses acid to solubilize metals for removal from soils, a technique akin to that used in the mining industry. After extraction with an acid, such as hydrochloric, sulfuric, or nitric, the soil is separated from the acid, rinsed with water to remove excess acid and metals, dewatered, and neutralized. The extracted metals are separated from the solution during acid regeneration, and the acid solution is recycled back to the process. The extracted metals can be precipitated and recovered.

Soil Washing. Soil washing is as simple a process as the name implies. Soil is excavated, physically broken up, washed with a water-based solution, and separated into size fractions (sand/gravel and silt/clay). The goal of soil washing is to transfer pollutants to the wash water and to isolate the finer soil fraction that preferentially adsorbs organics and metals. Various surfactants, extractants, or detergents are added to the wash water to enhance removal of organic contaminants from the soil particles. Soil fines and contaminants are removed from the wash water, which is then recycled back to the process.

The finer soil fraction contains adsorbed organics, small metallic particles, and bound ionic metals. This fraction may be treated further to remove the contaminants, or it may be incinerated or landfilled. The "clean" coarse fraction may contain some residual metallic fragments. With metal contamination, both the fine and coarse soil fractions may be leached with an acid solution to remove the metals (Benker 1995).

Solvent Extraction. Organic contaminants are removed from soil by extraction with solvents such as hexane, triethylamine, acetone, liquified propane, or even supercritical carbon dioxide. After extraction, the solvent is separated from the soil and treated by processes such as microfiltration and distillation before recycling back to the process. The soil may be further treated by vapor extraction to remove most of the residual solvent, followed by biological treatment. Organics that are amenable to solvent extraction from soils include pesticides, PCBs, petroleum hydrocarbons, chlorinated hydrocarbons, polynuclear aromatic hydrocarbons, polychlorinated dibenzo-p-dioxins, and polychlorinated dibenzofurans.

Thermal Desorption. In this process, heat is applied either directly or indirectly to contaminated soil to transfer volatile compounds to an air (or gas) stream. Heat is controlled to prevent combustion or incineration. The volatilized compounds are usually removed by carbon adsorption, destroyed by thermal oxidation, or condensed as recovered materials.

Several innovative thermal desorption systems have been tested or used in field applications for the treatment of volatile and semivolatile organics, PCBs, organometallic complexes, and total petroleum hydrocarbons (EPA 1994b, EPA 1994c, EPA 1992). Soils are heated to temperatures ranging from 200°C to 600°C, and volatilized organics and metals are removed in a gas stream, which is then treated. The ECO LOGIC process uses a hydrogen-rich carrier gas to transport the volatilized contaminants to a chemical reduction reactor. Prior to introduction into the reactor, the gas stream passes through a molten metal bath to remove a portion of the volatilized metals.

LAND-BASED SYSTEMS

Concern about soil and ground water contamination and inadequate waste treatment has given rise to stringent control of land-based systems such as landfills, surface impoundments, land treatment units, and underground injection wells. Federal regulations have been developed under RCRA and its amendments covering the design, operation, and closure of land-based systems.

In addition, there are federal standards restricting the management of hazardous wastes in land-based units unless the waste is treated in a particular way or achieves minimum treatment levels. These restrictions on hazardous waste, which are found at 40 CFR §268, are referred to as land bans or land disposal restrictions (LDRs). There are three types of LDRs: those that address total waste concentration, extracted or leachate waste concentration, and specified technology. There are different LDRs for wastewaters and non-wastewaters. Wastewaters are defined (with some exceptions) as wastes containing less than 1% by weight total organic carbon (TOC) and less than 1% by weight total suspended solids (TSS).

Landfills

Major federal regulations covering landfill design and operation are the Subpart N standards of 40 CFR §§264 and 265 for hazardous waste and the standards for municipal landfills at 40 CFR §258. The standards for hazardous waste landfills cover:

- Design and operation of liner, leachate collection, and storm water systems;
- Action leakage rates and response plan for leak detection systems;
- Monitoring and inspection of liners, covers, and leachate collection systems;

- Surveying and recordkeeping for each landfill cell;
- Closure of the landfill or individual landfill cells and post-closure care;
- Requirements for ignitable, reactive, and otherwise incompatible wastes; and
- Requirements for bulk and containerized liquids, overpacked drums (lab packs), containers, and certain dioxin-contaminated listed wastes.

The standards for municipal waste landfills under §258 cover:

- Location restrictions regarding airports, floodplains, wetlands, fault areas, seismic impact zones, and unstable areas;
- Operating criteria for hazardous waste prohibition, cover material, disease vector control, explosive gas control, air pollution, access, storm water, surface water discharges, liquid waste restrictions, and record-keeping;
- Design criteria for liners and leachate collection;
- Ground water monitoring and corrective action for ground water con-tamination;
- Closure and post-closure; and
- Financial assurance.

The following paragraphs deal primarily with landfill construction and design. The reader is referred to §§258, 264, and 265 of 40 CFR for details on specific regulatory requirements.

Basic Landfill Components. The basic components of a landfill are a liner, a leachate collection system, and a final cover. Many different types and com-binations of materials are used for these components. In addition, inasmuch as both liners and covers are designed to prevent liquid migration, many of the same materials are used for both.

Common landfill materials include clay or low-permeable soil materials, geosynthetics (geogrids, geonets, geotextiles, geomembranes, flexible mem-brane liners), and synthetic clay liners. Brief descriptions of these materials follow:

- *Clay or soil material*—Normally, natural clays with a hydraulic conduc-tivity of at least 1×10^{-7} cm/sec. More permeable soils may be mixed with clay or other additives to create a soil material with sufficient im-permeability.
- *Geogrid*—A geotextile used for additional structural stability for steep slopes and liner support under differential settlement.
- *Geonet*—A geotextile used as a drainage net.

- *Geosynthetic*—A general term referring to a group of synthetic materials that includes geogrids, geonets, geotextiles, and geomembranes.
- *Geotextile*—A thin, fabric-type material that is used as a particle filter, drainage net, or geocell containment. Also referred to as *geofabric*. When used for drainage, the term *geonet* is often used.
- *Synthetic clay liner*—A manufactured material made of a thin layer of sodium montmorillonitic clay sandwiched between two geotextile layers.
- Synthetic membrane—A thin, flexible, plastic-type material used as an impermeable layer. Also called geomembrane or flexible membrane liner (FML).

The main purpose of synthetic membranes is to prevent liquids from migrating into or out of a landfill. In final covers, synthetic membranes prevent precipitation from entering a landfill. In landfill liners, these membranes prevent leachate from migrating through to secondary landfill liners or outside the landfill into the soil and ground water.

Synthetic membranes are made from common polymers such as butyl rubber, chlorinated polyethylene, chlorosulfonated polyethylene, ethylene-propylene rubber, high-density polyethylene, medium- to very-low density polyethylene, and polyvinyl chloride. Other polymers that may be used in synthetic membranes are nylon, polyester, neoprene, ethylene propylene diene monomer, and urethane.

Key factors in choosing synthetic membranes are compatibility with the waste leachate, permeability, and mechanical properties. Compatibility with the waste leachate is important because the membrane can be chemically weakened or damaged and fail to contain the leachate. Permeability is important because a membrane must be a barrier not only to water, but to other leachate constituents such as hydrocarbons. Mechanical properties are important because they determine how easy the membrane is to install, how well it resists tears and punctures, and how well it holds up under temperature extremes and sunlight.

Liners. Landfill liners are meant to prevent waste leachate from migrating into the soil and ground water outside the landfill. They range in complexity from natural liners (where the existing soil acts as a barrier) to single, double, and multiple liners. Natural liners were common many years ago before soil and ground water contamination problems became widely known. Today, new landfill construction includes at least a single liner; hazardous waste landfills must have at least a double liner.

The materials used for landfill liners include synthetic membranes, clay, soils that have been mixed with additives to improve their characteristics (amended soils), and alternate barrier materials such as synthetic clay liners.

Leachate Collection. Leachate originates from a mixture of precipitation falling on a landfill and the liquids present in landfilled wastes. As this mixture percolates through the landfill, it picks up, or leaches, contaminants from the wastes. Leachate is removed from a landfill to reduce the hydraulic head on liners and the amount of leachate that could breach the liner and contaminate soil and ground water.

A leachate collection system usually consists of a gridwork of perforated pipe placed above the landfill liner. The piping grid is covered with a layer of sand and gravel that allows easy drainage of the leachate into the pipe and prevents the pipe from being crushed by the weight of the waste above it. As an alternative to a piping grid, a leachate collection system may also be constructed with geonets. An example of a primary leachate collection system is shown in Figure 5.6.

Covers. A final cap or cover is placed over the waste when a landfill is closed. The primary purpose of this final cover is to minimize the infiltration of precipitation and thus minimize the generation of leachate. A cover also helps to maintain the integrity of the landfill while improving the final appearance of the disposal site.

Final covers are actually multiple layers. Typically, the order in which these layers are placed over the waste is as follows: a grading layer, a gas-venting layer, a barrier layer, a drainage layer, a protective layer, and a topsoil/vegetative layer.

The grading layer is placed directly over the waste to even out the contours of the landfill surface, providing a stable and smooth base for the next layer. The grading layer may also serve as the gas-venting layer if a coarse-grained material such as sand or gravel is used.

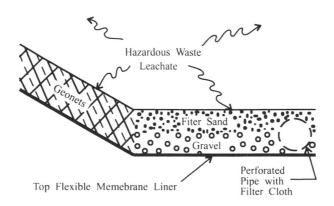

Source: Adapted from EPA (1989).

Figure 5.6 Cross-sectional view of a primary leachate collection system.

Gases evolve from a landfill as wastes decompose. Gas vents or extraction wells may be installed for various reasons: to prevent rupture of the final cover; to prevent stress on the vegetation cover, which can lead to erosion; to meet air quality criteria; to minimize health risks to the local population; or to burn for heating value. The type and quantity of gases generated depend on the type of waste. When putrescible (biodegradable) wastes are landfilled, gas venting is considered essential (Bagchi 1994). Municipal landfills, which contain a high percentage of biodegradable wastes, can produce significant quantities of methane and carbon dioxide. Unless gas extraction wells are installed in the landfill, a venting layer is needed as part of the final cover (LaGrega et al. 1994). Coarse-grained materials such as sand and gravel are used to allow these gases to pass freely to the vents. Vents may be isolated or connected in a grid system.

A barrier layer is placed over the grading and gas-venting layers. The barrier layer minimizes the infiltration of precipitation and storm water. Barriers are typically constructed of clay, or of a synthetic membrane over a low-permeable material such as clay. Landfills that are lined with synthetic membranes typically use synthetic membranes in the barrier layer. This prevents the buildup of excessive hydraulic head on the liner, because the permeabilities of the barrier and the liner will be the same. Synthetic clay liners may also be used as barriers.

A drainage layer placed over the barrier layer diverts water laterally away from the barrier, preventing the buildup of hydraulic head. When synthetic membranes are used in the barrier layer, a drainage layer over the membrane prevents the saturation of the upper soil cover, which otherwise can lead to failure and erosion. The drainage layer also protects synthetic membranes from damage when final soil covers are placed. Drainage layers may be constructed of sand and gravel, geonets, or geocomposites.

Topsoil is placed over the drainage layer. The type of soil used must be cohesive enough to maintain the contours of the landfill, minimize erosion, and support the vegetative cover. Together with the drainage layer, topsoil protects the barrier layer from developing cracks that can result from drying and freeze/thaw cycles. If the topsoil is sandy and has a permeability or 1×10^{-3} cm/sec or more, then a separate drainage layer may not be needed (Bagchi 1994). Many different types of vegetative (grass) covers can be used, but it is important to choose a vegetative cover that grows well in the area and is easy to maintain. Natural vegetation may be allowed to grow as the final cover; however, until the natural vegetation is fully established, additional seeding or covering will be needed to prevent the final cover from eroding.

Hazardous Waste Landfills. Hazardous waste landfills constructed after January 29, 1992, must have a double liner and a leachate collection system above each liner. An example of a double liner, leachate collection system is shown in Figure 5.7. The double liner requirements may be waived, however, for landfills containing only wastes from foundry furnace emission controls or metal

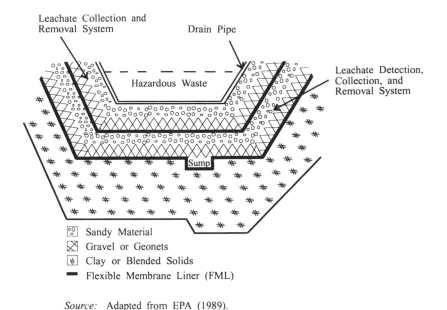

Source: Adapted from EPA (1989).

Figure 5.7 Example of a landfill with liners and a leachate management system.

casting molding sand (monofills) and certain replacement landfill units. The top liner must be designed to prevent hazardous constituents from migrating into the liner, not only while the landfill is in operation, but throughout the post-closure care period, which may be 30 years or longer. The bottom liner must be a composite liner of at least two components. The upper component, like the top liner, must be designed to prevent the migration of hazardous constituents. The lower component of the bottom liner must have at least 3 feet of compacted soil material with a hydraulic conductivity of 1×10^{-7} cm/sec or less.

The top liner and upper component of the bottom liner are usually geomembranes. The compacted soil material for the lower component of the bottom liner is usually a clayey soil or one that has been amended to meet the hydraulic conductivity criterion.

There must be a leachate collection system above the top liner and one above the bottom liner. This system must be constructed with either a granular drainage material such as gravel and sand or a synthetic material such as a geonet. If a granular layer is used, it must be at least 12 inches thick and have a hydraulic conductivity of at least 1×10^{-2} cm/sec. If a geonet is used, it must have a transmissivity of at least 3×10^{-5} m²/sec. The bottom slope of the leachate collection system must be at least 1 percent.

The final cover for a hazardous waste landfill must provide long-term minimization of liquid migration, require minimum maintenance, promote drainage while minimizing erosion or abrasion, maintain its integrity with settling and subsidence of the landfill contents, and have a premeability less than or equal to that of any bottom liner or natural subsoils.

Surface Impoundments

Surface impoundments, or ponds, are commonly used for biological wastewater treatment and for impounding storm water. Impoundments are used more often for nonhazardous wastewaters, because the hazardous waste regulations for this type of unit are very strict. Impoundments provide temporary storage for storm water prior to discharge or treatment. In arid areas, storm water basins may be used as evaporation ponds in lieu of creating a discharge that must have a wastewater permit.

Surface impoundments used to manage hazardous wastewaters must meet the technical standards given at 40 CFR Parts 265 and 264 as well as land disposal restrictions in Part 268. Although there are some exceptions, hazardous waste impoundments are subject to minimum technology requirements (MTRs). MTRs include two or more liners and a leachate collection and removal system between liners. Requirements for liner components and the leachate system are similar to those for hazardous waste landfills except that granular drainage materials must have a hydraulic conductivity of at least 1×10^{-1} cm/sec and geonets must have a transmissivity of at least 3×10^{-4} cm/sec. LDRs require that treatment residues generated in hazardous waste impoundments be removed at least once a year unless the residues meet applicable treatment or prohibition levels.

Land Application

In land application or treatment, waste is mixed into the upper soil layer, where microorganisms degrade organic compounds and metals are adsorbed. Land application of municipal wastewater biosolids is common because such materials have value as fertilizers (nitrogen, phosphorus) and as soil amendments. Land treatment of hazardous wastes has been significantly curtailed because of land disposal restrictions on hazardous wastes.

Land application of biosolids are regulated under 40 CFR Part 503. These regulations contain limits for arsenic, cadmium, chromium, copper, lead, mercury, molybdenum, nickel, selenium, and zinc. There are limits for both the metal content in the biosolids (ceiling and high quality limits), as well as maximum loading rates on the land (annual and total cumulative). Metal concentrations in biosolids must not exceed the ceiling or maximum concentration limits. Cumulative loading rates apply to biosolids that do not meet the high-quality limits. Biosolids meeting the high-quality limits are exempted from certain requirements.

Land treatment of hazardous waste is regulated under 40 CFR Parts 264, 265, and 268. The land treatment standards of Parts 264 and 265 cover treatment demonstrations, design and operating requirements, restrictions regarding food-chain crops; monitoring of the unsaturated zone, recordkeeping, closure and post-closure, and special requirements for ignitable, reactive, incompatible, and dioxin contaminated wastes. Part 268 imposes land disposal restrictions prior to the application of hazardous wastes to the land treatment unit.

Underground Injection

Underground injection wells are used for the disposal of aqueous solutions containing acids, heavy metals, organics, or inorganics that are either more difficult or expensive to treat by other methods. Typically, wastewaters are pretreated prior to injection to remove oils or solids that could plug the injection zone.

Injection wells are regulated under the Safe Drinking Water Act (SDWA) and RCRA. Hazardous waste injection wells are subject to land disposal restrictions at 40 CFR Part 268. Pretreatment systems for hazardous waste are also subject to RCRA technical standards at 40 CFR Parts 264 and 265.

Hazardous waste subject to land disposal restrictions cannot be disposed of in an injection well unless the waste meets the treatment standards of Part 268 or is exempted by the regulatory authority. To obtain this exemption, the injection system must meet stringent criteria defined at 40 CFR Part 148, which includes a *no migration* standard. Under this standard, injected wastes cannot migrate upward out of the injection zone or laterally within the zone to a point of discharge, or interface with an underground source of drinking water for 10,000 years. In addition, before the waste can migrate out of the zone or to a point of discharge or interface with a drinking water source, it must be rendered nonhazardous by attenuation, transformation, or immobilization of its hazardous constituents. Information required to support an exemption petition must include a description of the waste treatment methods and a monitoring plan.

REFERENCES

American Petroleum Institute, "In Situ Air Sparging: Evaluation of Petroleum Industry Sites and Considerations for Applicability, Design, and Operation," Publication No. 841-46090, Washington, DC, 1995.

Bagchi, Amalendu, *Design, Construction, and Monitoring of Landfills*, 2d ed., John Wiley & Sons, New York, 1994.

Baker, Ralph S., "Bioslurping LNAPL Contamination," *Pollution Engineering*, Vol. 28/No. 3, Cahners Publishing Co., Highlands Ranch, CO, 1996.

Benker, Keith W., "Removing Metals from Soil," *Civil Engineering*, Vol. 65/No. 10, American Society of Civil Engineers, New York, 1995.

Bouwer, E.J., and G.D. Cobb, "In Situ Groundwater Treatment Technology Using Biodegradation," U.S. Army Toxic and Hazardous Materials Agency, Report AMXTH-TE-CR-88023, 1987.

Brubaker, Gaylen R., "The Boom in In Situ Bioremediation," *Civil Engineering*, Vol. 65/No. 10, American Society of Civil Engineers, New York, 1995.

Chambers, C.D., et al., *In Situ Treatment of Hazardous Waste-Contaminated Soils*, Noyes Data Corporation, Park Ridge, NJ, 1991.

Cheremisinoff, Paul N., "Selecting Hazardous Waste Treatment Systems: Thermal Classification Is Key," *The National Environmental Journal*, Vol. 4/No. 1, Columbus, GA, 1994.

Clifton, C.E., *Introduction to Bacterial Physiology*, McGraw-Hill, New York, 1957.

Corbin, Michael H., Nancy A. Metzer, and Michael F. Kress, "Project Summary: Field Investigation of Effectiveness of Soil Vapor Extraction Technology," EPA/600/SR-94/142, U.S. Environmental Protection Agency, Cincinnati, OH, 1994.

Costa, Michael J., "Case Study: Vacuum Extraction Harnessed for Emergency Clean-up," *Environmental Protection*, Vol. 6/No. 4, Stevens Publishing, Waco, TX, 1995.

Eckenfelder, W. W., Jr., *Industrial Water Pollution Control*, McGraw-Hill, New York, 1966.

Eckenfelder, W. Wesley, Jr., and Jack L. Musterman, *Activated Sludge Treatment of Industrial Wastewater*, Technomic Publishing Company, Lancaster, PA, 1995.

Elizardo, Kelly, "Biotreatment Techniques Get Chemical Help," *Pollution Engineering*, Vol. 25/No. 11, Cahners Publishing Co., Highlands Ranch, CO, 1993.

Fagan, Michael R., "Peroxygens Enhance Biological Treatment," *Environmental Protection*, Vol. 5/No. 9, Stevens Publishing, Waco, TX, 1994.

Hart, D.P. and R. Couvillion, *Earth-Coupled Heat Transfer*, National Water Well Association, Dublin, OH, 1986.

Hasbach, A., "Moving Beyond Pump-and-Treat," *Pollution Engineering*, Vol. 25/No. 6, 1993.

Hazen, Terry C., "Savannah River Site: A Test Bed for Cleanup Technologies," *Environmental Protection*, Vol. 6/No. 4, Stevens Publishing, Waco, TX, 1995.

Huling, S.G., et al., "Enhanced Bioremediation Utilizing Hydrogen Peroxide as a Supplemental Source of Oxygen: A Laboratory and Field Study," EPA/600/2-90/006, U.S. Environmental Protection Agency, Washington, DC, 1990.

Hutchins, et al., "Nitrate for Biorestoration of an Aquifer Contaminated with Jet Fuel, Report for June 1988–Sept. 1990," U.S. Environmental Protection Agency in co-operation with Solar Universal Technologies, Inc., NSI Technology Services Corp., Traverse Group, Inc., and 9th Coast Guard District, 1991.

Jensen, Ric, "Successful Treatment with Supercritical Water Oxidation," *Environmental Protection*, Vol. 5/No. 6, Stevens Publishing, Waco, TX, 1994.

Kemblowski, M.W., et al., "Fate and Transport of Residual Hydrocarbon in Ground Water: A Case Study," Petroleum Hydrocarbons and Organic Chemicals in Ground Water: Prevention, Detection, and Restoration; Conference and Exposition, National Water Well Association and American Petroleum Institute, November 17–19, 1987.

King, R.B., et al., *Practical Environmental Bioremediation*, Lewis Publishers, Boca Raton, FL, 1992.

Krishnan, E. Radha, et al., *Recovery of Metals from Sludges and Wastewaters*, Noyes Data Corporation, Park Ridge, NJ, 1993.

LaGrega, Michael D., Philip L. Buckingham, and Jeffrey C. Evans, *Hazardous Waste Management*, McGraw-Hill, New York, 1994.

Matso, Kalle, "Mother Nature's Pump and Treat," *Civil Engineering*, Vol. 65/No. 10, American Society of Civil Engineers, New York, 1995.

Miller, R.N., "A Field-Scale Investigation of Enhanced Petroleum Hydrocarbon Biodegradation in the Vadose Zone Combining Soil Venting as an Oxygen Source with Moisture and Nutrien Addition," Ph.D. diss., Civil and Environmental Engineering Department, Utah State University, Logan, UT, 1990.

Morgan, P., and R.J. Watkinson, "Factors Limiting the Supply and Efficiency of Nutrient and Oxygen Supplements for the In Situ Biotreatment of Contaminated Soil and Groundwater," *Water Research*, Vol. 26, 1992.

Musterman, Jack L., and Victor J. Boero, "Granular Activated Carbon vs. Macroreticular Resins," *Industrial Wastewater*, Vol. 3/No. 2, Water Environment Federation, Alexandria, VA, 1995.

Nelson, Christopher H., "Using Ozone to Speed Up Air Sparging," *Environmental Protection*, Vol. 6/No. 4, Stevens Publishing, Waco, TX, 1995.

Noyes, Robert, *Unit Operations in Environmental Engineering*, Noyes Publications, Park Ridge, NJ, 1994.

Pheiffer, T., "Case Study of the Eglin Air Force Base Ground Water Bioremediation System," in *Innovative Operational Treatment Technologies for Application to Superfund Sites: Nine Case Studies*, U.S. Environmental Protection Agency, Washington, DC, 1990a.

Pheiffer, T., "Case Study of Ground Water Extraction with Air Stripping, Des Moines TCE Site," in *Innovative Operational Treatment Technologies for Application to Superfund Sites: Nine Case Studies*, U.S. Environmental Protection Agency, Washington, DC, 1990b.

Pheiffer, T., "Case Study of Ground Water Extraction with Air Stripping, Eau Claire Municipal Well Field Site," in *Innovative Operational Treatment Technologies for Application to Superfund Sites: Nine Case Studies*, U.S. Environmental Protection Agency, Washington, DC, 1990c.

Pitter, Pavel, and Jan Chudoba, *Biodegradability of Organic Substances in the Aquatic Environment*, CRC Press, Boca Raton, FL, 1990.

Raymond, R.L., et al., "Field Report, Field Application of Subsurface Biodegradation of Gasoline in a Sand Formation," Project No. 307-77, Committee on Environmental Affairs, American Petroleum Institute, Washington, DC, 1976.

Reidy, P.J., et al., "Assessing UST Corrective Action Technologies: Early Screening of Cleanup Technologies for the Saturated Zone," EPA/600/2-90/027, U.S. Environmental Protection Agency, 1990.

Roeck, Douglas R., and Alfred Sigg, "Carbon Injection Proves Effective in Removing Dioxins," *Environmental Protection*, Vol. 7/No. 1, Stevens Publishing, Waco, TX, 1996.

Russell, D.L., *Remediation Manual for Petroleum-Contaminated Sites*, Technomic Publishing Co., Lancaster, PA, 1992.

Russell, H.H., et al., "TCE Removal from Contaminated Soil and Ground Water," Ground Water Issue, EPA/540/S-92/002, U.S. Environmental Protection Agency, Washington, DC, 1992.

Semprini, L., et al., "A Field Evaluation of In Situ Biodegradation for Aquifer Restoration," U.S. Environmental Protection Agency, EPA/600/2-87/096, Washington, DC, 1987.

Sewell, G.W., et al., "Anaerobic In Situ Treatment of Chlorinated Ethenes," U.S. Environmental Protection Agency, EPA/600/D-90/204, Washington, DC, 1990.

Sims, J.L., et al., "In Situ Bioremediation of Contaminated Ground Water," U.S. Environmental Protection Agency, EPA/540/S-92/003, Washington, DC, 1992.

U.S. Environmental Protection Agency, "Manual: Ground Water and Leachate Treatment Systems," EPA/625/R-94/005, Washington, DC, 1995a.

U.S. Environmental Protection Agency, "Project Summary: Assessing UST Corrective Action Technologies: Lessons Learned About In Situ Air Sparging at the Denison Avenue Site, Cleveland, Ohio," EPA/600/SR-95/040, Cincinnati, OH, 1995b.

U.S. Environmental Protection Agency, "SITE Technology Capsule: IITRI Radio Frequency Heating Technology," EPA/540/R-94/527a, Cincinnati, OH, 1995c.

U.S. Environmental Protection Agency, "SITE Technology Capsule: KAI Radio Frequency Heating Technology," EPA/540/R-94/528a, Cincinnati, OH, 1995d.

U.S. Environmental Protection Agency, "SITE Demonstration Bulletin: ZenoGem™ Wastewater Treatment Process, ZENON Environmental Systems," EPA/540/MR-95/503, Cincinnati, OH, 1995e.

U.S. Environmental Protection Agency, "Development Document for Proposed Effluent Limitations Guidelines and Standards for the Pharmaceutical Manufacturing Point Source Category," EPA/821/R-95/019, Washington, DC, 1995f.

U.S. Environmental Protection Agency, "EPA and Industry Sign CRADA to Develop Innovative 'Lasagna' Process," in *Bioremediation in the Field*, Center for Environmental Research Information, Cincinnati, OH, 1994a.

U.S. Environmental Protection Agency, "Eco Logic International Gas-Phase Chemical Reduction Process, The Thermal Desorption Unit: Applications Analysis Report," EPA/540/AR-94/504, Washington, DC, 1994b.

U.S. Environmental Protection Agency, "SITE Demonstration Bulletin, Thermal Desorption System: Clean Berkshires, Inc.," EPA/540/MR-94/507, Cincinnati, OH, 1994c.

U.S. Environmental Protection Agency, "SITE Emerging Technology Summary: Cross-Flow Pervaporation for Removal of VOCs from Contaminated Wastewater," EPA/540/SR-94/512, Cincinnati, OH, 1994d.

U.S. Environmental Protection Agency, "CAV-OX® Cavitation Oxidation Process, Magnum Water Technology, Inc.: Application Analysis Report," EPA/540/SR-93/520, Washington, DC, 1994e.

U.S. Environmental Protection Agency, "Emerging Technology Summary: Laser Induced Photochemical Oxidative Destruction of Toxic Organics in Leachates and Groundwaters," EPA/540/SR-92/080, Washington, DC, 1993a.

U.S. Environmental Protection Agency, "EPA Signs License Agreement for TCE-Degrading Microorganisms," in *Bioremediation in the Field*, Center for Environmental Research Information, EPA/540/N-93/001, Cincinnati, OH, No.8, 1993b.

U.S. Environmental Protection Agency, "Demonstration Bulletin: CAV-OX® Ultraviolet Oxidation Process, Magnum Water Technology," EPA/540/MR-93/520, Washington, DC, 1993c.

U.S. Environmental Protection Agency, "Perox-pure™ Chemical Oxidation Technology Peroxidation Systems, Inc.: Applications Analysis Report," EPA/540/AR-93/501, Washington, DC, 1993d.

U.S. Environmental Protection Agency, "Low Temperature Thermal Treatment (LT^3®) Technology, Roy F. Weston, Inc.: Applications Analysis Report," EPA/540/AR-92/019, Washington, DC, 1992.

U.S. Environmental Protection Agency, "Treatment Technology Background Document," PB91-160556, Washington, DC, 1991.

U.S. Environmental Protection Agency, "Best Demonstrated Available Technology (BDAT) Background Document for Aniline Production Treatability Group (K103,

K104)," vol. 7, proposed, EPA/530/SW-88/0009g, Washington, DC, 1988.

U.S. Environmental Protection Agency, "Compendium of Technologies Used on the Treatment of Hazardous Waste," EPA/625/8-87/014, Center for Environmental Research Information, Cincinnati, OH, 1987.

U.S. Environmental Protection Agency, "Requirements for Hazardous Waste Landfill Design, Construction and Closure," EPA/625/4-89/022, U.S. EPA, Cincinnati, OH, 1989.

Ware, P.J., "Supplemental Air to Reduce Remediation," *The National Environmental Journal*, Vol. 3/No. 4, Columbus, GA, 1993.

Water Environment Federation, "Ironing Out Groundwater Contamination," Water Environment & Technology, Vol. 6/No. 6, Alexandria, VA, 1994.

Water Environment Federation (formerly Water Pollution Control Federation), *Hazardous Waste Treatment Processes*, Manual of Practice FD-18, prepared by Task Force on Hazardous Waste Treatment, Alexandria, VA, 1990.

Weber, Walter J., Jr., *Physiochemical Processes for Water Quality Control*, John Wiley & Sons, New York, 1972.

Woodside, Gayle, *Hazardous Materials and Hazardous Waste Management*, John Wiley & Sons, New York, 1993.

BIBLIOGRAPHY

Adams, Carl E., Jr., Davis L. Ford, and W. Wesley Eckenfelder Jr. (1981), *Development of Design and Operational Criteria for Wastewater Treatment*, Enviro Press, Inc., Nashville, TN.

American Petroleum Institute (1995), *Modeling Aerobic Biodegradation of Dissolved Hydrocarbons in Heterogenous Geologic Formations*, Publication No. 848-00200, Washington, DC.

American Society of Civil Engineers and American Water Works Association (1990), *Water Treatment Plant Design*, 2d ed., McGraw-Hill, New York.

Bausmith, David S., Darcy J. Campbell, and Radisav D. Vidic (1996), "In Situ Air Stripping," *Water Environment & Technology*, Vol. 8/No.2, Water Environment Federation, Alexandria, VA.

Bouley, Jeffrey (1993), "Liners & Lining Systems," *Pollution Engineering*, Vol. 25/No. 4, Cahners Publishing Co., Highlands Ranch, CO.

Chaudhry, G. Rasul, ed. (1994), *Biological Degradation and Bioremediation of Toxic Chemicals*, Dioscorides Press, Portland, OR.

Dietrich, Jonathan A. (1995), "Membrane Technology Comes of Age," *Pollution Engineering* Vol. 27/No. 7, Cahners Publishing Co., Highlands Ranch, CO.

Flood, Donald R. (1994), "Synthetic Linings for Hazardous Wastes," *The National Environmental Journal*, Vol. 4/No. 3, Columbus, GA.

Gray, N.F. (1990), *Activated Sludge: Theory and Practice*, Oxford Science Publications, New York.

Hartley, Robert P. (1992), *Surface Impoundments: Design, Construction, and Operation*, Noyes Data Corporation, Park Ridge, NJ.

Jenkins, David, Michael G. Richard, and Glen T. Daigger (1993), *Manual on the Causes and Control of Activated Sludge Bulking and Foaming*, Lewis Publishers, Boca Raton, FL.

Metcalf & Eddy, Inc. (1972), *Wastewater Engineering: Collection, Treatment, Disposal*, McGraw-Hill, New York.

Miller, Michael E. (1995), "Bioremediation on a Big Scale," *Environmental Protection*, Vol. 6/No. 7, Stevens Publishing, Waco, TX.

National Research Council (1994), *Alternatives for Ground Water Cleanup*, National Academy Press, Washington, DC.

———— (1993), *In Situ Bioremediation: When Does It Work?*, National Academy Press, Washington, DC.

Paff, Stephen W., Brian E. Bosilovich, and Nicholas J. Kardos (1994), "Emerging Technology Summary: Acid Extraction Treatment System for Treatment of Metal Contaminated Soils," EPA/540/SR-94/513, U.S. Environmental Protection Agency, Cincinnati, OH.

Russell, David L. (1996), "Using Horizontal Wells as a Remediation Tool," *Environmental Protection*, Vol. 7/No. 1, Stevens Publishing, Waco, TX.

Santos, Ed, (1995), "Standardized and Customized pH Neutralization Systems," *Environmental Protection*, Vol. 6/No. 12, Stevens Publishing, Waco, TX.

U.S. Environmental Protection Agency (1995), "Project Summary: Removal of PCBs from Contaminated Soil Using the CF Systems® Solvent Extraction Process: A Treatability Study," EPA/540/SR-95/505, Cincinnati, OH.

U.S. Environmental Protection Agency (1995), "SITE Technology Capsule: Terra-Kleen Solvent Extraction Technology," EPA/540/R-94/521a, Cincinnati, OH.

———— (1995), "Geosafe Corporation In Situ Vitrification: Innovative Technology Evaluation Report," EPA/540/R-94/520, Washington, DC.

———— (1994), "SITE Emerging Technology Bulletin: Volatile Organic Compound Removal from Air Streams by Membranes Separation, Membrane Technology and Research, Inc.," EPA/540/F-94/503, Cincinnati, OH.

———— (1994), "SITE Technology Capsule: Clean Berkshires, Inc. Thermal Desorption System," EPA/540/R-94/507a, Cincinnati, OH.

———— (1993), "Membrane Treatment of Wood Preserving Site Groundwater by SBP Technologies, Inc.: Applications Analysis Report," EPA/540/AR-92/014, Washington, DC.

———— (1993), "Resources Conservation Company B.E.S.T.® Solvent Extraction Technology: Applications Analysis Report," EPA/540/AR-92/079, Washington, DC.

———— (1993), "SITE Demonstration Bulletin: Microfiltration Technology, EPOC Water, Inc.," EPA/540/MR-93/513, Cincinnati, OH.

———— (1992), "Mercury and Arsenic Wastes: Removal, Recovery, Treatment, and Disposal," Noyes Data Corporation, Park Ridge, NJ.

———— (1992), "Silicate Technology Corporation's Solidification/Stabilization Technology for Organic and Inorganic Contaminants in Soils: Applications Analysis Report," EPA/540/AR-92/010, Washington, DC.

Water Environment Federation (formerly Water Pollution Control Federation) and American Society of Civil Engineers (1988), *Aeration: A Wastewater Treatment Process*, Manual of Practice No. FD-13, Alexandria, VA, and New York.

Weymann, David (1995), "Biosparging and Subsurface pH Adjustment: A Realistic Assessment," *The National Environmental Journal*, Vol. 5/No. 1, Campbell Publishing, Columbus, GA.

6

WASTEWATER AND WATER QUALITY

Wastewater and water quality are inextricably linked. Without proper treatment, wastewater discharges into our nations waterways degrade the water we use for drinking, swimming, and fishing. Dissolved inorganic solids, such as salts, can make drinking water unpalatable. Toxic pollutants can endanger aquatic life and the people who eat the fish and drink the water from affected areas. Even biodegradable pollutants and nutrients such as nitrogen and phosphorus can impact water quality by causing algal blooms and lowering oxygen levels, which affect the health and type of fish that live in the water.

Water quality is protected by setting standards for water bodies and limiting the types and amounts of pollutants in wastewaters that are discharged. Wastewater limits are based on either instream water quality standards or levels that treatment technology is capable of achieving, whichever set of criteria is the most restrictive. Technology-based limits have been established by the EPA for many industries in the form of national effluent limitations and standards.

This chapter discusses water quality standards, effluent limitations and standards, storm water permits, and pretreatment programs for publicly owned treatment works (POTWs). To help in understanding these concepts, example calculations are given for wastewater limits based on water quality standards, wastewater limits based on national categorical limitations, and local limits for pretreatment programs.

WATER QUALITY STANDARDS

Water quality standards are designed to protect water resources and the animals and people that contact or use them. The federal water quality standards program is defined by regulatory requirements at 40 CFR Part 131. Standards may be established on a state or federal level. State standards, although customized by each state, must meet the general requirements of Part 131. For states that have not established adequate water quality standards, EPA has set "national" standards for priority toxic pollutants specific to these states under the National Toxics Rule, which was promulgated in 1992 (57 FR 60910). Table 6.1 lists these 126 priority pollutants.

The federal requirements at 40 CFR Part 131 are divided into four subparts (A–D). Subpart A covers general provisions such as scope, purpose, definitions, minimum requirements for state program submittals, and special considerations of Indian tribes. Subpart B contains the requirements for state water quality standards, including designated uses of water resources, water quality criteria, and an antidegradation policy. Subpart C sets out the requirements for reviewing and revising state water quality standards every three years. Subpart D contains the national standards established for states that EPA believes do not have a standards program that meets the requirements of the Clean Water Act (CWA).

Designated uses of a stream, river, lake, bay, or other water resource determine the water quality standards that must be developed to protect those uses. Designated uses may include public water supply; protection and propagation of fish, shellfish, and wildlife; recreation; agriculture; industry; and navigation. Other less commonly considered uses may include aquifer protection, ground water recharge, coral reef preservation, hydroelectric power, and marinas.

Water quality standards contain both narrative and numeric criteria. Narrative criteria are statements such as the following, suggested by EPA in its "Water Quality Standards Handbook" (EPA 1994b):

All waters, including those within mixing zones, shall be free from substances attributable to wastewater discharges or other pollutant sources that:

1. Settle to form objectional deposits;
2. Float as debris, scum, oil, or other matter forming nuisances;
3. Produce objectionable color, odor, taste, or turbidity;
4. Cause injury to, or are toxic to, or produce adverse physiological responses in humans, animals, or plants; or
5. Produce undesirable or nuisance aquatic life (54 FR 28627, July 6, 1987).

Numeric water quality criteria are specific limits on pollutants such as metals and toxic organics or instream parameters such as dissolved oxygen, pH, chlorides, and sulfates. Numeric criteria are usually developed as aquatic life and human health criteria, depending on the type of water body and designated use. Aquatic life criteria are developed separately for fresh-and saltwater systems and for short-term (acute) and long-term (chronic) exposure conditions. Human health criteria are based on water, fish, and shellfish consumption in freshwater systems and fish and shellfish consumption in saltwater systems; they may also be based on contact and noncontact recreation.

TABLE 6.1 Priority Toxic Pollutants

Metals

Antimony	Chromium	Mercury	Silver
Arsenic	Copper	Nickel	Thallium
Beryllium	Lead	Selenium	Zinc
Cadmium			

Miscellaneous Pollutants

Cyanide	Asbestos	2,3,7,8-TCDD (dioxin)

Volatile Organics

Acrolein	1,2-Dichloropropane
Acrylonitrile	1,3-Dichloropropylene
Benzene	Ethylbenzene
Bromoform	Methyl bromide
Carbon tetrachloride	Methyl chloride
Chlorobenzene	Methylene chloride
Chlorodibromomethane	1,1,2,2-Tetrachloroethane
Chloroethane	Tetrachloroethylene
2-Chloroethylvinyl ether	Toluene
Chloroform	1,2-trans-Dichloroethylene
Dichlorobromomethane	1,1,1-Trichloroethane
1,1-Dichloroethane	1,1,2-Trichloroethane
1,2-Dichloroethane	Trichloroethylene
1,1-Dichloroethylene	Vinyl chloride

Acid Extractables

2-Chlorophenol	4-Nitrophenol
2,4-Dichlorophenol	3-Methyl-4-chlorophenol
2,4-Dimethylphenol	Pentachlorophenol
2-Methyl-4,6-dinitrophenol	Phenol
2,4-Dinitrophenol	2,4,6-Trichlorophenol
2-Nitrophenol	

Base/Neutral Extractables

Acenaphthene	Diethyl phthalate
Acenaphthylene	Dimethyl phthalate
Anthracene	Di-n-butyl phthalate
Benzidine	2,4-Dinitrotoluene
Benzo(a)anthracene	2,6-Dinitrotoluene
Benzo(a)pyrene	Di-n-octyl phthalate

TABLE 6.1 *Continued*

Benzo(b)fluoranthene	1,2-Diphenylhydrazine
Benzo(ghi)perylene	Fluoranthene
Benzo(k)fluoranthene	Fluorene
Bis(2-chloroethoxy)methane	Hexachlorobenzene
Bis(2-chloroethyl)ether	Hexachlorobutadiene
Bis(2-chloroisopropyl)ether	Hexachlorocyclopentadiene
Bis(2-ethylhexyl)phthalate	Hexachloroethane
4-Bromophenyl phenyl ether	Indeno(1,2,3-cd)pyrene
Butylbenzyl phthalate	Isophorone
2-Chloronaphthalene	Naphthalene
4-Chlorophenyl phenyl ether	Nitrobenzene
Chrysene	n-Nitrosodimethylamine
Dibenzo(a,h)anthracene	n-Nitrosodi-n-propylamine
1,2-Dichlorobenzene	n-Nitrosodiphenylamine
1,3-Dichlorobenzene	Phenanthrene
1,4-Dichlorbenzene	Pyrene
3,3'-Dichlorobenzidine	1,2,4-Trichlorobenzene

Pesticides and PCBs

Aldrin	Dieldrin	PCB-1242
alpha-BHC	alpha-Endosulfan	PCB-1254
beta-BHC	beta-Endosulfan	PCB-1221
gamma-BHC	Endosulfan sulfate	PCB-1232
delta-BHC	Endrin	PCB-1248
Chlordane	Endrin aldehyde	PCB-1260
4,4'-DDT	Heptachlor	PCB-1016
4,4'-DDE	Heptachlor epoxide	Toxaphene
4,4'-DDD		

Source: Information from 40 CFR Section 131.36(b)(1)

Although not nearly as well developed as aquatic and human health criteria, other types of water quality criteria may include sediment criteria, biological criteria, wildlife criteria, and numeric criteria for wetlands. Sediment and wildlife criteria will be EPA's next areas of focus.

Antidegradation

Federal regulations outline three levels or tiers of antidegradation of water quality, as discussed in EPA's "Water Quality Standards Handbook" (EPA 1994b). Tier 1 maintains and protects existing uses and the water quality necessary for these uses. Tier 2 protects waters whose quality is better than that necessary to protect fishable, swimmable uses. Tier 3 protects waters that are designated as outstanding national resource waters (ONRWs), having exceptionally high water quality or ecological significance.

Tier 1 is the base or absolute minimum level of protection for all waters. Under Tier 1, water quality in a water body may be lowered only if existing uses will still be fully protected and existing water quality and downstream water quality standards are not exceeded. An additional requirement is that a better use of the water body (which would require higher water quality) cannot be achieved.

Tier 2 protects high-quality waters from antidegradation. The quality of these waters exceed that necessary to protect aquatic life, wildlife, and recreational uses. Water quality may be lowered in these waters only after an antidegradation review shows that it is necessary for important economic and social development, that there has been adequate intergovernmental coordination and public participation, that point sources meet the highest pollutant control requirements, and that nonpoint sources are controlled by best management practices.

Tier 3 protects the highest-quality waters by prohibiting any degradation of water quality. This means that no new or increased discharges are allowed in ONRWs or in their tributaries. However, states may allow limited activities if they degrade water quality only temporarily over a short time period (a few weeks or months).

Aquatic Life Criteria

Aquatic life criteria are developed for acute and chronic exposure of aquatic organisms in both fresh- and saltwater (marine) systems. Criteria are derived from toxicity tests of single chemicals on a variety of aquatic plants and animals. Toxicity tests for acute criteria are usually 48- to 96-hour tests that measure lethality or immobilization of the organism. Toxicity tests for chronic criteria are run for much longer periods, often greater than 28 days, and measure effects on survival, growth, or reproduction. Toxic concentrations are referred to as the criterion maximum concentration (CMC) for acute exposure and the criterion continuous concentration (CCC) for chronic exposure.

EPA recommends limiting exposure to the acute CMC to a 1-hour period (EPA 1994b) although, in practice, regulators use 1-day periods for setting waste load allocations and permit limits (EPA 1991). EPA recommends limiting exposure to the chronic CCC to a 4-day period (EPA 1994b), which can then be statistically translated into a monthly average permit limit (EPA 1991). Many states, however, establish chronic criteria as 7-day averages.

Criteria for certain chemicals or types of chemicals may include other water quality parameters such as pH (ammonia, pentachlorophenol) and hardness (metals). Metals criteria are often given as total recoverable metals, which represents both the particulate and dissolved fractions. Because the dissolved form is believed to represent the form of the metal that is most available to the organism biologically (bioavailable), such as through adsorption on the gill surface, EPA recommends that compliance with water quality standards be based on the dissolved fraction (EPA 1994b).

Human Health Criteria

Human health criteria are designed to protect against chronic effects from long-term exposure. For carcinogenic chemicals, the exposure period is a person's lifetime, assumed to be equal to 70 years. Criteria are usually based on consumption of water and contaminated fish or shellfish, depending on the type and designated uses of the water source. For example, estuarine and marine waters are not used for drinking water, so the human health criteria for salt water would be based on food consumption alone. Consumption of contaminated organisms is of particular concern where a chemical bioaccumulates, or concentrates in the organism, as do PCBs in the lipid or fatty tissues of fish.

EPA's equation for deriving a human health criterion, C, for a noncarcinogenic chemical is as follows (EPA 1994b):

$$C \text{ (mg/L)} = \frac{(RfD \times WT) - (DT + IN)WT}{WI + (FC \times L \times FM \times BCF)} \tag{6.1}$$

where RfD is the oral reference dose in mg toxicant/kg human body weight/day, WT is the weight of an average human adult (70 kg), DT is the dietary exposure other than fish in mg toxicant/kg human body weight/day, IN is the inhalation exposure in mg toxicant/kg human body weight/day, WI is the average human adult water intake (2 L/d), FC is the daily fish consumption in kg/day, L is the ratio of lipid fraction of fish tissue consumed to 3%, FM is the food chain multiplier, and BCF is the bioconcentration factor for fish with 3% lipid content in mg toxicant/kg fish divided by mg toxicant/L water.

EPA's equation for deriving a human health criterion, C, for a carcinogenic chemical is as follows (EPA 1994b):

$$C \text{ (mg/L)} = \frac{(RL \times WT)}{q_1^*(WI + FC \times L \times FM \times BCF)} \tag{6.2}$$

where RL is the risk level (number of additional cancers per population), usually in the range of 10^{-4} to 10^{-6}, q_1^* is the carcinogenic potency factor in kg day/mg, and all other variables are as defined for Equation 6.1.

Sediment Criteria

EPA is currently developing sediment quality criteria, initially focusing on nonionic organic chemicals and metals for aquatic life protection. EPA is considering various applications of sediment criteria: assessing risks of contaminated sediments, site monitoring, ensuring control of contaminants in discharges, maintaining uncontaminated sediments, early warning of potential problems, establishing permit limits for point source discharges and target levels for nonpoint sources, and aiding remediation activities (EPA 1994b). EPA's contaminated sediment management strategy is designed to:

- Prevent further sediment contamination that may cause unacceptable ecological or human health risks;
- When practical, clean up existing sediment contamination that adversely affects the nation's water bodies or their uses, or that causes other significant effects on human health or the environment;
- Ensure that sediment dredging and dredged material disposal continue to be managed in an environmentally sound manner; and
- Develop and consistently apply methodologies for analyzing contaminated sediments (EPA 1994a).

Criteria for nonionic organic chemicals will likely be based on the equilibrium partitioning (EqP) approach, which assumes that a chemical partitions between the organic carbon in the sediment and the pore water in the interstices of the sediment and is at equilibrium. Criteria for metals will likely be based on acid volatile sulfides (AVS) and other binding factors.

Sediment criteria based on the EqP approach depend on the properties of the particular chemical as well as the type of sediment. The following equation is being considered by EPA to calculate sediment quality criteria (SQC) for nonionic organic chemicals (EPA 1993); units for SQC are in µg/g sediment:

$$SQC = \frac{K_p FCV}{1,000} \tag{6.3}$$

K_p is the partition coefficient (L/kg sediment), which is the ratio of chemical concentration in the sediment and in the pore water at equilibrium. *FCV*, or final chronic value (µg/L), is derived from the water quality criteria for the chemical and is the final acute value (FAV) divided by the final acute-chronic ratio.

Because sediments differ from place to place, a more useful form of the sediment criterion is SQC_{oc}, which is normalized to be independent of sediment type (defined by organic fraction). SQC_{oc} is derived from Equation 6.3 and the following calculation for K_p:

$$K_p = f_{oc} K_{oc} \tag{6.4}$$

where f_{oc} is the mass fraction of organic carbon in the sediment representing the sorption capacity for the chemical and K_{oc} is the partition coefficient for sediment organic carbon. Substituting for K_p in Equation 6.3 and normalizing SQC by dividing by the organic carbon fraction, f_{oc}, a sediment criterion can be derived that is independent of sediment organic carbon:

$$SQC_{oc} = K_{oc} FCV \tag{6.5}$$

The partition coefficient, K_{oc}, is calculated from the octanol/water coefficient, K_{ow}, using the following equation:

$$\log_{10}K_{oc} = 0.00028 + 0.983\log_{10}K_{ow} \tag{6.6}$$

Mixing Zones

A mixing zone is an area where a wastewater discharge is allowed to initially mix with a receiving water without having to meet a water quality standard. The idea behind a mixing zone is that a small area around an outfall where a standard may be exceeded is not harmful to the overall receiving water. A mixing zone must be carefully delineated so that:

- It does not impair the integrity of the water body as a whole;
- There is no lethality to organisms passing through the mixing zone; and
- There are no significant health risks, considering likely pathways of exposure (EPA 1994b).

The type of mixing zone relates to the applicable water quality criteria (acute aquatic, chronic aquatic, or human health). The immediate area around or downstream of the discharge is the zone of initial dilution (ZID). Acute aquatic criteria must be met at the edge of the ZID, although within the ZID the criteria may be exceeded. Beyond the ZID is the secondary mixing zone. Chronic aquatic criteria may be exceeded within this zone, and compliance is determined at the edge of the mixing zone. A mixing zone may also be defined for human health criteria; the edge of this zone is usually considered to be the point of complete (100%) mixing with the receiving water.

Federal regulations allow states to incorporate mixing zones into their water quality standards. Although many do, states are not obligated to use mixing zones. Without mixing zones, water quality criteria apply directly to the wastewater at the point of discharge (end-of-pipe). Because water quality criteria for many pollutants are quite restrictive, lack of a mixing zone can result in discharge limits that are very difficult and expensive to meet.

Mixing zones may be defined by specific size areas around or at distances from the discharge point and translated into dilution factors. For example, the mixing zone may be one fourth of the stream width for an acute aquatic criterion, representing a dilution factor of 25%. In this case, the effluent would have to meet the acute aquatic criterion at the edge of the ZID, whereas the permit limit would be four times the acute criterion (the acute criterion divided by 0.25). Where multiple criteria apply to a discharge (acute aquatic, chronic aquatic, human health), the most stringent criterion determines the permit limit (see Example 6.1).

Whole Effluent Toxicity (WET)

Whole effluent toxicity (WET) testing directly measures the toxicity of an effluent on aquatic organisms. An organism is exposed to wastewater effluent in diluted or undiluted form, depending on the type of WET test, and evaluated for growth, survival, fecundity, or other characteristics. The WET test attempts to evaluate toxicity through a more holistic approach, as it incorporates all the effluents characteristics; hence, the term *whole effluent*. In this way it differs from numeric criteria, which are based on studies using clean water samples spiked with a single toxic chemical. WET testing is one way of ensuring that narrative criteria to protect water quality are being met.

Wastewater discharge permits include WET testing requirements. Typically, two species must be tested, one a feeder type such as the crustacean, *Ceriodaphnia dubia*, or water flea, and a fish such as *Pimephales promelas*, or fathead minnow. Both of these are freshwater species and are used only to test effluents that discharge into fresh water. Marine or estuarine species are used for effluents discharging into marine or estuarine environments. Saltwater species often used for WET testing are *Mysidopsis bahia*, or mysid shrimp, *Menidia beryllina*, or Atlantic silverside, and *Cyprinodon variegatus*, or sheepshead minnow.

States may set WET limits in wastewater permits that are analogous to numeric permit limits for a specific chemical. Other permits may include only requirements for conducting the WET test and reporting the results. Test frequency varies and usually depends on how often the test results demonstrate no toxicity. For example, a discharger may have to test monthly at the beginning of the permit, then be allowed to reduce testing to a quarterly or semiannual basis if no toxicity has been shown. If repeated tests show toxicity, the discharger will be required to investigate the cause and implement changes to eliminate the toxicity by means of a toxicity reduction evaluation (TRE).

EXAMPLE 6.1
Water Quality-Based Permit Limits for An Organic Chemicals Plant

Where wastewater technology-based permit limits do not protect water quality or the designated uses, additional water quality-based effluent limits and/or conditions are included in NPDES (National Permit Discharge Elimination System) permits. For a proposed organic chemicals plant, calculate permit limits for benzene and hexachlorobenzene based on protection of water quality. The plant is expected to discharge an average of 1.5 million gallons per day (MGD) into a tidally influenced river with an average flow of 2,500 MGD.

Mixing zone dilution factors are 0.1 for acute criteria, 0.01 for chronic criteria, and 0.001 for human health criteria. Saltwater quality criteria for benzene are 2.7 mg/L for marine acute aquatic toxicity, 1.35 mg/L for marine chronic aquatic toxicity, and 0.0125 mg/L for human health consumption of fish and shellfish. There are no aquatic criteria for hexachlorobenzene, but the human health criterion is very stringent (0.00000025 mg/L). To determine which of the criteria (acute, chronic, human health) is most restrictive, you must first convert each criterion to a long-term average. The conversion factors are 0.32 for the acute criterion and 0.53 for the chronic criterion; the human health criterion is already a long-term average and needs no conversion. To convert long-term averages into maximum monthly averages and maximum daily values for permit limits, use the following factors: for aquatic criteria, 1.31 for the monthly average and 3.11 for the daily maximum; for human health criteria, 1.0 for the monthly average and 2.38 for the daily maximum. Calculate monthly average and daily maximum permit limits in lbs/d.

Solution. First convert the criteria into long-term averages, incorporating the dilution factors based on mixing zones. The long-term averages are as follows:

Benzene, acute aquatic

$$\frac{2.7 \text{ mg/L}}{0.1} \times 0.32 = 8.64 \text{ mg/L}$$

Benzene, chronic aquatic

$$\frac{1.35 \text{ mg/L}}{0.01} \times 0.53 = 71.55 \text{ mg/L}$$

Benzene, human health

$$\frac{0.0125 \text{ mg/L}}{0.001} \times 1 = 12.5 \text{ mg/L}$$

Hexachlorobenzene, human health

$$\frac{2.5 \times 10^{-7} \text{ mg/L}}{0.001} \times 1 = 0.00025 \text{ mg/L}$$

The minimum long-term average for benzene is 8.64 mg/L, based on the acute aquatic criterion. The monthly average and daily maximum permit loads for benzene would therefore be:

$$PL_{mon} = 1.31 \times 8.64 \text{ mg/L} \times 1.5 \text{ MGD} \times \frac{8.34 \text{ lbs}}{\text{MG·mg/L}} = 142 \text{ lbs/d}$$

$$PL_{day} = 3.11 \times 8.64 \text{ mg/L} \times 1.5 \text{ MGD} \times \frac{8.34 \text{ lbs}}{\text{MG·mg/L}} = 336 \text{ lbs/d}$$

The monthly average and daily maximum permit loads for hexachlorobenzene would be:

$$PL_{mon} = 1.0 \times 0.00025 \text{ mg/L} \times 1.5 \text{ MGD} \times \frac{8.34 \text{ lbs}}{\text{MG·mg/L}} = 0.00313 \text{ lbs/d}$$

$$PL_{day} = 2.38 \times 0.00025 \text{ mg/L} \times 1.5 \text{ MGD} \times \frac{8.34 \text{ lbs}}{\text{MG·mg/L}} = 0.00744 \text{ lbs/d}$$

Great Lakes Water Quality Guidance

In 1995, EPA published regulations at 40 CFR Part 132 outlining its guidance for water quality in the Great Lakes system. These regulations identify minimum water quality standards, antidegradation policies, and implementation procedures to protect human health, aquatic life, and wildlife in the Great Lakes system. The Great Lakes system is defined as all the streams, rivers, lakes, and other bodies of water within the drainage basin of the Great Lakes. States that are subject to the regulations are Illinois, Indiana, Michigan, Minnesota, New York, Ohio, Pennsylvania, and Wisconsin; Indian tribes within the drainage basin of the Great Lakes are also subject to the regulations. These states and tribes were required to submit criteria, methodologies, policies, and procedures for water quality standards to EPA by September 23, 1996. The majority of the Part 132 regulations is contained in six appendices outlining the Great Lakes antidegradation policy, implementation procedures, and methodologies for developing criteria for aquatic life, human health, wildlife, and bioaccumulation factors. Wildlife criteria focus on bioaccumulative chemicals of concern (BCCs). A wildlife BCC is one that is persistent in the environment and tends to accumulate in the aquatic food chain, which may cause adverse effects on the reproduction, development, and population survival of birds and mammals.

The Great Lakes water quality program is very similar to, yet at the same time quite different from, the national water quality program. The Great Lakes program includes three antidegradation categories similar to Tiers I–III of the national water quality program covering (1) waters not attaining designated uses, (2) high-quality waters, and (3) outstanding national or international resource waters. The Great Lakes guidelines also provide for mixing zones; however, mixing zones for BCCs must be eliminated by the year 2005, although exceptions will be allowed. The Great Lakes rule includes minimum water quality criteria for 29 pollutants. When developing new or additional numbers for their own programs, the states are required to use either Tier I or Tier II methodologies. Tier I is used to calculate numeric *criteria*. When there are not enough data to satisfy Tier I requirements, Tier II is used to calculate numeric *values*. Because Tier II methodologies depend on less data, they incorporate a high degree of conservatism and generally produce numbers more stringent than those of Tier I.

EPA guidance for the Great Lakes water quality program incorporates bioaccumulation factors (BAFs) in human health and wildlife criteria (EPA 1995). BAFs reflect the net accumulation of a substance by an aquatic organism from water, sediment, and food. The accumulation of toxic chemicals in aquatic organisms can increase to levels that are harmful to the wildlife and humans that eat them. A BAF represents the ratio of a substances concentration in the tissue of aquatic organisms to its concentration in the ambient water in situations where both the organism and its food are exposed and the ratio does not change substantially over time. BAFs can be determined by several different methods. EPA recommends four methods, in the following order of preference:

- A BAF measured in the field, in fish collected at or near the top of the food chain.
- A BAF derived from a biota-sediment accumulation factor (BSAF). A BSAF relates the concentration of a substance in the tissue of an aquatic organism (normalized by lipid content) to the concentration of the substance in the sediment (normalized by organic carbon content).
- A BAF calculated by multiplying a bioconcentration factor (BCF) measured in the laboratory, preferably on an indigenous fish species, by a food chain multiplier (FCM) factor. A BCF represents the net accumulation of a substance by an aquatic organism from the ambient water only through gill membranes or other external body surfaces. An FCM is a multiplying factor that accounts for the biomagnification of a chemical through trophic levels in the food chain.
- A BAF calculated by multiplying a BCF, calculated from the octanol-water partition coefficient, by an FCM.

BAFs and the methodologies for deriving BAFs are not limited to the Great Lakes, as EPA will be incorporating them into wildlife and human health criteria for other bodies of water.

EFFLUENT DISCHARGE LIMITATIONS AND STANDARDS

EPA has developed national limitations and standards for wastewater for 51 specific industrial categories. These categories, listed in Table 6.2, cover a diverse range of industries including food processing, metal manufacturing, electrical components, inorganic and organic chemicals, plastics, and mining.

These effluent guidelines apply to both direct and indirect dischargers. A direct discharge is one that goes into a receiving body of water, such as a creek, stream, river, lake, bay, or ocean, and must have a permit under the NPDES or a state-equivalent NPDES permit. An indirect discharge is one that goes to a publicly owned treatment works (POTW) such as a municipal wastewater plant.

There are several classes of effluent guidelines. There are separate effluent guidelines for direct and indirect dischargers, as well as different guidelines for existing and new dischargers. There are effluent guidelines that cover common pollutants such as biochemical oxygen demand (BOD), total suspended solids (TSS), and pH and others that cover toxic pollutants. The following is a generalized description of these different types of guidelines and limitations for wastewater discharges.

TABLE 6.2 Industrial Categories with Federal Effluent Standards

CFR Section	Industrial Category
40 CFR §467	Aluminum Forming
40 CFR §427	Asbestos Manufacturing
40 CFR §461	Battery Manufacturing
40 CFR §431	Builders Paper and Board Mills
40 CFR §407	Canned and Preserved Fruits and Vegetables Processing
40 CFR §408	Canned and Preserved Seafood Processing
40 CFR §458	Carbon Black Manufacturing
40 CFR §411	Cement Manufacturing
40 CFR §434	Coal Mining
40 CFR §465	Coil Coating
40 CFR §468	Copper Forming
40 CFR §405	Dairy Products
40 CFR §469	Electrical and Electronic Components
40 CFR §413	Electroplating
40 CFR §457	Explosives Manufacturing
40 CFR §412	Feedlots
40 CFR §424	Ferroalloy Manufacturing
40 CFR §418	Fertilizer Manufacturing
40 CFR §426	Glass Manufacturing
40 CFR §406	Grain Mills
40 CFR §454	Gum and Wood Chemicals Manufacturing
40 CFR §460	Hospitals
40 CFR §447	Ink Formulating
40 CFR §415	Inorganic Chemicals Manufacturing
40 CFR §420	Iron and Steel Manufacturing
40 CFR §425	Leather Tanning and Finishing
40 CFR §432	Meat Products
40 CFR §433	Metal Finishing
40 CFR §464	Metal Molding and Casting
40 CFR §436	Mineral Mining and Processing
40 CFR §421	Nonferrous Metals
40 CFR §471	Nonferrous Metals Forming and Metal Powders
40 CFR §435	Offshore Oil and Gas Extraction
40 CFR §440	Ore Mining and Dressing
40 CFR §414	Organic Chemicals, Plastics, and Synthetic Fibers
40 CFR §446	Paint Formulating
40 CFR §443	Paving and Roofing Materials
40 CFR §455	Pesticide Chemicals Manufacturing
40 CFR §419	Petroleum Refining
40 CFR §422	Phosphate Manufacturing
40 CFR §439	Pharmaceutical Manufacturing
40 CFR §459	Photographic Processing
40 CFR §463	Plastics Molding and Forming
40 CFR §466	Porcelain Enameling
40 CFR §430	Pulp, Paper, and Paper Board
40 CFR §428	Rubber Processing

TABLE 6.2 *Continued*

CFR Section	Industrial Category
40 CFR §417	Soaps and Detergents
40 CFR §423	Steam Electric Power Generating
40 CFR §409	Sugar Processing
40 CFR §410	Textile Mills
40 CFR §429	Timber Products

For direct dischargers:

- BPT—Effluent limits for existing dischargers that are based on the *best practicable technology* for wastewater treatment in a particular industrial category. These limits apply to conventional pollutants, which are defined at 40 CFR §401.16 as BOD, TSS, fecal coliform bacteria, pH, and oil and grease, and nonconventional pollutants such as chemical oxygen demand (COD) and ammonia.

- BCT—Effluent limits for existing dischargers that are based on the *best conventional technology* for conventional pollutants (usually BOD and TSS). BCT limits may be the same or more stringent than BPT limits. They are more stringent if the cost of providing a higher level of treatment is *reasonable*, which is decided by a two-part cost/benefit test.

- BAT—Effluent limits for existing dischargers that are based on the *best available technology*. These limits apply to toxic pollutants and nonconventional pollutants.

- NSPS—*New source performance standards* (effluent limits) for new direct discharges. New sources are specifically defined at 40 CFR §122.2 for direct dischargers and at 40 CFR §403.3 for indirect dischargers. In general, they are wastewater discharges that are generated by new construction or major modifications of facilities after effluent limits are proposed for a particular industrial category. NSPS are usually more stringent than BPT, BCT, and BAT limits because new facilities have an opportunity to install more efficient pollution control and treatment technology. NSPS standards apply to all types of pollutants: conventional, nonconventional, and toxic.

For indirect dischargers:

- PSES—*Pretreatment standards for existing sources* from indirect dischargers. These standards cover all types of pollutants: conventional, nonconventional, and toxic. PSES are not usually developed for the conventional pollutants BOD and TSS because they are assumed to be readily treated at POTWs and adequately controlled by local POTW limits or restrictions. PSES are analogous to BPT and BAT limits for direct dischargers.

- PSNS—*Pretreatment standards for new sources* from indirect dischargers. These standards cover all types of pollutants: conventional, nonconventional, and toxic. Like PSES, PSNS are not usually developed for the conventional pollutants BOD and TSS because they are assumed to be readily treated at POTWs and adequately controlled by local POTW limits or restrictions. PSNS are analogous to NSPS for direct dischargers.

Other types of effluent limits:

- BMP—*Best management practices*. These are procedures that usually address maintenance and good housekeeping to minimize spills and pollutant concentrations in wastewaters.
- BPJ—Effluent limits based on *best professional judgment*. BPJ is used when limits cannot be set entirely according to categorical effluent guidelines. For example, BPJ is used to set limits for wastewaters that are a mixture of categorical streams and utility wastewaters from cooling towers, boilers, and demineralizer units. BPJ is also used to set limits for pollutants not covered by effluent guidelines.

Effluent limitations guidelines and standards are revised periodically. The general trend is toward more stringent limits for all types of pollutants (conventional, nonconventional, toxic) and development of new limits for toxic pollutants. Effluent permit limits are set as either concentration limits or mass loads based on manufacturing production levels or wastewater flow. The following examples show how these types of limits are applied to direct dischargers.

EXAMPLE 6.2
Effluent Limits for Paper Mill

A new paper mill is planning to produce 4 million pounds per day of unbleached kraft linerboard and corrugating medium. The mill will be subject to NSPS effluent guidelines under 40 CFR Part 430, Subpart V, EPA Effluent Guidelines and Standards for Pulp, Paper, and Paperboard, Unbleached Kraft and Semi-Chemical Subcategory. Calculate the daily average (monthly) and daily maximum limits for BOD and TSS for this mill.

Solution. Permit limits for paper mills are based on production. The NSPS guidelines in Subpart V are 2.1 pounds BOD per thousand pounds of production (lbs/1,000 lbs) for the daily average and 3.9 lbs/1,000 lbs for the daily maximum. The TSS guidelines are 3.8 lbs/1,000 lbs (daily average) and 7.3 lbs/1,000 lbs (daily maximum). The permit limits for BOD are:

$$PL_{mon} = (2.1)(4,000) = 8,400 \text{ lbs/d}$$

$$PL_{day} = (3.9)(4,000) = 15,600 \text{ lbs/d}$$

The permit limits for TSS are:

$$PL_{mon} = (3.8)(4,000) = 15,200 \text{ lbs/d}$$

$$PL_{day} = (7.3)(4,000) = 29,200 \text{ lbs/d}$$

EXAMPLE 6.3
Effluent Limits for Organic Chemicals Plant

The organic chemicals plant in Example 6.1 has applied for a wastewater discharge permit for a new process facility it is planning to build. The process wastewater flow will be 1.5 MGD and will be treated in an activated sludge treatment plant. The plant is subject to NSPS effluent guidelines under 40 CFR Part 414, EPA Effluent Guidelines and Standards for Organic Chemicals, Plastics, and Synthetic Fibers. The new facility will produce both commodity organic chemicals (75%) and bulk organic chemicals (25%). Calculate the NSPS limits for this new facility. For the toxic pollutants, calculate only the multiplying factor to determine mass loads and the specific mass loads for benzene and hexachlorobenzene. Would water quality standards require more stringent limits for benzene and hexachlorobenzene?

Solution. The NSPS effluent limits for commodity organic chemicals and bulk organic chemicals are found at 40 CFR Subparts F and G, respectively. Subpart I, which is referenced in both of these subparts, contains the limits for toxic pollutants for direct discharges from end-of-pipe biological treatment (e.g., activated sludge treatment). The monthly and daily maximum limits, respectively, for BOD are 30 mg/L and 80 mg/ L for Subpart F and 34 mg/L and 92 mg/L for Subpart G. The monthly and daily maximum limits, respectively, for TSS are 46 mg/L and 149 mg/L for Subpart F and 49 mg/L and 159 mg/L for Subpart G. Limits for pH are the same for both Subparts; the pH must be within a range of 6 to 9 S.U. (standard units) at all times.[1] The BOD and TSS concentrations for the monthly and daily permit limits are calculated using the production percentages in each product subcategory. These concentrations are as follows:

BOD concentrations

$$C_{mon} = 0.75(30) + 0.25(34) = 31 \text{ mg/L}$$

$$C_{day} = 0.75(80) + 0.25(92) = 83 \text{ mg/L}$$

---

[1]If a direct discharge is monitored continuously for pH, then 40 CFR §401.17 allows excursions outside the 6 to 9 range for limited periods during the month. The pH may be outside the range for no more than 60 minutes at any one time and no more than a total of 7 hours and 26 minutes for a given month. The 7 hours, 26 minutes, represent 1% of the time during the month; hence, this provision is also referred to as the 1% rule.

TSS concentrations

$$C_{mon} = 0.75(46) + 0.25(49) = 46.75 \text{ mg/L}$$
$$C_{day} = 0.75(149) + 0.25(159) = 151.5 \text{ mg/L}$$

The permit limits for the organic chemicals category must be set as mass loads. Therefore, the BOD and TSS permit limits are:

BOD permit limits

$$PL_{mon} = 1.5(8.34)(31) = 388 \text{ lbs/d}$$
$$PL_{day} = 1.5(8.34)(83) = 1,038 \text{ lbs/d}$$

TSS permit limits

$$PL_{mon} = 1.5(8.34)(46.75) = 585 \text{ lbs/d}$$
$$PL_{day} = 1.5(8.34)(151.5) = 1,895 \text{ lbs/d}$$

In respect to toxic pollutants, the plant would receive permit limits for all 56 organic pollutants in Subpart I. It would also receive permit limits for any of the five metals and cyanide listed in Subpart I for that portion of the process flow that contains these pollutants. For simplicity in this example, it is assumed that none of the process streams contain metals or cyanide, so no permit limits are required for these pollutants. The multiplier for the organic pollutants would be the flow times 8.34 (to give mass limits in lbs/d). The monthly and daily concentrations for benzene from Subpart I are 0.037 mg/L and 0.136 mg/L, respectively. The monthly and daily concentrations for hexachlorobenzene are 0.015 mg/L and 0.028 mg/L, respectively. The monthly and daily mass loads for benzene and hexachlorobenzene are as follows:

Benzene mass loads

$$PL_{mon} = 1.5(8.34)(0.037) = 0.463 \text{ lbs/d}$$
$$PL_{day} = 1.5(8.34)(0.136) = 1.701 \text{ lbs/d}$$

Hexachlorobenzene mass loads

$$PL_{mon} = 1.5(8.34)(0.015) = 0.188 \text{ lbs/d}$$
$$PL_{day} = 1.5(8.34)(0.028) = 0.350 \text{ lbs/d}$$

Because the mass loads for benzene based on Subpart I are less than those based on water quality standards (Example 6.1), the permit limits would be based on Subpart I. However, water quality-based loads for hexachlorobenzene calculated in Example 6.1 are more stringent than Subpart I, so permit limits for this pollutant would be based on water quality standards.

EXAMPLE 6.4
Effluent Limits for a Petroleum Refinery

An existing petroleum refinery with petrochemical production is subject to effluent guidelines for petroleum refineries at 40 CFR Part 419, Subpart C, Petrochemical Sub-category. Total feedstock is 300 thousand (300K) barrels per day (bbl/day). Crude processes include vacuum crude distillation (150K bbl/day), crude desalting (300K bbl/day), and atmospheric crude distillation (300K bbl/day). Cracking and coking processes include fluid catalytic cracking (135K bbl/day), hydrocracking (45K bbl/day), and delayed coking (30K bbl/day). Calculate the monthly and daily maximum permit limits for BOD, TSS, and oil and grease for this facility.

Solution. Calculation of effluent limits for petroleum refineries involves many steps, but is still rather simple and straightforward. An example calculation is provided in the guidelines at 40 CFR §419.42(b)(3). The first step is to calculate the process configuration factor based on the throughput for each process. Each process contributes a portion of the configuration factor, which is its throughput ratio (relative to total feedstock) times a weighting factor. The weighting factor for crude processes is 1 and for cracking/coking processes is 6. The process configuration factor for each unit is calculated as follows:

Vacuum crude distillation

$$\frac{150}{300} \times 1 = 0.50$$

Crude desalting

$$\frac{300}{300} \times 1 = 1.0$$

Atmospheric crude distillation

$$\frac{300}{300} \times 1 = 1.0$$

Fluid catalytic cracking

$$\frac{135}{300} \times 6 = 2.7$$

Hydrocracking

$$\frac{45}{300} \times 6 = 0.9$$

Delayed coking

$$\frac{30}{300}\times 6 = 0.6$$

The sum of the unit process configuration factors, 6.7, is the total refinery process configuration factor. The configuration factor is used to look up the process factor from tables given in the guidelines for each subpart. For Subpart C, a process configuration factor of 6.7 gives a process factor of 1.08. The size factor, based on total feedstock, is another factor used in the calculation of effluent limits and is determined from tables in the guidelines. From Subpart C, the size factor for a 300K facility is 1.13. Effluent limits are calculated by multiplying the guideline number, given in pounds per thousand barrels of feedstock (lbs/1,000 bbl), by the process factor and size factor. The guideline numbers for Subpart C for conventional pollutants for existing dischargers are found at 40 CFR §419.34(a); these guidelines are BCT-based. Permit limits for monthly averages and daily maximums based on these guidelines are calculated as follows:

BOD

$PL_{mon} = (6.5 \text{ lbs BOD/K bbl/d})(300)(1.08)(1.13) = 2,380 \text{ lbs/d}$

$PL_{day} = (12.1)(300)(1.08)(1.13) = 4,430 \text{ lbs/d}$

TSS

$PL_{mon} = (5.25)(300)(1.08)(1.13) = 1,922 \text{ lbs/d}$
$PL_{day} = (8.3)(300)(1.08)(1.13) = 3,039 \text{ lbs/d}$

Oil and grease

$PL_{mon} = (2.1)(300)(1.08)(1.3) = 769 \text{ lbs/d}$
$PL_{day} = (3.9)(300)(1.08)(1.13) = 1,428 \text{ lbs/d}$

The preceding calculations are only for effluent loads from the process wastewater. Allocations for ballast water and storm water are also contained in the guidelines.

STORM WATER

Storm water associated with industrial activity and certain size municipal systems must have a discharge permit according to federal regulations at 40 CFR §122.26. Industrial activity is defined very specifically in these regulations. Municipal storm water discharges that must be permitted are those classified as medium to large. Generally speaking, large systems are those with a population of 250,000 or more and medium systems are those with a population between 100,000 and 250,000. Facilities that discharge storm water from industrial activity into storm water systems of medium or large municipalities must notify the operator of the municipal system and make their storm water pollution prevention plans available to the operator upon request.

Various types of storm water permits are issued by EPA. An individual permit can be issued that is tailored to a specific facility. The baseline general permit, as the name implies, sets forth general requirements that EPA considers fundamental to proper storm water management. Because there are so many storm water discharges that must be permitted, EPA has made the baseline general permit easy to obtain in order to promote its use. The requirements set forth by the baseline general permit are not trivial, however, as discussed later in this section. Multisector permits are similar to the baseline general permit except that they have been tailored to specific industrial and facility sectors. Table 6.3 lists the 29 sectors for which EPA has developed multisector permits.

Industrial Activity

EPA's definition of industrial activity in respect to storm water discharges is lengthy and subject to interpretation to a certain degree. In defining industrial activity, EPA focuses on storm water that can become contaminated with chemicals, oils, and dirt at industrial facilities and construction sites. Areas that are likely to generate storm water from industrial activity include plant yards, access roads and rail lines, product and chemical storage areas, waste management areas, material handling equipment areas, loading/unloading and shipping/receiving areas, tank storage areas, and large construction sites. There are 11 types of facilities and activities that are defined as industrial activity in 40 CFR §122.26(b)(14):

- Facilities subject to storm water effluent limitations guidelines, new source performance standards, or toxic pollutant effluent standards under 40 CFR subchapter N. Certain standard industrial classifications (SIC) with toxic pollutant standards are exempted (refer to the SICs at the end of this list).
- Facilities under SIC 24 (except 2434), 26 (except 265 and 267), 28 (except 283), 29, 311, 32 (except 323), 33, 3441, and 373.
- Facilities in the mineral industry under SIC 10 through 14, including active or inactive mining operations and oil and gas exploration, production, processing, or treatment operations, or transmission facilities that discharge storm water contaminated by contact with or that has come into contact with, any overburden, raw material, intermediate products, finished products, by-products, or waste products located on the site of such operations. Certain mining operations that have been reclaimed or have been released from reclamation requirements are not considered industrial activity.
- Hazardous waste treatment, storage, or disposal facilities, including those that are operating under interim status or a permit under subtitle C of RCRA.

TABLE 6.3 Industrial and Other Facilities Sectors with Multisector Storm Water Permits

Asphalt paving and roofing materials manufacturing and lubricant manufacturing
Automobile salvage yards
Chemical and allied products manufacturing
Coal mines and coal mining-related facilities
Electronic and electrical equipment and components, photographic and optical
  goods manufacturing
Fabricated metal products
Food and kindred products
Glass, clay, cement, concrete, and gypsum products manufacturing
Hazardous waste treatment, storage, or disposal facilities
Landfills and land application sites
Leather tanning and finishing
Metal mining (ore mining and dressing)
Mineral mining and processing
Oil and gas extraction
Paper and allied products manufacturing
Primary metals
Printing and publishing facilities
Publicly owned treatment works
Rubber, miscellaneous plastic products, and miscellaneous manufacturing
Scrap and waste recycling facilities
Ship and boat building or repairing yards
Steam electric power generating facilities, including coal handling areas
Textile mills, apparel, and other fabric products manufacturing
Timber products
Transportation equipment, industrial, or commercial machinery manufacturing
Vehicle maintenance areas and/or equipment cleaning operations at water
  transportation facilities
Vehicle maintenance areas, equipment cleaning areas, or deicing areas located at air
  transportation facilities
Vehicle maintenance or equipment cleaning areas at motor freight transportation
  facilities, passenger transportation facilities, petroleum bulk oil stations and
  terminals, rail transportation facilities, petroleum bulk oil stations and terminals,
  rail transportation facilities, and the U.S. Postal Service transportation facilities
Wood and metal furniture and fixture manufacturing

Source: Information from 60 FR 50808-9, September 29, 1995.

- Landfills, land application sites, and open dumps that receive or have received any industrial wastes from industrial activities, including those that are subject to regulation under subtitle D of RCRA.
- Recycling facilities, including metal scrapyards, battery reclaimers, salvage yards, and automobile junkyards, including but limited to those classified as SIC 5015 and 5093.
- Steam electric power generating facilities, including coal-handling sites.

- Transportation facilities classified as SIC 40, 41, 42 (except 4221-25), 43, 44, 45, and 5171 that have vehicle maintenance shops, equipment cleaning operations, or airport deicing operations.
- Domestic sewage treatment plants or any other sewage sludge or wastewater treatment device or system used in the storage, treatment, recycling, and reclamation of municipal or domestic sewage, including land dedicated to the disposal of sewage sludge, that are located within the confines of the facility, with a design flow of 1.0 MGD or more, or required to have an approved pretreatment program under 40 CFR Part 403. Facilities that are not included in this category are farm lands, domestic gardens or lands used for sludge management where sludge is beneficially reused and which are not physically located in the confines of the facility, or areas that are in compliance with section 405 of the CWA (disposal or use of sewage sludge).
- Construction activities, including clearing, grading, and excavation, extending over five or more acres. Areas of less than five acres are not considered industrial activity unless they are part of a larger common plan of development or sale.
- Facilities under SIC 20, 21, 22, 23, 2434, 25, 265, 267, 27, 283, 285, 30, 31 (except 311), 323, 34 (except 3441), 35, 36, 37 (except 373), 38, 39, and 4221-25 and which are not otherwise included in any of the previously mentioned industrial activity categories. Storm water from these facilities (SIC 20, etc.) is not considered to be from industrial activity, however, if the storm water has not come in contact with industrial materials or activities. In contrast, storm water from the other categories of industrial activity are assumed to be associated with industrial activity regardless of actual exposure.

Baseline General Permit

Baseline general permits from EPA are available only to facilities in those states and territories that do not have authorized NPDES programs, although states with authorized NPDES programs may develop their own types of general permits.

There are two types of EPA baseline general permits: one for construction sites and one for other types of industrial activity. Either type of permit may be obtained by submitting a notice of intent (NOI). In a similar fashion, facilities may terminate their general permits by submitting a notice of termination (NOT).

An important element of general permits is the storm water pollution prevention plan. Such plans must be prepared prior to submitting the NOI and updated as conditions change at the site.

Storm water pollution prevention plans for construction activities must include a description of the site, stabilization and other practices to prevent erosion and sediment runoff, management of storm water, waste disposal, other state or local requirements, maintenance of sediment control measures, inspections, and a list of any non-storm-water discharges that are combined with the storm water.

Storm water pollution prevention plans for other industrial activities are more extensive than those for construction sites. A pollution prevention team must be created and identified in the plan. The plan must also identify potential pollutant sources, storm water management controls, and a comprehensive site compliance evaluation. There are additional requirements for storm water that is discharged through municipal systems serving a population of 100,000 or more, as well as for those facilities that are subject to EPCRA Section 313 requirements and for those facilities that have salt storage piles.

Facilities other than construction sites must sample and analyze their storm water discharges either annually or semiannually, depending on the type of facility. Facilities that must monitor semiannually must also submit their results once a year to EPA; facilities with annual monitoring requirements need only retain their data onsite. Monitoring and reporting frequencies for different types of facilities are shown in Table 6.4. The pollutants that must be monitored depend on the type of facility; however, common pollutants that are monitored include oil and grease, BOD, COD, TSS, and pH.

TABLE 6.4   General Permit Monitoring Requirements for Non-Construction Sites

Semiannual Monitoring with Annual Reporting

- *Battery reclaimers:* storm water discharges from areas for storage of lead acid batteries, reclamation products, or waste products, and areas used for lead acid battery reclamation
- *EPCRA, Section 313 facilities subject to reporting requirements for water priority chemicals:* storm water discharges that come in contact with any equipment, tank, container, or other vessel or area used for storage of a Section 313 water priority chemical, or located at a truck or rail car loading or unloading area where a Section 313 water priority chemical is handled
- *Industrial facilities with coal piles:* storm water discharges from coal pile runoff
- *Land disposal units, incinerators, boilers and industrial furnaces (BIFs):* storm water discharges from active or inactive land disposal units without a stabilized cover that have received any waste from industrial facilities other than construction sites; and storm water discharges from incinerators and BIFs that burn hazardous waste
- *Primary metal industries (SIC 33):* all storm water discharges associated with industrial activity
- *Wood treatment facilities:* storm water discharges from areas that are used for wood treatment, wood surface application, or storage of treated or surface protected wood; facilities that use chlorophenolic formulations; facilities that use creosote formulations; and facilities that use chromium-arsenic formulations

Annual Monitoring with Records Kept Onsite

- *Airports with more than 50,000 flight operations per year:* storm water discharges from aircraft or airport deicing areas

TABLE 6.4 *Continued*

- *Animal handling and meat packing facilities:* storm water discharges from animal handling areas, manure management areas, production waste mangement areas exposed to precipitation at meat packing plants, poultry packing plants, facilities that manufacture animal and marine fats and oils

- *Automobile junkyards:* storm water discharges exposed to (a) more than 250 auto/truck bodies with drivelines, 250 drivelines, or any combination thereof; (b) more than 500 auto/truck units; more than 100 units dismantled per year where automotive fluids are drained or stored

- *Cement manufacturing facilities and cement kilns:* all storm water discharges associated with industrial activity except those from material storage piles that are not eligible for coverage under the general permit

- *Chemical and allied product manufacturers and rubber manufacturers (SICs 28 and 30):* storm water discharges that come in contact with solid chemical storage piles

- *Coal-fired steam electric facilities:* storm water discharges from coal handling sites, other than runoff from coal piles, which is not eligible for coverage under the general permit

- *Lime manufacturing facilities:* storm water discharges that have come in contact with lime storage piles

- *Oil-fired steam electric power generating facilities:* storm water discharges from oil-handling sites

- *Ready-mix concrete facilities:* all storm water discharges associated with industrial activity

- *Ship building and repairing facilities:* all storm water discharges associated with industrial activity

Source: Federal Register, vol. 57, pp. 41248–41250.

Multisector Permits

Multisector storm water permits are similar to baseline general permits except that EPA has specifically tailored them to certain types of industries and facilities (Table 6.3). Because of time constraints, EPA issued the baseline general permit before multisector permits.

Multisector permits are available from EPA for 29 sectors; however, like the baseline general permit, they are available only in those states or areas that do not have authorization from EPA to administer the NPDES program. EPA encourages states that do have NPDES authorization to use the multisector permit as a model for their own storm water permits. Like the baseline general permit, multisector permits are obtained by submitting an NOI and require a storm water pollution prevention plan.

Multisector permits have fewer requirements than the baseline general permit because EPA had time to tailor the general requirements to each sector. The rubber manufacturing sector is an exception, in that its multisector permit requires BMPs and monitoring for zinc. Monitoring requirements for multisector permits are based on the potential for pollutants to contaminate storm water at a particular facility. For facilities with low potential, only visual monitoring of storm water discharges is required. Other facilities must sample and analyze their discharges.

EPA has set numeric limits for only a few types of facilities, as shown in Table 6.5; several states have added other numeric limits for facilities in their domain. Monitoring for pollutants that do not have numeric limits must be done quarterly in the second year of the five-year permits. If the average value exceeds the benchmark value for that pollutant, monitoring must be repeated in the fourth year to see whether pollution controls have improved storm water quality.

PRETREATMENT REGULATIONS

Regulations implementing the federal pretreatment program were first published in 1978 when EPA issued its national pretreatment strategy. Since then, the pretreatment regulations have been amended several times. The federal pretreatment regulations at 40 CFR 403 require publicly owned treatment works (POTWs) to control the discharge of nondomestic wastewaters. POTWs are typically wastewater treatment plants operated by a city, town, or other municipality. They may, however, also include facilities that are specifically designed to treat industrial wastewaters, if they have been so structured to meet the legal definition of a POTW. Nondomestic wastewaters, although not strictly defined in the regulations, are wastewaters that are not normally thought of as domestic, that is, originating from residential dwellings. Key terms of the federal pretreatment regulations are listed in Table 6.6.

TABLE 6.5   Storm Water Discharges with Numeric Limits in Multisector Permit

Coal pile runoff (TSS, pH)

Contaminated runoff from phosphate fertilizer manufacturing facilities (total phosphorus, fluoride)

Runoff from asphalt paving and roofing emulsion production areas (TSS, oil and grease, pH)

Runoff from material storage piles at cement manufacturing facilities (TSS, pH)

Mine dewatering discharges from crushed stone, construction sand and gravel, and industrial sand mines located in Texas, Louisiana, Oklahoma, New Mexico, and Arizona (TSS, pH)

Louisiana special requirements

    All storm water discharges (TOC, oil and grease, chlorides)

    Oil and gas exploration and production facilities (COD, TOC, oil and grease)

Texas special requirements

    All storm water discharges (whole effluent toxicity (WET), arsenic, barium, cadmium, chromium, copper, lead, manganese, mercury, nickel, selenium, silver, zinc)

Source: 60 FR 50827, 51118–51173.

TABLE 6.6   Common Terms in the Federal Pretreatment Regulations

Term	Definition
Approval authority	Director in an NPDES state with an approved state pretreatment program and the appropriate regional administrator in a non-NPDES state or NPDES state without an approved state pretreatment program.
Control authority	POTW, if it has an approved pretreatment program, or the approval authority if the pretreatment program has not yet been approved.
Indirect discharge	Introduction of pollutants into a POTW from any nondomestic source regulated under Section 307(b), (c), or (d) of the Clean Water Act.
Interference	Effect of a discharge which, alone or in conjunction with discharges from other sources, both (1) inhibits or disrupts the POTW, its treatment processes or operations, or its sludge processes, use, or disposal; and (2) therefore is a cause of a violation of any requirement of the POTW's NPDES permit (including an increase in the magnitude or duration of a violation) or the prevention of sewage sludge use or disposal in compliance with the following statutory provisions and regulations or permits issued thereunder (or more stringent state or local regulations): Section 405 of the Clean Water Act, the Solid Waste Disposal Act (SWDA) (including Title II, more commonly referred to as the Resource Conservation and Recovery Act, and including state regulations contained in any state sludge management plan prepared pursuant to Subtitle D of the SWDA), the Clean Air Act, the Toxic Substances Control Act, and the Marine Protection, Research and Sanctuaries Act.
Local limits	Site-specific numeric limits developed by a POTW to implement the narrative discharge prohibitions of the federal pretreatment regulations.
Pass-through	Effect of a discharge which exits the POTW into waters of the United States in quantities or concentrations which, alone or in conjunction with a discharge or discharges from other sources, is a cause of a violation of any requirement of the POTW's NPDES permit, including an increase in the magnitude or duration of a violation.
POTW	Publicly owned treatment works, as defined by Section 212 of the Clean Water Act, which is owned by a state or municipality, as defined by Section 502(4) of the Act. This term includes any devices and systems used in the storage, treatment,

(Continued)

TABLE 6.6   *Continued*

Term	Definition
	recycling, and reclamation of municipal sewage or industrial wastes of a liquid nature. It also includes sewers, pipes, and other conveyances only if they convey wastewater to a POTW treatment plant. The term also means the municipality, as defined in Section 502(4) of the Clean Water Act, that has jurisdiction over the indirect discharges to and the discharges from such a treatment works.
Pretreatment	Reduction of the amount of pollutants, the elimination of pollutants, or the alteration of the nature of pollutant properties in wastewater prior to or in lieu of discharging or otherwise introducing such pollutants into a POTW. The reduction or alteration may be obtained by physical, chemical, or biological processes, by process changes, or by other means, except as prohibited by the federal pretreatment regulations. Appropriate pretreatment technology includes control equipment, such as equalization tanks or facilities, for protection against surges or slug loadings that might interfere with or otherwise be incompatible with the POTW.
Significant industrial user	Discharger to a POTW that (1) is subject to categorical pretreatment standards under 40 CFR Chapter I, Subchapter N; or (2) discharges an average of 25,000 gallons per day or more of process wastewater (excluding sanitary, noncontact cooling and boiler blowdown wastewater); contributes a process wastestream that makes up 5% or more of the average dry weather hydraulic or organic capacity of the treatment plant; or is designated as such by the control authority on the basis that the industrial user has a reasonable potential for adversely affecting the POTW's operation or for violating any pretreatment standard or requirement.

Source: Information from 40 CFR Part 403.

The term *industrial discharger*, which is used in the regulations to describe dischargers of nondomestic wastewater, is somewhat of a misnomer. Although industrial discharges from manufacturing processes are clearly non-domestic, there are numerous other sources of nondomestic wastewater from commercial, trade, and institutional establishments. For example, waste-waters from restaurants are generally considered nondomestic because they contain more organics, oils, and greases than household wastes.

Nondomestic dischargers are also sometimes referred to as indirect dischargers. An indirect discharge is any nondomestic discharge to a POTW that is regulated under §§307(b), (c), or (d) of the Clean Water Act (sections that provide for the establishment of national pretreatment standards).

POTWs control nondomestic discharges through highly structured pretreatment programs required by the federal regulations. However, only certain POTWs are required to establish pretreatment programs. POTWs that must develop pretreatment programs are those with a total flow (domestic and nondomestic) greater than 5 million gallons per day (MGD). If multiple POTWs are controlled by the same authority, and their combined flow is greater than 5 MGD, then they too must develop pretreatment programs. In addition, the nondomestic wastewater must contain discharges that are either subject to federal pretreatment standards or contain pollutants that will cause problems for the POTW. In reality, almost any city with a wastewater flow greater than 5 MGD will be large enough to have some industrial development that will require the city to establish a pretreatment program. If a POTW has a total flow of 5 MGD or less, it may still be required to establish a pretreatment program if there are treatment problems, wastewater permit violations, contamination of the treatment sludge, or discharges that warrant stricter control.

POTWs must submit their pretreatment programs to EPA for approval. To be approved, pretreatment programs must contain several required elements. The POTW must demonstate that it has the proper legal authority to carry out its pretreatment regulations, including issuing pretreatment permits or equivalent means of discharge control, and assessing civil and criminal penalties for noncompliance. The POTW must inspect and sample the discharges of its users, track their compliance status and publish the names of those with significant noncompliances, and develop discharge limits based on the characteristics of the treatment facility (local limits).

The POTW must enforce its pretreatment program and is routinely monitored in this regard by either EPA or the state. The state performs the monitoring function in lieu of EPA if the state has received authorization for the pretreatment program from EPA. The federal pretreatment regulations set out the requirements that states must satisfy in order to receive authorization.

Purpose of a Pretreatment Program

The primary purpose of a pretreatment program is to prevent the discharge of pollutants that pass through inadequately treated or that interfere with the operation of the treatment facility. Pass-through occurs when the treatment facility cannot remove or degrade a compound sufficiently to meet its discharge permit limits. Interference is broadly defined in the regulations. It includes damage to or obstruction of the sewer system, conditions in the sewer that endanger workers or can cause explosions, negative impacts on process units such as excessive oil loadings or inhibitory effects on the biological units, and degradation of the biological sludge (biosolids) quality that prevents its use for land application or composting.

An additional goal of a pretreatment program is to increase opportunities to reuse wastewaters and biosolids. The fewer pollutants that these materials contain, the more opportunities there are for reuse. Thus, preventing pass-through and interference of pollutants is key to meeting this goal.

Elements of a Pretreatment Program

POTW Requirements. POTWs must demonstrate that they have the legal authority to carry out their pretreatment programs. They must have the legal authority to enforce compliance by their users, to deny or set conditions on discharges, and to issue pretreatment permits to users or control their discharges in some similar fashion. In addition, they must have legal authority to require users to establish compliance schedules and to submit notices and discharge monitoring reports, and to inspect and monitor users.

Pretreatment programs must include certain procedures. Procedures must be established to identify all users subject to the pretreatment program and to characterize their discharges, to notify these users of their requirements, to review discharge monitoring reports and notices sent in by users, to sample and analyze user discharges, to investigate noncompliances, and to publish the names of users in significant noncompliance at least annually in the largest daily local newspaper. There are special requirements regarding what are called significant industrial users (SIUs; see definition in Table 6.6). Stated simply, SIUs are those users that have sufficiently high flows or pollutant loadings to impact the treatment facility. Of special distinction are SIUs that are categorical users, those that have federal pretreatment standards established for their particular industrial category (see Table 6.2 for a list of these categories and the citation of where their pretreatment standards are found in the federal regulations). The POTW must prepare a list of SIUs, notify SIUs of their SIU status, inspect SIU facilities and sample their discharges annually, and evaluate at least every two years whether these SIUs must develop slug control plans to control spills and accidental discharges.

POTWs must show that they have adequate resources and personnel to implement their pretreatment programs. They must also prepare an enforcement response plan that details how they will investigate noncompliances, what enforcement actions will be taken, and how these actions will be linked to the severity of the noncompliance (notice of violation, administrative order, suspension of service, termination of permit).

Each year, POTWs must submit to the approval authority a status report on the pretreatment program. At a minimum, the POTW must include a current list of users and the discharge standards that are applicable to each one (categorical, local limits, other); the compliance status of each user over the last year; a summary of compliance, enforcement, and inspection activities; and any other information relevant to the pretreatment program.

Of particular importance to the POTW is the development of local limits. Local limits are discharge limits for users based on the treatment capacity of the POTW facility and the disposal method for biosolids. A more detailed discussion of local limits and how they may be calculated is discussed later in this section.

EPA has published a number of guidance manuals to help POTWs develop their pretreatment programs, as listed in Table 6.7.

Discharger Requirements. Dischargers are subject to wastewater limitations and requirements for monitoring, notification, and reporting.

Dischargers are subject to three types of limitations. The first type comprises discharge prohibitions, which include both numeric limitations as well as narrative restrictions (see Table 6.8). There is a general prohibition against any discharge that causes pass-through or interference. In addition, there are several specific prohibitions. It is prohibited to discharge a wastewater that can create a fire or explosion hazard; the flash point must always be 140°F (60°C) or greater. Corrosive wastewaters are prohibited, including any wastewater with a pH less than 5. High temperature wastewaters are prohibited if they will inhibit the treatment facility's biological processes or raise the tem-

TABLE 6.7 EPA Guidance Manuals for POTW Pretreatment Programs

"Control of Slug Discharges to POTWs Guidance Manual"
"Guidance for Developing Control Authority Enforcement Response Plan"
"Guidance Manual on the Development and Implementation of Local Discharge Limits Under the Pretreatment Program"
"Guidance Manual for the Identification of Hazardous Wastes Delivered to POTWs by Truck, Rail, or Dedicated Pipe"
"Guidance Manual for Implementing the Total Toxic Organics (TTO) Pretreatment Standards"
"Guidance Manual for POTWs to Calculate the Economic Benefit of Noncompliance"
"Guidance Manual for Preparation and Review of Removal Credit Applications"
"Guidance Manual for Preventing Interference at POTWs"
"Guidance Manual for the Use of Production-Based Standards and the Combined Wastestream Formula"
"Guidance to Protect POTW Workers from Toxic and Reactive Gases and Vapors"
"Guides to Pollution Prevention, Municipal Pretreatment Programs"
"Industrial User Inspection and Sampling Manual for POTWs"
"Industrial User Permitting Guidance Manual"
"Model Pretreatment Ordinance"
"RCRA Information on Hazardous Wastes for Publicly Owned Treatment Works"
"Supplemental Manual on the Development and Implementation of Local Discharge Limitations Under the Pretreatment Program: Residential and Commercial Toxic Pollutant Loadings and POTW Removal Efficiency Estimation "

TABLE 6.8   Discharge Prohibitions in the Federal Pretreatment Regulations

General Prohibition

- Discharges that cause pass through or interference

Specific Prohibitions

- Pollutants that create a fire or explosion hazard in the POTW including, but not limited to, wastestreams with a closed cup flashpoint of less than 140°F (60°C) using the test methods specified in 40 CFR §261.21
- Pollutants that will cause corrosive structural damage to the POTW, but in no case discharges with pH lower than 5.0, unless the POTW is specifically designed to accomodate such discharges
- Solid or viscous pollutants in amounts that will cause obstruction in the flow in the POTW, resulting in interference
- Any pollutant, including oxygen demanding pollutants (BOD, etc.), released in a discharge at a flow rate and/or pollutant concentration that will cause interference with the POTW
- Heat in amounts that will inhibit biological activity in the POTW, resulting in interference, but in no case heat in such quantities that the temperature at the POTW treatment plant exceeds 40°C (104°F) unless the Approval Authority, upon request by the POTW, approves alternate temperature limits
- Petroleum oil, nonbiodegradable cutting oil, or products of mineral oil origin in amounts that will cause interference or pass-through
- Pollutants that result in the presence of toxic gases, vapors, or fumes within the POTW in a quantity that may cause acute worker health and safety problems
- Any trucked or hauled pollutants, except at discharge points designated by the POTW

Source: Information from 40 CFR Part 403.

perature at the headworks above 104°F (40°C). Also prohibited are wastewaters that can obstruct the flow in the sewer, contain petroleum or mineral oils that will cause interference, or generate emissions that endanger the health or safety of workers. The last specific prohibition refers to trucked- or hauled in wastewaters, unless the POTW has approved the discharge at a designated point into the treatment facility.

The second type of limitation includes the national categorical pretreatment standards for specific industrial categories (see Table 6.2). These limitations are detailed in subchapter N of 40 CFR Chapter I and are included in the pretreatment regulations only by reference. Users that are subject to this type of limitation are called categorical users and, by definition, are considered SIUs.

Categorical limitations for a particular categorical user may be modified by the combined wastestream formula, removal credits, intake credits, or a fundamentally different factors variance. The combined wastestream formula is used where a categorical wastewater is combined with a noncategorical wastewater. The discharge limit is calculated by flow-weighting the concentration or mass of pollutant in each waste stream; because the noncategorical waste streams usually have lower levels of the pollutant, the combined discharge limit is typically lower than the categorical limit. A POTW can increase a categorical limit by granting a removal credit; the credit is based on the amount of the pollutant the POTW can remove (degrade or settle out) within its own treatment facility. A categorical limit can also be increased by an intake credit when the pollutant is found in the user's water supply; the limit can be increased up to the amount in the intake water. A variance from a categorical limit, because a user is fundamentally different from those facilities used to develop the categorical standard, may be requested by the user *or* by another interested party; the variance can either lower or raise the categorical limit.

Local limits constitute the third type of limitation. As the term implies, local limits are specific to each POTW and are based on the POTW's treatment capabilities and the regulations that affect the POTW's operation. A local limit for a pollutant must satisfy all of the following criteria, as applicable: the POTW's wastewater permit limit, the water quality standard in the receiving water, prevention of treatment process upset or inhibition, sludge (biosolids) disposal limit, and protection of worker safety and health. The lowest concentration that satisfies all of these criteria becomes the local limit for that pollutant.

All SIUs are required to sample and analyze (self-monitor) their discharges. SIUs must submit self-monitoring reports summarizing their results at least once every six months. In addition to these periodic six-month reports, categorical SIUs are required to submit a one-time baseline monitoring report (BMR) and a categorical compliance deadline report. A BMR is required after a federal categorical pretreatment standard becomes effective; its purpose is to provide a baseline description of the discharge flow and pollutants. The compliance deadline report is similar to a BMR; however, it must be submitted 90 days after the date for final compliance with the categorical standard. Self-monitoring and reporting for users who are not SIUs are based on the type and size of discharge.

Users are required to provide notification of certain events to the POTW. A user must notify the POTW immediately if the user's discharge could cause problems such as pass-through or interference. A user must notify the POTW prior to making a substantial change in the volume or character of its discharge. If a user unintentionally bypasses any part of its pretreatment facilities, the user must notify the POTW orally within 24 hours, and follow with a written report within five days. If a user anticipates a bypass, for example, for equipment maintenance, the user should notify the POTW at least 10 days in advance, if possible. Categorical users must notify the POTW of violations of a categorical limit within 24 hours. A user must also notify the POTW of any discharge that would be a hazardous waste, if it were not exempted by the domestic sewage exclusion (DSE) within 180 days after the discharge starts.[2] Users discharging 15 kilograms per month or less of nonacute hazardous wastes are exempt from this notification requirement.

Development of Local Limits. Local limits on users' discharges are based on the characteristics of the individual POTW. Such characteristics normally include the types of wastewater treatment and sludge management processes, effluent limits in the POTW's wastewater discharge permit, water quality standards for the water body that receives the POTW's discharge, sludge disposal regulations, and worker health and safety standards.

Each POTW must develop local limits or demonstrate that they are not necessary; however, a POTW is unlikely to be able to exempt all pollutants. At a minimum, EPA expects a POTW to develop local limits for cadmium, chromium, copper, lead, nickel, and zinc because they are common to industrial discharges and have the potential to impact the POTW. In addition to these six metals, EPA recommends that the POTW develop local limits for arsenic, cyanide, mercury, and silver. Beyond these ten basic pollutants, the POTW uses its own judgment to add any other pollutants to its local limits list.

The first step in developing local limits is to identify the pollutants of concern to the POTW's operation. The POTW samples its influent wastewater, effluent wastewater, and treatment sludge to identify the pollutants that enter the system and whether or not these pollutants are degraded, removed with the sludge, or discharged in the effluent. The proportion of each pollutant that is degraded, removed, or discharged depends on the individual pollutant. Although all three of these processes apply to organics, metals exit the treatment plant only through sludge removal or effluent discharge. To obtain additional information on pollutants that may enter the treatment plant, the POTW conducts a survey of its major users. The POTW reviews these users' discharge data from their self-monitoring and the POTW's own sampling. The POTW may also review other information from these users that identifies pollutants that are or may be discharged, such as a description of the manufacturing process and chemicals used in the manufacturing process and any pretreatment facility.

The second step in developing local limits is to calculate how much of the pollutant can be contained in the wastewater influent (allowable headworks loading) before process problems occur or the POTW's effluent or sludge becomes noncompliant with standards or requirements. An allowable headworks loading is calculated for each criterion (permit limit, water quality stan-

---

[2]Under the DSE, a mixture of domestic sewage and other waste that passes through the sewer to a POTW is not a solid waste under RCRA. Because the mixture is not a *solid waste*, it cannot be defined as a hazardous waste subject to the requirements of RCRA. The DSE was included in RCRA because the federal pretreatment program was thought to adequately regulate the discharge of hazardous nondomestic waste to POTWs, and requiring additional pretreatment under RCRA would be redundant.

dard, process inhibition, sludge quality standard); the lowest loading among these is chosen. Because each criterion may be based on a different time period (daily maximum, monthly average, etc.), the loadings should be converted to the same basis (for example, long-term average) before choosing the lowest load. Afterward, the load is converted to the time basis of the local limits, usually a daily maximum. Calculations to determine the lowest allowable headworks loading are described in the following paragraphs.

Wastewater Permit Limit. The allowable headworks loading for a wastewater permit limit depends on how much of the pollutant is removed in the treatment system. For example, if the permit limit is 100 µg/L and the treatment system removes 90 percent of the pollutant from the influent, the allowable influent concentration is 1,000 µg/L. Thus, the equation for the allowable headworks loading for a wastewater permit limit is:

$$L_{permit} = \frac{8.34 F_{permit} C_{permit} Q_{potw}}{1 - R_{potw}} \tag{6.7}$$

where

L_{permit} = allowable headworks loading based on wastewater permit limit, lbs/d

C_{permit} = wastewater permit limit, mg/L

F_{permit} = factor to convert the wastewater permit limit to the same common time scale as the other criteria

Q_{potw} = POTW flow, MGD

R_{potw} = fraction of pollutant removed through treatment system

8.34 = unit conversion factor, (lbs/d)/(MG·mg/L)

EXAMPLE 6.5
Allowable Headworks Loading for Permit Limit

Calculate the allowable headworks loading for arsenic if the wastewater permit limit is 0.1 mg/L. The permit limit is a daily maximum limit (24-hour average) and the loading will be calculated as a long-term average. The ratio of the long-term average to daily maximum is 0.322. The POTW flow is 8.3 MGD, and the removal efficiency through the POTW is 45%.

Solution. Using Equation 6.7, the allowable headworks loading is:

$$L_{permit} = \frac{(8.34)(0.322)(0.1)(8.3)}{(1 - 0.45)}$$

$$= 4.05 \text{ lbs/d}$$

Water Quality Standard. The allowable headworks loading for a water quality standard is calculated much like the loading for a permit limit; however, the mixing effects of the receiving water and the background concentration of the pollutant in the receiving water must also be considered. There can be as many as three water quality standards for a pollutant: an acute aquatic life standard, a chronic aquatic life standard, and a human health standard (see the section "Water Quality Standards" earlier in this chapter). The most restrictive standard of the three is used to calculate the allowable headworks loading. However, these standards reflect different time periods, and they must be converted to a common time scale before choosing the lowest standard to use in the loading calculation. The equations to convert these criteria to the same time scale are as follows:

$$C_{wqa} = F_{wqa} C_{acute} \tag{6.8}$$

$$C_{wqc} = F_{wqc} C_{chronic} \tag{6.9}$$

$$C_{wqhh} = F_{wqhh} C_{hh} \tag{6.10}$$

where

$C_{acute}, C_{chronic}, C_{hh}$ = water quality standards for acute aquatic life, chronic aquatic life, and human health, respectively

$F_{wqa}, F_{wqc}, F_{wqhh}$ = factors to convert acute aquatic life, chronic aquatic life, and human health standard, respectively, to the same common time scale as the other criteria

$C_{wqa}, C_{wqc}, C_{wqhh}$ = water quality standards for acute aquatic life, chronic aquatic life, and human health, respectively, and on same time scale

EXAMPLE 6.6
Water Quality Standards Converted to Same Time Scale

Convert the water quality standards for arsenic to long-term average concentrations representing each standard. The acute, chronic, and human health standards for arsenic are 0.069 mg/L, 0.036 mg/L, and 0.0014 mg/L, respectively. The acute standard is a daily maximum value; the ratio of the acute standard to its long-term concentration is 0.468. The chronic standard is a four-day average value; the ratio of the chronic standard to its long-term concentration is 0.644. The human health standard is a long-term average value.

Solution. Using Equations 6.8, 6.9, and 6.10, the long-term average concentrations representing the acute, chronic, and human health standards are 0.0323 m/L, 0.0232 mg/L, and 0.0014 mg/L, respectively.

$$C_{wqa} = (0.468)(0.069) = 0.0323 \text{ mg/L}$$

$$C_{wqc}=(0.644)(0.036)=0.0232 \text{ mg/L}$$
$$C_{wqhh}=(1)(0.0014)=0.0014 \text{ mg/L}$$

The next step is to determine the allowable concentration in the wastewater effluent for each of these standards, taking into account the conditions in the receiving water. These conditions include the background or ambient concentration of the pollutant, the fraction of the pollutant that is available to affected organisms, and dilution with the receiving water. Allowable concentrations in the wastewater effluent based on each standard are calculated as follows:

Acute aquatic life

$$C_{effa}=\frac{C_{wqa}-C_b(1-DF_{wqa})}{F_{ava}DF_{wqa}} \tag{6.11}$$

Chronic aquatic life

$$C_{effc}=\frac{C_{wqc}-C_b(1-DF_{wqc})}{F_{avc}DF_{wqc}} \tag{6.12}$$

Human health

$$C_{effhh}=\frac{C_{hh}-C_b(1-DF_{hh})}{F_{avhh}DF_{hh}} \tag{6.13}$$

$C_{effa}, C_{effc}, C_{effhh}$ = allowable effluent concentration based on acute aquatic, chronic aquatic, and human health standards, respectively, mg/L

$DF_{wqa}, DF_{wqc}, DF_{hh}$ = dilution fraction or fraction of effluent after mixing in the receiving water for acute aquatic, chronic aquatic, and human health standards, respectively

$F_{ava}, F_{avc}, F_{avhh}$ = fraction of pollutant that is available to affected organisms

C_b = background concentration in receiving water, mg/L

EXAMPLE 6.7
Allowable Effluent Concentration for Water Quality Standard

What is the most restrictive effluent concentration for arsenic based on the long-term averages in Example 6.6? The dilution fraction is 0.00312 and is the same for all three water quality standards inasmuch as the outfall is an ocean discharge. Assume that the fraction available is the same for all three standards and is equal to 1.0 and that the background concentration is zero.

Solution. Because the dilution fraction and the available fraction are the same for all three standards, it is clear that the human health standard will give the most restrictive allowable effluent concentration. The allowable concentration, using Equation 6.13, is therefore:

$$C_{efhh} = \frac{0.0014 - 0(1 - 0.00312)}{1(0.00312)}$$

$$= 0.449 \text{ mg/L}$$

In freshwater systems, the dilution fraction and available fraction will differ among the standards and the allowable effluent concentration will have to be calculated for each standard. The lowest of the three allowable effluent concentrations is then used in the following equation to calculate the allowable headworks loading based on water quality standards.

$$L_{wq} = \frac{8.34 C_{wq} Q_{potw}}{1 - R_{potw}} \tag{6.14}$$

where

L_{wq} = allowable headworks loading based on water quality standard, lbs/d
C_{wq} = minimum allowable effluent concentration satisfying all acute aquatic, chronic aquatic, and human health standards

EXAMPLE 6.8
Allowable Headworks Loading for Water Quality Standard

What is the allowable heading works loading for arsenic for the most restrictive water quality standard in Example 6.7?

Solution. The most restrictive effluent concentration is 0.449 mg/L, based on the human health standard. Therefore, with a POTW flow of 8.3 MGD and a removal rate of 45%, the allowable headworks loading for water quality standards is:

$$L_{wq} = \frac{8.34(0.449)(8.3)}{(1 - 0.45)}$$

$$= 56.5 \text{ lbs/d}$$

Process Inhibition. The allowable headworks loading to protect against inhibition or upset of the process units is based on the maximum concentration a unit can receive without developing problems and how much of the pollutant is removed prior to the unit. The main focus is on the biological units, particularly the aeration basin. The allowable headworks loading for secondary treatment units such as activated sludge is:

$$L_{pi} = \frac{8.34 F_{pi} C_{pi} Q_{potw}}{1 - R_{pri}} \qquad (6.15)$$

where

L_{pi} = allowable headworks loading based on process inhibition, lbs/d
C_p = minimum concentration causing process inhibition, mg/L
F_{pi} = factor to convert the process inhibition concentration to the same common time scale as the other criteria
R_{pri} = compound removal through the treatment system prior to the unit

For the protection of digestor units, the following equation is used:

$$L_{pi} = \frac{8.34 F_{pi} C_{pi} Q_{potw}}{R_{potw}} \qquad (6.16)$$

The lowest loading that protects against process inhibition is used with compare to the other loadings.

EXAMPLE 6.9
Allowable Headworks Loading for Process Inhibition

What is the allowable headworks loading for arsenic to protect against process inhibition of an activated sludge unit if the limiting concentration is 0.1 mg/L and the removal rate up to that point is 24%? The limiting concentration is a 24-hour average, and the ratio of the long-term concentration to the 24-hour average is 0.468.

Solution. The allowable headworks loading that would protect against process inhibition is 4.26 lbs/d, as shown here:

$$L_{pi} = \frac{8.34(0.468)(0.1)(8.3)}{(1 - 0.24)}$$

$$= 4.26 \text{ lbs/d}$$

Sludge Quality Standard. Standards for the use and disposal of sewage sludge produced by POTWs are found in 40 CFR Part 503. These standards include pollutant limits and management practices for land application, surface disposal, landfilling, and incineration. Because limits on metal content are stringent, they often determine the local limits that POTWs must set for their nondomestic influents.

The allowable headworks loading to ensure sludge quality is based on how the POTW uses or disposes of the sludge and how much of the pollutant is contained in the sludge. If the POTW uses more than one method to dispose of its sludge, for example, land application and incineration, the standard that gives the lowest allowable headworks loading is used. Standards for land application, a common method of disposal for POTWs, are listed in Table 6.9.

Land application standards are given as sludge concentrations in mg/kg, as cumulative loading rates in kg/hectare, or as annual loading rates in kg/hectare/yr. The allowable headworks loading based on sludge concentration is calculated by the following equation:

$$L_{sl} = \frac{F_{sl} C_{sl} Q_{sl}}{R_{potw}} \times \frac{\text{yr}}{365 \text{ days}} \times \frac{10^{-6} \text{kg}}{\text{mg}} \tag{6.17}$$

where

L_{sl} = allowable headworks loading based on sludge quality standard, lbs/d

F_{sl} = factor to convert the sludge quality standard to the same common time scale as the other criteria

C_{sl} = sludge quality standard, mg/kg

Q_{sl} = annual dry weight sludge application rate, lbs/yr

Equation 6.17 can also be used with the cumulative and annual sludge loading standards after these standards are converted to equivalent sludge concentrations, as follows:

Cumulative loading standard

$$C_{csl} = \frac{Q_c A}{Y Q_{sl}} \times \frac{2.205 \text{ lb}}{\text{kg}} \times \frac{10^6 \text{ mg}}{\text{kg}} \tag{6.18}$$

where

C_{csl} = cumulative sludge loading standard converted to mg/kg

Q_c = cumulative sludge loading standard, kg/hectare

A = land application area, hectares

Y = life of land application area, years

Annual loading standard

$$C_{asl} = \frac{Q_s A}{Q_{sl}} \times \frac{2.205 \text{ lb}}{\text{kg}} \times \frac{10^6 \text{ mg}}{\text{kg}} \tag{6.18}$$

TABLE 6.9 Standards for Land Application of Sewage Sludge

Constituent	Ceiling Concentration (mg/kg)	Monthly Concentration (mg/kg)	Cumulative Loading (kg/hectare)	Annual Loading (kg/hectare)
Arsenic	75	41	41	2
Cadmium	85	39	39	1.9
Copper	4,300	1,500	1,500	75
Lead	840	300	300	15
Mercury	57	17	17	0.85
Molybdenum	75	—	—	—
Nickel	420	420	420	21
Selenium	100	100	100	5
Zinc	7,500	2,800	2,800	140

Source: 40 CFR §503.13.

where

C_{asl} = annual sludge loading standard converted to mg/kg

Q_s = annual sludge loading standard, kg/hectare/yr

EXAMPLE 6.10
Allowable Headworks Loading for Sludge Quality Standard

Calculate the allowable headworks loading for arsenic for a sludge standard of 41 mg/kg (limit for the monthly average concentration). The ratio of the long-term average to the monthly average standard is 0.84, and the sludge application rate is 7,056,000 dry pounds a year.

Solution. The allowable loading is calculated using Equation 6.17 as follows:

$$L_{sl} = \frac{(0.84)(41)(7,056,000)(10^{-6})}{(0.45)(365)}$$

$$= 1.48 \text{ lbs/d}$$

If the sludge is incinerated, the sludge concentration used in Equation 6.17 is calculated from emission standards for certain metals (40 CFR 503.43). The calculation differs depending on the particular metal. For arsenic, cadmium, chromium, and nickel, the following equation is used:

$$C_{inc} = \frac{RSC}{DF(1-CE)SF} \times \frac{86,400 \text{ s}}{d} \times \frac{\text{metric ton}}{1,000 \text{ kg}} \times \frac{1,000 \text{ mg}}{gm} \quad (6.20)$$

where

C_{inc}=sludge concentration limit for incineration, mg/kg
RSC=risk specific concentration, μg/m³
DF=dispersion factor, μg/m³/(gm/s)
CE=incinerator control efficiency, fraction
SF=sludge feed rate, metric tons/d

EXAMPLE 6.11
Sludge Concentration Limit for Arsenic Based on Incineration

How much arsenic can a sludge contain before it exceeds the incineration standard? In this case, the risk specific concentration is 0.023 μg/m³, the dispersion factor is 58.24 μg/m³/(gm/s), the incinerator control efficiency is 0.96, and the sludge feed rate is 8.77 metric tons/d.

Solution. Using Equation 6.20, the concentration limit is:

$$C_{inc}=\frac{(0.023)(86,400)(1,000)}{(58.24)(1-0.96)(8.77)(1,000)}$$

$$=97.3 \text{ mg/kg}$$

For lead, the following equation is used:

$$C_{inc}=\frac{0.1 NAAQS}{DF(1-CE)SF} \times \frac{86,400 \text{ s}}{d} \qquad (6.21)$$

where

$NAAQS$=national ambient air quality standard for lead, μg/m³

EXAMPLE 6.12
Sludge Concentration Limit for Incineration (Lead)

What is the limiting sludge concentration of lead if the standard for lead emissions is 1.5 μg/m³ and the control efficiency is 0.67? Assume the same dispersion factor and sludge feed rate as in Example 6.11.

Solution. Using Equation 6.21, the limiting lead concentration in the sludge is:

$$C_{inc}=\frac{(0.1)(1.5)(86,400)}{(58.24)(1-0.67)(8.77)}$$

$$=76.9 \text{ mg/kg}$$

For mercury, the following equation is used based on the regulations at 40 CFR 61, Subpart E:

$$C_{inc} = \frac{NES_{hg}}{Q_{sl}} \times \frac{365\ \text{d}}{\text{yr}} \times \frac{2.205\ \text{lbs}}{\text{kg}} \times \frac{1,000\ \text{mg}}{\text{gm}} \tag{6.22}$$

where

NES_{hg} = national emissions standard for mercury, gm/d

EXAMPLE 6.13
Sludge Concentration Limit for Incineration (Mercury)

What is the limiting sludge concentration of mercury if the standard for mercury emissions is 3,200 gm/d? Assume that the sludge application rate is the same as in Example 6.11.

Solution. Using Equation 6.22, the sludge limit for mercury is:

$$C_{inc} = \frac{(3,200)(365)(2.205)(1,000)}{7,056,000}$$

$$= 365\ \text{mg/kg}$$

For beryllium, incineration emissions must comply with the 40 CFR 61 Subpart C standard. The calculation for beryllium is similar to that for mercury except that the control efficiency of the incinerator is included (zero capture efficiency is assumed in the preceding equation for mercury). The following equation is used to convert the emission standard for beryllium to a sludge concentration:

$$C_{inc} = \frac{NES_{be}}{(1-CE)Q_{sl}} \times \frac{365\ \text{d}}{\text{yr}} \times \frac{2.205\ \text{lbs}}{\text{kg}} \times \frac{1,000\ \text{mg}}{\text{gm}} \tag{6.23}$$

where

NES_{be} = national emissions standard for beryllium, gm/d

EXAMPLE 6.14
Sludge Concentration Limit for Incineration (Beryllium)

What is the limiting sludge concentration of beryllium if the emissions standard for beryllium is 10 gm/d and the control efficiency is 0.99? Assume the same sludge application rate as in Example 6.11.

Solution. Using Equation 6.23, the sludge limit for beryllium is:

$$C_{inc} = \frac{(10)(365)(2.205)(1,000)}{(1-0.99)(7,056,000)}$$

$$= 114 \text{ mg/kg}$$

Lowest Allowable Headworks Loading. The lowest of all of the previously discussed loadings (permit, water quality, process inhibition, sludge quality) is selected as the total allowable headworks loading, representing the combined load from domestic and nondomestic sources.

EXAMPLE 6.15
Lowest Allowable Headworks Loading

What is the most restrictive headworks loading for arsenic, as calculated from the preceding examples?

Solution. The lowest loading is 1.48 lbs/d, which is based on protection of sludge quality for land application.

Allocation of Loading. The third step in developing local limits is to allocate the allowable load among the nondomestic users. First, the total allowable load is reduced by a safety factor to account for the uncertainty in assumptions and to set aside capacity for future users. Then, the domestic load is subtracted from the total load, leaving the allowable load that can be allocated to nondomestic users. The allowable nondomestic load is calculated as follows:

$$L_i = F_{ll}[(1-SF)L_{min} - 8.34 F_d C_d F_q Q_{potw}] \tag{6.24}$$

where

L_i = allowable nondomestic headworks load, lbs/d

F_{ll} = factor to convert load to time scale assumed for local limit (for example, daily maximum)

SF = fraction of total load to use as safety factor

L_{min} = minimum of allowable loads calculated for permit limits, water quality standards, process inhibition, and sludge standards, lbs/d

F_d = factor to convert the domestic background concentration to the same common time scale as the other criteria

C_d = domestic background concentration, mg/L

F_q = fraction of POTW influent from domestic sources

EXAMPLE 6.16
Allowable Nondomestic Load

Calculate the allowable arsenic load that can be allocated to nondomestic users if 10% of the total allowable load is reserved for future capacity. Assume that domestic wastewater contains no arsenic. In addition, because the local limit will be set as a daily maximum limit (24-hour average), you will be converting the long-term average load to a daily maximum. Assume that the ratio of daily maximum to long-term average is 3.11.

Solution. Using Equation 6.24, the allowable nondomestic load that may be allocated among the industrial users is:

$$L_i = (3.11)(1-0.1)(1.48) = 4.14 \text{ lbs/d}$$

The POTW can allocate its allowable influent loading among its non-domestic users in any fashion, subject, of course, to other regulatory restrictions. For example, the POTW cannot give a categorical user a discharge limit that is higher than the categorical limit. EPA suggests four ways in which a POTW could allocate the loading: uniform concentration, contributory flow basis, mass proportional, and selected reduction (EPA 1987).

With use of the uniform concentration method, every nondomestic user gets the same concentration limit. The local limit is calculated by dividing the allowable load by the total nondomestic flow.

$$C_u = \frac{L_i}{8.34(1-F_q)Q_{potw}} \tag{6.25}$$

where

C_u = uniform concentration local limit, mg/L

EXAMPLE 6.17
Uniform Concentration Local Limit

What would the local limit be for arsenic based on the allowable nondomestic load calculated in Example 6.16 and the uniform concentration method? Flow to the POTW is 66% domestic and 34% nondomestic.

Solution. Using Equation 6.25, the local limit for arsenic, as a daily maximum limit, would be:

$$C_u = \frac{4.14}{(8.34)(1-0.66)(8.3)}$$

$$= 0.18 \text{ mg/L}$$

The contributory flow method is similar to the uniform concentration method, except that only the flow from users that discharge the pollutant is used. Each user discharging the pollutant gets the same concentration limit; however, this limit applies only to those users with that pollutant in their discharge. A local limit based on contributory flows is calculated as follows:

$$C_c = \frac{L_i}{8.34Q_c} \tag{6.26}$$

where

C_c = local limit based on flows from users discharging the pollutant, mg/L
Q_c = flow from only those users discharging the pollutant, MGD

In the mass proportional method, each user receives an allocation in proportion to what it is currently discharging. A local limit based on the mass proportional method is calculated as follows:

$$C_{m,x} = \frac{L_x}{L_t} \times L_i \times \frac{1}{8.34Q_x} \tag{6.27}$$

where

$C_{m,x}$ = local limit based on mass loads currently discharged by a user x (each user gets its own local limit), mg/L
L_x = current load discharged by a user x, lbs/d
L_t = total current load in POTW influent, lbs/d
Q_x = flow discharged by user x, MGD

Notice in the preceding equation that if the total current load in the POTW influent exceeds the allowable load, each user's allocation is *reduced* in proportion to its current loading.

In the selected reduction method, the current POTW influent load is greater than the allowable load and the POTW determines which users must reduce their loads and by how much. Users that are required to reduce their loads are those that are likely to have the greatest impact or load on the POTW. Reductions are based on the type of wastewater generated by the user and the capabilities of pretreatment technologies. A local limit calculated by the selected reduction method is calculated as follows:

$$C_{r,x} = \frac{(1-R_x)L_x}{8.34Q_x} \tag{6.28}$$

where

$C_{r,x}$ = local limit for a particular user based on selected reduction, mg/L
R_x = required pollutant reduction fraction for user x

In conjunction with any of the allocation methods described here, the POTW may account for losses of the pollutant in the sewer through degradation or volatization, which serve to increase a user's limit. Volatilization in the sewer may in itself become a health or safety issue, as discussed in the following paragraphs.

Worker Health and Safety. The POTW and its workers must be protected against toxic gases and explosions in the sewer and at the treatment facility. Local limits are calculated for those pollutants that volatilize, forming toxic or flammable gases in confined spaces. Instead of an allowable headworks loading, the local limit to protect worker health and safety is calculated directly as a wastewater concentration, based on the vapor pressure and the Henry's law constant for the pollutant. The following equation below assumes that steady state conditions exist in the confined spaces and that each pollutant acts independently of the other; these assumptions are quite simplistic, but conservative.

$$C_{hs} = \frac{F_{hs} C_{vap}}{H} \tag{6.29}$$

where

C_{hs} = local limit based on worker health and safety, mg/L
C_{vap} = vapor phase concentration limit, mg/m^3
F_{hs} = factor to convert C_{vap} to time scale assumed for local limit (for example, daily maximum)
H = Henry's law constant, (mg/m^3)/(mg/L)

EPA recommends using the ACGIH TLV-TWA (threshold limit value, time-weighted average) values for C_{vap} to protect against toxicity and 10% of the lower explosive limit (LEL) to protect against explosivity (EPA 1992). TLV-TWAs are given in mg/m^3; however, the vapor concentration for explosivity must be converted as follows:

$$C_{vap} = 0.001 LEL \frac{1 \text{ atm}}{\left(\frac{0.08206 \text{ atm·L}}{\text{mol·K}}\right) 298.15 \text{ K}} \times \frac{10^3 \text{ L}}{\text{m}^3} \times MW \times \frac{10^3 \text{ mg}}{\text{gm}} \tag{6.30}$$

where

LEL = lower explosive limit, %

MW = molecular weight of pollutant, gm/mol

EXAMPLE 6.18
Worker Health and Safety Limit for Benzene

Calculate a local limit for benzene to protect worker health and safety. Assume that the TLV-TWA value is equivalent to a daily maximum. Use 225 (mg/m^3)/(mg/L) for Henry's law constant for benzene.

Solution. The vapor concentration, C_{vap}, is calculated first, using Equation 6.30. The lower explosive limit for benzene is 1.4% and the molecular weight of benzene is 78. Therefore, the vapor concentration is:

$$C_{vap} = \frac{(0.001)(1.4)(78 \times 10^6)}{(0.08206)(298.15)}$$

$$= 4,463 \text{ mg/m}^3$$

Next, the local limit is calculated using Equation 6.29.

$$C_{hs} = \frac{(1)(4,463)}{225} = 20 \text{ mg/L}$$

REFERENCES

U.S. EPA, "Water Quality Guidance for the Great Lakes System: Supplementary Information Document (SID)," EPA/820/B-95/001, Washington, DC, 1995.

U.S. EPA, "EPA's Contaminated Sediment Management Strategy," EPA/823/R-94/001, Washington, DC, 1994a.

U.S. EPA, "Water Quality Standards Handbook," EPA/823/B-94/005a, Washington, DC, 1994b.

U.S. EPA, "Technical Basis for Deriving Sediment Quality Criteria for Nonionic Organic Contaminants for the Protection of Benthic Organisms by Using Equilibrium Partitioning," EPA/822/R-93/011, Washington, DC, 1993.

U.S. EPA, "Guidance to Protect POTW Workers from Toxic and Reactive Gases and Vapors," EPA/812/B92/001, Washington, DC, 1992.

U.S. EPA, "Technical Support Document for Water Quality-Based Toxics Control," EPA/505/2-90-001, Washington, DC, 1991.]

U.S. EPA, "Guidance Manual on the Development and Implementation of Local Discharge Limits Under the Pretreatment Program," EPA/833/B87/202, Washington, DC, 1987.

BIBLIOGRAPHY

Federal Register (1995), "Final National Pollutant Discharge Elimination System Storm Water Multi-Sector General Permit for Industrial Activities; Notice."

—— (1992), "Final NPDES General Permits for Storm Water Discharges Associated with Industrial Activity; Notice."

—— (1992), "Final NPDES General Permits for Storm Water Discharges from Construction Sites; Notice."

—— (1990), "40 CFR Parts 122, 123, and 124, National Pollutant Discharge Elimination System Permit Application Regulations for Storm Water Discharges; Final Rule."

U.S. EPA (1994), "Industrial User Inspection and Sampling Manual for POTWs," EPA/831/B94/001, U.S. Environmental Protection Agency, Washington, DC.

—— (1993), "Guidelines for Deriving Site-Specific Sediment Quality Criteria for the Protection of Benthic Organisms," EPA/822/R-93/017, Washington, DC.

—— (1993), "Guides to Pollution Prevention, Municipal Pretreatment Programs," EPA/625/R-93/006, Washington, DC.

—— (1992), "Model Pretreatment Ordinance," EPA/833/B92/003, U.S. Environmental Protection Agency, Washington, DC.

—— (1992), "Storm Water Management for Industrial Activities: Developing Pollution Prevention Plans and Best Management Practices," Government Institutes, Inc., Rockville, MD.

—— (1991), "Control of Slug Discharges to POTWs Guidance Manual," EPA/21W/4003, U.S. Environmental Protection Agency, Washington, DC.

—— (1991), "Storm Water Guidance Manual for the Preparation of NPDES Permit Applications for Storm Water Discharges Associated with Industrial Activity," Government Institutes, Inc., Rockville, MD.

—— (1991), "Supplemental Manual on the Development and Implementation of Local Discharge Limitations Under the Pretreatment Program: Residential and Commercial Toxic Pollutant Loadings and POTW Removal Efficiency Estimation," EPA/21W/4002, U.S. Environmental Protection Agency, Washington, DC.

—— (1990), "Guidance Manual for POTWs to Calculate the Economic Benefit of Noncompliance," U.S. Environmental Protection Agency, Washington, DC.

—— (1989), "Industrial User Permitting Guidance Manual," EPA/833/B89/001, U.S. Environmental Protection Agency, Washington, DC.

—— (1989), "Guidance for Developing Control Authority Enforcement Response Plan," U.S. Environmental Protection Agency, Washington, DC.

—— (1987), "Guidance Manual for Preventing Interference at POTWs," EPA/833/B89/201, U.S. Environmental Protection Agency, Washington, DC.

—— (1987), "Guidance Manual for the Identification of Hazardous Wastes Delivered to POTWs by Truck, Rail, or Dedicated Pipe," U.S. Environmental Protection Agency, Washington, DC.

—— (1985), "Guidance Manual for the Use of Production-Based Standards and the Combined Wastestream Formula," EPA/833/B85/201, U.S. Environmental Protection Agency, Washington, DC.

_____ (1985), "RCRA Information on Hazardous Wastes for Publicly Owned Treatment Works," EPA/833/B85/202, U.S. Environmental Protection Agency, Washington, DC.

_____ (1985), "Guidance Manual for Preparation and Review of Removal Credit Applications," EPA/833/B85/200, U.S. Environmental Protection Agency, Washington, DC.

_____ (1985), "Guidance Manual for Implementing the Total Toxic Organics (TTO) Pretreatment Standards," EPA/440/1-85/009g, U.S. Environmental Protection Agency, Washington, DC.

7

STORAGE AND CONTAINMENT

This chapter discusses the storage of materials in containers, tanks, and containment buildings and secondary containment for these units designed to contain spill and leaks. There are major regulations covering the storage of hazardous substances, petroleum products, and hazardous wastes in these types of units. These regulations are important in establishing minimum design standards and sound operational practices. The discussion of tank storage includes aboveground and underground tank systems for hazardous substances, petroleum products, and hazardous wastes. In addition to regulatory requirements for tanks, information is provided on corrosion and tank integrity tests, construction materials, for tanks, and compatibility of chemicals with tank materials. The management of hazardous wastes in containers is discussed, including types of containers, storage area requirements, and general management practices. The discussion of containment buildings for managing hazardous wastes includes design standards and general operation requirements.

REGULATIONS

It is important to classify a material or waste properly, according to the regulation (or regulations) that apply. Facilities are periodically inspected by regulatory personnel, and one of the most common mistakes that a facility makes is to classify a material incorrectly, which results in its being cited for noncompliance with regulations that apply to that classification. Quite often, the wrong labels are used for containers, for example, using a "Hazardous Waste" label may be used when the waste is really nonhazardous. To avoid such mis-

169

takes, stored materials should always have an appropriate label. If a waste is not regulated by any specific regulations, a label such as the one shown in Figure 7.1 may be used. If the waste is regulated but is not hazardous, a label such as that shown in Figure 7.2 may be used. Hazardous waste labels are discussed in the following section.

Hazardous Waste

Minimum requirements for the management of hazardous waste in tanks, containers, and containment buildings are set out in the RCRA regulations under 40 CFR. States that have authorization from EPA for their own hazardous waste programs must establish requirements that are at least as stringent as the federal requirements.

Tanks, containers, and containment buildings have specialized definitions under the hazardous waste regulations. A *tank* is a stationary device, designed

NON-REGULATED WASTE

GENERATOR INFORMATION (Optional)

SHIPPER _____

ADDRESS _____

CITY, STATE, ZIP _____

PROPER D.O.T. SHIPPING NAME _____

_____

UN or NA NO. _____

CONTENTS _____

NON-REGULATED WASTE

QS1584 ©EMED Co., Bflo., NY 14240 • 1-800-442-3633

Figure 7.1 Example label for nonregulated waste.

to contain an accumulation of hazardous waste, that is constructed primarily of nonearthen materials such as wood, concrete, steel, or plastic, which provide structural support. A *tank system* includes a tank and its associated ancillary equipment (piping, etc.) and containment system. A *container* is any portable device in which a hazardous waste is stored, transported, treated, disposed of, or otherwise handled. A *containment building* is a completely enclosed, self-supporting structure that has sufficient strength to withstand the stresses of daily operation from personnel and heavy equipment, has a primary barrier against waste migration, controls for fugitive dust emissions, and prevents tracking of materials outside the building.

Units that are used to accumulate waste temporarily have fewer requirements than units that are used routinely and must be permitted. A *temporary unit* is either a satellite accumulation area or what is called a "90-day unit." These temporary units are briefly discussed in the following paragraphs.

NON-HAZARDOUS WASTE

OPTIONAL INFORMATION

SHIPPER _____

ADDRESS _____

CITY, STATE, ZIP_____

CONTENTS_____

_____

NON-HAZARDOUS WASTE

QS 1583 MFD. BY: Emed Co: Bflo, NY. 14240

Figure 7.2 Example label for nonhazardous waste.

Satellite Accumulation Areas. A hazardous waste generator is allowed to accumulate up to 55 gallons of hazardous waste, or up to one quart of acutely hazardous waste[1] without a hazardous waste permit if certain conditions are met. The waste must be accumulated in a container at or near the point of generation where the waste initially accumulates, which is under the control of the operator of the process generating the waste. The container label must say "Hazardous Waste" or bear another description that identifies what is stored in the container. An example of such a label is shown in Figure 7.3. The container must be maintained in good condition, and if found leaking, the waste must be transferred to a non-leaking, structurally sound container. The waste must be compatible with the container materials so that the container is not damaged by reaction with the waste. The container must be kept closed at all times, except when waste is added or removed. Any amount that accumulates over the allowed limit must be removed from the area within three days; otherwise, the generator becomes subject to additional requirements.

Temporary Storage (90 Days or Less). Hazardous waste may be accumulated by the generator in containers, tanks, or containment buildings for up to 90 days without a hazardous waste permit if certain conditions are met. The generator must have a system that minimizes hazards to health and safety, train its personnel in the management of the waste, and have a plan for emergencies and other contingencies. If a generator is treating hazardous waste in these 90-day units in order to meet land disposal restrictions, the generator must have a written waste analysis plan describing such procedures. Each container and tank must be labeled, "Hazardous Waste." An example of a hazardous waste container label is shown in Figure 7.4. If the waste is placed in containers, the generator must comply with standards for structural integrity, location restrictions, compatibility of materials, compatibility of wastes, general operation, inspections, and control of air emissions. Each container must be marked with the date upon which accumulation begins (note that the label in Figure 7.4 contains a space for recording the accumulation date). If tanks are used, the generator must comply with standards for structural integrity, design and installation, leak detection and containment, location restrictions, general operation, inspections, compatibility of wastes, control of air emissions, and closure of the unit. If the waste is managed in a containment building, the building must be certified as complying with certain design standards and there must be procedures and documentation that the 90-day limit is met for each waste volume or that the building is emptied at least once every 90 days.

Small-quantity generators, those generating more than 100 kilograms but less than 1,000 kilograms per month of hazardous waste, may accumulate up to 6,000 kilograms of waste for up to 180 days without a permit if they meet cer-

---

[1]Commercial chemical products that become acutely hazardous wastes when discarded are listed at 40 CFR 261.33(e). Examples are acrolein, arsenic pentoxide, and carbon disulfide.

Figure 7.3 Example label for satellite accumulation.

tain requirements. This period can be extended to 270 days if the waste has to be shipped at least 200 miles for offsite treatment, storage, or disposal. The requirements that the small quantity generator must meet to qualify for this permit exemption are similar to, but less extensive than, those for 90-day units.

Permitted Units. The standards for containers, tanks, and containment buildings that require hazardous waste permits are found in 40 CFR §264. The subparts of 264 that directly address units of this type are:

- Subpart I—Use and Management of Containers
- Subpart J—Tank Systems
- Subpart CC—Air Emissions for Tanks, Surface Impoundments, and Containers
- Subpart DD—Containment Buildings

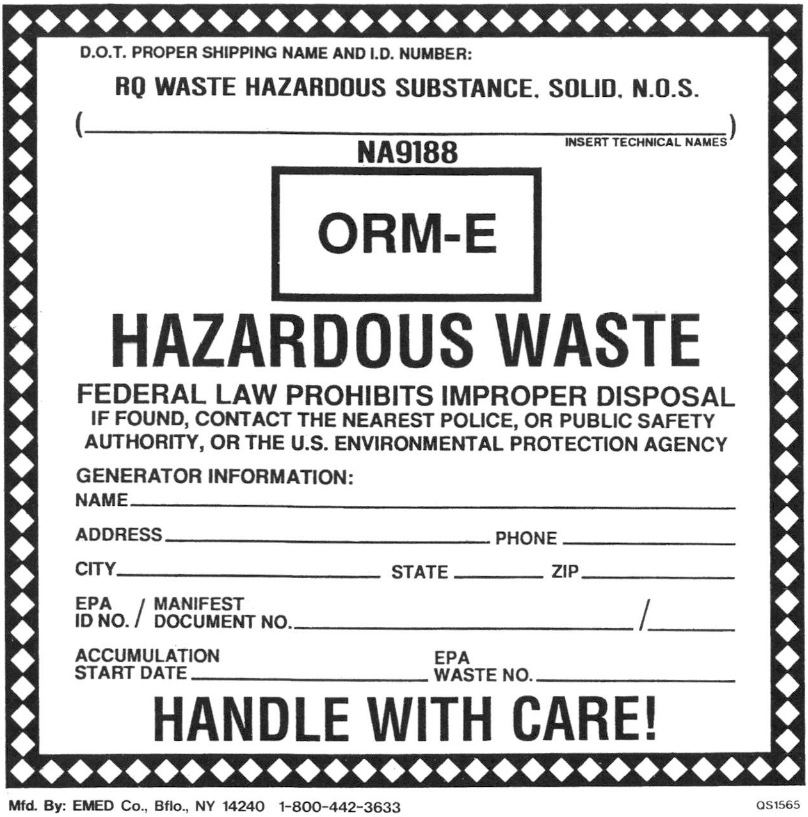

D.O.T. PROPER SHIPPING NAME AND I.D. NUMBER:

RQ WASTE HAZARDOUS SUBSTANCE. SOLID. N.O.S.

(_____)

NA9188   INSERT TECHNICAL NAMES

ORM-E

HAZARDOUS WASTE

FEDERAL LAW PROHIBITS IMPROPER DISPOSAL
IF FOUND, CONTACT THE NEAREST POLICE, OR PUBLIC SAFETY AUTHORITY, OR THE U.S. ENVIRONMENTAL PROTECTION AGENCY

GENERATOR INFORMATION:
NAME_____

ADDRESS_____ PHONE _____

CITY_____ STATE _____ ZIP_____

EPA   / MANIFEST
ID NO. / DOCUMENT NO._____ /_____

ACCUMULATION   EPA
START DATE _____ WASTE NO._____

HANDLE WITH CARE!

Mfd. By: EMED Co., Bflo., NY 14240 1-800-442-3633   QS1565

Figure 7.4   Example label for hazardous waste storage in containers.

The topics covered by these standards are listed in Table 7.1. The details of these standards are discussed later in this chapter in the individual sections on tanks, containers, and containment buildings.

Hazardous Substances

Hazardous substances that are managed in underground storage tanks must comply with 40 CFR §280. Hazardous substances are those that are defined in Section 101(14) of CERCLA, excluding those identified as hazardous wastes. Certain types of tanks are excluded from Part 280 requirements; these exclusions are discussed in the section "Underground Tanks."

Petroleum Products

Petroleum and petroleum-based materials that are managed in underground storage tanks must comply with 40 CFR §280. Part 280 defines these materials to include, but not to be limited to, those substances consisting of a complex

TABLE 7.1   Hazardous Waste Standards for Containers, Tanks, and Containment Buildings (40 CFR §264, Subparts I, J, CC, and DD)

Section	Description
Subpart I—Use and Management of Containers	
§264.170	Applicability
§264.171	Condition of containers
§264.172	Compatibility of waste with containers
§264.173	Management of containers
§264.174	Inspections
§264.175	Containment
§264.176	Special requirements for ignitable or reactive waste
§264.177	Special requirements for incompatible wastes
§264.178	Closure
§264.179	Air emission standards
Subpart J—Tank Systems	
§264.190	Applicability
§264.191	Assessment of existing tank system's integrity
§264.192	Design and installation of new tank systems or components
§264.193	Containment and detection of releases
§264.194	General operating requirements
§264.195	Inspections
§264.196	Response to leaks or spills and disposition of leaking or unfit-for-use tank systems
§264.197	Closure and post-closure
§264.198	Special requirements for ignitable or reactive wastes
§264.199	Special requirements for incompatible wastes
§264.200	Air emission standards
Subpart CC—Air Emission Standards for Tanks, Surface Impoundments, and Containers	
§264.1080	Applicability
§264.1081	Definitions
§264.1082	General standards
§264.1083	Waste determination procedures
§264.1084	Tank standards
§264.1086	Container standards
§264.1087	Standards for closed-vent systems and control devices
§264.1088	Inspection and monitoring requirements
§264.1089	Recordkeeping requirements
§264.1090	Reporting requirements
§264.1091	Alternative control requirements for tanks
Subpart DD—Containment Buildings	
§264.1100	Applicability
§264.1101	Design and operating standards
§264.1102	Closure and post-closure care

blend of hydrocarbons derived from crude oil through processes of separation, conversion, upgrading, and finishing. Examples include motor fuels, jet fuels, distillate fuel oils, residual fuel oils, lubricants, petroleum solvents, and used oils. Certain types of tanks are excluded from Part 280 requirements; these exclusions are discussed in the "Underground Tanks" section.

The storage of oil at non-transportation related facilities is subject to 40 CFR §112 if the volume of underground storage exceeds 42,000 gallons or the volume above ground exceeds 660 gallons (single vessel) or a combined total of 1,320 gallons. Oil is defined in Part 112 as oil of any kind or in any form, including, but not limited to, petroleum, fuel oil, sludge, oil refuse, and oil mixed with wastes other than dredged spoil. Part 112 requires what is called a Spill Prevention Control and Countermeasure Plan (SPCC Plan).

Used oil is subject to RCRA standards at 40 CFR §279. Part 279 defines used oil as any oil that has been refined from crude oil, or any synthetic oil, that has been used and as a result of such use is contaminated by physical and chemical properties. As specified in Part 279 standards, the storage of used oil is subject to 40 CFR §112 and, if underground, is also subject to 40 CFR §280.

TANKS

Tanks are usually classified as either aboveground tanks or underground tanks for the purpose of regulation. Whether a tank is above or below ground would appear to be a simple question; however, these classifications are defined strictly in some regulations. For example, in the hazardous waste regulations, an aboveground tank is defined as one whose entire surface area is completely above the plane of the adjacent surrounding surface and can be visually inspected (this includes the tank bottom). An example of an aboveground tank with secondary containment is shown in Figure 7.5. An underground tank, as cited in the hazardous waste regulations, is one whose entire surface area is totally below the surface of and is covered by the ground.[2] Underground tanks that are used for hazardous substances and petroleum are defined by the volume of material contained below the surface (10% or more). The volume of material that is contained in the tank's underground piping must be included in this calculation.

Hazardous waste managed in tanks is subject to Subpart J of 40 CFR §265 (interim status units) or Subpart J of 40 CFR §264 (permitted units). The storage of hazardous substances and petroleum products in underground tanks is regulated under 40 CFR §280. Aboveground tanks used for process materials and products are not regulated under RCRA, but may be subject to other federal, state, or local regulations.

---

[2]There is no term in the hazardous waste regulations for those tanks that either sit directly on or are partly submerged in the ground.

Figure 7.5 Example of an aboveground tank with secondary containment.

Basic Design

In designing a tank system, general items for consideration include materials of construction, structural support, operating pressure, corrosion protection, and overfill and spill prevention. Material selection is based on compatibility with wastes or process or product materials. A general overview of materials and chemical compatibilities is provided later in this chapter.

In designing structural support for a tank, important items to consider include soil-load-bearing data, piling requirements, foundation properties, tank wall thickness requirements, integrity of seams (particularly in raised systems), and support of ancillary piping and connections. Other items affecting structural design are local and climatic conditions such as the risk of earthquakes, hurricanes, and flooding and extreme temperatures. For underground tanks, important considerations include the weight of vehicular traffic, the weight of structures above the tanks, subsurface saturation characteristics, depth to ground water, and depth of the freeze zone.

To protect tanks against overpressurizing, pressure controls should be supplemented with devices such as rupture disks, pressure relief valves, and automatic shutoffs at preset pressure changes. To prevent overfilling, liquid level controls for loading and unloading operations should include a preset high-level shutoff. To prevent negative pressures, there should be vacuum relief valves. Alarms for these and other abnormal operating conditions are also appropriate.

There are various factors affecting the corrosion of an underground tank system. These include soil properties such as pH, moisture content, sulfide level, and the presence of metal salts. In addition, corrosion rates can be affected by other metal structures and interference (stray) electric current. Corrosion can occur inside a metal tank if the tank metallurgy is not adequate for the material stored.

Aboveground Tanks

Types of Aboveground Storage Tanks. There are several broad categories of aboveground storage tanks, including atmospheric tanks, low pressure tanks, pressure tanks, and cyrogenic tanks. Atmospheric tanks operate at atmospheric pressure, although some field applications operating at pressures less than 0.5 pounds per square inch gauge (psig) are referred to as atmospheric applications. Low pressure tanks can operate at pressures ranging from 0.5 to 15 psig. Pressure tanks operate at pressures greater than 15 psig.

Tank designs for atmospheric tanks include the open top tank, fixed roof tank, and the floating roof tank. An example of a double-contained open atmospheric tank is shown in Figure 7.6. The open top tank is often used in wastewater treatment applications for equalization or neutralization. The fixed roof tank is used for storing materials with low vapor pressure, such as methanol, ethanol, and kerosene. Floating roof tanks are used mainly for petroleum products such as crude oil and gasoline. The floating roof design reduces evaporative losses while providing increased fire protection.

Low pressure tank designs include hemispheroid, noded hemispheroid, and spheroid tanks. Low pressure tanks are used for storage of volatile materials, which are gases at ambient temperature and pressure, such as pentane. In addition, combustible chemicals such as benzene and large-volume volatile liquids such as butane can be stored under low pressure.

Pressure tanks are typically cylindrical and are supported in either a vertical or horizontal position. These tanks are used to store high-volatility materials such as liquified petroleum gas.

Cryogenic tanks are made of specialty steels with rigid low-temperature specifications and are heavily insulated. Examples of materials stored in cryogenic tanks include ammonia, liquid nitrogen, liquid propane, and other compounds that are gaseous at atmospheric temperature and pressure.

Design Concepts. In addition to any regulatory standards, there are several other design concepts that can be incorporated into any new aboveground double-containment project to maximize environmental protection (Langlois, Bauer, and Woodside 1990):

- Maximum ease of inspecting and testing,
- Greater than 100% capacity of secondary containment for indoor applications, as well as outdoor applications,

Figure 7.6 Example of a double-contained open atmospheric tank.

- Specialized leak detection, and
- Elimination of penetrations through containment.

These concepts are not detailed specifically in current regulations, but their use is good management practice where economically achievable.

Inspectability/Testability. Installing tanks and ancillary piping above ground is the first step toward making a tank system easy to inspect and test. Raising the primary tank on beams or saddles inside the secondary containment allows for daily inspections under the tank. If the secondary containment is also raised, the system becomes 100% inspectable.

Likewise, if ancillary piping is run above ground whenever possible, leaking joints or material failures can be easily detected. For cases where "pipe-in-a-pipe" or double-walled piping is installed, ports with leak detectors and drain valves can be installed every 50 to 200 feet for assessment of primary pipe integrity.

Tanks and piping can be tested for leaks using the methods described in elsewhere in this chapter. A good engineering practice is to set up a schedule for periodic testing of all tanks at the facility. The length of time between tests will vary, depending on application and materials of construction.

Secondary Containment Greater Than 100%. There are hazardous waste tank requirements for greater than 100% containment for outdoor liners and vaults used as secondary containment. In addition to these requirements, environmental protection can be maximized if indoor secondary containment is sized for catastrophic failure of the primary unit, plus spillage from ancillary piping. An example of an indoor tank with secondary containment is shown in Figure 7.7. A capacity of 125% containment is typically adequate. Use of this design concept can negate the possibility of spilled materials overflowing secondary containment and penetrating through the building floor.

Specialized Leak Detection. Standard leak detection systems, such as electronic moisture sensors and sensor floats, are used widely in double containment. Although usually adequate for indoor applications, outdoor leak detection can be maximized with specialized systems that differentiate chemical leakage from rainwater. Examples of these devices are detectors that signal an alarm based on pH, oxidation/reduction potential, or conductivity.

Eliminating Penetrations Through Secondary Containment. Joints and penetrations are usually the weakest points in secondary containment and are the most likely to leak should catastrophic failure occur. Locating pumps inside the secondary containment is one way to minimize penetrations. In addition, overhead trestles can be used to route piping over, instead of through, secondary tank walls.

Figure 7.7   Example of an indoor tank with secondary containment.

Optimum Containment for Tanks, Pipes, and Pumps. As discussed earlier, optimum systems include inspectable and testable primary and secondary containment. An example of this type of tank system is a tank in a tank, with both tanks raised on beams. A tank in a pan, with tank and pan raised on beams, is another option. Maximum environmental protection is provided by double-contained piping options including pipe-in-a-pipe on an overhead trestle. Another piping option is pipe on a rack inside a pan, building, or trench.

As mentioned previously, the optimum design is to have pumps placed inside the containment system. When this design is not practical, pump drip pans can be used and are particularly useful for smaller applications. Drip pans are not adequate, however, for failures of pressurized pipes or other major failures such as diaphragm tears or broken seals. They are also less useful for outdoor applications, because rainwater must be drained or absorbed after each rain.

EXAMPLE 7.1
Secondary Containment Calculations

Design a secondary containment dike area for three aboveground storage tanks, based on 125% of the largest tank volume. All three tanks will be located within the same dike. Two of the tanks are the same size, 10 feet in diameter with a capacity of 10,000 gallons. The third tank is larger, 15 feet in diameter with a capacity of 25,000 gallons. If the dike wall is to be 3 feet high, what is the required length of the dike perimeter if 6 inches of freeboard are required for the dike?

Solution. The largest tank volume is 25,000 gallons. The required volume that must be retained within the dike is

$$1.25 \times 25{,}000 \text{ gal} \times \frac{\text{ft}^3}{7.48 \text{ gal}} = 4{,}178 \text{ ft}^3$$

The dike must be sized to include the area taken up by the three tanks. The area of each of the smaller tanks is

$$(5 \text{ ft})^2 \pi = 78.5 \text{ ft}^2$$

The area of the larger tank is

$$(7.5 \text{ ft})^2 \pi = 177 \text{ ft}^2$$

Therefore, the total area taken up by the three tanks is

$$2(78.5 \text{ ft}^2) + 177 \text{ ft}^2 = 334 \text{ ft}^2$$

The total dike area needed is therefore

$$\frac{4{,}178 \text{ ft}^3}{(3 - 0.5) \text{ ft}} + 334 \text{ ft}^2 = 2{,}005 \text{ ft}^2$$

If the sides of the dike are the same length, they would be approximately 45 feet.

Underground Tanks

As noted earlier in this section, the definition of an underground tank depends on which regulation applies to the material being stored in the tank. Underground tanks used for hazardous waste are those that are completely below the ground surface and are covered by the ground. Underground tanks used for hazardous substances (not hazardous wastes) and petroleum and petroleum-based materials are those whose volume (including piping) is at least 10% underground. Standards for hazardous waste tanks are found in Subpart J of both 40 CFR §265 (for nonpermitted units and those awaiting a permit (interim status tanks)) and 40 CFR §264 (permitted tanks). Standards for hazardous substances and petroleum products are found at 40 CFR §280.

Certain tanks are excluded from these regulations. Wastewater treatment tanks that are part of a National Pollutant Discharge Elimination System (NPDES) treatment facility or discharge to a publicly owned treatment works (POTW) are exempt from the hazardous waste standards for tanks. Underground tanks that are excluded from Part 280 requirements include:

- Those containing hazardous wastes;
- Wastewater treatment tanks that are part of an NPDES system or discharge to a POTW;
- Equipment or machinery that contains hazardous substances or petroleum products, such as hydraulic lift tanks and electrical equipment tanks;
- Tank systems that are of 110 gallons or less;
- Tank systems containing only a *de minimis* concentration of hazardous substances or petroleum products; and
- Any emergency spill or overflow containment system that is emptied expeditiously after use.

Tanks that are not defined as "underground tanks" in relation to 40 CFR 280 include:

- Farm or residential tanks of no more than 1,100 gallons that contain motor fuel for noncommercial purposes;
- Tanks containing heating oil for use on the premises;
- Septic tanks;
- Pipeline facilities (including gathering lines) that are regulated under the Natural Gas Pipeline Safety Act of 1968 or the Hazardous Liquid Pipeline Safety Act of 1979, or that are intrastate facilities regulated under state laws comparable to either of these two Acts;
- Storm water or wastewater collection systems;
- Flow-through process tanks;
- Liquid traps or associated gathering lines directly related to oil or gas production and gathering operations; and

- Storage tanks situated in an underground area (such as a basement, cellar, mine-working, drift, shaft, or tunnel) if they are on or above the surface of the floor.

Underground tanks are typically used for high-volume flammable materials such as gasoline, diesel fuel, and heating oils. They are normally horizontal cylindrical tanks, ranging from 5,000 to 20,000 gallons. Underground tanks are especially suited for flammable materials because diurnal temperature changes and tank breathing losses are minimized. Other applications for underground tanks include storage of flammable product or process chemicals.

Basic Design. An integral part of underground tank design is cathodic protection to prevent surface corrosion. Guidelines for cathodic protection are published by several standards organizations and include:

- American Petroleum Institute (API) Publication 1632 (1987)—Cathodic Protection of Underground Petroleum Storage Tanks and Piping Systems, 2d ed.
- National Association of Corrosion Engineers (NACE) Standard RP0285 (1985)—Control of External Corrosion on Metallic Buried, Partially Buried, or Submerged Liquid Storage Systems
- NACE Publication 2M363(1963)—Recommended Practice for Cathodic Protection
- Underwriters Laboratories (UL) Standard 1746(1989)—Corrosion Protection Systems for Underground Storage Tanks

These guidelines address design considerations, coatings, cathodic protection voltage criteria, and installation and testing of cathodic protection systems.

Corrosion is an electrochemical phenomenon. In an electrochemical cell, the positive terminal—the anode—undergoes an oxidation reaction, while the negative terminal—the cathode—undergoes a reduction reaction. Corrosion is an oxidation reaction. Cathodic protection is a standard engineering method used to prevent external corrosion on the surface of a buried, partially buried, or submerged metal tank by allowing the metal surface to become the cathode of the electrochemical cell. A protective array of anodes is set in the soil at calculated distances from the tank to protect against corrosion.

Two recognized types of cathodic protection systems are galvanic anode and impressed current anode. Galvanic anodes are usually made of magnesium or zinc and, depending on soil properties, may require a special backfill material such as mixtures of gypsum, bentonite, and sodium sulfate (NACE, RP0285 1985). The anodes are made of a less noble metal than the tank so that they will be more susceptible to undergoing an oxidation reaction. Once installed, the system provides sacrificial protection to the tank by inducing a continuous current source to inhibit electrical corrosion. Galvanic anode cathodic protection systems are limited in electrical current output.

Thus, applications for these systems are limited to tanks well insulated by a nonconductive coating that minimizes the exposed surface of the tank. Impressed current anodes systems are made of higher-grade materials such as graphite, platinum, and steel. As with galvanic anode systems, soil properties may require the use of special backfill material such as coke breeze and calcined petroleum coke (NACE, RP0285 1985). A direct current source is used. The anodes are connected to a positive terminal of the power source, and the structure is connected to the negative terminal.

Verifying proper operation of a cathodic protection system should include measurements such as structure-to-soil potential, anode current, structure-to-structure potential, piping-to-tank isolation (if protected separately), and effect on adjacent structures. A regulated system must be tested within six months after it is installed and at least every three years thereafter. Some state agencies and standards organizations require or recommend more frequent testing.

Underground tanks for petroleum products are often constructed of carbon steel that has been coated at the factory. Desirable characteristics of a coating include resistance to deterioration when exposed to products stored in the tank, high dielectric resistance, and resistance to moisture transfer and penetration. Testing using American Society of Testing and Materials (ASTM) methods for determining cathodic disbonding of coatings can be performed on coated tanks and piping. In addition to being tested for disbonding, the coating should be tested for pinholes or other defects. Furthermore, precautions must be taken by the construction personnel to ensure that the protective coat is not damaged during tank installation.

Hazardous Substances and Petroleum Products. Performance standards under 40 CFR §280 for the management of hazardous substances and petroleum products in underground tanks include the following:

- Tanks must be properly designed and constructed, and any portion underground that routinely contains product must be protected from corrosion.
- Piping that routinely contains product and is in contact with the ground must be properly designed, contructed, and protected from corrosion.
- Spill and overfill prevention equipment associated with product transfer must be designed to prevent a release to the environment during transfers (including the use of catch basins for hoses, automatic shutoffs, flow restriction, alarms, and other equipment).
- All tanks and piping must be properly installed in accordance with a nationally recognized code of practice. The installation must be performed by a certified installer and completed properly.

Standards for proper design and installation of underground tanks and piping are documented in several standards, including:

- API RP 1615 (1987)—Installation of Underground Petroleum Storage Systems, 4th ed.
- American Society of Testing and Materials (ASTM) D 4021-86—Standard Specification for Glass-Fiber-Reinforced Polyester Underground Petroleum Storage Tanks
- UL Standard 1316 (1983)—Standard for Glass-Fiber-Reinforced Plastic Underground Storage Tanks for Petroleum Products

Additional requirements are defined for tanks that contain CERCLA hazardous substances. These include secondary containment performance requirements established to prevent releases to the environment at any time during the operational life of the tank.

Maximizing Environmental Protection. EPA's guidance manual entitled *Detecting Leaks: Successful Methods Step by Step* (1989) provides information on design concepts for underground storage tanks (UST) that maximize environmental protection. Information is provided on acceptable types of secondary containment, interstitial monitoring, and underground piping. Release detection methods are also addressed.

Secondary Containment. Secondary containment should be designed to provide an outer barrier, between the tank and the soil and backfill material, that is capable of holding the material long enough that a release can be detected. For tanks containing hazardous substances, prevention of leak migration is also a requirement. A double-walled tank, a jacketed tank, and fully enclosed external liners can provide this type of protection. In addition, the secondary containment must have enough sufficient capacity to hold the contents of the tank (or the largest tank if more than one tank is included in the system) and should prevent precipitation or ground water intrusion.

To maximize environmental protection for double-walled tanks, the second tank should be made of the same material as the primary tank. For jacketed tanks or tanks with liners, the material used should be compatible with the product. The material should also be sufficiently thick and impermeable to direct a release to the monitoring point and permit its detection. The permeation rate of the stored substance through the jacket or liner should be 10^{-6} centimeters per second (cm/sec) or less.

Interstitial Monitoring. An interstitial monitoring system should be designed to detect any leak from the underground tank under normal operating conditions. In most cases, a leak detection system does not measure the leak rate, but only the presence of a leak. Interstitial monitoring systems operate to detect leaks through on one of several mechanisms, such as electrical conductivity, pressure sensing, fluid sensing, hydrostatic monitoring, manual inspection, and vapor monitoring. Conductivity, fluid sensing, manual inspection, and vapor monitoring are suitable for all applications. Pressure sensing and hydrostatic monitoring are applicable only to double-walled installations.

Because of the inability to physically inspect tanks that are underground, the interstitial monitoring system becomes the primary means of verifying tank integrity. For this reason, potential problems with the monitoring system should be identified before installation and reflected in the design. These potential problems might include ground water or aboveground runoff penetration into the secondary containment from pinhole leaks in the secondary containment system or from inadequate lining. Further, faulty electrical installation can render some monitoring devices inoperable.

Other Release Detection Methods. Besides interstitial monitoring, other release detection methods include inventory control, vapor monitoring, ground water monitoring, manual or automatic tank gauging, and tank tightness testing using volumetric tests.

Product inventory control is a technique effective in finding leaks of more than one gallon per hour. Because of its low sensitivity, however, it must be combined with tank tightness testing. In general, this method requires careful tracking of inputs, withdrawals, and amounts still remaining in the tank. API RP 1621, *Recommended Practice for Bulk Liquid Stock Control at Retail Outlets* (API 1987), provides guidance for meeting inventory control requirements.

Vapor monitoring is used predominantly with USTs that store petroleum products. Success of this type of release detection system is dependent on site geology and soil types. A vapor monitoring system consists of a vapor monitoring well and a vapor sensor. If a product leaks from a UST, the liquid and vapors spread throughout the surrounding soil. If the vapor reaches a sensor in a concentration above a predetermined set point, the system responds with an alarm.

Ground water monitoring is a release detection method effective for petroleum and other products that have a specific gravity of less than 1.0. The method can be used in areas where the ground water table is very near to the excavation zone of the tank. Permanent observation wells are placed close to the tank, and the wells are checked periodically for evidence of free phase product on top of the water in the well.

Manual and automatic tank gauging and tank tightness testing using volumetric tests are also acceptable methods of leak detection. These methods are described elsewhere in this chapter.

Design of Undergound Piping. Double-contained piping with interstitial monitoring maximizes environmental protection in underground systems and is required for uses with hazardous substances. For pressurized lines, the use of automatic flow restrictors to restrict flow in the event of a leak and automatic shutoff devices to stop the flow when a pressure drop occurs also provides environmental protection.

Underground double-contained piping options include the use of trench liners as well as double-walled pipe. For the trench liner application, the pipe trench is lined with a flexible membrane that is imprevious to the stored product. Often, the liners are thermosplastic or polymeric sheets and are at least 50 mils thick. The trench is designed sloping away from the tank so that pipe leaks can be differentiated from tank leaks.

For applications of double-walled piping, the inner and outer pipes can be made of the same material or the outer pipe can be made of a cheaper material, such as fiberglass-reinforced plastic. Fiberglass-reinforced plastic is not used for high-pressure applications, but if compatible with the product, can work under nonpressurized or low-pressure conditions. Double-walled piping is usually sloped to a containment structure, sump, or observation well that can be monitored for the presence of liquids or vapors.

Hazardous Waste Tanks

The requirements for managing hazardous waste in tanks found at 40 CFR §264 are discussed in this section. Part 264 standards cover both aboveground and underground tanks, although the requirements may differ depending on the type of tank. These standards are for hazardous waste tanks that require hazardous waste permits; however, they are essentially the same as the Part 265 requirements for interim status units that have applied for, but not yet received, a permit. Portions of the Part 265 requirements are also applied to temporary storage tanks (see the discussion earlier in this chapter)

Part 264 requirements for managing hazardous waste in tanks include structural integrity assessment, installation inspection, specifications for backfill material, tightness testing, proper support of ancillary piping, corrosion protection, certification statements by tank system experts, and secondary containment and leak detection systems that meet specific design criteria.

Requirements for secondary containment and leak detection systems include the following:

- The system must be designed, installed, and operated to prevent any migration of wastes or accumulated liquid out of the system and into the environment.
- The system must be capable of detecting and collecting releases and accumulated liquids until the collected material is removed.
- The secondary containment for tanks must include either a liner external to the tank, a vault, a double-walled tank, or an equivalent device.

- A liner or vault must have enough capacity to contain the contents of the largest tank when full, plus precipitation from a 25-year, 24-hour rainfall.
- Liners, vault systems, and double-walled tanks must meet specified design criteria.
- Ancillary equipment must be provided with full, secondary containment such as a trench, jacketing, double-walled piping, or other containment method that prevents migration of wastes into the environment and provides leak detection. Some exceptions apply to this requirement, for example, if the aboveground piping is inspected daily and piping systems are equipped with automatic shutoff devices.
- Tanks that manage hazardous wastes containing volatile compounds must comply with air emission control requirements.

Any existing tank system that fails must be retrofitted to meet the established standards before the tanks can be put back into service. Other existing tanks must be retrofitted to the standards by the time they reach 15 years of age. All new tank systems must meet the requirements before being placed into service.

Testing Methods

Testing tanks periodically for corrosion resistance can indicate the need for repairs and aid in averting catastrophic failure through early warning of material thinning or fatigue. Likewise, tank integrity testing, if performed according to a preset periodic schedule, can allow for early detection of leaks or pending failure. The following paragraphs discuss methods for testing corrosion resistance and tank integrity.

Corrosion Resistance. There are numerous testing procedures for determining the corrosion resistance of materials commonly used in tank systems. Some of these testes may be suitable for field use to determine the extent of corrosion in existing equipment. Others are useful for screening materials during the system design phase. Tank manufacturers usually perform a variety of corrosion resistance and other tests on tank materials and make the test data available to potential customers.

Pitting and Crevice Corrosion Resistance of Stainless Steels and Related Alloys. ASTM G 48-76 is used to test pitting and crevice corrosion resistance of stainless steels and related alloys. This ASTM standard describes two test methods, Method A and Method B. Method A specifies total immersion in ferric chloride to determine the relative pitting resistance of stainless steels and other alloys, particularly those that have been heat treated or have had surface finishes applied. Method B, a ferric chloride crevice test, is used to determine both pitting and crevice corrosion resistance of these same types of materials.

In both tests, standard specimens are immersed in a 10% ferric chloride solution in both tests for approximately 72 hours. For crevice testing, the specimens are attached to TFE-fluorocarbon blocks with O-rings or rubber bands to form crevices at the points of contact. The test specimens are evaluated by visual inspection, measurement of pit and crevice depth, and weight loss.

Corrosion Coupon Tests in Plant Equipment. ASTM G 4-84 is a guide that defines a method for evaluating corrosion of engineering materials under the variable conditions occurring in actual service. The method typically is used to evaluate materials of construction for use in plant equipment or as replacement or modification materials.

The size and shape of test specimens are dependent on the specific test application. The duration of exposure is based on known rates of corrosion of the materials in use, or determined by the convenience at which plant operations can be interrupted to introduce and remove test specimens. Evaluation of the specimen includes microscopic inspection for etching, pitting, tarnishing, scaling, and other defects. The depth of the pits can be measured, and number, size, and distribution noted. A metallographic examination for intergranular corrosion or stress-corrosion cracking may also be performed.

Standard Test Method for Chemical Resistance of Protective Linings. ASTM C 868-85 offers a procedure for testing and evaluating the chemical resistance of a protective lining applied to steel or other metals. The liner is applied and cured, and then immersed in the service solution for six months. As necessary to simulate actual conditions, heat may be applied. Color, surface gloss, surface texture, and blisters are all visually evaluated before, at interim times, and after the completion of the test. In addition, lining thickness is measured before and after the test to quantify the chemical attack on or the dissolution of the lining material.

Standard Methods of Testing Vulcanizable Rubber Tank and Pipe Lining. ASTM D 3491-85 provides a procedure for testing and evaluating the chemical resistance of vulcanizable rubber tank and pipe lining. A test specimen of a rubber component applied to a steel plate is immersed in the service solution and heated, if required. The duration of the test should be a minimum of six months, with inspections performed every month. At the end of the test period, a visual inspection is made for changes in surface texture, evidence of cracking, blistering, swelling, delaminating, or permeation. In addition, substrate attack or corrosion, such as rusting or metal darkening, is noted. Because of the length of time required for this test, prescreening of lining materials can be performed using ASTM D 471-79(1991), which is a more convenient method.

X-Ray Fluorescence. ASTM A 754-79(1990) offers an X-ray fluorescence (XRF) test method for determining coating thickness. A radiation detector that can discriminate between the energy levels of all radiations is used to measure the thickness of a coating and substrate exposed to an intense beam of radiation generated by a radioisotope source or an X-ray tube. The combined interaction of the coating and substrate with the beam of radiation generates X rays of well-defined energy, which are singularly characteristic of that element.

If XRF thickness testing is performed in the field, environmental factors such as temperature, humidity, and surface cleanliness must be taken into account. Other factors, including specimen size, specimen uniformity, radiation source, and radiation detector, can also affect the test results.

Other ASTM Standards. Other ASTM standards, guides, and practices that address corrosion testing are documented in Volume 3.02, "Wear and Erosion: Metal Corrosion," *Annual Book of ASTM Standards* (see References). Examples of these standards include:

- G 1-90—Recommended Practice for Preparing, Cleaning, and Evaluating Corrosion Test Specimens
- G 15-90—Terminology Relating to Corrosion and Corrosion Testing
- G 46-76(1986)—Recommended Practice for Examination and Evaluation of Pitting Corrosion
- G 50-76(1984)—Recommended Practice for Conducting Atmospheric Corrosion Tests on Metals
- G 78-89—Guide for Crevice Corrosion Testing of Iron Base and Nickel Base Stainless Alloys in Seawater and Other Chloride-Containing Aqueous Environments
- G 82-83(1989)—Guide for Development and Use of a Galvanic Series for Predicting Galvanic Corrosion Performance
- G 96-90—Practice for On-Line Monitoring of Corrosion in Plant Equipment (Electrical and Electrochemical Methods)
- G 104-89—Test Method for Assessing Galvanic Corrosion Caused by the Atmosphere

Integrity Testing. Once a tank system is installed, periodic integrity testing is a necessary part of the operation and maintenance program. A guide for selecting leak-testing methods for tank and material testing and other applications is presented in ASTM E 432-91. In addition, commonly used nondestructive test methods are addressed in the following resources (see References):

- ASTM, *Annual Book of ASTM Standards*, "Nondestructive Testing," vol. 3.03
- American Society of Mechanical Engineers (ASME), *ASME Boiler and Pressure Vessel Code*, Section V, "Nondestructive Testing"
- API, *Design and Construction of Large, Welded, Low-Pressure Storage Tanks*, Standard 620, 8th ed.
- EPA, *Detecting Leaks: Successful Methods Step by Step*, EPA/530/UST-89/012

Most methods for tank system testing are applicable to aboveground tanks. Tests for underground tanks are limited because of inaccessibility.

Holiday Test. The holiday test, defined in ASTM G 62-87, is used to detect pinholes, voids, or small faults that allow current drainage through protective coatings on steel pipe or polymeric precoated corrugated steel pipe. Although this test method defines the holiday test for pipeline coatings, it may also be used to test the walls and bottom of a coated tank. A highly sensitive electrical device is used in conjunction with water or another electrically conductive wetting agent to locate pinholes and thin spots in coatings (defined as holidays). If electrical contact is made on the metal surface through a holiday, an alarm is activated to alert the operator of the coating flaw.

There are two methods defined for performing the test, based on the thickness of the coating. For thin-film coatings of 1 to 20 mils, a low-voltage holiday detector is used, which has an electrical energy source of less than 100 V d-c. This method detects pinholes and other voids, but does not detect thin spots in the coating. Where coating thicknesses are greater than 20 mils, a high-voltage detector that has an energy source of 900 to 20,000 V d-c can be used to detect both pinholes and thin spots in the coating.

Magnetic Particle Testing. Magnetic particle testing is used to detect cracks and other discontinuities near the surface of ferromagnetic materials. Applications include tank walls, tank bottoms, and welds. The area to be tested is cleaned and then magnetized. Magnetic particles are applied to the surface. The particles form patterns where there are disturbances in the normal magnetic field and indicate cracks or flaws. The method is sensitive to very small discontinuities. ASTM standards that relate to this method include the following:

- A 275/A 275M-90—Method for Magnetic Particle Examination of Steel Forgings
- E 125-63(1985)—Reference Photographs for Magnetic Particle Indications on Ferrous Castings
- E 709-80(1985)—Practice for Magnetic Particle Examination

Ultrasonic Testing. Ultrasonic testing is defined in ASTM E 1002-86 and in the ASME Code, Section V, Article 5. This test can be used to locate pressurized gas leaks and estimate leak rates. In general, it is considered a screening tool to be used prior to other more sensitive and time-consuming tests. This test method uses an acoustic leak detection system to detect impulsive signals that are much stronger than background noise level. The ultrasonic test system provides for detection of acoustic energy in the ultrasonic range and translates energy into an audible signal that can be heard through the use of speakers or earphones. The detected energy is indicated on a meter readout. Leak rates can be approximated from a formula that uses the maximum detection distance at calibrated sensitivity.

Acoustic Emission Testing. This test method can be used to monitor vessels and piping during operation to detect defects such as flaws and cracks. The tank or pipe is put under pressure, which results in a stress concentration, causing the defect to enlarge. This enlargement generates sound vibrations that can be detected by sensors located along the surface of the structure. Arrival times at the sensors are used to pinpoint the defect. For proper evaluation of the acoustical vibrations, the tank should be in quiescent state. ASTM standards that pertain to this test include:

- E 750-88—Practice for Measuring Operating Characteristics of Acoustic Emission Instrumentation
- E 976-84(1988)—Guide for Determining the Reproducibility of Acoustic Emission Sensor Response
- E 1067-89—Practice for Acoustic Emission Testing of Fiberglass Reinforced Plastic Resin (FRP) Tanks/Vessels
- E 1139-87—Practice for Continuous Monitoring of Acoustic Emission from Metal Pressure Boundaries
- E 1211-87—Practice for Leak Detection and Location Using Surface-Mounted Acoustic Emission Sensors

Hydrostatic Leak Testing. Hydrostatic testing, documented in ASTM E 1003-84(1990) and API Standard 620, is a method for testing tanks with a pressurized liquid. The test requires that a component be filled completely with a liquid, preferably water. Pressure is applied slowly to the liquid until the required pressure, usually between 75% and 150% of the design operating pressure, is reached. The pressure is held for a designated period of time. Leakage can be determined by visual inspection or pressure drop. Because liquid may clog small leaks, this method of testing is performed after pneumatic testing. In addition, the test liquid temperature must be equal to or above ambient temperature; otherwise, condensation can form on the outside of the tank, making visual detection of a leak difficult.

Pneumatic Pressure Testing. Pneumatic pressure testing is described fully in the ASME Code. The method outlines a procedure for testing tanks and metal piping with air pressure. The test is not appropriate for fiberglass or other low-pressure materials. The empty tank or pipe is pressurized to the design operating pressure and held for a specified period of time, during which any drop in pressure is noted. Because air at high pressure can be an explosive hazard if a contaminant such as oil vapor is present, inspections should be made at a reasonable distance from the tank, with the use of field glasses, as required.

Liquid Penetrant Testing. Liquid penetrant test methods and practices are described in several ASTM standards and in the ASME Code, Section V, Article 6. ASTM standards include:

- E 165-91—Practice for Liquid Penetrant Inspection Method
- E 433-71(1985)—Reference Photographs for Liquid Penetrant Inspection
- E 1208-91—Test Method for Fluorescent Liquid Penetrant Examination Using the Lipophilic Post-Emulsification Process
- E 1209-91—Test Method for Fluorescent Penetrant Examination Using the Water-Washable Process
- E 1210-91—Test Method for Fluorescent Penetrant Examination Using the Hydrophilic Post-Emulsification Process
- E 1219-91—Test Method for Fluorescent Penetrant Examination Using the Solvent-Removable Process
- E 1220-91—Test Method for Visible Penetrant Examination Using the Solvent-Removable Process

Liquid penetrant testing allows for visible examination for penetrants, so as to detect discontinuities such as cracks, opening in seams, and isolated porosity. The test involves spreading a liquid penetrant, typically a light oil with visible or fluorescent dyes in it, over the surface to be tested. The penetrant is given time to enter open discontinuities. Excess penetrant is removed, and the liquid in any discontinuity is drawn out. The discontinuity and near surfaces are stained by the penetrant in the process. The surface can then be visually inspected for indications of surface discontinuities.

Partial-Vacuum Tests. Partial vacuum tests on closed tanks are defined in API Standard 620. These tests are performed to ensure that the tank walls and roof meet design specifications. During the test, water is withdrawn from the tank with all vents closed until the design partial vacuum is developed at the top of the tank. Observations are made as to when the vacuum relief valves start to open. These valves should open before the design pressure is reached. These tests are performed with the tank full, half full, and empty. Partial-vacuum testing using a vacuum box can be performed on welds of a tank bottom that rests directly on the ground. This test is accomplished by applying a solution film at the joints and pulling a partial vacuum of at least 3 psig. For this application, the vacuum box must have a transparent top.

Tank Gauging. Manual tank gauging, commonly called static testing, is defined in EPA's guidance manual for release detection (EPA 1989). The test is effective for small-volume tanks of less than 550 gallons. The liquid level is measured in a quiescent tank at the beginning and end of a 36-hour period or other specified length of time. Any change in liquid level can be used to calculate the change in volume. Unless dramatic temperature changes occur, the liquid level change can be compared against established guidelines to determine whether any differences in the measurements are significant enough to indicate a leak. Automatic gauging systems can be permanently installed in tanks to provide both tank integrity testing and inventory information. The system can measure the change in product level within the tank over time and can detect drops in level not associated with tank withdrawals.

Volumetric Tank Testing. This test, defined in EPA's guidance manual (EPA 1989) is applicable mainly to underground storage tanks. A known volume of liquid, usually water, is placed in the tank for a period of time. The tank level is monitored for any changes that might indicate leakage. For maximum sensing of level change, the tank should be overfilled so that the liquid reaches the fill tube or standpipe located above grade. Because the level changes occur in a small area, small changes in volume can be detected readily with gauging equipment. Temperature variations must be taken into account during testing because they can affect the volume. In addition, structural deformation resulting from filling the tank can occur, so a waiting period must be observed to ensure conditions have stabilized.

Materials of Construction

In selecting material for a tank system, chemical compatibility and material cost are the two key considerations. Material selection must be made carefully for each application and must include expected variations in chemical or waste solutions and storage temperature and pressure ranges. Other factors to consider include normal atmospheric conditions, as well as hazardous climatic conditions such as the possibility of earthquakes and hurricanes.

Materials selected for the primary tank should be as optimum as the budget will allow. To keep costs down, the secondary containment, which is used only in emergencies, can often be made of less expensive materials.

Material selection for tank systems must be based on the parameters of the individual application. Test data for compatibility of specific chemicals with various metal and nonmetallic materials have been compiled by the National Association of Corrosion Engineers (NACE) in two volumes entitled *Corrosion Data Survey* (NACE 1975, NACE 1985). Included in metal tests are:

- Iron-based metals such as carbon steel and stainless steel;
- Copper-based metals and alloys;
- Nickel alloys such as nickel-chromium-iron and nickel-chromium-molybdenum; and
- Other metals and alloys including aluminum, silver, or titanium.

Nonmetal tests cover materials such as carbon, glass, synthetic and natural rubber, epoxy fiberglass and other fiberglass materials, and some plastics such as polyethylene, polypropylene, and polyvinyl chloride. The temperature range for use of these materials is generally 70°F to 140°F.

A general overview of types of chemical families that are typically used in industry is given in the following paragraphs, including a general discussion of compatibility requirements for these chemical types, based on NACE data. Both metals and nonmetallic materials are addressed. All material selections should be investigated thoroughly and/or tested before being put into service.

Acids. The iron-based metals—cast iron, carbon steel, and stainless steel—are not recommended for weak acids. Austenitic stainless steels, such as AISI 316 and 317, and several nickel-based alloys with molybdenum are suitable for some acids at varying concentrations and at low temperatures (<200°F). Other alloys and metals, such as gold and platinum alloys, silver, tantalum, titanium, and hastelloy, are also resistant to specific acids at somewhat higher temperatures. These alloys and metals typcially are not used in large applications because of the expense. Gaskets, valves, and other small but critical parts, however, often are made of these stronger materials because fittings and joints are particularly subject to chemical attack.

Acid-compatible nonmetals include polyester-fiberglass, glass-lined steel, and synthetic (butyl and fluorine) and natural rubber; selection of these materials depends on the application. Epoxy fiberglass and carbon are also compatible for some applications. Plastics such as polyethylene and polypropylene are adequate for use with some weak acids. These materials are not recommended for use with high concentrations of strong acids such as sulfuric acid, nitric acid, and hydrofluoric acid.

Alcohols. Carbon and stainless steel, cast iron, aluminum, and other metal alloys are compatible with most alcohols at low temperatures. Synthetic or natural rubber can sometimes be applied with success, but the type of rubber varies according to specific alcohol type. If the rubber is incompatible, the alcohol will dissolve it. Isopropyl alcohol is most compatible with fluorine rubber, but butyl and natural rubber can also be used at low temperatures. Other acceptable nonmetals vary significantly according to type of alcohol.

Aldehydes. Generally, stainless steels of AISI 304 and higher designations are compatible with aldehydes, depending on the specific chemical. Alloys such as nickel-based alloys and copper-based alloys can also be used successfully in most applications. Nonmetallic materials are not as suitable. Furfuryl alcohol-glass, glass-lined steel, and epoxy-asbestos-glass are compatible for various applications, but material costs are generally prohibitive. Compatibility with other nonmetals varies. Plastics such as polypropylene soften at higher concentrations and temperatures. Applications for synthetic and natural rubber are very limited.

Ammonium Solutions. Ammonium solutions are not compatible with copper-based metals. Stainless steel and nickel-based alloys are compatible with most ammonium solutions. Many solutions are compatible with aluminum. Synthetic and natural rubber are resistant to almost all ammonium solutions at low temperatures. Glass-lined steel is resistant to the more aggressive solutions, such as ammonium fluoride. Polychloroprene, polyethylene, and polypropylene are compatible for many less aggressive solutions. Ammonia is compatible with most metals and is commonly handled in carbon steel. Urea is compatible with AISI 304 stainless steel.

Caustics. Carbon steel and AISI 304 and 316 stainless steel are acceptable for most caustic solutions, particularly at low temperatures. Nickel and nickel-based alloys can also be used. Natural and butyl rubbers are generally compatible with caustics. Other materials such as polychloroprene, polypropylene, epoxy fiberglass, and polyvinyl chloride also are generally acceptable.

Petroleum Distillates and Off-Shore Applications. Most petroleum distillates are compatible with a wide variety of metals, with carbon steel being the preferred material for tank systems throughout a petroleum facility. Synthetic and natural rubbers generally are inadequate. Polyethylene and polypropylene also are inadequate because the aromatic hydrocarbons tend to cause varying degrees of material swelling, softening, and stress cracking. Some glass, epoxy, and fiberglass materials have proven compatible, but petroleum distillate tanks are normally too large for these materials from a structural standpoint. Offshore piping and other applicatons must be resistant to seawater; cupronickel alloys are compatible for these uses.

CONTAINERS

Hazardous Waste

Hazardous waste that is managed in containers is subject to either the standards in Subpart I of 40 CFR §265 (interim status units) or those of 40 CFR §264 (permitted units). The main difference between the two types of standards is that permitted units are subject to additional standards for containment areas and closure. Because they are more extensive, this section discusses only the requirements for permitted units.

Types of Containers. Containers used to manage hazardous waste must be made of materials that do not react with and are compatible with the waste. This requirement also applies to container liners.

Containers and container liners are considered "empty" of hazardous waste and are not subject to the container standards if:

- All material has been removed that can normally be removed for that type of material, and

- There is less than one inch of residue on the bottom.

Or, alternatively:

- If the container volume is 110 gallons or less, the residue is no more than 3% by weight of the total container capacity, or
- If the container volume is more than 110 gallons, the residue is no more than 0.3% by weight of the total container capacity.

If the container held a compressed gas, it is considered empty if the inside pressure is nearly atmospheric. If the waste in the container was an acute hazardous waste, it is considered empty if:

- The container has been triple-rinsed with a solvent capable of removing such material,
- Another equivalent cleaning method has been used, or
- The liner has been removed. In this case, only the outer container is considered empty of hazardous waste.

Even though a container may be classified as "empty" for regulatory purposes, the residual material in the container may still present a hazard. Containers storing such materials can be labeled with instructions on handling or cleaning when empty. An example of such a label is shown in Figure 7.8.

Storage Area Design. The design requirements for hazardous waste container storage areas depend on the type of waste. There are more requirements if the waste is a liquid, contains free liquids, or is one of certain listed hazardous wastes containing dioxins (F020, F021, F023, F026, and F027). For wastes of this type, the container storage area must have an imprevious base and secondary containment. The base must be sloped or otherwise designed to drain and collect liquids from leaks, spills, and precipitation, unless the containers themselves capture spills from liquids (for example, by elevating them above the base). There must be no cracks or gaps in the base. The secondary containment system must be large enough to hold 10% of the total volume in the containers or the volume of the largest container, whichever is the greater. Storm water must be prevented from entering the storage area unless the storage area has the capacity to contain such waters. If there are no wastes containing free liquids or the previously mentioned F-listed wastes stored in the area, the only requirements are that the area must be sloped or otherwise designed to drain and remove precipitation, or the containers must be elevated to prevent contact with such.

Waste Management. Incompatible wastes and materials cannot be placed in the same container because of the danger of violent reactions, explosions, fire, or container damage or corrosion. Containers must be washed before reuse if the waste to be stored is incompatible with the previously stored waste. Incom-

WARNING

THIS CONTAINER DANGEROUS WHEN EMPTY

MAY HAVE RESIDUAL CHEMICALS

HANDLE WITH CARE

DO NOT REUSE UNTIL THOROUGHLY CLEANED

DO NOT STORE NEAR HEAT OR FLAMES

ECWD66 © Emed Co.. Bflo.. NY 14240

Figure 7.8 Example label for empty container.

patible wastes stored nearby, in other containers or storage units, must be separated or protected from contact with each other by a wall, dike, berm, or other device. Ignitable or reactive wastes must be at least 50 feet from the facility's property line.

Containers must be kept closed at all times, except when waste is being added or removed. Containers must be handled properly so that they are not damaged or pierced. The storage area and containers must be inspected at least once a week to check for leakage or deterioration. If a leaking container or one in poor condition is found, the waste must be transferred to another container or managed in some way to satisfy the standards.

Air emissions must be controlled in accordance with Subpart CC of §264. Hazardous wastes not subject to these requirements are those whose average volatile organic concentration at the point of generation is less than 100 parts per million by weight (ppmw) or whose organic content has been reduced by a process that meets certain specific requirements. In addition, containers that are 0.1 m^3 (approximately 26 gallons) or smaller are not subject to Subpart CC.

When the storage area is no longer in use, it must be closed. All hazardous waste and residues must be removed. Containers, liners, bases, and soil contaminated with hazardous waste must be decontaminated or removed.

Secondary Containment

Secondary containment of portable chemical and waste containers both during storage and when in use is becoming a widespread practice in industry. Containment options vary widely, but 100% inspectable/testable secondary containment typically is not achievable. For most drum storage areas, such as container storage rooms and chemical distribution centers, curbed and coated concrete can be used. For these applications, a coating should be selected not only for chemical compatibility, but also for its ability to withstand fork truck and other equipment loadings. Physical inspections of the coating should be made regularly to ensure that cracking or delaminating has not occurred.

Smaller staging areas for chemical and waste containers can use bermed and coated concrete or a containment pan. For these applications, fork truck loadings are not usually a factor. The containment area should be large enough to contain, at a minimum, spillage from the largest container.

Routine barrel or drum pumping operations should be contained using, as a minimum, a drip pan under the barrel and, preferably, a pan sized to contain at least the contents of one barrel. Relatively inexpensive open-grated portable containment pallets are available that hold up to four drums and have enough spill capacity to easily contain a drum failure. These pallets can be moved with a fork truck. Another portable containment unit that holds one or two drums and is moved easily from one job site to another is also available. Specially designed and very durable, portable chemical containers that hold approximately 220 to 440 gallons (four to eight 55-gallon drums) are available as well. Although spill containment per se is not part of the container design, some containers are enclosed with a metal casing to ensure that dropping or excessive bumping does not crack the container. This design can reduce the risk of spills during transport.

CONTAINMENT BUILDINGS

Containment buildings that are used to manage hazardous waste are subject to 40 CFR §265 standards if they are 90-day units or interim status units (awaiting a hazardous waste permit). Containment buildings that operate under a hazardous waste permit are subject to 40 CFR §264. The requirements are essentially the same for both Part 265 and Part 264.

Design

The containment building must have a floor, walls, and a roof to completely enclose the waste so as to prevent exposure to weather. Floors and walls must be sturdy enough to prevent collapse of the building and to withstand the rigors of day-to-day operation with people and equipment. Building materials must be chemically compatible with any waste they contact. Industry standards, such as those from the American Concrete Institute (ACI) and ASTM, are used by EPA to judge the structural integrity of the building. Prior to starting operations in the building, the building design must be certified by a qualified registered professional engineer.

A containment building must have a primary barrier. If the waste contains free liquids or is treated with free liquids, the primary barrier must be designed to prevent the migration of hazardous constituents into the barrier, for example, by using a geomembrane covered with a concrete surface for wear protection. The floor must be sloped so that any liquids falling on the barrier will drain to a collection system where they can be removed. When liquids are being handled in the building, there must also be a secondary containment system, including a secondary barrier with leak detection system. The leak detection system must be constructed of at least 12 inches of granular material with a minimum hydraulic conductivity of 1×10^{-2} cm/sec, or of a geonet drainage material with a minimum transmissivity of $3 \times 10^{-5} m^2$/sec. The bottom slope of the leak detection system must be at least 1%. Secondary containment may be waived if the only free liquids that are used are limited amounts for dust suppression, used to meet occupational health and safety requirements. Under certain conditions, a containment building can serve as secondary containment or as an external liner system for tanks placed within.

Operation

The primary barrier of a containment building must be maintained to be free of large cracks or gaps, corrosion, or other deterioration. Hazardous waste cannot be placed higher than any containment wall. The tracking of waste outside the building must be minimized, and there must be an area for equipment decontamination and rinsate collection. There cannot be any visible fugitive dust emissions through openings such as doors, windows, vents, or cracks. Waste treatment must be conducted so that liquids, wet materials, or liquid aerosols are not released to other parts of the building.

At least once every seven days, inspections must be conducted of monitoring equipment, the leak detection system, the building itself, and the area immediately outside the building for any evidence of hazardous waste leaks. Inspections must be recorded in the operating record.

When a condition is detected that could cause or has caused a release of hazardous waste, it must be corrected promptly. The problem must be recorded in the operating record, and the affected area must be immediately removed from service. A schedule identifying specific actions to correct the problem must be established. The owner or operator must notify EPA of the problem within 7 days and, within 14 working days, provide EPA with the written schedule to correct the problem. Correction of the problem must be verified to EPA in writing, signed by a qualified registered professional engineer.

REFERENCES

American Petroleum Institute, "Design and Construction of Large, Welded, Low-Pressure Storage Tanks," 8th ed., API Standard 620, Washington, DC, 1990.

American Petroleum Institute, "Recommended Practice for Bulk Liquid Stock Control at Retail Outlets," 4th ed., API RP 1621, 1987.

American Society of Mechanical Engineers, *ASME Boiler and Pressure Vessel Code*, Section V, "Non-Destructive Testing," New York, 1986.

American Society for Testing and Materials, "Nondestructive Testing," *Annual Book of ASTM Standards*, Vol. 3.03, Philadelphia, 1992.

American Society for Testing and Materials, "Wear and Erosion: Metal Corrosion," *Annual Book of ASTM Standards*, Vol. 3.02, Philadelphia, 1992.

Langlois, Kelvin E., Chris Bauer, and Gayle Woodside, "Double Contained Wastewater Treatment Tanks," paper presented at the Water Pollution Control Federation Annual Conference, Washington, DC, October 1990.

National Association of Corrosion Engineers, "Control of External Corrosion on Metallic Buried, Partially Buried, or Submerged Liquid Storage Systems," NACE Standard RP0285-85, Houston, TX, 1985.

National Association of Corrosion Engineers, *Corrosion Data Survey*, Metals Section, 6th ed., Houston, TX, 1985.

National Association of Corrosion Engineers, *Corrosion Data Survey, Nonmetals Section*, 5th ed., Houston, TX, 1975.

U.S. Environmental Protection Agency, "Detecting Leaks: Successful Methods Step by Step," EPA/530/UST-89/012, Washington, DC, 1989.

BIBLIOGRAPHY

Arbuckle, J. Gordon, et al (1991), *Environmental Law Handbook*, Government Institutes, Inc., Rockville, MD.

ASME (1986), *ASME Boiler and Pressure Vessel Code*, Section VIII, "Rules for Construction of Pressure Vessels," American Society of Mechanical Engineers, New York.

ASME/ANSI (1987), "Chemical Plant and Petroleum Refinery Piping," ASME/ANSI Standard B31.1, American Society of Mechanical Engineers, New York, and American National Standards Institute, New York.

ASTM (1991), *Acoustic Emission: Current Practice and Future Directions*, Special Technical Publication 1077, Sachse, Roget, and Yamaguchi, eds., American Society of Testing and Materials, Philadelphia.

———— (1990), *Corrosion Testing and Evaluation: Silver Anniversary Volume*, Special Technical Publication 1000, Baboian and Dean, eds., American Society of Testing and Materials, Philadelphia.

———— (1989), *Effects of Soil Characteristics on Corrosion*, Special Technical Publication 1013, Chaker and Palmer, eds., American Society of Testing and Materials, Philadelphia.

———— (1988), *Galvanic Corrosion*, Special Technical Publication 978, H.P. Hack, ed., American Society of Testing and Materials, Philadelphia.

Cole, Mattney G. (1992), *Underground Storage Tank Installation and Management*, Lewis Publishers, Boca Raton, FL.

De Renzo, D.J. (1985), *Corrosion Resistant Materials Handbook*, 4th ed., Noyes Data Corporation, Park Ridge, NJ.

Ecology and Environment, Inc., and Whitman, Requardt, and Associates (1985), *Toxic Substance Storage Tank Containment*, Noyes Data Corporation, Park Ridge, NJ.

Gangadharan, et al. (1988), *Leak Prevention and Corrective Action for Underground Storage Tanks*, Noyes Data Corporation, Park Ridge, NJ.

Jawad, Maan H., and James R. Farr (1989), *Structural Analysis and Design of Process*, John Wiley & Sons, New York.

LeVine, Richard, and Arthur D. Little, Inc. (1988), *Guidelines for Safe Storage and Handling of High Toxic Hazard Materials*, American Institute of Chemical Engineers, Center for Chemical Process Safety, New York.

NACE (1975), "Control of Internal Corrosion in Steel Pipelines and Piping Systems," National Association of Corrosion Engineers, NACE Standard RP0175, Houston, TX.

Rizzo, Joyce A. (1991), *Underground Storage Tank Management: A Practical Guide*, Government Institutes, Rockville, MD.

Rizzo, Joyce A., and Albert D. Young (1990), *Aboveground Storage Tanks: A Practical Guide*, Government Institutes, Rockville, MD.

Schwendeman, Todd G., and H. Kendall Wilcox (1987), *Underground Storage Systems: Leak Detection and Monitoring*, Lewis Publishers, Boca Raton, FL.

U.S. EPA (1989), "Volumetric Tank Testing: An Overview," EPA/625/9-89/009, U.S. Environmental Protection Agency, Washington, DC.

_____ (1989), "Soil Vapor Monitoring for Fuel Tank Leak Detection: Data Compiled for Thirteen Case Studies," prepared by On-Site Technologies for Environmental Monitoring Systems Laboratory, U.S. Environmental Protection Agency, Washington, DC.

_____ (1988), "Standard Practice for Evaluating Performance of Underground Storage Tank External Leak/Release Detection Components and Systems," prepared by Radian Corporation for Environmental Monitoring Systems Laboratory, U.S. Environmental Protection Agency, Washington, DC.

_____ (1988), "Review of Effectiveness of Static Tank Testing," prepared by Midwest Research Institute for the Office of Underground Storage Tanks, Office of Solid Waste and Emergency Response, U.S. Environmental Protection Agency, Washington, DC.

_____ (1988), "Evaluation of Volumetric Leak Detection Methods for Underground Fuel Storage Tanks," vol. 1, EPA/600/2-88/068a, prepared by Vista Research, Inc., for U.S. Environmental Protection Agency, Washington, DC.

_____ (1988), "Common Human Errors in Release Detection Usage," prepared by Camp Dresser & McKee, Inc., for U.S. Environmental Protection Agency, Washington, DC.

_____ (1988), "Analysis of Manual Inventory Reconciliation," prepared by Midwest Research Institute for the Office of Underground Storage Tanks, Office of Solid Waste and Emergency Response, U.S. Environmental Protection Agency, Washington, DC.

_____ (1987), "Soil-Gas Measurement for Detection of Subsurface Organic Contamination," Environmental Monitoring Systems Laboratory, U.S. Environmental Protection Agency, Washington, DC.

_____ (1986), "Underground Tank Leak Detection Methods: A State-of-the-Art Review," EPA/600/2-88/001, prepared by IT Corporation for Hazardous Waste Engineering Research Laboratory, Office of Research and Development, U.S. Environmental Protection Agency, Washington, DC.

Whitlow, R. (1990), *Corrective Response Guide for Leaking Underground Storage Tanks*, Government Institutes, Rockville, MD.

Woodside, Gayle (1993), *Hazardous Materials and Hazardous Waste Management: A Technical Guide*, John Wiley & Sons, New York.

_____ (1991), "Aboveground Storage: Double Containment Strategies for Today and Tomorrow," *Proceedings from Anuual Air and Waste Management Association Annual Conference*, Vancouver, June 1991.

Woodside, Gayle, and John J. Prusak (1992), "Aboveground Storage: State-of-the-Art Systems," *Proceedings from Annual Air and Waste Management Association Annual Conference*, Kansas City, June 1992.

8

STATISTICAL APPLICATIONS

Statistical analysis is a common element of the data-oriented environmental field. For example, statistics are used in verifying site cleanup, setting wastewater permit limits, assessing ground water compliance, projecting treatment system performance, and judging analytical and laboratory performance, to name but a few areas. This chapter describes some of the most common statistical techniques used in the environmental field and demonstrates their use through practical examples. The equations behind these techniques are also provided to improve the understanding and use of graphics and statistical tools in computer programs.

BASIC STATISTICS

Choosing a Distribution

When one speaks of the distribution of a data set, usually one means how the data plot in comparison with well-known statistical functions. It is not always necessary, but it is helpful to plot data before picking a distribution since the shape of the data can suggest which distribution to use as well as highlight anomalies that would have significant effects on statistical calculations. For example, the normal distribution looks like a bell-shaped curve when plotted as a histogram, and as a straight line on a probability plot.

Histograms. Histograms can be used to decide which statistical function fits a set of data. A histogram is a plot of frequency of occurrence over the range of data. Histograms may also be referred to as frequency diagrams or distributions. To prepare a histogram, first divide the range of data values into intervals. Next, count the number of data points that fall within each interval. For example, if the data range from 0 to 479, divide the data into 10 intervals (the number of intervals depends on the number of points and the degree of refinement desired in the histogram), starting with 0 to 49, then 50 to 99, and so on. Plot each interval as a frequency count on the vertical axis; the horizontal axis will show the range of data. Alternatively, the frequency can be plotted as a percentage by dividing the interval count by the total number of data points.

EXAMPLE 8.1
Plotting a Histogram

The organic strength and biodegradability of wastewaters are often described as biochemical oxygen demand (BOD). Plot the following BOD data for a wastewater effluent in a histogram. Based on the shape of the histogram, what distribution might you use to fit the data?

BOD (mg/L)

5	7	8	8	9	9	10	10	11
12	12	12	13	14	14	15	15	15
16	16	16	17	17	17	18	18	18
18	19	19	20	20	21	21	22	22
22	23	23	24	24	24	24	25	25
26	26	26	27	27	28	28	30	31
33	34	34						

Solution. The data range from 5 to 34 mg/L. A convenient interval spacing would be 5; for example, 5 to 9, 10 to 14, 15 to 19, and so on. The number and percentage of data points in each interval are:

Interval	5–9	10–14	15–19	20–24	25–29	30–35
Count	6	9	15	13	9	5
%Total	10%	16%	26%	23%	16%	9%

The histogram is plotted in Figure 8.1. Note the vertical axis is scaled for both frequency count and percentage.

Probability Plots. Two of the most common distributions are the normal distribution and its close relation, the lognormal distribution. These two distributions appear frequently in environmental data. Therefore, the discussion in this section focuses on these distributions.

If the data are likely to follow a normal-type distribution, the data should be plotted in a probability plot. If the data in a probability plot fall in a fairly straight line, the normal distribution is a good fit. Figures 8.2 and 8.3 show two sets of data in probability plots. The data in Figure 8.2 have an S-curve shape and, therefore, do not fit the normal distribution well. The data in Figure 8.3 plot as a staight line and, therefore, fit the normal distribution well.

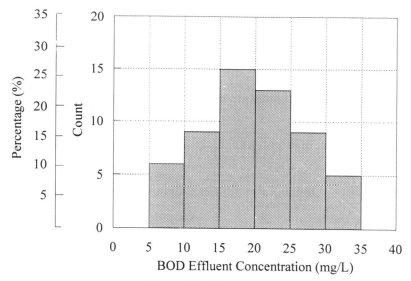

Figure 8.1 Histogram.

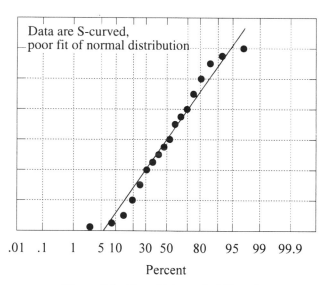

Figure 8.2 Poor fit in probability plot.

To draw a probability plot, first rank the data from least to greatest, that is, from $i=1$ to n. Then the percentile rank, P, for each point is calculated by Equation 8.1. Each data point is plotted versus percentile in the probability plot.

$$P = 100(i - \frac{1}{2})/n \qquad\qquad (8.1)$$

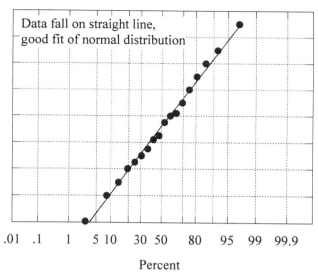

Figure 8.3 Good fit in probability plot.

EXAMPLE 8.2
Probability Plot to Test Normal Distribution

Prepare a probability plot of the influent wastewater data that follow. Based on the shape of the plot, would you use the normal distribution for these data?

Benzene Concentration in Influent Wastewater (mg/L)

0.67	0.33	1.47	0.52	0.43	1.88	0.25	0.30	0.73
0.31	0.42	0.34	0.31	1.10	0.57	0.46	0.80	0.59
0.32	0.35	1.41	0.56	0.55	0.36	0.67	0.22	0.40
0.90	0.73	0.27	0.43	0.79	0.37	0.42	0.50	0.36
0.39	0.32	0.30	0.21	0.19	0.88	0.38	0.21	0.16
0.47								

Solution. The data are ranked and their percentiles are calculated using Equation 8.1. The data and percentiles are plotted in a probability plot in Figure 8.4. Because the data do not fall even roughly in a straight line (note the large degree of curvature), the normal distribution would not be used with these data. However, transformation of the data by the logarithmic function may help straighten out the curvature and produce a normal distribution. If the data are lognormally distributed, they will plot as a straight line if the vertical axis of a probability plot is changed to log-scale.

EXAMPLE 8.3
Probability Plot with Lognormal Data

Plot the data in Example 8.2 in a probability plot using a log-scale for the vertical axis. Would the lognormal distribution be a good fit for these data?

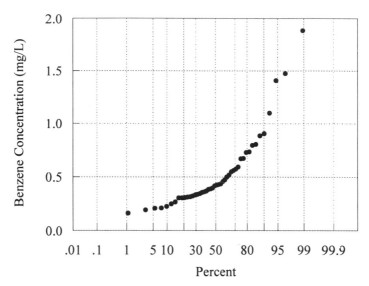

Figure 8.4   Probability plot of benzene data.

Solution. A log-scale probability plot of the data is prepared with the same percentiles calculated in Example 8.2. By changing the vertical axis to log-scale, you can avoid taking the logarithms of the data values. Based on the probability plot in Figure 8.5, the data fit the lognormal distribution better because they fall in a straight line on the probability plot using log-scale.

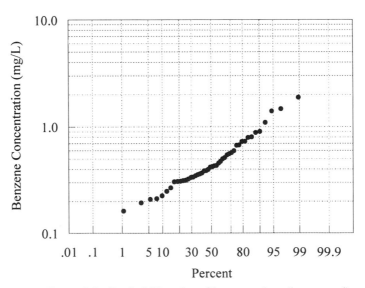

Figure 8.5   Probability plot of benzene data (lognormal).

Normality Tests. There are a number of statistical procedures for testing how well the normal distribution fits a set of data. These procedures include the chi-squared test, the Kolmogorov-Smirov test, and the Shapiro-Wilk's test (EPA 1989). All three of these tests require fairly large sample sizes (20 for the chi-squared test and 50 for the other two tests). For most environmental data, however, it is sufficient to determine the goodness-of-fit with a simple probability plot that does not require a minimum number of samples and is more intuitive than calculations made with normality tests.

Deviations from normality that are seen in probability plots are reflected in the coefficient of variation, a parameter that can be used as a quick numerical test for normality. When data are initially analyzed, it is common to calculate the mean and standard deviation. The coefficient of variation is then calculated from the mean and standard deviation (see Equations 8.7, 8.9, and 8.10 later in this chapter). As a general rule of thumb, a normal distribution should have a coefficient of variation less than 100% (EPA 1989). Values greater than 100% indicate a large degree of variability in the data that may be caused by outliers or extreme values that do not fit the normal distribution well. Limiting the coefficient of variation to 100% also assures that values in the distribution are greater than zero, which is the general case for environmental data.

Data Transformations. There are many statistical tests available for data when they fit the normal distribution. That is why the normal distribution is usually the first choice when trying to fit environmental data. If the data do not fit a normal distribution, they can sometimes be transformed so that they do. A special case in point is the lognormal distribution. The data are transformed by taking their logarithms. Usually, the natural or base e logarithm is used, written as ln. This transformation is so common with many data types that it has its own special name, that is, the lognormal distribution.

There are other functions that can be used to transform data so that they will fit the normal distribution. The following are examples of common transformation functions, showing x' as the transformed value of the original value, x.

Logarithm (natural, base e)

$$x' = \ln x \tag{8.2}$$

Logarithm (base 10)

$$x' = \log_{10} x \tag{8.3}$$

Reciprocal or inverse

$$x' = \frac{1}{x} \tag{8.4}$$

Square root

$$x' = \sqrt{x} \tag{8.5}$$

Nonparametric Method. There will be times when the data do not fit a convenient distribution such as the normal distribution or any other "parametric" distribution. A parametric distribution is one that is defined by a function with parameters (variables). For example, the parameters of a normal distribution are its mean and standard deviation. These two parameters can be used to calculate the value of the distribution at any percentile point, such as the 95th percentile.

The nonparametric method is not defined by functions with parameters. Instead, percentile values are calculated directly from the ranked data.

EXAMPLE 8.4
Percentile Values using Nonparametric Method

In the following ranked data set, what is the 90th percentile value?

Data	21	32	39	45	57	59	65	71	83	97
Rank	1	2	3	4	5	6	7	8	9	10

Solution. Multiply the percentile by the number of data points to find the rank of the closest data point that matches the percentile.

$$0.90(10) = 9\text{th value out of } 10$$

The 90th percentile value is 83. This is, of course, a very simple example. In a larger data set, if the calculated rank is not a whole number, you would round the rank to the nearest whole number.

Nonparametric methods can be used when environmental data do not fit a normal distribution even after data transformation. Nonparametric methods can be used to set permit limits, such as for wastewater discharges, although the normal and lognormal distributions are by far the most common and most preferred methods. EPA has used the nonparametric method to propose wastewater effluent limits for a number of industries, including pulp and paper, pesticides, and pharmaceutical manufacturing (EPA 1980, EPA 1982, Eynon and Valdés 1983).

The following equation is used to calculate a percentile value at a 50% confidence level using a nonparametric method:

$$1 - \sum_{i=r}^{n} \frac{n!}{i!(n-i)!} p^i (1-p)^{n-i} \geqslant 0.5 \tag{8.6}$$

where n is the number of data points, r is the rank of the value in the data set representing a given percentile with a 50% confidence, and p is the pth percentile (for example, $p = 0.99$ represents the 99th percentile). Notice the difference in how the rank is calculated between Equation 8.6 and Example 8.4. The difference arises from the addition of a confidence level to Equation 8.6. In order to use Equation 8.6, there must be at least 69 values (hint: set r equal to n in Equation 8.6 and solve for n).

Descriptive Statistics

In the following paragraphs, some of the most common descriptive parameters for data sets are discussed, illustrated by example calculations. These parameters are the mean, variance, standard deviation, coefficient of variation, and median.

Mean or Average. The mean or average, $\bar{x}$, of n data points $(x_1 \ldots x_n)$ is:

$$\bar{x} = \frac{1}{n}\sum x \tag{8.7}$$

The mean of a lognormal distribution, or log-mean, is calculated the same way, using the logarithms of the data points.

EXAMPLE 8.5
Mean or Average

Calculate both the mean and log-mean of the following TSS data for a wastewater effluent.

TSS (mg/L)

| 15 | 23 | 28 | 21 | 18 | 10 | 45 | 38 |

Solution. The mean is the sum of the values divided by the number of data points.

$$\bar{x} = \frac{198}{8} = 24.75$$

To calculate the log-mean, first take the logarithm of each value. Using the natural logarithm (base e), the log-transformed data are:

ln TSS

| 2.708 | 3.135 | 3.332 | 3.044 | 2.890 | 2.302 | 3.807 | 3.638 |

The log-mean, the sum of the preceding values divided by the number of values, is

$$\bar{x}_{\ln} = \frac{24.856}{8} = 3.107$$

Variance and Standard Deviation. The variance, s^2, and standard deviation, s, of n data points are:

$$s^2 = \frac{\sum (x - \bar{x})^2}{n - 1} \tag{8.8}$$

and

$$s = \sqrt{s^2} = \left(\frac{\sum (x - \bar{x})^2}{n - 1} \right)^{\frac{1}{2}} \tag{8.9}$$

The calculations are the same for a lognormal data set, using the logarithms of the data points. For a lognormal distribution, the mean in the preceding equations is the log-mean.

EXAMPLE 8.6
Variance and Standard Deviation

Using the TSS data in Example 8.5, calculate the variance and standard deviation for both the original data and the log-transformed data.

Solution. The differences between each value and the mean are:

x	-9.75	-1.75	3.25	-3.75	-6.75	-14.75	20.25	13.25
$x_{\ln}$	-0.399	-0.028	0.225	-0.063	-0.217	-0.805	0.700	0.531

The variance of x is the sum of the differences squared, divided by $n - 1$:

$$s^2 = \frac{971.5}{7} = 138.79$$

The standard deviation of x is the square root of the variance, or 11.78. The variance of $x_{\ln}$ is:

$$s^2 = \frac{1.682}{7} = 0.2402$$

The standard deviation of $x_{\ln}$ is the square root of the variance, or 0.4901.

Coefficient of Variation and Relative Standard Deviation. The ratio of the standard deviation to the mean is called the coefficient of variation (COV).

$$COV = \frac{s}{\bar{x}}$$ (8.10)

When expressed as a percentage, the COV may also be referred to as the relative standard deviation (RSD).

The COV is a measure of the scatter in the data set and can also be used as a quick test of normality (see "Normality Tests" earlier in this chapter).

EXAMPLE 8.7
Coefficient of Variation

Calculate the COV of the TSS data in Example 8.5, first for the original data and then for the log-transformed data. Based on the COVs, which distribution (normal, lognormal) do you think would fit the data better? How would you confirm it?

Solution. The COV is the standard deviation divided by the mean. For the original, untransformed data, the COV is

$$COV = \frac{11.78}{24.75} = 0.48$$

For the log-transformed data, the COV is

$$COV = \frac{0.4901}{3.107} = 0.16$$

Because the COV of the lognormal distribution is smaller (even though the COV of the normal distribution is less than the 100% guideline), the lognormal distribution might be a better fit. The probability plots in Figures 8.6 and 8.7 confirm that the lognormal distribution is a better fit.

Median. The median of a data set is the middle value of a ranked set of data. When there is an odd number of data points, the median is exactly the middle value. For example, in a set of 5 data points, the median would be the third value. If there is an even number of data points, the median is the average of the two middle values. For example, in a set of 10 data points, the median would be the average of the 5th and 6th values.

The median can be used instead of the mean when the data set contains one or more extreme values that distort or skew the distribution. Because the median is based only on the position of the data points, not their values, it is

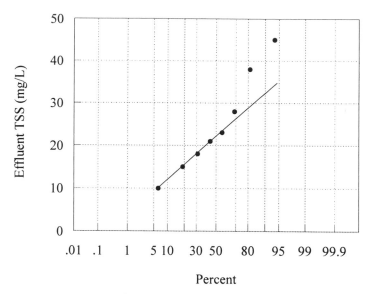

Figure 8.6   Probability plot of TSS data.

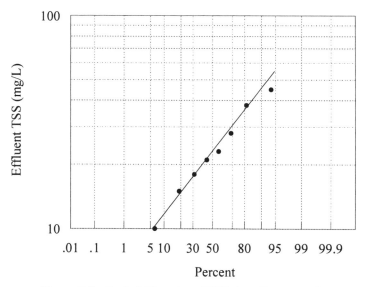

Figure 8.7   Probability plot of TSS data (lognormal).

not influenced by extreme points. This makes the median useful to determine the "average" value (in this sense, average meaning typical) without the distortion of a few anomalous values. The median is the middle value, or 50th percentile, of a data set.

EXAMPLE 8.8
Median of a Data Set

Calculate the median of the following MLVSS concentrations for a wastewater treatment plant. The data have already been ranked from least to greatest. Mixed liquor
volatile suspended solids (MLVSS) is a measure of the concentration of microorganisms in a biological treatment plant. What is a typical operating level based on
these data? Is the mean or median a better measure of a typical level?

MLVSS (mg/L)

2420	2440	2450	2480	2490	2500	2510	2520
2570	2580	2650	2670	2710	2780	2790	2800
2820	2830	2850	2870	2880	2900	2920	2940
5500	5700						

Solution. There are 26 values, so the median will be the average of the 13th and
14th values.

$$\frac{2710 + 2780}{2} = 2745$$

The mean is 2,907, which is higher than the median because of the two extreme values.
Although the difference between the mean and the median is not great in this case, the
median would be a better choice for a typical operating level because the last two values
are clearly outside the normal range of the data.

Percentile Estimates. Percentile estimates for normal distributions, such as the
95th or 99th percentile, are calculated from the mean and standard deviation,

$$x_{1-\alpha} - \bar{x} \perp t_{1-\alpha,\nu} s \qquad (8.11)$$

where t is the value of the Student or t normal distribution and α is the tail area
probability of the distribution (see Figure 8.8). The t value is a measure of how
far a percentile is from the mean, using the standard deviation as the unit of
measure. For example, the 95th percentile is 1.645 standard deviations from
the mean for large data sets. The value of α is a measure of the area under the
curve, between the "endpoint" of the distribution and a particular percentile.
For the 95th and 99th percentiles, α is 0.05 and 0.01, respectively. Both t and α
can be measured from either side of the mean. When percentiles less than 50%
are calculated, the minus sign, as in Equation 8.11 is used. The upper percentiles between the 90th and 99th are of most interest in environmental data,
however.

The value of t depends on the percentile and on the degrees of freedom, ν.
The degrees of freedom for Equation 8.11 for a data set of n values, is $n-1$.

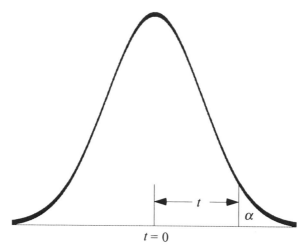

$t = 0$

Figure 8.8   Relationship of α and Student's t in the normal distribution.

Table 8.1 lists the t values for the most common values of α used with environmental data. Interpolation can be used for values that are not specifically given in the table.

EXAMPLE 8.9
95th Percentile of a Normal Distribution

Calculate the 95th percentile of the following influent wastewater flow data.

Influent Wastewater Flow (millions of gallons per day, MGD)

9.8	8.5	6.9	8.1	9.0
7.8	8.6	8.3	9.4	8.8
10.1	7.4	8.2	8.8	9.1

Solution. The data fit the normal distribution as seen in the probability plot in Figure 8.9. The mean and standard deviation are 8.59 and 0.858, respectively. From Table 8.1, t is 1.761 for $\alpha=0.05$ and $\nu=14$. The 95th percentile is

$$x_{0.95}=8.59+1.761\times0.858$$

$$=10.1$$

Notice how well the calculated percentile agrees with the value read from the probability plot. When the data fit the normal distribution well, the calculated and graphical percentile will be close.

TABLE 8.1 Values of t for Student's Distribution

	α				
ν	0.1	0.05	0.025	0.01	0.005
1	3.078	6.314	12.706	31.821	63.657
2	1.886	2.920	4.303	6.965	9.925
3	1.638	2.353	3.182	4.541	5.841
4	1.533	2.132	2.776	3.747	4.604
5	1.476	2.015	2.571	3.365	4.032
6	1.440	1.943	2.447	3.143	3.707
7	1.415	1.895	2.365	2.998	3.499
8	1.397	1.860	2.306	2.896	3.355
9	1.383	1.833	2.262	2.821	3.250
10	1.372	1.812	2.228	2.764	3.169
11	1.363	1.796	2.201	2.718	3.106
12	1.356	1.782	2.179	2.681	3.055
13	1.350	1.771	2.160	2.650	3.012
14	1.345	1.761	2.145	2.624	2.977
15	1.341	1.753	2.131	2.602	2.947
16	1.337	1.746	2.120	2.583	2.921
17	1.333	1.740	2.110	2.567	2.898
18	1.330	1.734	2.101	2.552	2.878
19	1.328	1.729	2.093	2.539	2.861
20	1.325	1.725	2.086	2.528	2.845
21	1.323	1.721	2.080	2.518	2.831
22	1.321	1.717	2.074	2.508	2.819
23	1.319	1.714	2.069	2.500	2.807
24	1.318	1.711	2.064	2.492	2.797
25	1.316	1.708	2.060	2.485	2.787
26	1.315	1.706	2.056	2.479	2.779
27	1.314	1.703	2.052	2.473	2.771
28	1.313	1.701	2.048	2.467	2.763
29	1.311	1.699	2.045	2.462	2.756
30	1.310	1.697	2.042	2.457	2.750
40	1.303	1.684	2.021	2.423	2.704
60	1.296	1.671	2.000	2.390	2.660
120	1.289	1.658	1.980	2.358	2.617
∞	1.282	1.645	1.960	2.326	2.576

Source: Information from Box et al. (1978).

Percentiles for a lognormal distribution are calculated, using Equation 8.11, with the log mean and the log standard deviation. The result is then transformed back to its original scale. If logarithms are base e, which is common for environmental data, then percentiles are calculated by:

$$x_{1-\alpha} = e^{\bar{x}_{\ln} \pm t_{1-\alpha,\nu} s_{\ln}} \tag{8.12}$$

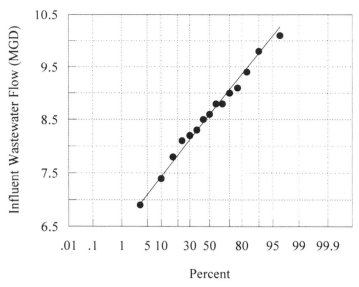

Figure 8.9   Probability plot of wastewater flow.

EXAMPLE 8.10
95th Percentile of a Lognormal Distribution

Calculate the 95th percentile value of the log-transformed data in Example 8.6.

Solution. There are 8 data points, so there are 7 degrees of freedom. The t value corresponding to 7 degrees of freedom and an α of 0.05 is 1.895. Therefore, the 95th percentile of the log-transformed data is

$$x_{0.95} = e^{3.107 + 1.895(0.4901)}$$
$$= e^{4.036}$$
$$= 57$$

Tolerance Intervals. Tolerance intervals are used to determine the range of values that represents a certain percentage of a data set. For example, saying that 95% of the data values lies between a and b is a way of stating a tolerance interval. Calculating a tolerance interval is similar to calculating percentiles, except that α is divided by 2, representing the lower and upper tails of the normal distribution (minimum and maximum tolerance limits, respectively, or x_L and x_U). Equations 8.13 and 8.14 are used to calculate tolerance intervals.

$$x_L = \bar{x} - k_{1-\alpha/2, \nu} s \tag{8.13}$$

$$x_U = \bar{x} + k_{1-\alpha/2, \nu} s \tag{8.14}$$

The tolerance factor, k, is similar to Student's t, except that k incorporates a confidence level or tolerance coefficient. A typical confidence level is 95%. For example, a 95%:90% tolerance interval means that one is 95% confident that 90% of the values will lie within the tolerance interval. Table 8.2 gives values of k for calculating tolerance intervals at a 95% confidence level. P in the table represents the percentage of the data set that lies within the tolerance interval and is equal to $1-\alpha$.

EXAMPLE 8.11
Tolerance Interval

Calculate a 95% tolerance interval at a 95% confidence level for the flow data in Example 8.9.

Solution. The k tolerance factor for a 95%:95% tolerance interval for $n = 15$ is 2.95. The upper and lower tolerance limits are 11.1 and 6.1, respectively.

$$x_U = 8.59 + 2.95 \times 0.858$$

$$= 11.1$$

$$x_L = 8.59 - 2.95 \times 0.858$$

$$= 6.1$$

Therefore, one would be 95% confident that 95% of the wastewater flows would lie between 6.1 and 11.1 million gallons per day.

TABLE 8.2   K Factors for Tolerance Interval at 95% Confidence Level

n	P / $\alpha/2$	0.95 / 0.025	0.99 / 0.005
2		37.67	48.43
3		9.92	12.86
4		6.37	8.30
5		5.08	6.63
6		4.41	5.78
7		4.01	5.25
8		3.73	4.89
9		3.53	4.63
10		3.38	4.43
15		2.95	3.88
20		2.75	3.62
25		2.63	3.46

Source: Information from Taylor (1990).

With some types of data, like ground water data, only the upper tolerance limit is of interest. In these cases, α represents only the upper tail of the distribution and k is adjusted accordingly. Values of k for a 95% confidence level are given in Table 8.3.

EXAMPLE 8.12
Upper Tolerance Limit

Calculate an upper 95% tolerance limit (one-sided) with 95% confidence for chromium with the following data from an upgradient ground water well.

Background Well, Chromium (μg/L)

3.4	1.6	4.3	2.1	2.5
1.0	4.7	2.7	1.3	1.7
3.6	5.0	3.9	3.0	4.6

Solution. The mean and standard deviation of the chromium values are 3.027 and 1.313, respectively. For $n = 15$, k is 2.566 from Table 8.3. Therefore, the upper 95% tolerance limit is:

$$x_U = 3.027 + 2.566 \times 1.313$$

$$= 6.40$$

Confidence Interval and Standard Error. A confidence interval describes the range of values within which the mean is expected to lie. An example of a 95%

TABLE 8.3   K Factors for Upper Tolerance Limit at 95% Confidence Level

n	P 0.95 α 0.05
3	7.655
4	5.145
5	4.202
6	3.707
7	3.399
8	3.188
9	3.031
10	2.911
15	2.566
20	2.396
25	2.292

Source: Information from EPA (1989).

confidence interval would be to state that one is 95% confident that the mean lies between a and b. The upper and lower limits of a confidence interval are, respectively:

$$\bar{x}_L = \bar{x} - t_{1-\alpha,v} \frac{s}{\sqrt{n}} \tag{8.14}$$

$$\bar{x}_U = \bar{x} + t_{1-\alpha,v} \frac{s}{\sqrt{n}} \tag{8.15}$$

Notice in the preceding equations that the standard deviation has been divided by the square root of n, n being the number of values that make up the mean. This term, $s/\sqrt{n}$ is called the standard error. Because a confidence interval is two-sided, the value of α in Equations 8.14 and 8.15 represents the upper and lower tails, respectively, of the Student's t distribution in Table 8.1. Thus, to calculate a 95% confidence interval, a value of 0.025 is used for α inasmuch as the sum of the two tails will equal 0.05.

EXAMPLE 8.13
Upper Confidence Limit

Calculate an upper limit for a 90% confidence interval for the mean of the benzene data in Example 8.2.

Solution. Because the benzene data are lognormally distributed, the data must be first transformed by taking the logarithm (base e) of each value. The log-mean and log-standard deviation are −0.7841 and 0.5471, respectively. The log-standard error is

$$\frac{s}{\sqrt{n}} = \frac{0.5471}{\sqrt{46}}$$

$$= 0.08066$$

With $n-1$, or 45, degrees of freedom, t is 1.68 for the upper 5% of the distribution from Table 8.1 (by interpolation). The upper confidence limit is

$$\bar{x}_U = e^{(-0.7841 + 1.68 \times 0.08066)}$$

$$= 0.52$$

Correlation Coefficient. The correlation coefficient, r, is used to determine whether there is a relationship between two variables, x and y.

$$r = \frac{n \sum xy - \sum x \sum y}{\sqrt{(n \sum x^2 - (\sum x)^2)(n \sum y^2 - (\sum y)^2)}} \tag{8.16}$$

The correlation coefficient can range from 1 to -1. Positive values indicate a positive correlation (x increases with y) with 1 representing perfect positive correlation. Negative values indicate a negative or inverse relationship (x increases as y decreases) with -1 representing perfect negative correlation. The midpoint value of 0 represents no correlation at all.

The significance of the correlation coefficient depends on the number of data points and can be tested by a simple t test.

$$t = \left| \frac{\sqrt{n-2}\, r}{\sqrt{1-r^2}} \right| \tag{8.17}$$

The value of t is compared with the critical value in Table 8.1 for $n-2$ degrees of freedom and the chosen significance level. If t is greater than the critical value, then the correlation coefficient is considered significant.

One should be careful when drawing conclusions from correlation coefficients. Sometimes data will appear to be related, but there is no real cause and effect between the two variables. In addition, some extreme values can make the correlation seem much better than it actually is. It is best to plot the data in conjunction with calculating the correlation coefficient to *see* whether the relationship is really there or if certain data points have too much influence on the relationship.

EXAMPLE 8.14
Correlation Coefficient

Using the following data, calculate the correlation coefficient between biochemical oxygen demand (BOD) and chemical oxygen demand (COD) in a wastewater effluent. Is the correlation coefficient significant at a 95% significance level (one-tail distribution)?

BOD (mg/L)	32	36	24	31	21	19	15	29	27
COD (mg/L)	85	101	70	78	67	65	72	84	81

Solution. Calculate the correlation coefficient with Equation 8.16.

$$r = \frac{9(18,799) - (234)(703)}{\sqrt{(9(6,454) - 234^2)(9(55,925) - 703^2)}}$$
$$= 0.85$$

To test the significance of this result, calculate the t value using Equation 8.17 and compare it with the critical t value in Table 8.1.

$$t = \frac{\sqrt{9-2}\,0.85}{\sqrt{1-0.85^2}}$$

$$= 4.3$$

Because 4.3 is greater than the critical value of 1.895, the correlation between BOD and COD is considered significant.

MORE COMPLEX STATISTICS

Outliers

An outlier is a data point that does not follow the same pattern as the rest of the data. For example, an outlier may be an extreme value within a distribution or a data pair in a linear regression that is far removed from the general slope of the rest of the data. A decision may be made to exclude outliers from data sets when they distort calculations or conclusions about the data. At the same time, on must be very cautious when deleting data. For example, environmental monitoring is often performed to determine if an extreme event has occurred that results in a noncompliance with a regulatory control permit.

Outliers can be identified visually by plotting the data or by a variety of statistical tests. In every case, however, data points should not be deleted indiscriminantly. Data points that are identified as potential outliers should be looked at again to see whether they resulted from a transcription or math error, or whether they truly represent extreme values that can occur and should be retained in the data set.

Two statistical tests that can be used to detect outliers are Grubbs' T test and Dixon's test of ratios. Youden's rank test can be used to detect outlying laboratories in interlaboratory studies. As noted in the following discussion some of these tests are based on the assumption that the data are normally distributed. If a data set is lognormally distributed, these tests can still be used if the data are first transferred to lognormal scale, that is, by taking the logarithms of the original values.

Grubbs' T Test. Grubbs' T test is based on the assumption that the data are normally distributed. T is calculated as:

$$T = \frac{|x - \bar{x}|}{s} \tag{8.18}$$

The value x is considered an outlier if T is greater than the corresponding value in Table 8.4, based on the number of data points and the chosen significance level.

TABLE 8.4 Critical Values for Grubbs' T Test (One-Sided Test)

n	P α	0.975 0.025	0.95 0.05
3		1.155	1.153
4		1.481	1.463
5		1.715	1.672
6		1.887	1.822
7		2.020	1.938
8		2.126	2.032
9		2.215	2.110
10		2.290	2.176
11		2.355	2.234
12		2.412	2.285
13		2.462	2.331
14		2.507	2.371
15		2.549	2.409
16		2.585	2.443
17		2.620	2.475
18		2.651	2.504
19		2.681	2.532
20		2.709	2.557
30		2.908	2.745
40		3.036	2.866
50		3.128	2.956
60		3.199	3.025
70		3.257	3.082
80		3.305	3.130
90		3.347	3.171
100		3.383	3.207
120		3.444	3.267
140		3.493	3.318

Source: Information from ASTM (1989).

EXAMPLE 8.15
T Test for Outliers

The flow data in Example 8.9 originally included a value of 16.2 MGD in addition to the other 15 data points. Use the T test with a 95% significance level (one-sided) to demonstrate that this value is an outlier.

Solution. The mean and standard deviation of the flow data, including the 16.2 MGD, are 9.0625 and 2.0758, respectively. T is:

$$T = \frac{|16.2 - 9.0625|}{2.0758}$$

$$= 3.4$$

The critical value of T from Table 8.4 is 2.443. Because T is greater than this, 16.2 MGD is considered an outlier. Although this value may be considered an outlier from the rest of the data, it is real (barring malfunction of the flow measurement device or transcription error) and signifies a particular flow event that may be necessary to investigate further.

Dixon's Test. Dixon's test is a quick test for small data sets because it does not require calculating the standard deviation. Dixon's test is based on the relative differences between values in a normal distribution. The data are ranked from smallest, x_1, to largest, x_n. Dixon's ratio, r, is then calculated for the suspected outlier (x_1 or x_n), depending on the number of data points as shown below.

		Suspected Outlier		
n	Ratio	x_n	x_1	
3–7	r_{10}	$\dfrac{(x_n - x_{n-1})}{(x_n - x_1)}$	$\dfrac{(x_2 - x_1)}{(x_n - x_1)}$	(8.19)
8–10	r_{11}	$\dfrac{(x_n - x_{n-1})}{(x_n - x_2)}$	$\dfrac{(x_2 - x_1)}{(x_{n-1} - x_1)}$	(8.20)
11–13	r_{21}	$\dfrac{(x_n - x_{n-2})}{(x_n - x_2)}$	$\dfrac{(x_3 - x_1)}{(x_{n-1} - x_1)}$	(8.21)
14–30	r_{22}	$\dfrac{(x_n - x_{n-2})}{(x_n - x_3)}$	$\dfrac{(x_3 - x_1)}{(x_{n-2} - x_1)}$	(8.22)

The value of r is compared with the value in Table 8.5 corresponding to n and the chosen significance level.

EXAMPLE 8.16
Dixon's Outlier Test

Use Dixon's test with a significance level of 5% on the flow data in Example 8.15.

Solution. There are 16 values, so Equation 8.22 for r_{22} is used.

$$r_{22} = \frac{(x_n - x_{n-2})}{(x_n - x_3)}$$
$$= \frac{(16.2 - 9.8)}{(16.2 - 7.8)}$$
$$= 0.762$$

TABLE 8.5   Critical Values of r for Dixon's Test

n	Significance Level	
	0.05	0.01
3	0.941	0.988
4	0.765	0.889
5	0.642	0.780
6	0.560	0.698
7	0.507	0.637
8	0.554	0.683
9	0.512	0.635
10	0.477	0.597
11	0.576	0.679
12	0.546	0.642
13	0.521	0.615
14	0.546	0.641
15	0.525	0.616
16	0.507	0.595
17	0.490	0.577
18	0.475	0.561
19	0.462	0.547
20	0.450	0.535
21	0.440	0.524
22	0.430	0.514
23	0.421	0.505
24	0.413	0.497
25	0.406	0.489
26	0.399	0.486
27	0.393	0.475
28	0.387	0.469
29	0.381	0.463
30	0.376	0.457

Source: Information from ASTM (1989).

The critical value from Table 8.5 is 0.507, which is less than 0.762. Therefore, 16.2 MGD would be considered an outlier.

Youden's Rank Test for Laboratories. Youden's ranking test is used to identify outlying laboratories in collaborative or round-robin interlaboratory tests. Suspect laboratories tend to have either high or low analytical results as compared with the other laboratories.

In an interlaboratory test, several samples with different concentrations are analyzed by multiple (n) laboratories. For each sample, each laboratory is given a rank, from 1 to n, with the laboratory with the largest analytical result given the value of 1, the next largest 2, and so on. If two or more laboratories tie in rank, each receives a rank equal to the average of the individual ranks. For example, if three laboratories tie for the 3rd, 4th, and 5th positions, each one would receive a rank of 4. After all samples have been ranked, each laboratory's cumulative score is calculated. A score that falls outside the range given in Table 8.6 indicates an outlying laboratory. Note, however, that it is possible

TABLE 8.6 Acceptable Range in Scores for Youden's Ranking Test

No. of Labs	3	4	5	6	7	8	9	10	11	12	13	14	15
3		4	5	7	8	10	12	13	15	17	19	20	22
		12	15	17	20	22	24	27	29	31	33	36	38
4		4	6	8	10	12	14	16	18	20	22	24	26
		16	19	22	25	28	31	34	37	40	43	46	49
5		5	7	9	11	13	16	18	21	23	26	28	31
		19	23	27	31	35	38	42	45	49	52	56	59
6	3	5	7	10	12	15	18	21	23	26	29	32	35
	18	23	28	32	37	41	45	49	54	58	62	66	70
7	3	5	8	11	14	17	20	23	26	29	32	36	39
	21	27	32	37	42	47	52	57	62	67	72	76	81
8	3	6	9	12	15	18	22	25	29	32	36	39	43
	24	30	36	42	48	54	59	65	70	76	81	87	92
9	3	6	9	13	16	20	24	27	31	35	39	43	47
	27	34	41	47	54	60	66	73	79	85	91	97	103
10	4	7	10	14	17	21	26	30	34	38	43	47	51
	29	37	45	52	60	67	73	80	87	94	100	107	114
11	4	7	11	15	19	23	27	32	36	41	46	51	55
	32	41	49	57	65	73	81	88	96	103	110	117	125
12	4	7	11	15	20	24	29	34	39	44	49	54	59
	35	45	54	63	71	80	88	96	104	112	120	128	136
13	4	8	12	16	21	26	31	36	42	47	52	58	63
	38	48	58	68	77	86	95	104	112	121	130	138	147
14	4	8	12	17	22	27	33	38	44	50	56	61	67
	41	52	63	73	83	93	102	112	121	130	139	149	158
15	4	8	13	18	23	29	35	41	47	53	59	65	71
	44	56	67	78	89	99	109	119	129	139	149	159	169
16	4	9	13	19									
	47	59	71	83									
17	5	9	14	20									
	49	63	76	88									
18	5	9	15	21									
	52	67	80	93									
19	5	10	15	21									
	55	70	85	99									
20	5	10	16	22									
	58	74	89	104									

Source: Information from Youden and Steiner (1975).

that an *outlying* laboratory may be the one whose results are actually closest to the true value and that all the other laboratories are merely consistent with each other. Youden's ranking test should be used to identify *suspect* laboratories whose data should be discarded only after careful review. Depending on the purpose of the round-robin study, it can be useful to share the results with the participating laboratories so that coscientious laboratories have the opportunity to evaluate and improve their performance.

EXAMPLE 8.17
Youden's Rank Test in an Interlaboratory Study

In 1984, EPA conducted extensive interlaboratory testing of the 600 series analytical methods (now found in 40 CFR 136, Appendix A) in order to develop regression equations for accuracy and precision. The following are the original data from the 15 laboratories that used Method 624 for volatiles to analyze six samples of industrial effluent spiked with benzene (EPA 1984). Use Youden's rank test to determine whether any of the laboratories should be considered outliers.

	Spiked Concentration in Industrial Effluent (µg/L)					
Lab	10.8	12.0	114.0	120.0	480.0	432.0
	Analytical Result (µg/L)					
1	9.9	11.1	102.7	114.3	428.7	410.3
2	12.2	17.2	135.8	153.9	745.5	604.9
3	9.4	39.5	125.7	146.3	388.8	362.1
4	17.9	13.3	120.8	190.8	465.8	603.9
5	0.0	0.0	76.8	93.8	446.8	276.8
6	8.8	10.7	98.1	77.5	298.0	291.8
7	8.3	9.5	103.0	97.7	185.0	181.0
8	14.8	15.6	144.8	135.2	443.2	415.2
9	12.8	14.9	144.0	149.0	288.0	296.0
10	10.6	11.6	106.3	99.7	289.7	247.2
11	11.4	11.1	74.3	97.2	422.0	361.0
12	10.8	12.5	104.0	147.0	600.0	441.0
13	10.9	11.8	108.8	0.0	353.5	265.5
14	10.5	13.1	122.3	113.7	443.1	304.0
15	9.0	12.5	118.3	96.9	370.8	307.5

Solution. The laboratory ranks for each sample are shown in the following table.

Lab			Individual Rank				Summed Rank
1	10	11.5	12	7	7	5	52.5
2	4	2	3	2	1	1	13
3	11	1	4	5	9	6	36
4	1	5	6	1	3	2	18
5	15	15	14	13	4	12	73
6	13	13	13	14	12	11	76
7	14	14	11	10	15	15	79
8	2	3	1	6	5	4	21
9	3	4	2	3	14	10	36
10	8	10	9	9	13	14	63
11	5	11.5	15	11	8	7	57.5
12	7	7.5	10	4	2	3	33.5
13	6	9	8	15	11	13	62
14	9	6	5	8	6	9	43
15	12	7.5	7	12	10	8	56.5

From Table 8.6, the allowable range in scores for 15 laboratories analyzing six samples is 18–78. Laboratory 2 is outside the range with a score of 13, indicating that this laboratory routinely obtained values on the high side. Laboratory 7 is also outside the range with a score of 79, indicating lower than normal analytical results. Both of these laboratories are considered potential outliers, and their data would be verified before deleting them from the data set.

Linear Regression

Linear regression is obtaining the best fit of a straight line through a set of data points. The linear function is written with x and y as the independent and dependent variables, respectively:

$$y = mx + b \tag{8.23}$$

Linear regression can also be done where the intercept, b, is forced to be zero, a logical condition for many sets of data when x is equal to zero.

The slope, m, the regression line is calculated by:

$$m = \frac{\sum xy - n\overline{xy}}{\sum x^2 - n\overline{x}^2} \tag{8.24}$$

The intercept, b, is calculated by:

$$b = \bar{y} - m\bar{x} \tag{8.25}$$

If the intercept is forced to be zero, then the slope is calculated by:

$$m = \frac{\sum xy}{\sum x^2} \tag{8.26}$$

The standard errors associated with the regression parameters are:

Slope (with intercept)

$$SE_m = \left[\frac{\sum (y-\bar{y})^2 - m^2 \sum (x-\bar{x})^2}{(n-2) \sum (x-\bar{x})^2} \right]^{0.5} \tag{8.27}$$

Slope (intercept equal to zero)

$$SE_m = \left[\frac{\sum (y-mx)^2}{n-1 \sum x^2} \right]^{0.5} \tag{8.28}$$

Intercept, b

$$SE_b = \left[\left(\frac{\sum (y-\bar{y})^2 - m^2 \sum (x-\bar{x})^2}{n-2} \right) \left(\frac{1}{n} + \frac{\bar{x}^2}{\sum (x-\bar{x})^2} \right) \right]^{0.5} \tag{8.29}$$

The statistical significance of the slope and intercept can be tested with their standard errors in a t test to evaluate the strength of the linear regression. For example, you can test to see whether the slope is significantly different from zero (i.e., no slope, no relationship) and likewise for the intercept.

It is always wise to plot the data with linear regression. Quite often, extreme values can make a regression appear much better mathematically than it is in reality. Take the case of the data plotted in Figure 8.10, which shows waste generation versus manufacturing production. The data are clustered randomly near the origin, with two extreme data points acting as anchors on the far end of the scale. When these two data points were removed and the rest of the data were replotted in Figure 8.11, the evidence of any trend in the data rapidly disappeared.

EXAMPLE 8.18
Linear Regression

You would like to develop a relationship between the return sludge concentration and the influent BOD load in your activated sludge plant in order to determine whether higher sludge concentrations can be maintained under increased BOD loads. The following data represent a period of normal, well-operated conditions.

Influent BOD (lbs/d)	Sludge Concentration (mg/L)
18,000	5,700
16,100	5,200
14,000	4,100
17,500	3,950
18,500	4,600
22,000	6,100
24,000	5,700
21,500	6,300
25,500	7,500
26,000	5,100
36,500	7,900
28,600	5,950
27,000	5,900
26,000	5,300
22,000	5,500
30,000	8,100
32,000	6,900
33,000	6,200
29,000	6,700
14,000	5,200

Solution. The plot of the data in Figure 8.12 shows that there is a linear relationship between the sludge concentration and the influent BOD load. The slope and intercept of the regression line are 0.132 and 2728, respectively. The correlation coefficient, R, is 0.75. The standard errors for the slope and intercept are 0.027 and 681, respectively.

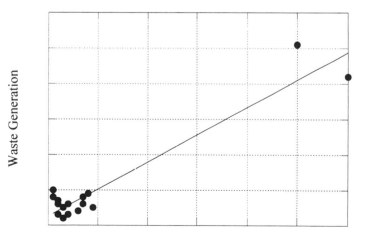

Manufacturing Production

Figure 8.10   Linear regression with extreme points.

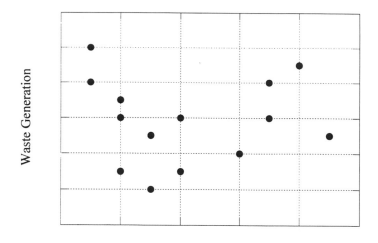

Manufacturing Production

Figure 8.11 Linear regression with extreme points removed.

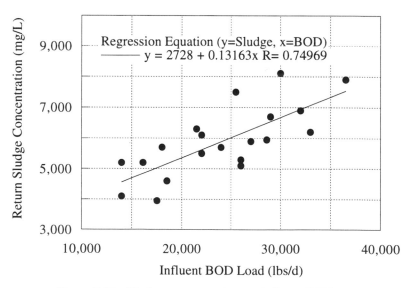

Figure 8.12 Sludge concentration vs. influent BOD.

To test the significance of R, use Equation 8.17.

$$t = \frac{\sqrt{20-2}\times0.75}{\sqrt{1-0.75^2}}$$

$$= 4.81$$

A significance level of 95% ($\alpha=0.05$) is a common test standard for statistical tests. The critical value of t from Table 8.1 with a significance level of 0.05 is 1.73. Because this is less than 4.81, the correlation is considered significant.

Next, the t for the slope is calculated using Equation 8.11. The assumption being tested is that the slope is not zero.

$$t = \frac{x_{1-\alpha}-\bar{x}}{s}$$

$$= \frac{0.132-0}{0.027}$$

$$= 4.89$$

Because t is greater than the critical value of 1.73, the slope is considered significantly greater than zero.

Finally, the intercept is tested to see whether it is significantly different from zero. If the intercept is not shown to be significantly different from zero, you might decide to use the regression equation with no intercept ($y=mx$). The t for the intercept is:

$$t = \frac{2728-0}{681}$$

$$= 4.00$$

Because t is greater than 1.73, the intercept is considered to be significantly greater than zero.

Nondetects and Censored Data

When environmental data are based on very low or trace concentrations, the data often include *nondetectable* or *less than* values, for example, <0.005 mg/L. These nondetects are called *censored*[1] data because their true values are not known.

Nondetects can be handled in a variety of ways, depending on how the data are being used. Several techniques were evaluated in a study by the National

---

[1] Censored data can be at the low (left-censored) or high (right-censored) end of a distribution; however, the problem with environmental data is almost always with nondetects (the low end).

Council of the Paper Industry for Air and Stream Improvement, Inc. (NCASI 1991). Some of the more common techniques are described in the following paragraphs.

Substitution with Zero or One Half the Detection Limit. A very simple technique is to replace the less than value with zero or one half the detection limit. The idea behind using zero is that if nothing was *detected*, the concentration should be zero. In reality, a less than value means that the real concentration can be anywhere between zero and the detection limit. Consequently, an alternative substitution is one half the detection limit. These substitutions should not be used indiscriminately, because they can distort both the mean and the variance. EPA recommends that one half the detection limit be used with data sets where no more than 15% of the values are nondetects (EPA 1989). NCASI does not recommend using the substitution technique to calculate variances that are used to make probabilistic statements, such as confidence intervals (NCASI 1991).

Median. The median can be used to estimate the mean if less than half of the data points are nondetects and the distribution is expected to be either normal or symmetric around the median. Using the median is a very simple way to estimate the mean; however, this technique does not provide a way of estimating the variance of the data.

Modified Delta-Lognormal Distribution. The modified delta-lognormal distribution has two parts: the fraction or delta portion, δ, of the data that are nondetects and the remaining detectable values that are assumed to follow a lognormal distribution. The expected mean, $E(x)$, and variance, $V(x)$, of this distribution are

$$E(x)=\delta D+(1-\delta)e^{\mu+0.5\sigma^2} \qquad (8.30)$$

and

$$V(x)=(1-\delta)e^{2\mu+\sigma^2}(e^{\sigma^2}-(1-\delta))+\delta(1-\delta)D(D-2e^{\mu+0.5\sigma^2}) \qquad (8.31)$$

where D is the detection limit of the nondetectable values and μ is the log mean and σ^2 is the log variance of the detectable values.

The modified delta-lognormal distribution was used by EPA to develop wastewater limits for the chemical and pesticide industries (EPA 1987, EPA 1993). One problem with this method, however, is that the detection limit must be the same (or assumed so) for all the nondetects. Consequently, EPA further modified the delta-lognormal distribution to include multiple detection limits (EPA 1995). In this latter case, the expected mean is

$$E(x)=\sum \delta_i D_i+(1-\delta)e^{\mu+0.5\sigma^2} \qquad (8.32)$$

where δ_i represents each individual fraction of nondetects at a particular detection limit, D_i, and the sum of the individual fractions is equal to the total fraction, δ.

Cohen's Method. This method assumes that the data are normally distributed. Its main disadvantage is that the detection limit must be the same for all the nondetects. In the case of multiple detection limits, it is possible to use the highest detection limit for all the nondetects and then apply Cohen's method; however, doing so effectively eliminates some of the accuracy in the data set. Cohen's method is fairly simple, but requires several steps and lookup tables.

Step 1. Calculate the mean, $\bar{x}_d$, and variance, s_d^2, of the m data points above the detection limit. For Cohen's method, the variance is calculated with m instead of $m-1$ in the denominator of Equation 8.8.

Step 2. Calculate the fraction, h, of the data points that are less than the detection limit where n is the total number of values and m is the number of values above the detection limit.

$$h = \frac{(n-m)}{n} \tag{8.33}$$

Step 3. Calculate γ where DL is the detection limit.

$$\gamma = \frac{s_d^2}{(\bar{x}_d - DL)^2} \tag{8.34}$$

Step 4. Lookup the value of λ in Cohen's tables (Cohen 1959, 1961), using h and γ. Cohen's estimates of the mean and the variance are then calculated by Equations 8.35 and 8.36, respectively.

$$\mu = \bar{x}_d - \lambda(\bar{x}_d - DL) \tag{8.35}$$

$$\sigma^2 = s_d^2 + \lambda(\bar{x}_d - DL)^2 \tag{8.36}$$

Youden Pairs and Plots

Youden pairs and plots (Youden and Steiner 1975) are useful in identifying differences between multiple data sets, such as performances of laboratories analyzing environmental samples. Youden's techniques were used by EPA in developing performance equations for the 600-series methods for volatile and semivolatile organic analyses found in Appendix A, 40 CFR 136.

A Youden pair comprises two samples that have different, but similar or close, concentrations. Usually, for an interlaboratory or round-robin test among laboratories, each laboratory is given a set of samples consisting of multiple Youden pairs representing a low, medium, and high concentration range. The data given in Example 8.17 are based on this type of study design. The samples are blinded; that is, the laboratories are not told the true concentrations, nor that the samples are Youden pairs. The advantage of using Youden pairs over duplicate samples is that the laboratories cannot unconsciously bias their analyses by "knowing" what concentration to expect in the second sample of a pair.

Laboratory results are plotted in a Youden plot for each sample pair, as illustrated in Figure 8.13. One of the sample concentrations is plotted on the horizontal axis, the other on the vertical axis. Each data point in the graph represents a laboratory. A circle is drawn on the graph with its origin at the data point representing the average sample concentrations found by the laboratories. The radius of the circle is related to the intralaboratory precision, which is the square root of the variance, calculated as follows:

$$s^2 = \frac{\sum (D_i - \bar{D})^2}{2(n-1)} \tag{8.37}$$

D_i is the difference between the two samples for each laboratory, $\bar{D}$ is the average of all the laboratory sample pair differences, and n is the number of sample pairs. The radius of the circle in Figure 8.13 is s times Student's t from Table 8.1 for a two-tailed 95% confidence area. To complete the Youden plot, a 45-degree line is drawn through the origin of the circle.

A Youden plot is interpreted visually as follows:

- *Precision*—A point that lies essentially on or very near the 45-degree line is a laboratory with good precision. The farther a point lies from the 45-degree line, the poorer its precision. The distance of a point from the 45-degree line is measured perpendicularly. The circle in the plot represents the pattern of results that would be expected from random error alone.

- *Accuracy*—Points closest to the circle center are laboratories that are closest to the mean accuracy (or recovery) of the laboratories in the study. The distance of any point to this center is measured from the point at which the perpendicular line discussed earlier intersects the 45-degree line. Points that lie farthest from the circle center are the extremes of the range in laboratory accuracies. If a point is farthest from the circle center and lies in quadrant I, it means that this laboratory gets consistently high recoveries relative to the other laboratories. If a point is farthest from the circle center in quadrant III, it means that the laboratory gets consistently low recoveries relative to the other laboratories. If the farthest point is in either quadrant II or IV, it means that the laboratory is inconsistent, getting high recoveries some of the time and low recoveries other times.

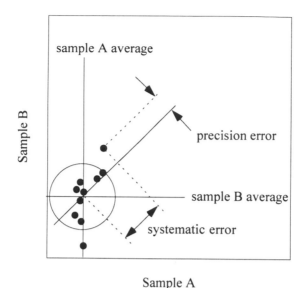

Figure 8.13 Youden plot of interlaboratory data.

- *Method performance*—If the true concentrations of the two samples are known and plotted on the same graph, the distance between the true point and the circle center is the systematic error in the analytical method. This distance is measured along either the vertical or the horizontal axis. If the plotted sample pairs are widely spaced, but close to the 45-degree line, all the laboratories have relatively small errors in precision as compared with systematic errors. It also means that the method may not be adequate in that the laboratories may be interpreting the procedures differently.

A minimum of six laboratories producing six usable sets of data is recommended for collaborative or interlaboratory testing (ASTM 1986). In EPA's interlaboratory studies for the 600-series analytical methods, 20 laboratories were tested.

PRACTICAL APPLICATIONS

Wastewater Permit Limits

EPA gives a detailed description of its guidance for setting permit limits in its *Technical Support Document for Water Quality-Based Toxics Control* (EPA 1991). Although the guidance is for toxic pollutants, the statistical principles are the same for conventional pollutants such as biochemical oxygen demand (BOD) and total suspended solids (TSS).

Although regulatory agencies differ in how they develop permit limits, the common approach follows EPA's guidance. The basic steps are as follows:

1. Determine whether the data fit a normal or lognormal distribution.
2. Identify and resolve any problems caused by outliers.
3. Calculate the distribution parameters (mean, standard deviation).
4. Calculate the daily maximum limit from the 99th percentile and the monthly average limit from the 95th percentile.

EXAMPLE 8.19
Wastewater Permit Limits

Your manufacturing facility has a wastewater treatment system whose permit is up for renewal. What permit limits might you expect based on the following BOD effluent data that was collected over a one-year period? Assume the daily maximum limit is based on the 99th percentile and the monthly average is based on the 95th percentile.

Wastewater Effluent BOD Concentrations (mg/L)

10	15	16	13	8	34	55	22
18	15	17	9	12	11	13	15
17	20	21	17	18	31	28	27
24	42	13	14	23	26	28	34

Solution. First, determine whether the data fit a normal or lognormal distribution. Judging from the probability plots in Figures 8.14 and 8.15, the lognormal distribution is a better fit because the data are curved in the normal-scale plot, but fall in a straight line in the log-scale plot. The lognormal distribution appears to be a good fit, and there are no apparent outliers in Figure 8.15. The log mean and log standard deviation are 2.933 and 0.4454, respectively. Using Equation 8.12 with a t value of 2.453 interpolated from Table 8.1 ($\alpha = 0.01$ and $v = 31$), the 99th percentile is

$$x_{0.99} = e^{2.933 + 2.453*0.4454}$$

$$= 56 \text{ mg/L}$$

Monthly average permit limits for wastewater are typically 95th percentiles based on a standard error and an n value of 30 (assumed as the number of days in a month). With a t value of 1.696 from Table 8.1, the monthly average limit would be:

$$x_{0.95} = e^{2.933 + 1.696*(0.4454)/\sqrt{30}}$$

$$= 22 \text{ mg/L}$$

Ground Water Monitoring

Routine statistical testing is required with ground water monitoring in hazardous waste permits. Detailed guidance is provided by EPA in its *Technical*

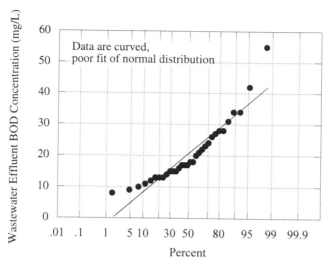

Figure 8.14   Normal probability plot of wastewater effluent BOD data.

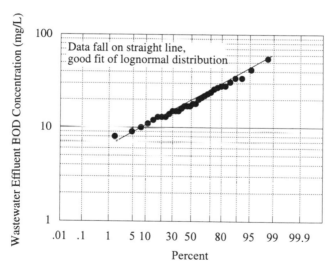

Figure 8.15   Lognormal probability plot of wastewater effluent BOD data.

Enforcement Guidance Document (TEGD) (EPA 1986a). Here is an example using the *t* test with a data set that contains censored or nondetectable values.

EXAMPLE 8.20
Ground Water Monitoring

You have just received the laboratory results from your quarterly ground water monitoring. Your hazardous waste permit requires that you test these results to see whether there is a significant increase over background concentrations, using the averaged replicate t test with Cohen's adjustment as necessary for nondetectable analyses. In the first year of quarterly ground water monitoring, you calculate an average background concentration of chromium of 3.0 µg/L with a standard deviation of 1.5 µg/L. Because you have one upgradient well with four quarterly results and three downgradient wells, your t value from EPA's TEGD is 6.297. Determine whether there has been a significant increase in downgradient well D1, having four replicate samples with concentrations of <10, 15, 14, and 17 µg/L. Use the following equation to calculate the maximum concentration that would be considered as background.

$$x_{max} = \bar{x} + ts \left(1 + \frac{1}{n}\right)^{0.5}$$

Solution. Calculate the maximum background concentration.

$$x_{max} = 3.0 + 6.297(1.5) \left(1 + \frac{1}{4}\right)^{0.5}$$

$$= 13.6 \text{ µg/L}$$

The first step in using Cohen's adjustment is to calculate the mean and variance of the values that are above the detection limit. For the three values in the set of four replicates, the mean and variance are 15.333 and 1.5275, respectively. Because one out of the four values was below the detection limit, use a value of 0.25 to calculate γ with Equation 8.34.

$$\gamma = \frac{2.3333}{(15.333 - 10)^2}$$

$$= 0.08204$$

For h=0.25 and γ=0.08204, Cohen's adjustment factor, λ, is 0.33350 (interpolated from Table 5, Appendix B of EPA's TEGD. Therefore, the adjusted mean calculated with Equation 8.35 is

$$\mu = 15.333 - 0.33350(15.333 - 10)$$

$$= 13.55 \text{ µg/L}$$

The statistical test indicates that the average concentration in well D1, even though it is higher than the average background concentration, is not a significant increase in concentration because it is within the normal range that one would expect for background concentrations.

Hazardous Characteristic (TCLP)

In its SW-846 manual (EPA 1986b), EPA suggests a statistical test for deciding whether a waste is characteristically hazardous using the Toxicity Characteristic Leaching Procedure (TCLP).[2] EPA suggests using the upper limit of an 80% confidence interval as the average concentration to compare with the TC limit. If the upper confidence limit is less than the TC limit, the waste is not considered hazardous for that particular TC compound. Because the statistic uses a confidence interval on the mean, the assumption here is that the TC is an average concentration limit.

EXAMPLE 8.21
Hazardous Characteristic (TCLP)

You want to determine whether the waste your facility generates is characteristically hazardous by the TCLP for benzene, using the method suggested by EPA in SW-846. The TCLP limit for benzene is 0.5 mg/L. You have collected weekly samples for 46 weeks (see data in Example 8.2). Based on these data and using EPA's method, what are your conclusions?

Solution. It was shown in Example 8.3 that the data are lognormally distributed. In Example 8.13, the log standard deviation was calculated as 0.08067. The critical value of t interpolated from Table 8.1 for 45 degrees of freedom and a significance level of 0.1 (corresponding to the upper tail of the 80% interval) is 1.301. The upper confidence limit is calculated by using Equation 8.15 and then transforming the lognormal value back to its original scale:

$$CL_U = e^{-0.78409 + 1.301(0.08067)}$$

$$= 0.507$$

The upper confidence limit is greater than the TC limit of 0.5 mg/L, which suggests that, based on its average characteristics, the waste is TC hazardous. However, because the upper limit is just slightly over the TC limit, you should go back and review the data to be certain of their accuracy before you conclude that the waste is hazardous.

Number of Samples to Collect

Many environmental projects involve extensive sampling, so there is an incentive to optimize the number of samples that have to be collected. Although it is obvious that collecting too few samples can undermine the reliability of your conclusions, it is not so obvious that beyond a certain point

---

[2]This is only one method suggested by EPA. Regional and local regulatory offices may choose different methods of applying the TC limits.

additional samples do not significantly increase the reliability of the data. Consider an extreme case: Would you collect 5,000 samples if your margin of error with just 20 samples was sufficient?

The estimation of the number of samples to collect is simply a different twist to the confidence limits calculated by Equations 8.14 and 8.15. To estimate the number of samples, you must make two assumptions: (1) how much the sample results will vary (a standard deviation) and (2) how great a difference from the true mean (based on an infinite number of samples) you are willing to tolerate. Let d be the difference between the true mean and upper (or lower) confidence limit based on n samples. The equation for the confidence limit can be rewritten as:

$$d = t_{1-\alpha/2} \frac{s}{\sqrt{n}}$$

Rearranging, and solving for n

$$n = \frac{t_{1-\alpha/2}^2 s^2}{d^2} \tag{8.38}$$

If the standard deviation is not known (based on population statistics or previous sampling), you must assume some value in order to use the equation. Afterward, you may revise your estimate of n using the standard deviation from the samples you have collected. This will also serve as a check that you collected the right number of samples. Use the value of t for $n = \infty$.

EXAMPLE 8.22
Number of Samples to Collect

You must collect enough samples of a waste to be reasonably confident that you have a reliable estimate of its average concentration. Based on previous sampling with a similar waste, you expect the individual concentrations to vary ± 25 mg/kg (as a standard deviation). You are willing to tolerate an error of 10 mg/kg from the true mean at a confidence level of 95% ($\alpha/2 = 0.025$). How many samples should you collect?

Solution. The number of samples to collect is 24, as shown by the calculation that follows. The value of t from Table 8.1 is 1.96.

$$n = \frac{(1.96)^2 (25)^2}{(10)^2}$$

$$= 24$$

Judging Laboratory Performance

To ensure that their environmental data are reliable, some clients audit the in-house procedures of the laboratories that analyze their samples. Another way of evaluating laboratory performance is to send blind samples for analyses. A blind interlaboratory study can be set up using Youden pairs where multiple laboratories are being considered or the client wants to develop an approved list of laboratories.

EXAMPLE 8.23
Judging Laboratory Performance Using Youden Pairs

You are the environmental coordinator in charge of several environmental remediation projects for your company. Because all of the sites involve fuel or gasoline contamination, you are primarily interested in analysis of the volatile organics, benzene, toluene, ethylbenzene, and xylenes (BTEX). Extensive sampling will be required, and the sites are in different locations throughout the United States. Therefore, you want to develop a list of approved laboratories that can be used for any of the samples. How would you set up an interlaboratory test using Youden pairs? Ground water sample concentrations are likely to range from detection limits for background and verification samples to 500 μg/L for contamination sites. According to EPA SW-846 method 8020 for purgeable volatile organic compounds, the detection limit for benzene, toluene, and ethylbenzene is 0.2 μg/L. Method 8020 does not list a detection limit for xylenes; however, for your projects you will require a detection limit of 2 μg/L.

Solution. Because ASTM recommends using a minimum of six laboratories in collaborative studies (ASTM 1986), solicit at least eight; the extra two will help if samples are broken or a laboratory's data are deleted as outliers.

Set up Youden pairs for three concentration levels. The first will be near the detection limits, the second near the upper limit of 500 μg/L, and the third midway between the other two. Because there is no specific requirement for how similar to make concentration pairs, you decide to use ±20% for the low concentration pair and ±10% for the middle and high concentration pairs. Your Youden pairs are as follows:

- Low concentration pair—(0.16 and 0.24 μg/L) for benzene, toluene, and ethylbenzene and (1.6 and 2.4 μg/L) for xylenes
- Middle concentration pair—(225 and 275 μg/L) for all four compounds
- High concentration pair—(450 and 550 μg/L) for all four compounds

Keep in mind that the prepared sample aliquots that will be sent to each laboratory do not have to match the aforementioned concentrations exactly, although whatever sample concentrations are prepared must be the same for each laboratory. Preparing samples with odd concentrations (543 μg/L as opposed to 550 μg/L) helps to prevent analysts from guessing the true concentration, as does preparing two similar, but not duplicate, samples. You may also decide to focus more on the concentrations near cleanup levels inasmuch as it is more critical to know when cleanup is complete than to know that a site is contaminated. In that case, you could add another low-concentration pair at 5 to 20 times the detection level, or substitute such a pair for the middle concentration pair (given in the preceding list).

Evaluating Wastewater Treatment Performance

The performance of a wastewater treatment system can be evaluated in any number of ways. The following example uses the median to define typical operating levels and correlation analyses to determine how much influence raw materials have on influent loads.

EXAMPLE 8.24
Evaluating Wastewater Treatment Performance

You have a year's worth of operating data for your wastewater treatment system, and you want to define operating levels that represent "average" conditions. There are a few erratic periods when unusual upsets occurred, as well as a few extreme values for some of the parameters. How would you identify outliers and define average operating levels? You also are interested in relating feedstock material to influent BOD load. You have monthly averages for two years of data. Does the quantity of feedstock material (tons/d) impact influent BOD loads (lbs/d)? If not, can you give possible explanations? To test the correlation coefficient, use a significance level of 95% for a two-tailed Student's t distribution.

Year 1	Feedstock	BOD	Year 2	Feedstock	BOD
Jan	12,500	14,000	Jan	13,000	18,500
Feb	17,000	20,500	Feb	12,500	26,500
Mar	14,500	21,000	Mar	16,500	27,000
Apr	15,500	16,500	Apr	13,000	31,000
May	12,500	19,000	May	14,000	31,000
Jun	13,500	17,500	Jun	16,000	25,500
Jul	14,000	22,000	Jul	12,500	21,000
Aug	17,500	25,000	Aug	13,000	20,500
Sep	14,000	23,000	Sep	13,500	25,500
Oct	12,000	19,500	Oct	16,000	19,000
Nov	17,000	23,500	Nov	14,000	22,500
Dec	16,500	18,000	Dec	14,000	28,500

Solution. Plotting each operating parameter as a time series will help you identify unusual periods and any parameters that correlate strongly with these periods. Probability plots will help to identify extreme values and average operating levels. Because you are interested in average performance for a well-operated treatment system, you delete those periods when either the operating parameters or influent/effluent loads were outside normal ranges or represent unusual events or poor management. Even with these deletions, you still have a few extreme points that you cannot relate to any particular cause, so you keep them in the data base. However, these extreme values will skew the mean, so you decide to use the median as a simpler way to calculate "average" operating levels. In doing so, you notice that for those parameters not having extreme points, the mean and median are essentially the same value. You conclude that using the median is a faster and easier way to calculate average conditions because you do not have make so many, and sometimes arbitrary, decisions about outliers.

The correlation coefficient for feedstock and influent BOD is 0.09. Because the correlation coefficient is so close to zero, you conclude that there is no apparent relationship between feedstock material and influent BOD load. This is confirmed by a t value of 0.41 (calculated using Equation 8.17), which is much lower than the critical t value of 2.074 from Table 8.1 (degrees of freedom = 22, α = 0.025). Influent BOD may not appear to be related to feedstock because most of the BOD load may be coming from a nonmanufacturing area, the method for measuring feedstock may not measure the organic portion very well, or the range in values for both parameters is too narrow to reveal any strong trends.

Demonstrating Cleanup Levels

Statistical tests for demonstrating when cleanup levels are met are site specific. The following example demonstrates cleanup levels with a t test based on background concentrations.

EXAMPLE 8.25
Demonstrating Cleanup Levels with the t Test

You are required to demonstrate compliance with cleanup levels for lead that are equivalent to background concentrations. You must use a t test with a 99% significance level (upper 1% of Student's t distribution) to verify that individual cleanup samples are consistent with background concentrations. What is the upper concentration in a cleanup sample that would still be considered within the range of values for background? Background is based on 10 samples with concentrations in mg/kg of 0.5, 1.5, 1.8, 1.1, 0.7, 0.6, 0.2, 0.9, and 0.1.

Solution. The mean and standard deviation of the background samples are 0.822 mg/kg and 0.567 mg/kg, respectively. The coefficient of variation is 69%, so you assume that the normal distribution is acceptable. With 9 degrees of freedom, the t value from Table 8.1 is 2.821. The upper concentration limit still representing background concentrations would be:

$$x = 0.822 + (2.821)(0.567)$$

$$= 2.4 \text{ mg/kg}$$

REFERENCES

American Society of Testing and Materials, "Standard Practice for Dealing with Outlying Observations," ASTM E 178-80, 1989.

American Society of Testing and Materials, "Standard Practice for Determination of Precision and Bias of Applicable Methods of Committee D-19 on Water," ASTM D 2777-86, 1986.

Box, George E.P., et al., *Statistics for Experimenters: An Introduction to Design, Data Analysis, and Model Building*, John Wiley & Sons, New York, 1978.

Cohen, A.C., Jr., "Tables for Maximum Likelihood Estimates: Singly Truncated and Singly Censored Samples," *Technometrics*, Vol. 3/No. 4, 1961.

Cohen, A.C., Jr., "Simplified Estimators for the Normal Distribution When Samples Are Single Censored or Truncated," *Technometrics*, Vol. 1/No. 3, 1959.

Eynon, Barrett P., and Alfonso J. Valdés, "Statistical Support for Pharmaceutical Rulemaking," SRI International, 1983.

National Council of the Paper Industry for Air and Stream Improvement, Inc., "Estimating the Mean of Data Sets That Include Measurements Below the Limit of Detection," Technical Bulletin No. 621, New York, 1991.

Taylor, John K., *Statistical Techniques for Data Analysis*, Lewis Publishers, Chelsea, MI, 1990.

Taylor, John K., *Quality Assurance of Chemical Measurements*, Lewis Publishers, Chelsea, MI, 1987.

U.S. Environmental Protection Agency, "Statistical Support Document for Proposed Effluent Limitations Guidelines and Standards for the Centralized Waste Treatment Industry," EPA 821-R-95-005, Office of Water, Washington, DC, 1995.

U.S. Environmental Protection Agency, "Development Document for Effluent Limitations Guidelines, Pretreatment Standards, and New Source Performance Standards for the Pesticide Chemicals Manufacturing Point Source Category," EPA 821/R-93/016, Washington, DC, 1993.

U.S. Environmental Protection Agency, "Technical Support Document for Water Quality-Based Toxics Control," Washington, DC, 1991.

U.S. Environmental Protection Agency, "Statistical Analysis of Ground Water Monitoring Data at RCRA Facilities: Interim Final Guidance," PB89-151047, Washington, DC, 1989.

U.S. Environmental Protection Agency, "Development Document for Effluent Limitations Guidelines and Standards for the Organic Chemicals, Plastics, and Synthetic Fibers Point Source Category," EPA 440/1-87/009, Washington, DC, 1987.

U.S. Environmental Protection Agency, "Resource Conservation and Recovery Act (RCRA) Ground Water Monitoring Technical Enforcement Guidance Document," PB87-107751, Washington, DC, 1986a.

U.S. Environmental Protection Agency, "Test Methods for Evaluating Solid Waste, Volume II: Field Manual, Physical/Chemical Methods, Chapter 9, Sampling Plan," Washington, DC, 1986b.

U.S. Environmental Protection Agency, "EPA Method Study 29, Method 624 Purgeables," prepared by Radian Corporation for Environmental Monitoring and Support Lab, Cincinnati, OH, 1984.

U.S. Environmental Protection Agency, "Development Document for Expanded Best Practicable Control Technology, Best Conventional Pollutant Control Technology, Best Available Technology, New Source Performance Technology, and Pretreatment Technology in the Pesticides Chemicals Industry," Washington, DC, 1982.

U.S. Environmental Protection Agency, "Development Document for Effluent Limitations Guidelines and Standards for the Pulp, Paper and Paperboard and the Builders Paper and Board Mills Point Source Categories, Proposed," EPA/440/1-80/025b, Effluent Guidelines Division, Washington, DC, 1980.

Youden, W.J., and E.H. Steiner, *Statistical Manual of the Association of Analytical Chemists*, Association of Official Analytical Chemists (now AOAC International), Arlington, VA, 1975.

BIBLIOGRAPHY

Bhattacharyya, Gouri K., and Richard A. Johnson (1977), *Statistical Concepts and Methods*, John Wiley & Sons, New York.

Garfield, Frederick M. (1991), *Quality Assurance Principles for Analytical Laboratories*, "Chapter 2, Statistical Applications and Control Charts," Association of Official Analytical Chemists (now AOAC International), Arlington, VA.

Harris, Robert L. et al. (1994), *Patty's Industrial Hygiene and Toxicology*, 3d ed., Vol. III, Part A, "Chapter 10, Statistical Design and Data Analysis Requirements," John Wiley & Sons, New York.

Keith, Lawrence H., ed. (1988), *Principles of Environmental Sampling*, American Chemical Society, Washington, DC.

Thompson, Steven K. (1992), *Sampling*, John Wiley & Sons, New York.

Wernimont, Grant T. (1985), *Use of Statistics to Develop and Evaluate Analytical Methods*, Association of Official Analytical Chemists (now AOAC International), Arlington, VA.

PART III

PRINCIPLES OF
SAFETY ENGINEERING

9

SAFETY MANAGEMENT

It is only in recent years that most safety professionals have been able to define their role in the safety work that is being accomplished. What they do has changed and will continue to evolve.

—(Dan Peterson, *Techniques of Safety Management*, 1989)

Safety management and motivation is a complex subject, essentially because of the human factor associated with it. Numerous accidents have been caused by humans committing unsafe acts—some of huge magnitude, such as the deadly toxic leak in Bhopal and the catastrophic oil spill at Prince William Sound. Thus, the human factor cannot be ignored. If human response were always predictable and scientifically quantifiable, then safety management and motivation could be "engineered" in much the same way that hazards are controlled through technology. Instead, the safety professional must rely on behavioral theories to make safety programs effective.

WORKER MOTIVATION

Hawthorne Experiment

One of the first attempts to define the effects of workplace conditions on productivity—in this case illumination—was made in 1924 at Western Electrics Hawthorne Works outside Chicago. The engineers performing the study were trying to ascertain the level of lighting in the workplace that optimized worker productivity. Experimental and control groups were set up, and illumination levels for the experimental group were carefully varied over time.

What was found from the experiment was seemingly unscientific. Both the experimental and the control groups increased their productivity. In addition, the experimental group continued to increase productivity even when lighting levels were lowered to the point of extreme dimness. In essence, the workers were responding to the attention they received from management and the researchers, and rather than to the lighting levels. This change in productivity as a result of such extra attention became known as the Hawthorne effect. Thus, it can be concluded that productivity, safety consciousness, quality of work, and other job performance indicators cannot be controlled merely by scientifically changing the physical workplace. Job performance must be approached from a behavioral standpoint as well.

Maslow's Hierarchy of Needs

Abraham Maslow had a profound effect on the field of human psychology with his theory of human motivation, first published in 1943 in *Psychological Review* as an article entitled "A Dynamic Theory of Human Motivation" and subsequently revised and expanded in his book *Motivation and Personality*, published in 1954. According to Maslow, once humans' basic physiological needs are met, other higher needs emerge. Then these other needs, rather than the physiological needs, dominate human beings, and when these are satisfied, new (and still higher) needs emerge. The concept of Maslow's hierarchy of needs is depicted in Figure 9.1.

The most basic needs are physiological and include such things as food, water, and sleep. The next layer of needs is concerned with one's safety, such as shelter and protection from danger. Further up the hierarchy are the social needs of friendship, belonging, and love. Next, needs of self-esteem must be met, which include self-confidence, independence, knowledge, status, personal recognition, achievement, and respect. The final step is self-actualization. A self-actualized individual is self-directed, accepts self and others, behaves creatively, and has a problem-solving orientation to life.

Maslow's assumption is that behavior and motivation are functions of needs satisfaction. Further, once a need is satisfied, it is no longer a motivator. Thus, according to this theory, for industrial managers to motivate employees, hierarchical needs must be recognized and can then be used as motivating factors. Typically, for most employees in a stable industrial work environment, at least physiological and safety needs will have been met and, for many, social needs as well. Using Maslow's theory, motivation would come from work that contains elements that make it possible to meet workers' needs for self-esteem and/or self-actualization.

Herzberg's Theory of Motivation-Hygiene

In the late 1950s, Frederick Herzberg proposed a theory about workers' motivational attitude, based on interviews with 200 engineers and accountants who represented a cross section of Pittsburgh industry. The theory is known as the

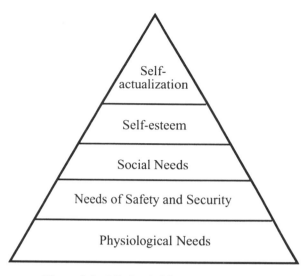

Figure 9.1 Maslow's hierarchy of needs.

Motivation-Hygiene theory. First published in *The Motivation to Work* (1959) and later in *Work and the Nature of Man* (1966), this theory views environmental—or hygiene—factors such as working conditions, wages, supervision, and company policies as factors that can produce short term changes in job attitudes, but that are rarely involved in long term motivational changes. These hygiene factors are, thus, termed "dissatisfiers" since they are factors that might prevent job dissatisfaction but do not, in and of themselves, motivate. Herzberg's research indicated that motivation forces, or "satisfiers," come from achievement, recognition of accomplishment, the work itself, responsibility, and advancement, the last three being of the greatest importance.

Herzberg points out that factors that produce job satisfaction are separate and distinct from factors that lead to job dissatisfaction. Job satisfaction and job dissatisfaction are not obverse of one another. Thus, the opposite of job satisfaction is *no* job satisfaction, not job dissatisfaction. Likewise, the opposite of job dissatisfaction is *no* job dissatisfaction. Thus, using this theory, it would follow that for safety to become a long-term, integral part of a worker's priority, it should involve elements such as individual responsibility and recognition of accomplishment, instead of only "good working conditions." Herzbergs theory is similar to Maslow's theory of needs hierarchy in that factors that create job satisfaction are commensurate with the needs of self-esteem and self-actualization.

McGregor's Theory X and Theory Y

About the same time that Maslow and Herzberg were publishing motivational theories that pertain to individuals, Douglas McGregor postulated that management style has a significant effect on motivation. In his book, *The Human Side of Enterprise* (1960), McGregor argued that conventional managers (at that time) allowed employees little freedom in performing their jobs because they believed that workers had an inherent dislike of work and would avoid it if possible. In addition, because of this dislike of work, these managers felt that most people must be coerced, controlled, directed, and threatened with punishment to get them to put forth adequate effort to meet organizational objectives. Further, McGregor thought that conventional managers believed that the average human being prefers to be directed, wishes to avoid responsibility, has relatively little ambition, and wants security above all. This managerial approach was termed Theory X.

As an alternative to Theory X, McGregor set forth the premise that employees work their best when given large amounts of autonomy and in an environment conducive to responsibility, growth, and recognition. McGregor referenced Herzberg's motivation-hygiene theory and alluded to Maslow's hierarchy of needs when he proposed that managers allow employees to seek and reach levels of self-esteem and self-actualization in their jobs. This second approach was termed Theory Y, which McGregor thought to be particularly applicable to managing managers and professional people.

McGregor states several assumptions about human behavior as the basis of Theory Y. He advocates that the expenditure of physical and mental effort in work is as natural as play or rest; thus, the average human being does not inherently dislike work. Other assumptions are that, under proper conditions, people will not only accept responsibility, but seek it, and that they will exercise self-direction and self-control in the service of objectives when they are committed to these objectives. Thus, the management style depicted in Theory X insults the worker's intelligence and stifles creativity. Because humans resist close supervisory control, it actually reduces worker effectiveness and harms employee motivation. A freer environment enables workers to fully utilize their intrinsic abilities. Using this theory, safety professionals would strive to ensure that their safety programs and procedures are not totally dictatorial and that they allow for worker input and participation.

The Great Jackass Fallacy

Harry Levinson published *The Great Jackass Fallacy* in 1973. This work discusses reasons for problems in worker motivation and cites such reasons as an executive's fear of losing control, managerial inexperience and lack of training, and the continued use of an ineffective management system of carrot and stick, also known as reward and punishment. This latter system Levinson calls "the great jackass fantasy." In management training sessions Levinson would ask executive participants to visualize a carrot and a stick and to describe the central image in that picture. Most frequently, the executives would respond that the central image was a jackass. Thus, according to Levinson, it becomes evident that management frequently feels that employees have "jackass" qualities such as stubbornness, stupidity, willfulness, and unwillingness to go where someone is driving them. When this perspective becomes obvious to workers through their observation of management's motivation techniques, they resist hearing the message and communications break down.

Further, Levinson argues, bureaucratic organizational structure compounds motivational problems. Based on a military model that assumes control from the top, every level is dependent on the higher level for management directives. Subordinates have little influence and no control over management directives and thus feel helpless and manipulated. In addition, when success is measured by one's ability to move up the pyramidal hierarchy, fewer and fewer people experience success along their career paths because there is room for fewer and fewer people to move up in the pyramid. While an employer is hoping for followers in the organization to be dedicated, committed, and innovative, the bureaucratic structure discourages these attributes in workers. Levinson contends that management must get rid of invalid motivational assumptions and revise the organizational structure wherever possible.

Perspectives of Peter Drucker

Early Writings. Peter F. Drucker has been writing about management for many decades. Some of his first management theories were set forth in his books *Concept of the Corporation, The New Society*, and *The Practice of Management.* In these works, Drucker suggests that managing workers by putting responsibility on them and by aiming at achievement is possible, but it places exceedingly high demands on both worker and manager. In a later work, *Management: Tasks, Responsibilities, Practices* (1973), Drucker suggests that management using McGregor's Theory Y (discussed earlier in this chapter) can be effective only if there are also direction or structure, objectives, and some controls. These would provide security of order and direction while still allowing workers to take on responsibility. Drucker further suggests that management using the carrot and the stick is ineffective—particularly the use of the stick, because employees do not fear that their ability to survive is tied to their job, as it was in the past. Although the carrot of monetary reward may have some effect on workers, over time this reward becomes expected and (as Herzberg pointed out and Drucker also emphasizes) managers must find means other than material rewards to motivate workers.

Management by Objectives. Drucker has always advocated setting objectives, priorities, and strategies, and his goal-setting theory later evolved and became popularized as the concept of Management by Objectives (MBO). His user's guide to MBO was first published in *Public Administration Review* (January-February 1976) and was later published in his book *Toward the Next Economics and Other Essays* (1981), although discussions of setting objectives, strategies, and priorities can be found very early in his writings.

MBO is typically used by public agencies, public service institutions, and very large and complex industries. In his user's guide, Drucker describes management by objectives as a two-part process—management by *objectives* must be addressed, as well as *management* by objectives. Management by *objectives* entails setting clear, measurable objectives, prioritizing (and sometimes eliminating) goals, and setting specific targets, timetables, and strategies. A clear definition of resources needed to obtain goals must be established. Performance measurements must also be established.

Management by objectives entails making informed decisions about objectives, strategies, priorities, resources, and measurements. *Management* by objectives brings out the basic views, dissents, and different approaches to the same task and same problems and allows the organization to maintain structure and direction.

Drucker explains that, if used correctly, MBO is at the core of planning and managing. It allows people to integrate themselves into the organization and direct themselves toward the organization's goals and purposes.

ACCIDENT PREVENTION

Heinrich's Theory of Accidents

Perhaps the most well-recognized accident theory is Heinrich's domino theory. Heinrich proposed that accidents are a result of five conditions, theoretically standing on end like dominoes. These conditions include social environment, fault of a person, unsafe act or condition, accident, and injury. If any one condition or "domino" is set in motion, the others will fall—unless one of the conditions can be removed. Thus, Heinrich theorizes that control of unsafe acts and conditions—particularly unsafe acts—can be the key to reducing accidents. In many cases, control of unsafe conditions can be accomplished through engineering design and technology. Control of unsafe acts is less scientific and includes education, personal involvement in safety, and other intangible factors.

Accident Statistics

The Centers for Disease Control and Prevention (CDC), which is part of the National Institute for Occupational Safety and Health (NIOSH), has published statistics pertaining to fatal injuries to workers for the decade of the 1980s (CDC 1993). Some key data from the NIOSH study are presented in Figures 9.2 to 9.4. The following statistics and information pertaining to work-related fatalities are summarized in the CDC study:

- The number of fatal accidents and the rate per 100,000 workers has declined by 23% and 37%, respectively.

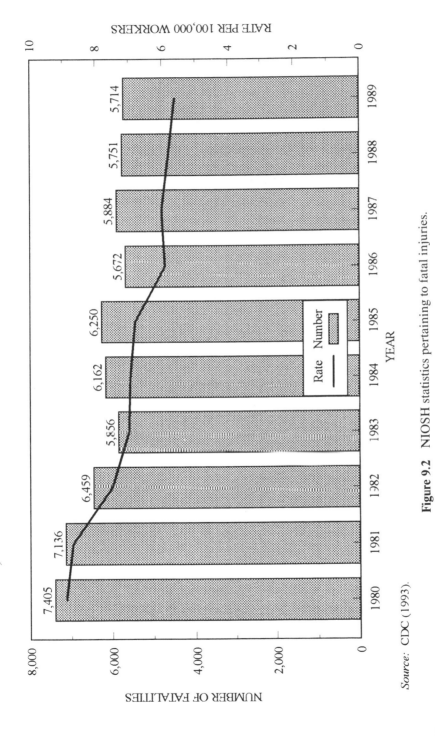

Figure 9.2 NIOSH statistics pertaining to fatal injuries.

Source: CDC (1993).

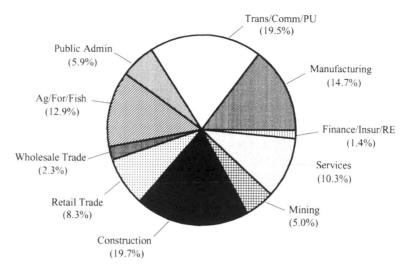

Source: Information from CDC (1993).

Figure 9.3 NIOSH statistics pertaining to fatal injuries.

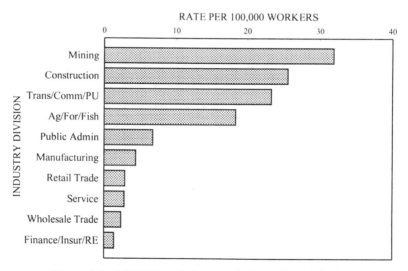

Figure 9.4 NIOSH statistics pertaining to fatal injuries.

- The leading causes of occupational injury death in the United States were motor vehicle crashes, machine-related incidents, homicides, falls, electrocution, and being struck by falling objects.
- The largest number of fatalities occurred in the construction, transportation/communication/public utilities, manufacturing, and agriculture/forestry/fishing industry divisions.

- The highest average annual fatality rate per 100,000 workers occurred in the mining industry, followed by construction, transportation/communication/public utilities, manufacturing, and agriculture/forestry/fishing industry divisions.
- The occupation division of transportation/material movers had the highest average annual fatality rate per 100,000 workers, followed by farmers/foresters/fishers, laborers, and precision production/craft/repair occupations.
- The age group with the largest number of occupational injury deaths was the 25- to 29-year-old group, followed by the 30- to 34-year old and 20- to 24-year old age groups.
- Workers 65 years old and older had the highest fatality rate per 100,000 workers of any age group.

Accident Investigation

Every accident should be investigated. The accident investigator should determine the cause of the accident, which could include any of the following:

- System failures, such as equipment malfunction, structural inadequacy, or design problems
- Procedural failures, such as poorly written task requirements, lack of necessary operating procedure, or inadequate training
- Employee error, such as performing job incorrectly or failure to use proper safety equipment

Once the cause of an accident is determined, corrective actions should be initiated. In some cases, the unsafe conditions at one location may exist elsewhere at the facility or job site. The prudent accident investigator will review areas with similar conditions (i.e., equipment or job tasks) to assess the potential for an occurrence of a similar accident in other areas.

Traditionally, the manager or immediate supervisor of the employee who has the accident performs the accident investigation. These investigations are often reviewed by the safety specialist and management during periodic safety meetings.

As a more proactive measure, the health and safety staff may decide to institute a method of tracking near miss accidents as well as those that actually occur. Campaign programs that reward employees for coming forward with examples of accidents that could occur in the workplace can be used to discover potential safety problems.

If an accident is serious, then a safety specialist or team of specialists should be called in to help with the investigation immediately. The involvement of an attorney during investigation of serious accidents may also be prudent, in case a lawsuit or other litigation arises in the future. As provided under 29 CFR 1904.8, an employer must report any accident to OSHA within eight hours when it involves a fatality or when three or more employees are hospitalized as a result of the accident.

Job Hazard Analysis

Another method commonly used by safety professionals to prevent accidents and injuries in the workplace is job hazard analysis. In conducting a job safety analysis, a specific job task is studied carefully, by steps of the job, in order to identify existing or potential job hazards. Candidates for jobs that might be analyzed using this method include those with the highest rates of accidents and injuries, as well as jobs in which close calls have occurred. After these jobs have been evaluated, other jobs, such as new jobs and jobs that have had changes made in their processes, can be addressed. Eventually, a job hazard analysis can be conducted for all jobs.

Elements of a job that may be reviewed include electrical hazards, use of tools, lighting, fire protection, personal protective equipment, ventilation, confined spaces, tripping hazards, ergonomic layout, and other safety and health elements. Once hazards are identified, improved job methods are developed to reduce or eliminate them. When possible, the source of a hazard should be eliminated. If the hazard cannot be eliminated, the danger should be reduced as much as possible. Employee involvement should be part of the process.

Once the job hazard analysis is completed, employees are trained on the proper procedures and the safer and more efficient work methods. In some cases, for instance, if the job requires the use of personal protective equipment (PPE), formal training and recordkeeping may be required.

SAFETY MANAGEMENT TODAY

There are many and varied approaches to safety management today; however, they often share common elements. There is an emphasis on the difference between behavior—something that can be observed and measured, such as wearing personal protective equipment—and personal factors that involve feelings and attitudes that cannot be observed and are difficult to measure. Safety managers focus on changing behavior, such as increasing the usage of PPE, by providing the proper environment, including equipment, engineering, physical layout, and so forth. They also consider the interplay between personal factors and behavior and environment.

Total quality management (TQM) concepts and techniques are also finding their way into safety management programs. Safety management programs today often include:

- Integration of behavior and personal approaches
- Avoidance of placing blame on individual
- Focus on the design of the process or activity in order to reduce accidents

- Observations of behavior and the sharing of these observations with workers
- Worker participation in defining and solving safety programs
- Recognition and incentives for workers

Like TQM, safety management today depends on the commitment of both management and employees to ensure that safety is fully integrated in the workplace.

Safety promotionals in the form of poster board campaigns, safety logos, safety publications, and other safety reminders have traditionally and continue to be a successful method of ensuring that safety remains a top focus in the workplace. Figure 9.5 shows some safety reminders that can be made into decals for hard hats or into posters for on site use. Figure 9.6 shows an example of a promotional signature flag for OSHA's Voluntary Protection Program, which is discussed in the next section of this chapter.

Figure 9.5   Safety promotional decals. (With permission from EMED Co., Buffalo, N.Y.)

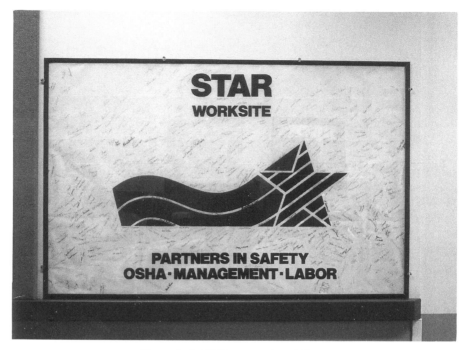

Figure 9.6   VPP promotional flag.

OSHA'S VOLUNTARY PROTECTION PROGRAM (VPP)

OSHA's VPP is authorized under the OSH Act and is intended to encourage employers and employees to reduce the number of occupational safety and health hazards at their places of employment, and to stimulate the improvement of programs for providing safe and healthful working conditions. The program recognizes and encourages worksites that have lower than average workplace injury rates and that have adopted comprehensive safety and health program management approaches that are considered the best in industry. VPP recognizes that good workplace safety and health programs go well beyond OSHA standards.

In addition to recognizing excellent safety and health programs, VPP supplements OSHA's enforcement effort so that VPP worksites are not subject to the same inspection requirements that other sites are. Once certified as a VPP site, participants are exempted from OSHA general inspections (from one to three years, depending on the level of certification), although employees rights to contact OSHA and to file complaints remain the same.

There is no one "best" design of a safety management system, inasmuch as operations and hazards of individual worksites are varied. Elements common to programs in approved **VPP** worksites include a comprehensive safety and health program; systematic hazard assessment and control; training programs; clear safety planning; work procedures, and rules; management commitment; and employee participation. Additional detail for each of these elements is presented in Table 9.1.

TABLE 9.1   Elements Common to Programs in Approved VPP Worksites

Safety and Health Program

- Documented required safety programs, such as hazard communication program, medical surveillance program, process safety management program, laser safety program, personal protective equipment program, lock-out/tag-out program, and others
- Specific safety and health training, including process equipment safety, electrical safety, personal protective equipment training, confined space training, training on processes containing highly hazardous chemicals, hazard communication training, and other task-related training, as required
- Safety manual accessible to managers and employees at worksite
- Job-specific or department-specific documented safety plans kept in the work area
- Total safety and health program evaluation system that provides for annual reports and includes evaluation of safety staffing, training, hazard assessment, safety rules and procedures, employee participation, and special emphasis programs
- Total injury rates and lost workday case rates for all employees averaged over three years are lower than the national average for the industry

Systematic Assessment and Control of Hazards

- Identification of system for identifying and controlling workplace hazards
- Program for conducting monthly self-assessments
- Documentation of findings during self-assessment and correction of hazards
- Documentation of environmental monitoring programs, including those for hazardous substances, noise, radiation, and others
- Training of personnel who conduct environmental sampling; verification that testing and analysis are performed in accordance with nationally recognized procedures
- System of communication to management of employee observations of unsafe or unhealthful conditions
- Accident investigation program, with documented results
- Preventative maintenance program
- Program for correcting identified hazards in timely fashion
- Medical program, including access to physician services and adequate first-aid provisions

TABLE 9.1 *Continued*

Training

- General periodic safety and health training such as safe lifting, good housekeeping practices, general chemical safety, hazard identification, and other general safety training
- Training program for new hires
- Management training on safety and health issues, policies, and procedures
- Training for emergency response and building or site evacuation

Safety Planning, Work Procedures, and Rules

- Safety and health planning, as integrated in the overall management system
- Documented safety rules and enforcement of safety rules and procedures
- Routine analysis—of jobs, processes, and/or construction crafts—for hazards
- Written personal protective equipment procedures
- Emergency planning documentation
- Identification of supervisors role in safety and health program

Management Commitment

- Written management commitment to worker safety and health protection
- Top management involvement in safety and health program
- Clear system of accountability for safety at the line level and clear assignment of safety responsibility
- Clearly established safety and health policies and results-oriented objectives
- Resources authorized for safety and health; certified safety professionals, industrial hygiene professionals, and medical personnel available on-site or through corporate or other means
- Contract workers provided with equivalent safety and health protection as provided for regular employees

Employee Participation

- Employees part of and proud of the safety and health programs
- Employees actively involved with safety and health programs

REFERENCES

Center for Disease Control and Prevention, *Fatal Injuries to Workers in the United States, 1980–1989: A Decade of Surveillance*, DHHS (NIOSH) Number 93-108, U.S. Department of Health and Human Services, Public Health Service, Centers for Disease Control and Prevention, National Institute for Occupational Safety and Health, Cincinnati, OH, 1993.

Drucker, Peter F., *Toward the Next Economics and Other Essays*, Harper & Row, New York, 1981.

Drucker, Peter F., *Management: Tasks, Responsibilities, Practices*, Harper & Row, New York, 1973.

Herzberg, Frederick, *Work and the Nature of Man*, World Publishing Company, Cleveland, OH, 1966.

Herzberg, Frederick, Bernard Mausner, and Barbara Bloch Snyderman, *The Motivation to Work*, John Wiley & Sons, Inc., New York, 1959.

Levinson, Harry, *The Great Jackass Fallacy*, Harvard University Press, Boston, 1973.

Maslow, A.H., *Motivation and Personality*, Harper & Brothers, New York, 1954.

Maslow, A.H., "A Dynamic Theory of Human Motivation," *Psychological Review*, 50, 370–396, 1943.

McGregor, Douglas, *The Human Side of Enterprise*, McGraw-Hill, New York, 1960.

Occupational Safety and Health Administration, *Job Hazard Analysis*, OSHA 3071, U. S. Department of Labor, Occupational Safety and Health Administration, Washington, DC, 1994.

Petersen, Dan, *Techniques of Safety Management: A Systems Approach*, 3d ed., Aloray, Inc., Goshen, NY, 1989.

Petersen, Dan, *Safety Management: A Human Approach*, 2d ed., Aloray, Inc., Goshen, New York, 1988.

BIBLIOGRAPHY

Argyris, Chris (1957), *Personality and Organization*, Harper, New York.

Colvin, Raymond J. (1991), *The Guidebook to Successful Safety Programming*, Lewis Publishers, Boca Raton, FL.

Curtis, Steven L. (1995), "Safety and Total Quality Management," *Professional Safety*, Vol. 40/No. 1.

Fahey, Robert G. (1992), "Accident Prevention Through Effective Safety Management," master's thesis, University of Florida, Gainesville, FL.

Geller, E. Scott (1994), "Ten Principles for Achieving a Total Safety Culture," *Professional Safety*, Vol. 39/No. 9.

Hidley, John H., and Thomas R. Krause (1994), "Behavior-based Safety: Paradigm Shift Beyond the Failures of Attitude-based Programs," *Professional Safety*, Vol. 39/No. 10.

Hannaford, E. (1976), *Supervisors Guide to Human Relations*, National Safety Council.

Martin, Desmond D., and Richard L. Shell (1980), "What Every Engineer Should Know About Human Resources Management," Marcel Dekker, Inc., New York.

Myers, Mike (1995), "Why Include Total Quality Management in Safety Programs?" *Water Environmental and Technology*, Vol. 7/No. 1.

Raouf, A., and B.S. Dhillon (1994), *Safety Assessment: A Quantitative Approach*, Lewis Publishers, Boca Raton, FL.

Trautlein, Barbara A. and David W. Milner (1994), "Work System Design: Creating a New Safety Paradigm," *Professional Safety*, Vol. 39/No. 12.

Woodside, Marianne, and Tricia McClam (1990), *An Introduction to Human Services*, Brooks/Cole, Pacific Grove, CA.

10

EQUIPMENT SAFETY

ELECTRICAL SAFETY

Understanding Electricity

Electric Current. Electric current is produced when free electrons in a wire or other conductive material are moved in the same direction. The flow of electrical energy is defined by Ohm's law:

$$I = V/R \tag{10.1}$$

where

  I is in amperes
  V is in volts
  R is in ohms

Simply put, the law states that electric current (I) is a function of electrical potential (V) between two points and the resistance (R) between them. Resistance is a function of the material over which electrons move. Electrical energy, also termed electrical charge, can move between the two points only where there is a difference in energy levels between two points. If there is more than one path between two points that have differing electrical energy levels, the electrons will move through the path of least resistance.

  All materials exhibit some resistance to the flow of electricity. Materials that allow electrons to flow easily are conductors. Examples include metals, water (except pure water), and other electrolytic fluids. Materials that do not allow electrons to flow easily are insulators. Examples include rubber, wood, polyethylene, and gases. The relative resistance of different metals as compared with copper (with copper equal to 1) is depicted in Figure 10.1. The relative conductance of these same materials is equal to 1/resistance, or the reciprocal of the relative resistance.

The resistance of a wire is directly proportional to its length and inversely proportional to its cross-sectional area. These relationships are shown graphically in Figure 10.2. As can be seen in the figure, the greater the cross-sectional area of the wire, the lower the resistance, and the longer the wire, the greater its resistance.

Temperature also has an effect on resistance. Normally, as the temperature of a material increases, the resistance increases, as with copper, aluminum, and nickel. As temperature increases, thermal vibration of the atoms reduces the mean free path and the electrical resistivity increases. There are some exceptions, such as with glasses, oxide ceramics (e.g., alumina and zircon porcelain), and semiconducting compounds (e.g., indium antimonite and zinc oxide), which experience a decrease in resistance as the temperature increases.

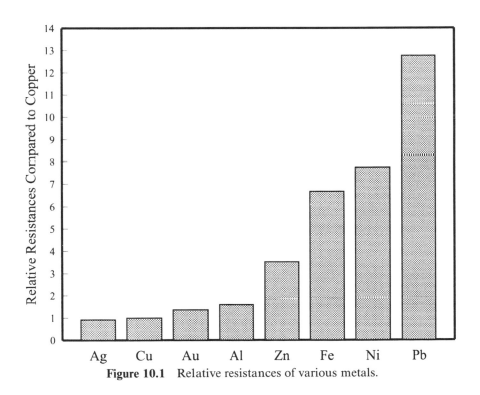

Figure 10.1   Relative resistances of various metals.

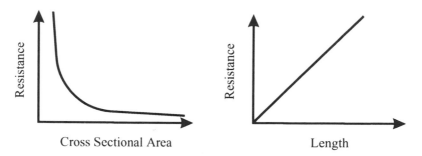

Figure 10.2   Relationships of resistance of wire to cross-sectional area and length.

Electric Circuits. Electric circuits are closed circuits made up of a power source, electric conductors in the form of wire, and a load. In most cases, the resistance of the load is much greater than that of the circuit wiring. There is also internal resistance within the power source of the circuit, but it is also very low as compared with the load. Loads can vary markedly in resistance. For instance, the element resistance of a toaster is approximately 9 ohms, whereas the motor winding resistance of an electric drill is 60 ohms or greater.

   The total resistance of a circuit, then, is the sum of the individual resistances of the power source, the circuit wiring, and the load. Typically, in applying Ohm's law, the resistances of the power source and the circuit wiring are ignored.

EXAMPLE 1

A 25-ohm resistor is used as the load in a circuit having a 200-volt battery as a voltage source, as shown in Figure 10.3a. If the current rating of the resistor is 10 amperes, will its rating be exceeded when the switch is closed?

Solution.   Use Ohm's law to find the current:

$$I = V/R = 200 \text{ volts}/25 \text{ ohms}$$

$$I = 8 \text{ amperes}$$

The current rating of 10 amperes is not exceeded when the switch is closed, so the use of the 25-ohm resistor is acceptable.

EXAMPLE 2

Determine the size of the resistor for the circuit shown in Figure 10.3b. The circuit has a current flow of 6 amperes, and the voltage source is a 30-volt battery.

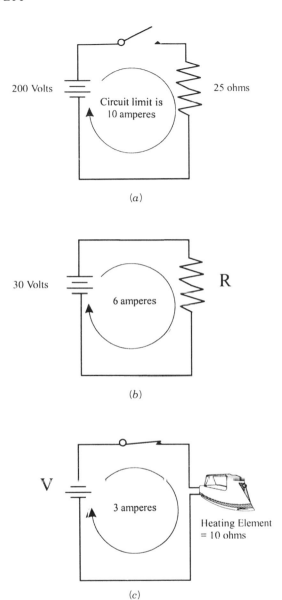

200 Volts

Circuit limit is
10 amperes

25 ohms

(a)

30 Volts

6 amperes

R

(b)

V

3 amperes

Heating Element
= 10 ohms

(c)

Figure 10.3   Example circuits.

Solution. Again, apply Ohm's law:

$$I = V/R$$
$$\text{so } R = V/I = 30 \text{ volts/6 amperes}$$
$$R = 5 \text{ ohms}$$

EXAMPLE 3

An electric iron has a heating element (resistor) of 10 ohms, and 3 amperes of current flow in the circuit when the switch is closed, as shown in Figure 10.3c. Size the voltage of the battery needed in this circuit.

Solution. This is the third relationship defined by Ohm's law:

$$I=V/R$$

$$\text{so } V=RI=3 \text{ amperes} \times 10 \text{ ohms}$$

$$V=30 \text{ volts}$$

Series Loads. When series loads are used, the total circuit resistance is the sum of the resistances of each individual load. Thus, if a circuit has five loads connected in series and each load is a 5-ohm resistor, the total circuit resistance is 5×5, or 25 ohms. A series circuit has only one path, and if the path is broken, current will not flow. Everything in that circuit that depends on the current is affected. The path can be broken by an open switch, a burned out fuse or resistor, or a broken wire.

EXAMPLE 4

What is the circuit current in Figure 10.4?

Solution. Find R_{TOT}; then apply Ohm's law.

$$R_{TOT}=R_1+R_2+R_3=10+7+13=30 \text{ ohms}$$

$$I=V/R=90 \text{ volts}/30 \text{ ohms}$$

$$I=3 \text{ amperes}$$

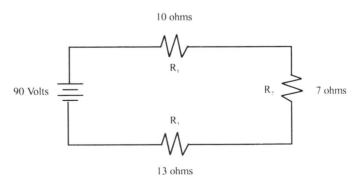

Figure 10.4   Example circuit.

Power. Power is the amount of work that can be done by a load in some standard amount of time. Power is measured in watts. A watt is defined as the power used when 1 ampere of current flows through a potential difference of 1 volt.

Power is calculated using the following equations:

$$P = I^2R \text{ and } P = VI \qquad (10.2)$$

When a circuit has a number of loads connected in series, the total power consumed in the circuit is the sum of the power consumptions of each load.

EXAMPLE 5

Find the power used in the circuit current in Figure 10.4.

Solution. Use the following equations:

$$P - I^2R$$

$$P_1 = 3 \times 3 \times 10 = 90 \text{ watts}$$

$$P_2 = 3 \times 3 \times 7 = 63 \text{ watts}$$

$$P_3 = 3 \times 3 \times 13 = 117 \text{ watts}$$

$$P_{TOT} = 90 + 63 + 117 = 270 \text{ watts}$$

Also,

$$P_{TOT} = VI = 90 \text{ volts} \times 3 \text{ amperes} = 270 \text{ watts}$$

Parallel Circuits. A parallel circuit is one in which there are one or more points where the current divides and follows different paths before recombining to flow back to the power source. Thus, for parallel circuits the current is not necessarily the same at every point. The total current is equal to the sum of the branch currents. A branch current is determined by the resistance of the branch and the voltage across it. Because all branches have the same voltage across them, the more resistance there is in a branch, the less current there will be.

Adding parallel loads to a circuit decreases the total circuit resistance and increases the total current. Effectively, parallel circuits increase the cross-sectional area of the conducting material.

In a parallel circuit, the total resistance is calculated by the following equation:

$$1/R_{TOT} = 1/R_1 + 1/R_2 + 1/R_3 \ldots \qquad (10.3)$$

There is another method of analyzing the circuit which, when visually displayed, adds clarity to the concepts involved with parallel circuits. If the parallel resistances are equal, as shown in Figure 10.5, then the total resistance equals one resistance divided by the number of resistances. If the resistances are unequal but in multiples of one another, any one resistor can be considered as two or more other resistors in parallel. For an illustration of this, refer to Figure 10.6. Figure 10.6a can be redrawn as shown in Figure 10.6b. The figures have equivalent resistances, and both methods for calculating the resistance are valid.

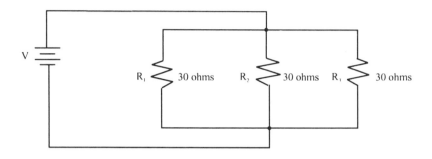

$R_{TOT} = 30/3 = 10$ ohms

Figure 10.5   Example circuit.

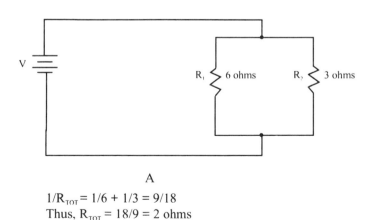

A

$1/R_{TOT} = 1/6 + 1/3 = 9/18$
Thus, $R_{TOT} = 18/9 = 2$ ohms

Figure 10.6   Example parallel circuit.

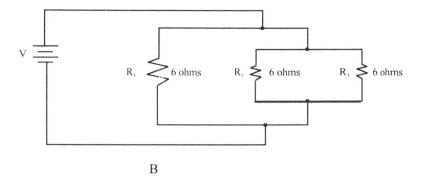

B

$R_{TOT} = 6/3 = 2$ ohms

Figure 10.6 *Continued*

EXAMPLE 6

Calculate the current in Figure 10.7.

Solution. Calculate resistance first, as follows:

$$1/R_{TOT}=1/R_1+1/R_2+1/R_3=1/20+1/10+1/40=7/40$$

thus,

$$R_{TOT}=40/7=5.714 \text{ ohms}$$

Then, using Ohm's law:

$$1=V/R=120/5.714=21 \text{ amperes}$$

In another approach to the same problem, take one branch at a time and consider it a series circuit. Then add all the branch currents to get the total current.

$$1=V/R_1+V/R_2+V/R_3=120/20+120/10+120/40$$

$$=6+12+3$$

$$I=21 \text{ amperes}$$

EXAMPLE 7

Find the total resistance of the circuit in Figure 10.8. What size battery is needed if the current is 10 amperes?

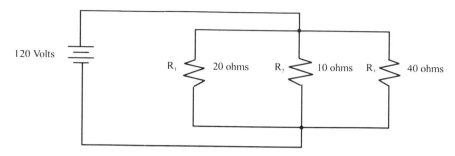

Figure 10.7   Example parallel circuits.

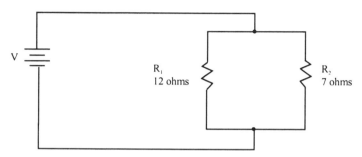

Figure 10.8   Example parallel circuit.

Solution. First, find the total resistance using the following equation:

$$1/R_{TOT} = 1/R_1 + 1/R_2$$
$$1/R_{TOT} = 1/12 + 1/7 = 19/84$$
$$R_{TOT} = 84/19 = 4.42$$

Next, use Ohm's law:

$$I = V/R; \text{ so}$$
$$V = RI = 10 \times 4.42 = 44.2 \text{ volts}$$

Because batteries do not come in this odd size, select the battery in the nearest size, but greater than 44.2 volts.

Electrical Hazards

Some employees work with electricity directly, as is the case with engineers, electricians, and electrical technicians who do wiring and circuit assembly. Others, such as manufacturing employees and even office workers, may work with it indirectly. As discussed at the beginning of this chapter, electricity travels in closed circuits, and its normal route is through a conductor.

Shock occurs when the human body becomes part of an electrical circuit. The current enters the body at one point and leaves at another. This dangerous and sometimes deadly condition can occur in any of three ways:

- A person comes in contact with both wires of an electric circuit.
- A person comes in contact with one wire of an energized circuit and the ground.
- A person comes in contact with a metallic part that is energized, or "hot," as a result of being in contact with an energized wire, and the person is also in contact with the ground.

The severity of shock is dependent on the amount of current flowing through the body, the path of the current through the body, and the length of time the body is in the circuit. Other secondary factors of shock severity include the frequency of the current, the phase of the heart cycle when shock occurs, and the general health of the person. Examples of the effects of electric current in the human body are presented in Table 10.1. As can be seen from the table, death can occur at current levels as low as 50 milliamperes.

TABLE 10.1 Effects of Electric Current in the Human Body

Current	Human Response
1 milliampere	Current is perceived as a tingle.
5 milliamperes	Slight shock is felt, but it is not painful.
6–25 milliamperes (women) 9–30 milliamperes (men)	Painful shock is felt and muscular control is lost; in some cases the accident victim cannot let go of the current.
50–150 milliamperes	Extreme pain is felt, respiratory functions are arrested, and the accident victim experiences severe muscular contractions; in almost all cases, the accident victim cannot let go unless the extensor muscles are excited by the shock and the person is thrown away from the current; death is possible.
1,000–4,300 milliamperes	Pumping of the heart ceases, and the accident victim goes into ventricular fibrillation; muscular contraction and nerve damage occur, and death is likely.
10,000 milliamperes	The accident victim experiences cardiac arrest and severe burns; death is probable.

Source: Information from OSHA 3075 (1991).

Correcting Electrical Hazards

Typically, electrical accidents are caused by faulty equipment, unsafe work-places, or unsafe work practices. Methods for protecting people from hazards caused by electricity include the use of insulation, guarding, grounding, electrical protective devices, and safe work practices.

Insulation. Insulation can provide a safeguard against energized wires and parts. Insulators such as rubber and plastic are used on conductors to prevent shock, fires, and short circuits. Insulation is typically color coded. Insulated wires used as grounding conductors are green or green with yellow stripes; grounded conductors that complete a circuit are generally covered with continuous white or natural gray-colored insulation. Ungrounded conductors, or "hot wires," are often black or red but can be any color other than green, white, or gray (OSHA 3075 1991).

Guarding. Live parts of electrical equipment operating at 50 volts or more must be guarded against accidental contact. OSHA considers it acceptable to guard these parts in any of the following ways (OSHA 3075 1991):

- The live parts can be located in a room, vault, or enclosure accessible only to qualified persons.
- The live parts are excluded from unqualified persons by use of permanent, substantial partitions or screens.
- The live parts are located on a balcony or platform elevated and arranged to exclude unqualified persons.
- The live parts are elevated 8 feet or more above the floor.

Entrances to rooms and other guarded locations containing exposed live parts must be marked with warning signs forbidding unqualified persons to enter. Indoor electric installations of more than 600 volts that are accessible to unqualified persons must be in a metal cabinet or enclosed in a vault that is locked.

Equipment Grounding. Equipment grounding is considered a secondary measure for protecting employees from electrical shock. Grounding is needed because the metal parts of electric tools and machines may become energized if there is a break in the insulation of a tool or machine wiring. A low-resistance path from the metallic case of the tool or machine to the ground is established through an equipment grounding conductor. The low-resistance conductor causes any unwanted current to pass directly to the ground instead of through the body of a worker. If the equipment grounding conductor is properly installed, the worker is protected. The system does not guarantee that the worker will not receive a shock or injury. It will, however, substantially reduce the possibility of this type of accident.

Electrical Protective Devices. Electrical circuit protection devices are designed to automatically limit or shut off the flow of electrical current in the event of a fault. Examples include fuses, circuit breakers, and ground-fault circuit interrupters (GFCI).

A GFCI is essentially a fast-acting circuit breaker that senses small imbalances in a circuit caused by current leakage to ground. In a fraction of a second, the device can shut off electricity. The GFCI continuously matches the amount of current going to an electrical device against the amount of current returning from the device along the electrical path. Whenever the amount going into the device differs from the amount returning by approximately 5 milliamperes, the GFCI interrupts the electrical power within as little as 1/40 of a second. This type of device does not, however, protect the employee from line-to-line contact hazards (OSHA 3007 1993). The GFCI is used in high-risk areas such as wet locations and construction sites.

Safe Work Practices. Safe work practices include deenergizing and locking out electric equipment before inspecting or making repairs, using electric tools that are in good repair, using good judgment when working near energized lines, and using appropriate protective equipment. To ensure that employees are aware of and use safe work practices, the employer must train workers on the electrical hazards to which they will be exposed. Training should include procedures for safety from electrical hazards and other safety-related topics.

OSHA Requirements in Respect to Electrical Safety

OSHA requirements in respect to electrical safety for general industry are promulgated under 29 CFR 1910 Subpart S. Topics covered under this Subpart S are delineated in Table 10.2.

Hazard (Classified) Locations. OSHA classifies hazard locations based on properties of flammable vapors, liquids, or gases, or combustible dusts or fibers that may be present at the locations. The locations are also classified by the likelihood that a flammable or combustible concentration or quantity is present. The six OSHA-defined designations are presented in Table 10.3.

Lockout/Tagout. The control of hazardous energy—typically termed lockout/tagout—is defined under 29 CFR Part 1910.147. In general, the rule requires that machines or equipment must be turned off and disconnected from the energy source before service or maintenance is performed. In addition, the energy-isolating device must be either locked or tagged out. Employers must develop written procedures for their lockout/tagout program and must specify locks or tags that are clearly identifiable. Training of employees and inspection of the use of the procedures are also part of the requirements.

TABLE 10.2   Topics Covered Under CFR 1910 Subpart S for Electrical Safety

Sections of Subpart S	Topic
1910.301	Introduction
1910.302	Electric utilization systems
1910.303	General requirements
1910.304	Wiring design and protection
1910.305	Wiring methods, components, and equipment for general use
1910.306	Specific purpose equipment and installations
1910.307	Hazard (classified) locations
1910.308	Special systems
1910.309–1910.330	Reserved for safety-related work practices
1910.331	Scope for safety related work practices
1910.332	Training
1910.333	Selection and use of work practices
1910.334	Use of equipment
1910.336–1910.3660	Reserved for safety-related maintenance requirements
1910.361–1910.380	Reserved for safety requirements for special equipment
1910.381–1910.398	Reserved for definitions
1910.399	Definitions applicable to Subpart S

HAND AND POWER TOOLS

Requirements for safety procedures and safeguards associated with use of hand and portable power tools in general industry are defined under 29 CFR 1910 Subpart P. The requirements address the use and guarding of these types of tools.

There are several basic safety practices that can help prevent accidents and unsafe use of hand and portable power tools. These include (OSHA 3080 1994):

- Keeping all tools in good condition and regularly maintained
- Using the right tool for the job
- Examining each tool for damage before use
- Operating a tool according to the manufacturer's instructions
- Providing and using the right protective equipment

Employers and employees must work together to establish and ensure use of safe working practices.

Hand Tools. Hand tools are nonpowered tools and include axes, hammers, screwdrivers, wrenches, and other similar tools. Typical hazards associated with these tools include misuse and improper maintenance.

TABLE 10.3   OSHA-Defined Hazard Locations

- *Class I, Division 1*—A location in which hazardous concentrations of flammable gases or vapors may exist under normal operating conditions; or in which hazardous concentration of such gases or vapors may exist frequently because of repair or maintenance operations or because of leakage; or in which breakdown or faulty operation of equipment or processes might release hazardous concentrations of flammable gases or vapors, and might also cause simultaneous ailure of electric equipment. This classification usually includes locations where volatile flammable liquids or liquefied flammable gases are transferred from one container to another; interiors of spray booths and areas in the vicinity of spraying and painting operations where volatile flammable solvents are used; locations containing open tanks or vats of volatile flammable liquids; drying rooms or compartments for the evaporation of flammable solvents; and other such hazardous applications defined in CFR 1910 Subpart S.

- *Class I, Division 2*—A location in which volatile flammable liquids or flammable gases are handled, processed, or used, but in which the hazardous liquids, vapors, or gases will normally be confined within closed containers or closed systems from which they can escape only in case of accidental rupture or breakdown of such containers or system, or in case of abnormal operation of equipment; or in which gases or vapors are normally prevented by positive mechanical ventilation and which might become hazardous through failure or abnormal operations of the ventilating equipment; or if the location is adjacent to a Class I, Division 1 location, and to which hazardous concentrations of gases or vapors might occasionally be communicated unless such communication is prevented by adequate positive-pressure ventilation from a source of clean air, and effective safeguards against ventilation failure are provided.

- *Class II, Division 1*—A location in which combustible dust is or may be in suspension in the air under normal operating conditions, in quantities sufficient to produce explosive or ignitible mixtures; or where mechanical failure or abnormal operation of machinery or equipment might cause such explosive or ignitible mixtures to be produced, and might also provide a source of ignition through simultaneous failure of electric equipment, operation of protection devices, or from other causes; or in which combustible dusts of an electrically conductive nature may be present.

- *Class II, Division 2*—A location in which combustible dust will not normally be in suspension in the air in quantities sufficient to produce explosive or ignitible mixtures, and dust accumulations are normally insufficient to interfere with the normal operation of electrical equipment or other apparatus; or dust may be in suspension in the air as a result of infrequent malfunction of handling or processing equipment, and dust accumulations resulting therefrom may be ignitible by abnormal operation or failure of electrical equipment or other apparatus.

- *Class III, Division 1*—A location in which easily ignitible fibers or materials producing combustible flyings are handled, manufactured, or used.

- *Class III, Division 2*—A location in which easily ignitible fibers are stored or handled, except in process of manufacture.

Power Tools. Types of power tools are determined by their power source, and include electric, pneumatic, liquid fuel, hydraulic, and powder-actuated. Typical hazards associated with these types of tools include cord damage, accidental start-up, improper maintenance, improper mounting, and improper transport of tools.

Electrical Tools. The most serious danger associated with electrical tools is electrocution. These tools must have a three-wire cord with ground and be grounded; or be double insulated; or be powered by a low-voltage isolation transformer. Double insulation is typically used because of its convenience.

Safeguards and safe practices that should be followed when using tools of this type include:

- Operation of tools within their design limitations
- Use of gloves and safety footwear when using these tools
- Ensuring that cords are inspected regularly and do not present a tripping hazard
- Ensuring that work locations are clean and dry

If an electric tool is used for grinding, cutting, polishing, or buffing—as is the case with abrasive wheel tools—then additional safeguards are needed. These include the use of eye protection, ensuring that the wheel is mounted properly and in good working condition, and ensuring proper use of safety guards.

Pneumatic Tools. Pneumatic tools are powered by compressed air. They include drills, sanders, nailers, staplers, and hammers. Dangers associated with the use of these tools include flying parts and dropped attachments. Noise is another hazard. Safeguards include positive locking devices for attachments, safety clips or retainers to prevent attachments from being unintentionally shot from the tool barrel, and hearing protection.

Liquid Fuel Tools. Liquid fuel tools are typically powered by gasoline. They include mowers, trimmers, generators, and other tools that are used outdoors or away from an electric power source. Hazards associated with these types of tools include dangerous exhaust fumes and the handling and transport of gasoline. If a tool is used in a closed area, proper ventilation or the use of respirators is required. In addition, fire extinguishers must be kept in the area.

Powder-Actuated Tools. Powder-actuated tools function like a loaded gun and should be operated only by employees who are appropriately trained in their use. The operator of a powder-actuated tool should select a powder charge that will meet the requirements of the work without resulting in excessive force. Hazards associated with these tools include accidental firing, misfiring, use by unauthorized persons, and defective parts. Safeguards and safe practices include ensuring that the muzzle end of the tool has a protective shield or guard centered perpendicularly on the barrel to confine any flying fragments or particles that might otherwise create a hazard when the tool is fired; use of ear, eye, and face protection; tagging and removing from service any tool that develops a defect during use; and proper operation of the tool.

Hydraulic Power Tools. Hydraulic power tools are powered by a fluid, such as hydraulic jacks and lifts. The fluid must be an approved fire-resistant fluid and must retain its operating characteristics at the most extreme temperatures to which it will be exposed. Hazards associated with these tools include material slippage, hose failure, and overloading. Safeguards and safe practices include blocking up the load once it is lifted; ensuring that the base of the tool rests on a firm and level surface; following manufacturers recommendation for safe operating pressure; and ensuring proper inspection and maintenance of hoses, valves, pipes, filters, and other fittings.

MACHINE SAFETY

Hazards associated with machines and equipment include rotating, reciprocating, and transversing motions; cutting, punching, shearing, and bending actions; and splashing of chemicals. These hazards are typically eliminated through guarding, safety and control devices, and splash protection. In addition, training on how to operate and maintain the machinery is an important part of any safety program.

Guarding

Guarding of machines in the workplace has been a common method of ensuring equipment safety for more than a century. Guards on machines serve several purposes, including keeping people and their clothing from coming in contact with moving parts; preventing flying particles from injuring a worker; and preventing excessive noise or dust.
  Typical guards include the following:

- Fixed guards—Designed to enclose a machine or otherwise protect the operator from moving parts and other hazards, these guards prevent accidents to fingers, hands, and other body parts, as well as prevent loose clothing from being caught by moving parts.
- Interlocked guards—These guards are designed to prevent operation of a machine if the guard is raised for purposes of machine setup, adjustment, or maintenance.
- Adjustable guards—These guards are designed to allow for a variety of production operations while maintaining hazard protection.
- Self-adjusting guards—Designed for a tool or machine, these guards "float" or otherwise self-adjust to allow for movement of materials while safeguarding the person operating the machine.

Advantages and limitations of each of these types of guards are presented in Table 10.4. Examples of machines that have interlocked guards are shown in Figures 10.9 and 10.10.

Safety Devices and Controls

In addition to guards, there are numerous safety devices and controls that are commonly used in industry to protect the worker from the hazards of machines and equipment, among which are the following:

- Moveable barriers—Designed to allow materials to be inserted or removed, a moveable barrier is interlocked with the machine so that the machine will not start unless the barrier is in place and stops if the barrier is opened.
- Automatic feed systems—These systems are designed to exclude contact of the operator with the hazardous parts of the machine.

TABLE 10.4   Advantages and Limitations of Various Types of Machine Guards

Fixed Guard

- Advantages—Can be constructed to suit many specific applications; in-plant construction is possible; can provide maximum protection; usually requires minimum maintenance; can be suitable to high production, repetitive operations.
- Limitations—May interfere with visibility; can be limited to specific operations; machine adjustment and repair often require its removal, thereby necessitating other means of protection for maintenance personnel.

Interlocked Guards

- Advantages—Can provide maximum protection; allows access to machine for removing jams without time consuming removal of fixed guards.
- Limitations—Requires careful adjustment, maintenance, and testing.

Adjustable Guards

- Advantages—Can be constructed to suit many specific applications; can be adjusted to admit varying sizes of stock.
- Limitations—Operator's hand may enter danger area, and protection may not be complete at all times; may require frequent maintenance and/or adjustment; may be made ineffective by the operator; the guard may interfere with visibility.

Self-Adjusting Guards

- Advantages—Off-the-shelf guards are often commercially available.
- Limitations—Does not always provide maximum protection; may interfere with visibility; may require frequent maintenance and adjustment.

Source: Information from OSHA (1992).

- Presence-sensing devices—These devices include mats, photoelectric sensors, and other devices that sense the presence of a person or body part in the zone of hazard and automatically shut off machine operation.
- Emergency stop control—This control is used in situations when the machine must be stopped immediately because of an emergency.
- Hand control devices—Control mechanisms such as pull out devices, restraint devices, and two- hand control devices are used to ensure that the operators hands are not at the point of operation.
- Low energy and inch controls—These controls are used during setup, cleaning, and maintenance activities to reduce the speed (and hazards) of the equipment.

Other controls to enhance machine safety include warning signs, antikickback devices, run controls, foot controls, and similar devices.

Splash Protection

Some operations that involve wet chemistry must have controls in place to guard against accidental splashes. These typically include process enclosure guarding with chemical-resistant panels, automatic feed and ejection systems

Figure 10.9 Example of interlock machine guard.

Figure 10.10 Example of interlock machine guard.

to minimize the operators exposure to the chemicals, use of proper personal protective equipment, and automatic shutoff controls at a convenient location.

Training

Training on proper equipment use, guarding, safety controls, proper clothing, and required personal protective equipment should be provided to each employee who will use machines and tools. All training should be given by a competent person and before any work is performed. In industrial facililities, equipment operating procedures are often used as a basis for training, as well as manufacturer's information and information from other sources.

Checklist to Ensure Machine Safeguarding

A checklist to help the safety professional ensure that machines and equipment are properly safeguarded is presented in Table 10.5. This can serve as a starting point for initial and periodic review of machine safety.

TABLE 10.5   Machine Safeguarding Checklist

Requirements for All Safeguards

	Yes	No
• Do the safeguards provided meet the minimum OSHA requirements (29 CFR 1910 Subpart O)?	____	____
• Do the safeguards prevent workers' hands, arms, and other body parts from making contact with dangerous moving parts?	____	____
• Are the safeguards firmly secured and not easily removable?	____	____
• Do the safeguards ensure that no objects will fall into the moving parts?	____	____
• Do the safeguards permit safe, comfortable, and relatively easy operation of the machine?	____	____
• Can the machine be oiled without removal of the safeguard?	____	____
• Is there a system for shutting down the machinery before safeguards are removed?	____	____
• Can the existing system be improved?	____	____

Mechanical Hazards

AT POINT OF OPERATION

	Yes	No
• Is there a point-of-operation safeguard provided for the machine?	____	____
• Does it keep the operator's hands, fingers, and body out of the danger area?	____	____
• Is there evidence that a safeguard has been tampered with or removed?	____	____
• Can you suggest a more practical, effective safeguard?	____	____
• Can changes be made on the machine to eliminate the point-of-operation hazard entirely?	____	____

POWER TRANSMISSION APPARATUS

	Yes	No
• Are there any unguarded gears, sprockets, pulleys, or flywheels on the apparatus?	____	____
• Are there any exposed belts or chain drives?	____	____
• Are there any exposed set screws, key ways, collars, etc.?	____	____
• Are starting and stopping controls within easy reach of the operator?	____	____
• If there is more than one operator, are separate controls provided?	____	____

OTHER MOVING PARTS

	Yes	No
• Are safeguards provided for all hazardous moving parts of the machine, including auxiliary parts?	____	____

TABLE 10.5 *Continued*

Nonmechanical Hazards

	Yes	No
• Have appropriate measures been taken to safeguard workers against noise hazards?	_____	_____
• Have special guards, enclosures, or personal protective equipment been provided, where necessary, to protect workers from exposure to harmful substances used in machine operation?	_____	_____

Electrical Hazards

• Is the machine installed in accordance with National Fire Protection Association and the National Electrical Code requirements?	_____	_____
• Are there loose conduit fittings?	_____	_____
• Is the machine properly grounded?	_____	_____
• Is the power supply correctly fused and protected?	_____	_____
• Do workers occasionally receive minor shocks while operating any of the machines?	_____	_____

Training

• Do operators and maintenance workers have the necessary training in how and why to use the safeguards?	_____	_____
• Have operators and maintenance workers been trained in where the safeguards are located, how they provide protection, and what hazards they protect against?	_____	_____
• Have operators and maintenance workers been trained in how and under what circumstances guards can be removed?	_____	_____
• Have workers been trained in the procedures to follow if they notice guards to be damaged, missing, or inadequate?	_____	_____

Protective Equipment and Clothing

• Is protective equipment required?	_____	_____
• If protective equipment is required, is it appropriate for the job, in good condition, kept clean and sanitary, and stored carefully when not in use?	_____	_____
• Is the operator dressed safely for the job (i.e., no loose-fitting clothing or jewelry)?	_____	_____

Machinery Maintenance and Repair

• Have maintenance workers received up-to-date instruction on the machines they service?	_____	_____

TABLE 10.5 *Continued*

	Yes	No
• Do maintenance workers lock out the machine from its power sources before beginning repairs?	_____	_____
• Where several maintenance persons work on the same machine, are multiple lockout devices used?	_____	_____
• Do maintenance persons use appropriate and safe equipment in their repair work?	_____	_____
• Is the maintenance equipment itself properly guarded?	_____	_____
• Are maintenance and servicing workers trained in the requirements of 29 CFR 1910.147, in regard to lockout/tagout hazard, and do the procedures for lockout/tagout exist before they attempt their tasks?	_____	_____

Source: OSHA (1992).

REFERENCES

Occupation Safety and Health Administration, "Hand and Power Tools," OSHA 3080, Washington, DC, 1994

Occupational Safety and Health Administration, "Ground-Fault Protection on Construction Sites," OSHA 3007, Washington, DC, 1993.

Occupational Safety and Health Administration, "Concepts and Techniques of Machine Safeguarding," PB93-129930, Washington, DC, 1992.

Occupational Safety and Health Administration, "Controlling Electrical Hazards," OSHA 3075, Washington, DC, 1991.

BIBLIOGRAPHY

ANSI (1993), *National Electrical Safety Code*, American National Standards Institute, New York.

Brauer, Roger L. (1990), *Safety and Health for Engineers*, Van Nostrand Reinhold, New York.

DOE (1993), *Electrical Safety Guidelines*, U.S. Department of Energy, Washington, DC.

NFPA (1996), *National Electrical Code Handbook*, National Fire Protection Agency, Quincy, MA.

NFPA (1995), *Electrical Safety Requirements for Employee Workplaces*, National Fire Protection Agency, Quincy, MA.

NFPA (1994), *Electrical Equipment Maintenance*, National Fire Protection Agency, Quincy, MA.

OSHA (1994), "Control of Hazardous Energy (Lockout/Tagout)," OSHA 3120, U.S. Department of Labor, Occupational Safety and Health Administration, Washington, DC, 1994.

Roadstrum, W.H. and D.H. Wolaver (1993), *Electrical Engineering for All Engineers*, 2d ed., John Wiley & Sons, New York.

Smith, R.J., and R.C. Dorf (1992), *Circuits, Devices, and Systems: A First Course in Electrical Engineering*, 5th ed., John Wiley & Sons, New York.

11

FIRE AND LIFE SAFETY

Fire protection systems are the main safety systems used at all facilities to mitigate fires, explosions, and other chemical emergencies. Life safety management, which is related to fire protection, includes means of egress, building occupancy management, and building construction. Useful references on these two subjects include the *Fire Protection Handbook* and *Life Safety Code®*, published by the National Fire Protection Association (NFPA), and the *Uniform Fire Code* and *Uniform Building Code*, published by the International Conference of Building Officials (ICBO). The *Fire Protection Handbook*, described by the NFPA as an encyclopedia of fire protection, contains a wealth of detailed information prepared by leading fire protection experts and includes a listing of official NFPA standards, codes, and recommended practices. The *Life Safety Code®* is also known as NFPA 101, "Code for Safety to Life from Fire in Buildings and Structures." Both the *Uniform Fire Code* and the *Uniform Building Code* are model codes developed to reduce the burden on jurisdictions for developing local regulations while helping to coordinate regulations on a national and international level.

This chapter introduces some of the basic elements in fire protection and life safety. Topics included are combustibility and flammability, classes of fires, fire extinguishing agents, portable fire extinguishers, material storage areas, classification systems for hazardous materials and building occupancy, means of egress, and other miscellaneous topics having to do with fire and life safety.

COMBUSTION AND FLAMMABILITY

There are various definitions of combustion and flammability that relate to standards and regulations for combustible and flammable materials. *Combustion* may be defined as any chemical process that involves oxidation sufficient to produce light or heat (ASSE 1988). When oxidation occurs relatively fast, combustion is accompanied by fire or flames and may be referred to as rapid oxidation (Meyer 1989). Spontaneous combustion occurs when heat accumulates from slow oxidation in poorly ventilated spaces, causing materials to auto-ignite (as with solvent-laden rags or wipes). A combustible material is one that is able to catch fire and burn (NSC 1992). *Flammable* refers to any substance that is easily ignited, burns intensely, or has a rapid rate of flame spread (NSC 1992).

The terms given here are general ones. Specific definitions are given in standards by the NFPA and others. NFPA 30, "Flammable and Combustible Liquids Code," defines a flammable liquid as one having a closed-cup flash point below 100°F and having a vapor pressure less than 40 pounds per square inch absolute (psia). Combustible liquids are those having a flash point greater or equal to 100°F. Classifications of flammable and combustible liquids are given in NFPA 321, "Basic Classification of Flammable and Combustible Liquids," and are shown in Table 11.1. Classification is based on flash point, vapor pressure, and boiling point. There are three classes of flammable liquids (Classes I-A, I-B, and I-C) and three classes of combustible liquids (Classes II, III-A, and III-B).

OSHA standards at 40 CFR 1910.106 use essentially the same definitions and classes of flammable and combustible liquids as those in NFPA 321. However, mixtures of different classes of liquids are defined further in the OSHA standard. If a mixture is made up of both low and high flash point components, and the higher flash point components make up at least 99% of the volume, the mixture is classified according to the higher flash point components.

The following are explanations of some common terms used in regard to combustible and flammable materials:

- *Flammable or explosive limits*—These limits define the range of concentration of a combustible material in air at which propagation of flame occurs on contact with an ignition source. Concentrations are expressed in percent vapor or gas in the air by volume. If the concentration is below the lower flammable limit (LFL) or lower explosive limit (LEL), the air mixture is considered too lean and flame propagation will not occur. Conversely, if the concentration is above the upper flammable limit (UFL) or upper explosive limit (UEL), the mixture is too rich and flame propagation will likewise not occur. Examples of upper and lower flammable limits for some common volatile chemicals are shown in Table 11.2.

TABLE 11.1   NFPA Classification of Flammable and Combustible Liquids

Classification	Flash Point	Boiling Point
Flammable Liquids (vapor pressure ⩽ 40 psia at 100°F)		
I-A	<73°F	<100°F
I-B	<73°F	⩾100°F
I-C	⩾73°F to <100°F	
Combustible Liquids		
II	⩾100°F to <140°F	
III-A	⩾140°F to <200°F	
III-B	⩾200°F	

Source: Information from NFPA 321.

TABLE 11.2   Flammable Limits and Flash Points for Selected Compounds

Compound	Flammable Limits Lower (% by Volume)	Upper	Flash Point (°F)
Acetone	2.5	13	−4
Benzene	1.3	7.1	12
Cyclohexane	1.3	8	−4
Ethylbenzene	1	6.7	59
Gasoline	1.4	7.6	−45
Hydrogen	4	74	Gas
Methane	5	15	Gas
Methanol	6	36	52
Toluene	1.4	6.7	40
Xylenes	1	7	81–90

Source: Information from NFPA (1991).

- *Flash point*—The lowest temperature at which a liquid gives off enough vapor to form an ignitable mixture with air near the surface of the liquid or within a test vessel, which is capable of flame propagation away from the source of ignition. Flash points for a few selected chemicals are shown in Table 11.2.
- *Explosion*—A rapid increase of pressure in a confined space followed by its sudden release due to rupture of the container, vessel, or structure; the increase in pressure is generally caused by an exothermic chemical reaction or overpressurization of a system (ASSE 1988). Bursting or rupture of a building or a container because of the development of internal pressure (NFPA 69).

- *BLEVE (boiling liquid expanding vapor explosion)*—An explosion caused when fire impinges on the shell of a bulk liquid container, tank, or vessel above the liquid level, causing loss of strength of the metal and consequently, an explosive rupture of the structure from internal pressure (ASSE 1988).
- *Deflagration*—An exothermic reaction that expands rapidly from burning gases to unreacted material by conduction, convection, and radiation, where the combustion zone progresses through the material at a rate that is less than the velocity of sound in the unreacted material (ASSE 1988). Propagation of a combustion zone at a velocity that is less than the speed of sound in the unreacted medium (NFPA 69).
- *Detonation*—An exothermic reaction characterized by the presence of a shock wave in a material that establishes and maintains the reaction, usually causing an explosion, where the reaction zone expands at a rate greater than the speed of sound in the unreacted material (ASSE 1988). Propagation of a combustion zone at a velocity that is greater than the speed of sound in the unreacted medium (NFPA 69).

FIRE PROTECTION

Classes of Fires

NFPA 10 classifies fires by type of material involved in a fire. Fires are classified as Classes A, B, C, and D. The following is a description of each class of fire:

- Class A—Fires in ordinary combustible materials such as wood, cloth, paper, rubber, and many plastics. Extinguishing the fire requires water or water solutions that absorb heat and cool, certain dry chemicals that coat materials and retard combustion, or dry chemicals or halogenated agents that interrupt the combustion chain reaction.
- Class B—Fires in flammable or combustible liquids, flammable gases, greases, oil, gasoline, paint, and similar materials. Extinguishing the fire requires excluding air/oxygen, inhibiting the release of vapors, or interrupting the combustion chain reaction.
- Class C—Fires in live electrical wiring and equipment. Extinguishing the fire requires the use of nonconductive materials such as dry chemicals or carbon dioxide. If the equipment is deenergized, extinguishers for Class A or Class B fires may be used, as appropriate.
- Class D—Fires in certain combustible metals, such as lithium, magnesium, powdered aluminum, potassium, sodium, titanium, zinc, and zirconium. Extinguishing the fire requires nonreactive, heat-absorbing materials.

Extinguishing Agents

There are a number of different types of materials used to suppress and extinguish fires. Selecting the right type of extinguishing agent is important because not all extinguishing agents react the same way in every type of fire. Using the wrong type of extinguishing agent may actually make the fire worse or increase the hazards associated with it.

Water is the most common extinguishing agent. The following are general types of extinguishing agents:

- Water and water-based agents
- Carbon dioxide
- Halogenated agents
- Dry chemicals
- Foam
- Agents for combustible metals

Table 11.3 lists these fire-extinguishing agents, where they are used, advantages and disadvantages in their use, and potential hazards.

Water. Water extinguishes a fire by cooling, smothering the fire by displacing or cutting off the oxygen source, emulsification, and dilution. Water cools the fire by absorbing heat energy, and it displaces oxygen available for combustion when it absorbs enough heat to convert to steam. However, water is not generally recommended for flammable liquids with flash points below 100°F because water cannot cool the liquids below the flash point to stop the production of flammable vapors that feed the fire. Water can smother a fire when enough water is converted to steam, which displaces the air around a fire. Flammable liquids with flash points above 100°F can be smothered by aqueous foam solutions; these solutions are applied gently to the surface of the liquid to avoid splashing and mixing. Fires of viscous liquids that are immiscible with water can be extinguished by the emulsion that forms when water is added. Emulsification has a cooling effect and sometimes a frothing effect that reduces the vaporization of the flammable liquid. Dilution with water may be used to extinguish fires of water soluble flammable materials such as alcohols; however, dilution is not practical where there is a danger of frothing or overflow that can spread the flammable material into other areas. NFPA standards related to water-based fire protection systems include the following:

- NFPA 13—Installation of Sprinkler Systems
- NFPA 13A—Inspection, Testing, and Maintenance of Sprinkler Systems
- NFPA 15—Water-Spray Fixed Systems for Fire Protection

TABLE 11.3 Fire Extinguishing Agents

Extinguishing Agent	Applications	Advantages/Disadvantages	Hazards
Water and water-based agents	Class A fires	Readily available. Have cooling effect (absorb heat). Steam displaces oxygen available for combustion. Not suitable for flammable liquids with flash points <100°F. Delivery systems subject to freezing in cold climates. Cannot blanket a fire because they run off quickly.	Danger of electric shock with live electrical equipment.
Carbon dioxide	Classes A, B, and C fires	Nonreactive. Nonconductive. Penetrates easily as a gas. Does not leave a residue.	Static electricity can build up on ungrounded nozzles. High concentrations in the atmosphere can cause death by asphyxiation.
Halon agents	Classes A, B, and C fires	Do not leave a residue.	Ozone-depleting potential will require replacement in the future. High concentrations can be irritating or toxic.
Dry chemicals	Classes A, B, and C fires	Ammonium phosphate is multipurpose agent for Classes A, B, and C fires (other dry chemicals are used for Classes B and C fires). These coat surfaces, but leave residue that is powdery or sticky and slightly corrosive.	Slightly corrosive. Ineffective if reignition occurs.

Foam agents	Classes A and B fires	Nontoxic and biodegradeable. Not suitable for water-reactive materials. Leave residue.	Potential for frothing and slopover in tank fires where bulk temperatures exceed 212° F. Electrically conductive.
Combustible metal extinguishing agents	Class D fires	Other common agents are not suitable because they are ineffective or intensify the fire. Leave residue.	Care must be used in the selection of an agent for each metal, so that it can be effective and to avoid adverse reactions.

- NFPA 16—Deluge Foam-Water Sprinkler and Foam-Water Spray Systems
- NFPA 18—Wetting Agents

Carbon Dioxide. Carbon dioxide is effective for many types of fires involving flammable liquids, electrically energized equipment, and ordinary combustibles such as paper and wood. It is not an effective extinguishing agent for materials that contain oxygen, such as cellulose nitrate, or for metals that react with and decompose carbon dioxide (magnesium, potassium, sodium, titanium, zirconium, and the metal hydrides).

Carbon dioxide works to extinguish fires by smothering (displacing oxygen) and cooling. The advantages of carbon dioxide are that it is generally nonreactive, as a gas it can penetrate and spread easily, it is electrically nonconductive, and it does not leave any residue. It is important, however, to ground all discharge nozzles to avoid the buildup and discharge of static electricity in potentially explosive atmospheres. In addition, because high levels of carbon dioxide can cause death by asphyxiation, adequate safety procedures must be designed into fire protection systems using carbon dioxide.

Guidance for carbon dioxide extinguishing systems is found in NFPA 12. NFPA 69, "Explosion Prevention Systems," provides guidance on using carbon dioxide for inerting or purging atmospheres.

Halogenated Agents. These fire-extinguishing materials are hydrocarbon-based compounds containing one or more of the following halogens: fluorine, chlorine, bromine, or iodine. Fluorine, chlorine, and bromine are the halogens used most often. Halon agents are identified by a simple numbering system based on the number of atoms of each element in the compound in this order: carbon, fluorine, chlorine, bromine, and iodine. For example, Halon 1301 refers to bromotrifluoromethane (C=1, F=3, Cl=0, and Br=1). Methyl iodine is Halon 10001, and dibromotetrafluoroethane is Halon 2402.

Halon agents may be used on Class B and Class C fires and on some Class A fires, but never on Class D fires. These agents extinguish fires by interrupting the flame chain reaction that propagates the fire. Like carbon dioxide, these agents vaporize and leave no residue, making them desirable for fires involving electrical equipment, computers, aircraft, ships, and vehicles and in special environments such as hospitals, museums, libraries, and vaults.

NFPA standards applicable to Halon agents include:

- NFPA 10—Portable Fire Extinguishers
- NFPA 12A—Halon 1301 Fire Extinguishing Agent Systems
- NFPA 12B—Halon 1211 Fire Extinguishing Systems
- NFPA 12CT—Halon 2402 Fire Extinguishing Agent Systems
- NFPA 75—Electronic Computer/Data Processing Equipment

Dry Chemicals. There are five basic types of dry chemical fire-extinguishing agents: monoammonium phosphate, potassium bicarbonate, potassium chloride, sodium bicarbonate, and urea-potassium bicarbonate. Other chemicals are added to these base chemicals to improve storage, flow, and water-repellent characteristics; common additives include metallic stearates, tricalcium phosphate, and silicones (NFPA 1991). Dry chemicals extinguish fires by smothering, heat shielding, and breaking the combustion chain reaction. Smothering of the fire occurs with the release of carbon dioxide from the decomposition of bicarbonate-based agents and the sticky residue left by decomposition of monoammonium phosphate. Some heat radiation shielding of the fuel is provided by the cloud of powder that results when dry chemicals are discharged. The major extinguishing action of dry chemicals, however, is the interference with the chain reaction propagated by free radicals.

Monoammonium phosphate is a multipurpose agent used for Classes A, B, and C fires. The other four basic dry chemicals are used on Classes B and C fires. The main disadvantage of dry chemical agents is the residue that is left behind, which is either powdery or sticky and slightly corrosive.

NFPA standards applicable to dry chemical agents include:

- NFPA 10—Portable Fire Extinguishers
- NFPA 17—Dry Chemical Extinguishing Systems

Foam. Fire-extinguishing foams are produced from aqueous solutions of liquid foaming agents. Foams extinguish a fire by smothering and cooling. Foams are used primarily to fight Class B fires of flammable and combustible liquids; however, some foams are useful with solid combustibles in Class A fires where the water in the solution can coat/penetrate and cool the material. They should not be used for Class C fires involving live electrical equipment because the foams are electrically conductive.

Foams are defined by their expansion ratio of foam volume to foam solution. They are divided into low-expansion foams with ratios up to 20:1, medium-expansion foams with ratios from 20 to 200:1, and high-expansion foams with ratios from 200 to 1,000:1 (NFPA 1991). Common types of foam include aqueous film-forming agents (AFFF), fluoroprotein foaming agents (FP), film-forming fluoroprotein agents (FFFP), protein foaming agents (P), low temperature foaming agents, alcohol-type foaming agents (AR), medium and high expansion foaming agents (SYNDET), vapor suppression foaming agents, and other synthetic hydrocarbon surfactant foaming agents.

According to the NFPA (1991), foams are most effective when:

- The burning liquid is below its boiling point at the ambient conditions of temperature and pressure.
- The bulk temperature of the burning liquid is less than the boiling point of water. When foams are applied to liquids at higher temperatures, an emulsion of steam, air, and fuel forms. This emulsion can greatly increase the volume in a tank fire and cause frothing or slop-over of the burning liquid.

- The burning liquid must not be capable of excessively degrading the foam, and the foam must not be highly soluble in the liquid.
- The burning liquid must not be reactive with water.
- The fire must be on a horizontal surface. Three-dimensional (falling fuel) or pressure fires cannot be extinguished by foam unless the material has a relatively high flash point and can be cooled to extinguishment by the water in the foam.

NFPA standards and codes applicable to foam extinguishing agents and systems include:

- NFPA 11—Low-Expansion Foam and Combined Agent Systems
- NFPA 11A—Medium- and High-Expansion Foam Systems
- NFPA 11C—Mobile Foam Apparatus
- NFPA 16—Deluge Foam-Water Sprinkler and Foam-Water Spray Systems
- NFPA 18—Wetting Agents
- NFPA 78—Lightning Protection Code
- NFPA 403—Aircraft Rescue and Fire Fighting Services at Airports
- NFPA 409—Aircraft Hangars
- NFPA 412—Aircraft Rescue and Fire Fighting Foam Equipment
- NFPA 414—Aircraft Rescue and Fire Fighting Vehicles
- NFPA 1901—Pumper Fire Apparatus

Combustible Metal Extinguishing Agents. In general, Class C fires involving combustible metals such as lithium, magnesium, potassium, sodium, titanium, and zirconium require specialized extinguishing agents because most normal agents for Classes A, B, and C fires will react with the metals and increase the intensity of the fire. For example, alkali metals such as lithium, potassium, and sodium react with water to form hydrogen gas, which is highly flammable.

Not all extinguishing agents for combustible metal fires are appropriate for all types of metals. Care must be taken in selecting an agent because some are ineffective on the fire or make it worse. Common agents include dry powders and materials such as sodium chloride, graphite, soda ash, lithium chloride, sand, powdered talc, dolomite, zirconium silicate, boron trifluoride, boron trichloride, inert gases, foundry flux, and cast iron borings (NFPA 1991). Special or proprietary agents include sodium chloride-based MET-L-X powder, sodium carbonate-based Na-X powder, graphite-based G-1 and Lith-X powders, TEC powder (a mixture of potassium chloride, sodium chloride, and barium chloride), Boralon (trimethoxyborane (TMB) and Halon 1211), and copper powder.

NFPA standards applicable to combustible metal fires and extinguishing agents include:

- NFPA 480—Storage, Handling, and Processing of Magnesium
- NFPA 481—Production, Processing, Handling, and Storage of Titanium
- NFPA 482—Production, Processing, Handling, and Storage of Zirconium
- NFPA 651—Manufacture of Aluminum and Magnesium Powder

Portable Fire Extinguishers

Standards for portable fire extinguishers are established in NFPA 10. This standard sets out requirements for the selection, installation, inspection, maintenance, and testing of portable fire extinguishers. Incorporated into NFPA 10 are the following testing and performance standards developed by the American National Standards Institute (ANSI) and Underwriters Laboratory (UL):

ANSI/UL 711, Rating and Testing of Fire Extinguishers
ANSI/UL 154 (carbon dioxide type extinguishers)
ANSI/UL 299 (dry chemical type extinguishers)
ANSI/UL 626 (water type extinguishers)
ANSI/UL 1093 (halogenated agent type extinguishers)
ANSI/UL 8 (foam type extinguishers)
ANSI/UL 1803 (third-party certified extinguishers)

Material Storage Areas

NFPA Requirements. NFPA has published a number of standards related to the storage of combustible materials; key standards are summarized in Table 11.4. In addition to these standards, NFPA's *Fire Protection Guide to Hazardous Materials* includes technical data from several key standards concerning the fire hazards of chemicals and other materials. NFPA's *Fire Protection Handbook* addresses storage practices for a variety of combustible materials, environments, and activities, including flammable and combustible liquids, gases, chemicals, explosives, solid fuels, plastics and rubber, and grain mill products. NFPA standards for general chemical storage and radioactive materials are discussed in the following sections.

Chemical Storage. In large chemical storage areas, segregation by chemical type or category provides the best fire protection. Separate rooms or buildings should be used to segregate oxidizing chemicals, flammable or combustible

TABLE 11.4   Selected NFPA Standards for Material Storage

NFPA 30	Flammable and Combustible Liquids Code
NFPA 30B	Manufacture and Storage of Aerosol Products
NFPA 40	Storage and Handling of Cellulose Nitrate Motion Picture Film
NFPA 40E	Storage of Pyroxylin Plastic
NFPA 43A	Storage of Liquid and Solid Oxidizing Materials
NFPA 43B	Storage of Organic Peroxide Formulations
NFPA 43C	Storage of Gaseous Oxidizing Materials
NFPA 43D	Storage of Pesticides in Portable Containers
NFPA 46	Storage of Forest Products
NFPA 55	Storage, Use, and Handling of Compressed and Liquefied Gases in Portable Cylinders
NFPA 58	Storage and Handling of Liquefied Petroleum Gases
NFPA 59	Storage and Handling of Liquefied Petroleum Gases at Utility Gas Plants
NFPA 59A	Production, Storage, and Handling of Liquefied Natural Gas (LNG)
NFPA 81	Fur Storage, Fumigation, and Cleaning
NFPA 122	Storage of Flammable and Combustible Liquids Within Underground Metal and Nonmetal Mines (Other than Coal)
NFPA 231	General Storage
NFPA 231C	Rack Storage of Materials
NFPA 231D	Storage of Rubber Tires
NFPA 231E	Storage of Baled Cotton
NFPA 231F	Storage of Roll Paper
NFPA 327	Cleaning or Safeguarding Small Tanks and Containers
NFPA 329	Underground Leakage of Flammable and Combustible Liquids
NFPA 386	Portable Shipping Tanks for Flammable and Combustible Liquids
NFPA 395	Storage of Flammable and Combustible Liquids on Farms and Isolated Construction Projects
NFPA 480	Storage, Handling, and Processing of Magnesium
NFPA 481	Production, Processing, Handling, and Storage of Titanium
NFPA 482	Production, Processing, Handling, and Storage of Zirconium
NFPA 490	Storage of Ammonium Nitrate
NFPA 495	Manufacture, Transportation, Storage, and Use of Explosive Materials
NFPA 1124	Manufacture, Transportation, and Storage of Fireworks

chemicals, unstable chemicals, corrosives, chemicals reactive to water or air, radioactive materials, and materials subject to self-heating. In addition, chemicals and other hazardous materials should be stored separately from nonchemical parts and other manufacturing supplies. Examples of fire extinguishing methods for different categories of chemicals are shown in Table 11.5.

TABLE 11.5 Fire Extinguishing Methods by Chemical Category

Chemical Category	Examples	Fire-Extinguishing Methods
Flammable/combustible	Gasoline, heavy oils, and sulfur/sulfides of sodium, potassium, and phosphorus	Large quantities of water or foam; sulfur and sulfide compounds can cause irritating gases.
Oxidizing	Potassium permanganate, potassium perchlorate, and sodium nitrate	Large quantities of water; fire fighters should wear self-contained breathing apparatus; area should be ventilated.
Unstable	Picric acid, TNT, and organic peroxides	Large quantities of water, preferably through automatic sprinklers because of explosion potential.
Corrosive	Sodium hydroxide, hydrochloric acid, sulfuric acid, and perchloric acid	Water spray in acid and alkali storage areas; toxic fumes can result in some cases; explosion potential exists in some cases.
Water reactive	Calcium carbide, chromyl chloride, and alkali metals	Graphite-based powder or inert materials for some chemicals.
Air reactive	Elemental sodium and yellow phosphorus	Water followed by dirt or sand.
Radioactive	Plutonium, radium, and uranium	Fire and explosion characteristics are not affected by radioactivity of material; automatic fire protection is necessary to avoid human exposure to radioactivity; preplanning for disposal of contaminated water is necessary.
Self-heating	Elemental sulfur and activated charcoal	Water or steam only on fire area; wet material is more susceptible to self-heating than dry material.

Source: Information from NFPA, *Fire Protection Handbook* (1992).

Drums or packages of chemicals may be stored on pallets and lifted into place on storage racks by forklifts or placed in racks by an automated system. The design of storage rack systems is presented in NFPA 231 Appendix A for general storage and NFPA 231C for rack storage of materials. Although this latter standard does not address chemical storage per se, it is a good guide on sprinkler design for materials of varying flammability.

In all cases, storage areas must be clean and orderly and fire protection equipment must be well maintained. Access to the racks and the load/unload areas must be roomy, with a clear path for egress. Aisle widths generally range from 4 to 8 feet, but should provide enough space for egress when fork trucks and other equipment are in use.

The design of a rack storage system inside a building is based generally on storage height and commodity class. Storage height is measured from floor level to the top of the storage container resting on the top rack. In design schemes, the most flammable products and packaging require in-rack sprinklers, as well as ceiling sprinklers of adequate water supply. An example of a drum rack storage system with in-rack sprinklers is shown in Figure 11.1. Horizontal barriers that cover the entire rack, including the flue spaces, also are required in many cases. Other considerations for storage areas include total water demand (gallons per minute per square foot), rack configuration, and storage to ceiling clearance. Specific design criteria should be researched thoroughly before sprinkler installation.

Radioactive Materials. Fire protection for facilities handling radioactive materials should be designed to minimize the spread of radioactive contamination during a fire. In addition, good adminstrative controls, which can minimize the impacts of fire at any facility, are especially important.

Fire protection practices for facilities handling radioactive materials have been developed by the NFPA in the following standards:

- NFPA 801—Recommended Fire Protection Practice for Facilities Handling Radioactive Materials
- NFPA 802—Nuclear Research and Production Reactors
- NFPA 803—Fire Protection for Light Water Nuclear Power Plants

These standards address key aspects of fire protection and fire response, including fire protection systems and equipment, general facility design, administrative controls, and other relevant topics.

Several design considerations are particularly important to facilities that handle radioactive materials: fire detection, life safety, electrical power, ventilation and duct systems, and fire drainage.

A fire detection system is important, such as a system that alarms at a fixed temperature or if there is an acclelerated rate of rise in temperature. This detection system will alert building occupants and allow immediate evacuation.

Figure 11.1 Drum rack storage with in-rack sprinklers.

The system also can be used by management as a signal to rapidly implement the fire emergency plan. Life safety design should include emergency lighting for means of egress, lighting protections, and interior finishes. Electrical operations equipment such as transformers, control panels, and main switches should be located well away from areas that handle radioactive and combustible materials.

Ventilation systems must include capabilities for heat removal, fire isolation, and filtration of radioactive gases and particles. Fresh air inlets should be located to avoid contact with radioactive contaminants. Fire-resistant or fire-retardant materials should be used in ventilation ducts. If shutdown of the ventilation system is allowed, fire dampers should be provided to resist the spread of contaminated smoke.

Drainage systems should be designed to remove liquids to a safe area for testing and proper disposal. The liquid-handling area should be designed to include the volume of a spill from the largest container, a 20-minute discharge from fire hoses or other suppressant system, and, for outdoor applications, a typical amount of rain or snow. The drains should be designed to minimize fire hazard and radioactive contamination of clean areas.

Administrative controls are a necessary part of a fire safety program. Included in a well-managed fire prevention program are a documented fire prevention program, a fire hazard analysis, a fire emergency plan, and an adequate testing, inspection, and maintenance program.

Proper personal protective equipment must be worn at all times, including face masks, clothing that prevents the entry of radioactive materials into the body, and a self-contained breathing apparatus. This protective equipment should be specified by a trained industrial hygienist or safety specialist.

Uniform Fire Code Requirements. Fire protection requirements for storage of a variety of materials are outlined in the *Uniform Fire Code.* A partial list of these materials and the corresponding Article number found in the *Code* are shown in Table 11.6. Hazardous materials addressed in the code include compressed gases (Article 74), cryogenic fluids (Article 75), explosive materials (Article 77), and flammable and combustible liquids (Article 79).

TABLE 11.6 Selected Articles in the Uniform Fire Code Related to Material Storage

Article	Material Type
27	Cellulose Nitrate Plastics (Pyroxylin)
28	Combustible Fibers
	Loose Storage, Baled Storage, Agricultural Products
29	Repair Garages, Flammable and Combustible Liquids
30	Exterior Lumber Storage
32	Tents, Canopies, and Temporary Membrane Structures
	Flammable and Combustible Liquids and LP-Gas
33	Cellulose Nitrate Motion Picture Film
34	Automobile Wrecking Yards, Tires
46	Fruit-Ripening Processes, Use of Ethylene
48	Magnesium
49	Welding and Cutting
	Cylinders, Calcium Carbide in Buildings
50	Manufacture of Organic Coatings
	Raw Materials and Finished Products
51	Semiconductor Fabrication Facilities
	Hazardous Production Materials (HPM)
52	Motor Vehicle Fuel-Dispensing Stations
	Flammable and Combustible Liquids
74	Compressed Gases
75	Cryogenic Fluids
77	Explosive Materials
78	Pyrotechnic Special Effects Material
79	Flammable and Combustible Liquids
80	Hazardous Materials
81	High-Piled Combustible Storage
82	Liquified Petroleum Gases
84	Motion Picture Projection, Film Storage
86	Pesticide Storage and Display
88	Aerosol Products

Hazardous materials are included as a discrete category in Article 80 of the code. This category includes additional requirements for the afore-mentioned hazardous materials, as well as storage requirements for other hazardous materials such as flammable solids, liquid and solid oxidizers, organic peroxides, pyrophoric materials, unstable and water-reactive materials, highly toxic solids and liquids, radioactive materials, corrosives, carcinogens, irritants, sensitizers, and other health hazard solids, liquids, and gases. For each category there is a defined exempt storage amount. Exemptions are typically based on type of storage and type of fire protection provided (i.e., unprotected by sprinkler or cabinet, stored within cabinet in unsprinklered building, stored in sprinklered building but not in cabinet, or stored in cabinet inside sprinklered building).

The next several paragraphs summarize some of the key requirements specified in the code. The summary is not all-inclusive, but should provide a general idea of the types of storage requirements for these materials. For inclusive information, the *Uniform Fire Code* should be reviewed.

Hazardous Materials. General storage requirements address spill control, drainage, containment, ventilation, separation of incompatible materials, construction of storage cabinets, fire extinguishing systems, and other relevant topics. Any special requirements related to a particular material are addressed separately in discussions of that material.

Compressed Gases. Compressed gases are required to be adequately secured to prevent gas cylinders from falling or being knocked over. Moreover, legible operating instructions must be maintained at the operating location. "No Smoking" and other warning signs must be posted near enclosures that hold the compressed gas cylinders. Additional items pertain to toxic gases and highly toxic gases. Requirements for these gases address ventilation, use of gas cabinets and gas detection systems, acceptable rates of release, treatment systems for exhausted gas cabinets, and related topics.

Cryogenic Fluids. Cryogenic fluids are classified in the code as flammable, nonflammable, corrosive/highly toxic, and oxidizer. These materials generally are not stored in storage rooms, but rather in aboveground, belowground, or in-ground containers. Requirements for each type of installation is addressed in the standard. In addition, minimum distances from tanks are specified for different classes of cryogenics.

Explosive Materials. Explosive materials must be carefully handled and stored. Smoking, matches, open flames, and other spark-producing devices are prohibited within 50 feet of magazines. Dried grass, leaves, trash, and debris must be cleared within 25 feet of magazines. Packages of damaged explosives cannot be unpacked or repacked within 50 feet of a magazine or inhabited building or in close proximity to other explosives. Any deteriorated explosive must be reported to the fire chief immediately and destroyed by an expert. Numerous other requirements pertain to explosive and blasting agents, and these requirements should be investigated thoroughly before such materials are brought onsite for storage.

Flammable and Combustible Liquids. Flammable and combustible liquids are addressed fully in the code. Requirements are grouped in following broad categories:

- General requirements
- Stationary aboveground tanks outside buildings
- Container and portable tank storage outside buildings
- Stationary aboveground tank storage inside buildings
- Container and portable tank storage inside buildings
- Underground tank storage

There are also numerous tables listing requirements for maximum sizes of containers and portable tanks, vent line diameters, minimum location and separation distances, sprinkler systems, and other components. An example of minimum location distances for tanks storing Class III-B liquids is shown in Table 11.7.

Flammable Solids. Indoor storage of these materials sometimes requires explosion venting or suppression. The size limit for each pile of material stored indoors is 1,000 cubic feet, and aisle widths between piles must be at least 4 feet or equal to the height of the piles, whichever is greater. Flammable solids cannot be stored in basements. For exterior storage, limits are 5,000 cubic feet for

TABLE 11.7   Minimum Location Distances for Aboveground Tanks Storing Class III-B Combustible Liquids Outside Buildings

Tank Capacity (C) (gallons)	Minimum Distance from Property Line of Property Which Is or Can Be Built Upon, Including the Opposite Side of a Public Way (feet)	Minimum Distance from Nearest Side of any Public Way or from Nearest Important Building on the Same Property (feet)
C≤12,000	5	5
12,000<C≤30,000	10	5
30,000<C≤50,000	10	10
50,000<C≤100,000	15	10
C>100,000	15	15

These requirements are for stable Class III-B liquids; requirements for unstable liquids are given elsewhere in the *Uniform Fire Code*. Other requirements apply if the tank storing a Class III-B liquid is located within a diked area or drainage path for tanks storing Class I or II liquids.
Source: Uniform Fire Code, Section 7902.2.2.6 (ICBO 1994b).

piles, with aisle widths at least 10 feet or one half the height of the piles, whichever is greater. In addition, for exterior storage, there must be at least 20 feet between the flammable solid and any building, property line, street, alley, public way, or exit to a public way, unless the storage area has an unpierced 2-hour fire-resistive wall extending not less than 30 inches above and to the sides of the storage area. Further requirements pertaining to ventilation, spill control, and other features are included.

Liquid and Solid Oxidizers. Examples of liquid oxidizers are bromine, hydrogen peroxide, nitric acid, perchloric acid, and sulfuric acid; solid oxidizers include chlorates, chromates, chromic acid, iodine, nitrates, perchlorates, and peroxides (ICBO 1994b). The code subdivides liquid and solid oxidizers into four classes, as shown in Table 11.8, with Class 4 being the most hazardous.

For indoor storage, amounts of these materials over exempt quantities require detached storage, as specified in the code. Detached storage buildings must be single story without a basement or crawl space. Requirements for distance from storage to any building, property line, or public way is dependent on the class of oxidizer. Class 4 oxidizers must be separated from other hazardous materials by at least 1-hour fire-resistive construction. Detached buildings for Class 4 oxidizers must be located at least 50 feet from other hazardous materials storage. For exterior storage, maximum quantities and arrangements are specified in the code. Also addressed are other requirements pertaining to safe distance to property boundary or public way, smoke detection systems, and ventilation systems.

Organic Peroxides. The code defines organic peroxides as flammable compounds that contain the double oxygen or peroxy (-O-O-) group and are subject to explosive decomposition. Organic peroxides can be found as liquids, pastes, or solids (usually finely divided powders). Organic peroxides are divided by the code into five classes and one "unclassified" group, as shown in Table 11.9.

For indoor storage, amounts of organic peroxides over exempted quantities require detached storage, depending on class. Separation distance of detached storage from any building, property line, or public way is also dependent on class. An approved supervised smoke detection system must be provided where Class I, II, III, or IV organic peroxides are stored, and a local alarm must be sounded when this detection system is activated. However, if there is an automatic fire-extinguishing system in a detached storage building, a smoke detection system is not required. In addition there are special storage conditions for organic peroxides. For example, 55-gallon drums cannot be stored more than one drum high, a minimum 2-foot clear space must be maintained between storage and uninsulated metal walls, containers and packages in storage areas must be closed, and there can be no bulk storage in piles or bins.

TABLE 11.8   Classification of Liquid and Solid Oxidizers According to Hazard

Class	Description	Examples
4	Can undergo an explosive reaction undergo an explosive reaction as a result of contamination or exposure to thermal or physical shock. In addition, the oxidizer will enhance the burning rate and may cause the spontaneous ignition of combustibles.	Ammonium perchlorate (>15 microns) Ammonium permanganate Guanidine nitrate Hydrogen peroxide (>91% solutions)
3	Will cause a severe increase in the burning rate of combustible materials with which it comes in contact or that will undergo vigorous self-sustained decomposition resulting from contamination or exposure to heat.	Ammonium dichromate Calcium hypochlorite (>50% weight) Chloric acid (10% maximum concentration) Hydrogen peroxide (>52%-91%) Potassium chlorate
2	Will cause a moderate increase in the burning rate or may cause spontaneous ignition of combustible materials with which it comes in contact.	Barium bromate Copper chlorate Hydrogen peroxide (>27.5%-52%) Zinc permanganate
1	Primary hazard is that it slightly increases the burning rate, but does not cause spontaneous ignition when it comes in contact with combustible materials.	Ammonium persulfate Inorganic nitrates and nitrites Hydrogen peroxide (>8%-27.5%) Sodium dichromate Zinc peroxide

Source: Section 2.1.5.3, Appendi VI-A in the Uniform Fire Code (ICBO 1994b).

Pyrophorics. Pyrophoric materials are those that can ignite spontaneously upon exposure to air. They can be in gaseous, liquid, or solid form. Examples of pyrophoric materials in gaseous form are diborane, phosphine, and silane; in liquid form, diethyl aluminum chloride, diethyl phosphine, and trimethyl gallium; and in solid form, cesium, hafnium, lithium, white or yellow phosphorous, plutonium, potassium, and sodium (ICBO 1994b).

Special storage requirements apply to pyrophoric materials that exceed exempt amounts. When stored indoors, piles of pyrophoric materials must be less than 100 square feet in area. The height of materials stored indoors cannot be greater than 5 feet. Individual containers cannot be stacked. Pyrophoric materials stored indoors must be separated from incompatible hazardous materials by 1-hour fire-resistive walls with protected openings, unless the materials are stored in storage cabinets that meet the code standards.

Unstable (Reactive) Materials. Materials of this type are divided into four classes in the code, as shown in Table 11.10. Group H occupancies (those

TABLE 11.9 Classification of Organic Peroxides According to Hazard

Class	Description	Examples
Unclassified	Capable of detonation. Present an extremely high explosion hazard through rapid explosive decomposition. Regulated in the *Uniform Fire Code* as explosive materials.	
Class I	Capable of deflagration, but not detonation. High explosion hazard through rapid decomposition.	Fulfonyl peroxide t-Butyl hydroperoxide (90%) Diisopropyl peroxydicarbonate (100%)
Class II	Burn very rapidly and present a severe reactivity hazard.	Acetyl peroxide (25%) t-Butyl peroxybenzoate (98%) Peroxyacetic acid (43%)
Class III	Burn rapidly and present a moderate reactivity hazard.	Benzoyl peroxide (78%) Benzoyl peroxide paste (55%) Cumene hydroperoxide (86%)
Class IV	Burn in same manner as ordinary combustibles and present a minimum reactivity hazard.	Benzoyl peroxide (70%) Benzoyl peroxide powder (35%) Laurel peroxide (98%)
Class V	Do not burn or present a decomposition hazard.	Benzoyl peroxide (35%) 2,5-di-(t-butyl peroxy) hexane (47%) 2,4-pentanedione peroxide (4% active oxygen)

Source: Section 2.1.6, Appendix VI-A in the *Uniform Fire Code* (ICBO 1994b).

involving manufacturing, processing, generation, or storage of materials that constitute a high fire, explosion, or health hazard) in indoor storage require detached buildings for solid and liquid materials exceeding 2 pounds if Class 4; 2,000 pounds if Class 3; and 50,000 pounds if Class 2. For gases, the limits are 20 cubic feet for Class 4; 2,000 cubic feet for Class 3; and 10,000 cubic feet for Class 2. For indoor storage, individual piles are limited to 500 cubic feet and the aisle width between piles must be at least equal to the height of the piles or 4 feet, whichever is greater. Unstable (reactive) materials cannot be stored in basements. For outdoor storage, individual piles are limited to 1,000 cubic feet and the aisle width between piles must be at least one half the height of the piles or 10 feet, whichever is greater. If the material is capable of deflagrating, outdoor storage must be more than 75 feet away from buildings, property lines, streets, alleys, public ways, or exits to a public way. If nondeflagrating, the distance requirement is 20 feet, unless an unpierced 2-hour fire-resistive wall extends 30 inches above and to the sides of the material.

TABLE 11.10   Classification of Unstable (Reactive) Materials According to Hazard

Class	Description	Examples
4	Readily capable of detonation or of explosive decomposition or explosive reaction at normal temperatures and pressures. Includes materials sensitive to mechanical or localized thermal shock at normal temperatures and pressures.	Acetyl peroxide Dibutyl peroxide Ethyl nitrate Peroxyacetic acid
3	Capable of detonation or of explosive decomposition or explosive reaction, but require a strong initiating source or heating under confinement before initiation. Includes materials sensitive to thermal or mechanical shock at elevated temperatures and pressures.	Hydrogen peroxide ($>$52%) Nitromethane Perchloric acid Tetrafluoroethylene monomer
2	Normally unstable and readily undergo violent chemical change, but do not detonate. Should include materials which can undergo chemical change with rapid release of energy at normal temperatures and pressures and which can undergo violent chemical change at elevated temperatures and pressures.	Acrolein Acrylic acid Hydrazine Styrene Vinyl acetate
1	Normally stable, but can become unstable at elevated temperatures and pressures.	Acetic acid Hydrogen peroxide (35%–52%) Paraldehyde Tetrahydrofuran

Source: Section 2.1.8, Appendix VI-A in the *Uniform Fire Code* (ICBO 1994b).

Water Reactive Materials. Water reactive materials are divided into three classes in the code, as shown in Table 11.11. Detached building storage is required for indoor storage for Group H occupancies if there are more than 2,000 pounds of Class 3 or 50,000 pounds of Class 2 water reactive materials. The floors of indoor storage areas must be liquid-tight and the rooms or areas must resist water penetration through the use of waterproof materials. Except for those in approved automatic fire-sprinkler systems, water pipes cannot be placed in indoor storage areas of water reactive materials. Class 3 materials must be kept in closed watertight containers if they are in areas protected by automatic fire-sprinkler systems. Indoor piles of water reactive materials must be less than 500 cubic feet each, with aisle clearances at least the height of the piles or 4 feet, whichever is greater. Class 3 materials cannot be stored in basements. This restriction also applies to Class 2 materials, unless they are stored in closed watertight containers or tanks. In order to be stored outdoors, water reactive materials must be placed in tanks or closed watertight containers. The minimum separation distance from buildings, property lines, streets, alleys, public ways, or exits to public ways is 75 feet for Class 3 materials and 20 feet for Class 1 and 2 materials. For Class 1 or 2 materials, the minimum distance is waived if there is an unpierced 2-hour fire-resistive wall extending at least 30 inches above and to the sides of the storage area. Pile size limits for outdoor storage are 100 cubic feet for Class 3 materials and 1,000 cubic feet for Class 1 and 2 materials. Aisle width between piles for all classes must be at least one half the height of the piles or 10 feet, whichever is greater.

Highly Toxic and Toxic Materials. These materials may be in gaseous, liquid, or solid form. Examples of highly toxic and toxic materials are given in the code; a selection of these is shown in Table 11.12. In indoor storage areas where a spill or another accidental emission can be expected to release highly toxic vapors, exhaust scrubbers or other systems must be provided. Highly toxic liquids and solids stored indoors must be separated from other hazardous materials by 1-hour fire-resistive construction or stored in approved hazardous material storage cabinets. Highly toxic and toxic liquids and solids that are stored outdoors must be kept more than 20 feet away from buildings, property lines, streets, alleys, public ways, and exits to a public way. The distance requirement is waived if an unpierced 2-hour fire-resistive wall extends at least 30 inches above and to the sides of the storage area. Outdoor storage must be in fire-resistive containers or be protected by an automatic, open head, deluge fire-sprinkler system or located under a noncombustible canopy protected by an automatic fire-sprinkler system. Outdoor piles of highly toxic materials are limited to 2,500 cubic feet each and must have aisle widths of at least one half the height of the piles or 10 feet, whichever is greater. Highly toxic liquids that can emit highly toxic vapors when spilled or released cannot be stored outdoors unless there is an effective collection and treatment system.

TABLE 11.11   Classification of Water Reactive Materials According to Hazard

Class	Description	Examples
3	Reacts explosively with water without heat or confinement.	Triethylaluminum Isobutylaluminum Bromine pentafluoride Diethylzinc
2	May form potentially explosive mixtures with water.	Calcium carbide Lithium hydride Potassium peroxide Sulfuric acid
1	May react with water with some release of energy, but not violently.	Acetic anhydride Sodium hydroxide Sulfur monochloride Titanium tetrachloride

Source: Section 2.1.9, Appendix VI-A in the *Uniform Fire Code* (ICBO 1994b).

TABLE 11.12   Examples of Highly Toxic and Toxic Materials

Gases	Liquids	Solids
	Highly Toxic Materials	
Arsine	Acrolein	Phenyl mercuric acetate
Chlorine trifluoride	Acrylic acid	4-Aminopyridine
Cyanogen	2-Chloroethanol	Arsenic pentoxide
Diborane	Hydazine	Arsenic trioxide
Fluorine	Hydrocyanic acid	Calcium cyanide
Germane	2-Methylaziridine	2-Chloroacctophenone
Hydrogen cyanide	2-Methyllactonitrile	Aflatoxin B
Nitric oxide	Methyl isocyanate	Decaborane (14)
Nitrogen dioxide	Nicotine	Mercuric bromide
Ozone	Tetranitromethane	Mercury (II) chloride
Phosphine	Tetraethylstannane	Pentachlorophenol
Hydrogen selenide		Methyl parathion
Stibene		Phosphorus (white)
		Sodium azide
	Toxic Materials	
Boron trichloride	Acrylonitrile	Acrylamide
Boron trifluoride	Allyl alcohol	Barium chloride
Chlorine	Alpha-chlorotoluene	Barium (II) nitrate
Hydrogen fluoride	Aniline	Benzidine
Hydrogen sulfide	1-Chloro-2,3-epoxy-propane	p-Benzoquinone
Phosgene	Chloroformic acid	Beryllium chloride
Silicon tetrafluoride	3-Chloropropene	Cadmium oxide
	o-Cresol	Hydroquinone
	Crotonaldehyde	Mercuric sulfate
	Phosphorus chloride	Oxalic acid
	Thionyl chloride	Phenol
		Sodium fluoride

Source: Section 2.2.1, Appendix VI-A in the *Uniform Fire Code* (ICBO 1994b).

Radioactive Materials. Indoor storage areas for radioactive materials must have a liquid-tight floor and detection equipment available that is suitable for determining surface level contamination for short-term hazard conditions. The maximum quantity and storage arrangement must be in compliance with Nuclear Regulatory Commission requirements as well as any state and local requirements. With the exception of some licensed, sealed sources, material stored in each individual container cannot exceed 2 millicuries for alpha emitters, 200 curies for beta emitters, or 0.1 curies for gamma emitters. Contaminated combustible materials must be kept in tightly closed noncombustible containers that do not contain other waste. If the combustible wastes are also contaminated with oxidizing materials subject to spontaneous heating, these wastes must be disposed of promptly.

Radioactive materials that are stored outdoors must be kept more than 20 feet away from property lines, streets, alleys, public ways, and exits to a public way. No minimum distance is required where there is an unpierced 2-hour fire-resistive wall extending at least 30 inches above and to the sides of the storage area. A minimum 20-foot separation distance from buildings is also required, unless the exterior walls of the building have at least a 1-hour fire-resistive rating. The minimum distance from building openings is 10 feet, and building openings that are less than 20 feet from outdoor storage areas must be protected by a fire assembly having a 45-minute fire-resistive rating. Radioactive material must be stored in fire-resistive containers, be protected by an automatic, open head, deluge fire-sprinkler system, or be located under a non-combustible canopy protected by an approved automatic fire-extinguishing system.

Corrosives. Corrosive materials can be acids, bases, or other materials. Examples of corrosive materials given in the code are listed in Table 11.13. Indoor storage areas for corrosive materials must have liquid-tight floors. Corrosive materials stored outdoors must be kept at least 20 feet from buildings, property lines, streets, alleys, public ways, and exits to a public way. No minimum distance is required if there is an unpierced 2-hour fire-resistive wall extending at least 30 inches above and to the sides of the storage area. Secondary containment is also required.

Carcinogens, Irritants, Sensitizers, and Other Health Hazard Materials. Irritants are chemicals that are not corrosive, but that cause a reversible inflammatory effect on living tissue by chemical action at the site of contact. Sensitizers are chemicals that cause a substantial proportion of exposed people or animals to develop an allergic reaction in normal tissue after repeated exposure. Other health hazard materials are those that affect target organs of the body such as the liver or kidney, damage the nervous system, decrease hemoglobin function, deprive the body tissue of oxygen, or affect reproduction (ICBO 1994b). Examples of these materials given in the code are listed in Table 11.14.

A liquid-tight floor and secondary containment are required for indoor storage of these materials. For outdoor storage, the minimum separation dis-

TABLE 11.13   Examples of Corrosive Materials

Acids	Bases	Other Corrosives
Chromic acid	Ammonium hydroxide (>10%)	Bromine
Formic acid	Calcium hydroxide	Chlorine
Hydrochloric acid (>15%)	Potassium hydroxide (>1%)	Fluorine
Nitric acid (>6%)	Sodium hydroxide (>1%)	Iodine
Perchloric acid (>4%)	Potassium carbonate	Ammonia
Sulfuric acid (>4%)		

Source: Section 2.2.3, Appendix VI-A in the *Uniform Fire Code* (ICBO 1994b).

TABLE 11.14   Examples of Carcinogens, Irritants, and Other Health Hazard Materials

Carcinogens*	Irritants	Other Health Hazard Materials
Asbestos	Diphenylaminechloroarsine	Asbestos
Benzene	Xylyl bromide	Carbon disulfide
Beryllium	Chloracetophene	Carbon monoxide
Carbon tetrachloride		Cyanides
Chloroform		Lead
Diazomethane		Mercury
p-Dioxane		Nitrosamines
Ethylene dichloride		Silica
Polychlorinated biphenyls		Uranium
Vinyl chloride		

*Known or suspected carcinogens.

Source: Section 2.2.4, Appendix VI-A in the *Uniform Fire Code* (ICBO 1994b).

tance is 20 feet from buildings, property lines, streets, alleys, public ways, and exits to a public way, unless there is an unpierced 2-hour fire-resistive wall extending at least 30 inches above and to the sides of the area. Individual piles stored outside cannot be larger than 2,500 cubic feet with a minimum aisle space between piles of one half the height of the piles or 10 feet, whichever is greater.

LIFE SAFETY

Life safety addresses the protection of human life from fire in buildings and structures. The key standard for life safety is NFPA 101, "Code for Safety to Life from Fire in Buildings and Structures," more commonly referred to as the Life Safety Code®. NFPA also publishes a companion reference, *Life Safety Code® Handbook*. Model standards for life safety for local building regulations are found in the *Uniform Building Code* published by the ICBO (see References).

Life Safety Code®

The NFPA Life Safety Code® addresses the following topics:

- Classification of occupancy and hazard of contents—Classifications include occupancies for assembly, education, health care, detention and correctional institutions; residences; mercantile, business, industry, storage, and special structures. Hazard of contents is classified as low, ordinary, or high.

- Means of egress (exit)—The Code addresses means, capacity, number, arrangement, illumination, and marking of egress components; travel discharge to exits; discharge from exits; emergency lighting; special provisions for high hazards; and mechanical equipment rooms, boiler rooms, and furnace rooms. The Code includes provisions to accommodate persons with mobility impairments to reflect the Americans with Disabilities Act (ADA).
- Features of fire protection—Information is provided on construction and compartmentation, smoke barriers, special hazard protection, and interior finish.
- Building service and fire protection equipment—Utilities; heating, ventilating, and air conditioning; and smoke control. Other topics covered include elevators, escalators, and conveyors; rubbish chutes, incinerators, and laundry chutes; fire detection, alarm, and communication systems; and automatic sprinklers and other extinguishing equipment.
- New and existing occupancies (by occupancy classification)—Topics vary by occupancy classification, but generally include general requirements, means of egress, protection, provisions special to the particular occupancy, and building services.

The following paragraphs address the Life Safety Code® provisions for classification of hazard of contents and means of egress. The information provided is meant to be only a general summary; the reader is referred to the Code itself for full details of the provisions.

Hazard of Content. The contents of any building or structure are classified as low, ordinary, or high hazard. Low-hazard contents are those that have such low combustibility that no self-progagating fire can occur. Ordinary hazard contents are those that are likely to burn with moderate rapidity or to give off a considerable volume of smoke. High hazard contents are those that are likely to burn with extreme rapidity or from which explosions are likely.

Industrial occupancies are subdivided according to the hazard of contents, type of building, and operation. General occupancy involves low- or ordinary-hazard manufacturing operations in conventionally designed buildings that are suitable for various types of manufacturing. Special-purpose occupancy involves low- or ordinary-hazard manufacturing operations in buildings designed for and suitable only for particular types of operations, which are characterized by relatively few employees and much of the area occupied by machinery or equipment. High-hazard occupancy involves those buildings having high-hazard materials, processes, or contents. Incidental high-hazard operations in low- or ordinary-hazard occupancies must be protected, but do not make the occupancy high hazard overall.

In high-hazard occupancies, no dead ends are permitted in the travel path for means of egress. Travel distance to exits must not exceed 75 feet in high hazard occupancies. In low- or ordinary-hazard general occupancies, the travel distance must not exceed 400 feet. In special-hazard occupancies involving low- or ordinary-hazard materials, the travel distance must not exceed 300 feet; however, if the building is protected by an automatic sprinkler system, the travel distance can be as long as 400 feet. High-hazard occupancies require automatic fire-extinguishing systems or other protection appropriate to the type of hazard, such as explosion venting or suppression so as to minimize the danger of explosion to occupants while they are escaping to safety.

Means of Egress. Components for egress structures include doors, stairs, smokeproof enclosures, horizontal exits, ramps, exit passageways, escalators and moving walks, fire escape stairs, fire escape ladders, slide escapes, alternating tread devices, and areas of refuge. General provisions for doors, stairs, and lighting are described in the following paragraphs. Additional provisions and exceptions to these regulations and others may be found in the Code.

Exit doors must be designed and constructed so that the way to exit is obvious and direct. Door openings must be side-hinged or of a pivoted-swinging type and have a clear width of at least 32 inches. Doors serving high-hazard areas must swing in the direction of exit travel. Door locks must not need a key to be opened from the inside of the building. Door latches must be provided with a means of release such as a knob, handle, or panic bar that can be seen under all lighting conditions. Power-operated doors must be designed so that in the event of power failure, they can be opened manually to permit egress or closed where necessary to safeguard means of egress.

Stairs in new construction must meet the size requirements shown in Table 11.15. Landings must have a minimum length in the direction of travel equal to the width of the stair, but where the stair has a straight run, the landing length cannot exceed 4 feet. If the slope of a new stair exceeds 1 in 15, handrails must be provided on both sides and within 30 inches of all portions of the stair width. Handrails must run the full length of each flight of stairs and continue through the turns between flights at landings. Handrails must be between 34 and 38 inches vertically above the surface of the tread. Handrails must have a circular cross section with an outside diameter between 1.25 and 2 inches and must be at least 1.5 inches from the wall. An example of stairs constructed outside a building is shown in Figure 11.2. Outside stairs must be designed to avoid any handicap to persons having a fear of heights.

Exit access, including stairs, aisles, corridors, ramps, escalators, and passageways, must be illuminated continuously as long as a means of egress is required. Floors of egress must be illuminated to at least 1 footcandle. Battery-operated lights are not allowed for primary illumination for egress but are acceptable for emergency lighting. Exit access must be provided with emergency lighting that activates without appreciable lag time upon changing lighting energy sources. If normal lighting fails, emergency lighting must be available for at least 1.5 hours. Emergency lighting must provide an average of 1 footcandle and a minimum of 0.1 footcandle at floor level. Emergency lighting

TABLE 11.15   Dimensional Criteria for New Stairs

Parameter	Required Dimension
Minimum width clear of all obstructions, except projections not exceeding 3.5 inches at or below handrail height on each side	44 inches
Height of risers	4 to 7 inches
Minimum tread depth	11 inches
Minimum headroom	6 feet, 8 inches
Maximum height between landings	12 feet

Source: Information from *Life Safety Code®* (1994).

must come on automatically when normal lighting is interrupted. Emergency generators powering lights must comply with NFPA 110, "Emergency and Standby Power Systems." Batteries in emergency lights must conform with NFPA 70, "National Electric Code®."

Uniform Building Code

The *Uniform Building Code* and its companion publication, *Uniform Building Code Standards*, also provide standards for building and life safety. These stan-

Figure 11.2   Outside stairs as means of egress for life safety.

dards include requirements for lighting, ventilation, sanitation, sprinkler and standpipe systems, fire alarms, construction type, construction height, allowable area, and special hazards. Other building standards in the code cover fire-resistive interior walls and ceilings, engineering pertaining to quality and design of materials of construction, and types of construction. The following paragraphs discuss some of the general requirements for occupancy classification, fire-resistant materials and constuction, interior finishes, smoke and heat control, and means of egress. The reader is referred to the code itself for more detail and exceptions to these requirements.

Classification of Occupancy. Buildings are classified by the types of occupancy shown in Table 11.16 and are subdivided into divisions within each classification. Group H, hazardous occupancy, is subdivided into seven divisions, as shown in Table 11.17.

Fire-Resistant Materials and Construction. A chapter of the *Uniform Building Code* covers fire-resistive materials and systems; protection of structural members, projections (cornices, overhangs, balconies), construction joints, insulation, fire blocks and draft stops, walls and partitions, floor-ceilings or roof-ceilings, shaft enclosures, usable space under floors, fire-resistive assemblies for protection of openings, and through-penetration fire stops. Standards of quality for fire-resistive materials recognized in the code are listed in Table 11.18.

TABLE 11.16   Classification of Building Code Occupancies

Group	Occupancy
A	Assembly
B	Business
E	Educational
F	Factory and Industrial
H	Hazardous
I	Institutional
M	Mercantile
R	Residential
S	Storage
U	Utility

Source: Information from *Uniform Building Code* (ICBO 1994a).

TABLE 11.17   Divisions of Group H, Hazardous Occupancies

Division	Type of Occupancy
1	Occupancies with materials exceeding exempt limits, which present a high explosion hazard. Certain exemptions apply to pyrotechnic special effect materials.
2	Occupancies where combustible dust is manufactured, used, or generated in such a manner that concentrations and conditions create a fire or explosion potential; occupancies with materials exceeding exempt limits, which present a moderate explosion hazard or a hazard from accelerated burning.
3	Occupancies where flammable solids, other than combustible dust, are manufactured, used, or generated. Also includes materials exceeding exempt limits, that present a high physical hazard.
4	Repair garages not classified as Group S, Division 3, Occupancies.
5	Aircraft repair hangars not classified as Group S, Division 5, Occupancies and heliports.
6	Semiconductor fabrication facilities and comparable research and development areas in which hazardous production materials (HPM) are used and the aggregate quantity exceeds exempt limits.
7	Occupancies with materials exceeding exempt limits that are health hazards.

Source: Information from *Uniform Building Code* (1994a).

TABLE 11.18   Standards of Quality for Fire-Resistive Materials and Systems

U.B.C. Standard 7-1	Fire Tests of Building Construction and Materials
U.B.C. Standard 7-2	Fire Tests of Door Assemblies
U.B.C. Standard 7-3	Tinclad Fire Doors
U.B.C. Standard 7-4	Fire Tests of Window Assemblies
U.B.C. Standard 7-5	Fire Tests of Through-Penetration Fire Stops
U.B.C. Standard 7-6	Thickness and Density Determination for Spray-Applied Fireproofing
U.B.C. Standard 7-7	Methods for Calculating Fire Resistance of Steel, Concrete Masonry, and Wood Construction
ASTM C 516	Vermiculite Loose-Fill Insulation
ASTM C 549	Perlite Loose-Fill Insulation
ANSI/NFPA 80	Standard for Fire Doors and Windows
ASTM C 587 and 588	Gypsum Base for Veneer Plaster and Gypsum Veneer
ASTM C 330 and 332	Lightweight Aggregates for Structural and Insulating Concrete
ASTM C 331	Lightweight Aggregates for Concrete Masonry Units
UL 555	Fire Dampers
UL 555C	Ceiling Dampers
UL 555S	Leakage Rated Dampers for Use in Smoke Control Systems
UL 33	Heat Response Links for Fire Protection Service
UL 353	Limit Controls

Source: Information from *Uniform Building Code* (ICBO 1994a).

Interior Finishes. Interior finishes are divided into three classes of materials, based on their flame-spread index as shown in Table 11.19. Standards in the code applicable to quality of interior finishes are:

- UBC Standard 8-1—Test Method for Surface-Burning Characteristics of Building Materials
- UBC Standard 8-2—Standard Test Method for Evaluating Room Fire Growth Contribution of Textile Wall Covering.

Requirements for interior finishes also include application of interior finishes, maximum allowable flame spread, textile wall coverings, insulation, and sanitation areas.

Smoke and Heat Control. The *Uniform Building Code* contains requirements for smoke control and smoke and heat venting. Topics covered in regard to smoke control are design methods, pressurization method, airflow method, exhaust method, design fire, equipment, power systems, detection and control systems, control air tubing, marking and identification, fire fighter's control panel, response time, and acceptance testing. Topics covered in regard to smoke and heat venting are occupancy requirements, types of vents, releasing devices, size and spacing of vents, and curtain boards.

Means of Egress. There are numerous requirements in the *Uniform Building Code* for means of egress. Topics covered by these requirements include occupant load, exits, doors, corridors and exterior exit balconies, stairways, ramps, exit illumination and signs, aisles, and many others. General descriptions of some of these requirements are given in the following paragraphs.

The cumulative exit width (in inches) of an occupancy must be at least the total occupant load multiplied by 0.3 for stairways and 0.2 for other exits. The width must be divided approximately equally among the individual exits. When only two exits are required, they must be separated by a distance at least one half the length of the maximum overall diagonal dimension of the building or area served. Additional separation distances are specified when there are more than two exits. Travel distance to an exit must not exceed 150 feet.

TABLE 11.19 Classification of Interior Finishes
Based on Flame-Spread Index

Class	Flame-Spread Index
I	0–25
II	26–75
III	76–200

Source: Information from *Uniform Building Code* (ICBO 1994a).

When an automatic sprinkler system is installed, the travel distance may be increased to 200 feet and by an additional 100 feet when the last 100 feet are entirely within a 1-hour fire-resistive corridor.

Exit passageways must not have any openings in the wall other than required exits from normally occupied spaces. Walls, floors, and ceilings of passageways must have at least a 1-hour fire resistance, but must also have at least the same fire resistance as other components in the building. Exit openings must be protected by fire assemblies having at least a 45-minute fire protection rating.

Exit signs must have lettering that stands out in contrast to the background color. Letters must be in block type at least 6 inches high with a stroke of at least 0.75 inches. If an exit sign is illuminated by an external light, the light must have an intensity of at least 5 footcandles on the sign face.

Accessible means of egress must be provided for those persons with physical disabilities or whose mobility is impaired. All spaces in the building that are required to be accessible must have at least one accessible means of egress. Exit stairways must have a clear width of 48 inches between handrails. When an accessible floor is at least four stories away from an exit, an elevator must be provided for accessible means of egress and standby power for the elevator is also required. Each area of refuge (a safe temporary area to await instructions or assistance) must be able to accommodate one wheelchair space (30 inches by 48 inches) for each 200 occupants or portion thereof. Each area of refuge must be identified by an AREA OF REFUGE sign and the international symbol of accessibility. This sign must be posted at each door to the area and must be illuminated. In addition, tactile signage must be located at each door to the area.

REFERENCES

American Society of Safety Engineers, *Dictionary of Terms Used in the Safety Profession*, ASSE, Des Plaines, IL, 1988.

International Conference of Building Officials, *Uniform Building Code*, Vol. 1, "Administrative, Fire- and Life-Safety, and Field Inspection Provisions," ICBO, Whittier, CA, 1994a.

International Conference of Building Officials, *Uniform Fire Code*, Vol. 1, ICBO, Whittier, CA, 1994b.

Meyer, Eugene, *Chemistry of Hazardous Materials*, Prentice-Hall, Englewood Cliffs, NJ, 1989.

National Fire Protection Association, *Life Safety Code® Handbook*, 6th ed., Ron Coté, ed., NFPA, Quincy, MA, 1994.

National Fire Protection Association, *Life Safety Code®*, NFPA 101, NFPA, Quincy, MA, 1994.

National Fire Protection Association, *Fire Protection Handbook*, NFPA, Quincy, MA, 1992.

National Fire Protection Association, *Fire Protection Guide on Hazardous Materials*, 10th ed., NFPA, Quincy, MA, 1991.

National Safety Council, *Accident Prevention Manual for Business & Industry: Engineering and Technology*, NSC, Chicago, 1992.

12

PROCESS AND SYSTEM SAFETY

Because of recent catastrophic releases of toxic, reactive, or flammable liquids and gases, national and international attention has been given to emergency planning and release mitigation. The Emergency Planning and Community Right-to-Know Act (EPCRA) of 1986 provided the framework for developing comprehensive emergency plans on the community level, and OSHA's standard for Process Safety Management of Highly Hazardous Chemicals, which was finalized in 1992, made process safety mandatory for major users of highly hazardous chemicals.

This chapter explores process and system safety in terms of OSHA's standard. In addition, system safety in terms of reliability is discussed and logic symbols and methods used for basic calculations of system reliability are provided. Finally, emergency planning, as required under EPCRA, is discussed.

PROCESS SAFETY MANAGEMENT

Section 304 of the Clean Air Act Amendments (CAAA) requires the Secretary of Labor, in conjunction with the Administrator of EPA, to promulgate, pursuant to the Occupational Safety and Health Act of 1970, a chemical process safety standard to prevent accidental releases of chemicals that could pose a threat to employees. This standard, delineated under 29 CFR 1910.119 and finalized on February 24, 1992, set forth requirements for an organized approach to process safety management (PSM).

Elements included in the standard, as specified in the CAAA, are process hazard assessment, employee education and training, documentation of operating and safety information, equipment maintenance and testing, emergency preparedness planning, and other management aspects. The standard mainly applies to manufacturing industries and other high-use chemical sectors. Examples of industries and other sectors that may be covered by the standard—depending on amounts of chemicals stored at the facility and other factors—are listed in Table 12.1. Covered employers are those companies that manufacture, store, or use any of more than 130 specific toxic and reactive chemicals in listed quantities. Flammable liquids and gases stored or used are covered if the amount is 10,000 pounds or greater. The standard does not apply to retail facilities, oil and gas well-drilling or servicing operations, normally unoccupied remote facilities, hydrocarbon fuels used solely for workplace consumption as fuel, and flammable liquids stored in atmospheric tanks, or transferred, that are kept below their normal boiling point without benefit of chilling or refrigerating and that are not connected to a process.

Basic Elements of the PSM Standard

Process Safety Information. All employers covered by the standard must compile written process safety information about the hazards of the highly hazardous chemicals used or produced by a process, information about the technology of the process, and information about the equipment used in the process. This information is a precursor to the hazard analysis required by the standard, since it provides the basis for identifying and understanding the hazards of a process. Types of information to be included are presented in Table 12.2.

Employee Involvement. Section 304 of the CAAA requires that employers involve employees in developing and implementing the process safety management program elements and hazard assessments. A written plan for methods of accomplishing this is required under the standard. Existing health and safety programs can be used as vehicles for keeping employees informed about hazards they may encounter, for training required under PSM, and for other obligations required under PSM (OSHA 1993).

TABLE 12.1   Examples of Manufacturing and Other Sectors Likely to Be Covered Under the Process Safety Management Standard

Manufacturing Industries	Other Sectors
• Chemical manufacturing	• Natural gas liquids
• Chemical products	• Farm product warehousing
• Transportation equipment	• Electric
• Primary metals	• Gas
• Fabricated metal products	• Sanitary services
• Others	• Wholesale trade
	• Pyrotechnics/explosives manufacturers

TABLE 12.2 Types of Information to Be Documented Under the PSM Standard

Hazard Information for Each Highly Hazardous Chemical

- Toxicity
- Permissible exposure limits
- Physical data
- Corrosivity data
- Thermal and chemical stability data
- Hazardous effects of inadvertent mixing of different materials

Information Pertaining to the Technology of the Process

- A block flow diagram or simplified process flow diagram
- Process chemistry
- Maximum intended inventory
- Safe upper and lower limits for items such as temperatures, pressures, flows, and compositions
- An evaluation of the consequences of deviations, including those affecting the safety and health of employees

Information Pertaining to Equipment Used in the Process

- Materials of construction
- Piping and instrument diagrams (P&IDs)
- Electrical classification
- Relief system design and design basis
- Ventilation system design
- Design codes and standards employed
- Material and energy balances for processes built after May 26, 1992
- Safety systems (e.g. interlocks, detection of suppression system)

Source: 29 CFR 1910.119(d).

Process Hazard Analysis. Process hazard analysis (PHA) is a thorough, organized, and systematic approach used to identify, evaluate, and control the hazards of processes involving highly hazardous chemicals. The analysis must address or include the following:

- Hazards of the process.
- Identification of any previous incident that had a potential for catastrophic consequences in the workplace.
- Engineering and administrative controls applicable to the hazards and their interrelationships, such as appropriate application of detection methodologies to provide early warning of releases. Acceptable detection methods may include process monitoring and control instrumentation with alarms, and detection hardware such as hydrocarbon sensors.

- Consequences of failure of engineering and administrative controls.
- Facility siting.
- Human factors.
- Qualitative evaluation of a range of the possible safety and health effects on employees in the workplace if there is a failure of controls.

Acceptable PHA methodologies include the what-if technique, the checklist technique, the what-if/checklist methodology, the hazard and operability study (HAZOP), the failure mode and effects analysis (FMEA), the fault tree analysis, or an equivalent methodology. Information pertaining to these methodologies are shown in Table 12.3.

OSHA suggests, in Appendix C of the standard, that the team conducting the PHA should understand the methodology used, and that the team leader should be fully knowledgeable about the methodology and the PHA process.

TABLE 12.3   Methodologies for Process Hazard Analysis

Methodology	Description
What-if	Used for relatively uncomplicated processes; "what-if" questions are used to evaluate the effects of component failures and procedural or other errors on various aspects of the process, from raw materials to product; useful for evaluating almost all parts of design and operation, including process development and conceptual design, pilot plant operation, engineering, construction/start-up, normal operation, expansion/modification, incident investigation, and shutdown.
Checklist	Used for more complex processes; during analysis, an organized checklist is developed and aspects of the process are assigned to committee members having the greatest experience or skill in evaluating those aspects; operator practices and job knowledge are audited in the field, and suitability of equipment and materials of construction is studied; chemistry and control systems are reviewed; operating and maintenance records are audited; useful for a stable process; often precedes use of a more sophisticated hazard assessment technique; useful for evaluation of design, pilot plant operation, engineering, construction/start-up, normal operation, expansion/ modification, and shutdown.

TABLE 12.3 *Continued*

Methodology	Description
What-if/Checklist	A comprehensive approach that combines the creative thinking of a selected team of specialists with the methodical focus of a prepared checklist; allows methodical examination of the operation from receipt of raw materials to delivery of the finished product to the customer's site; information acquired from a process review using this method can be used in training operating personnel; applicable to all parts of a facility, including equipment and operations of a process; useful for evaluation of design, pilot plant operation, engineering, construction/start-up, normal operation, expansion/modification, and shutdown.
Hazard and operability study (HAZOP)	A formal method of systematically investigating each element of a system for all potential deviations from design conditions; deviations in parameters such as flow, temperature, pressure, and time are reviewed in conjunction with material strength, locations of connections, and other design aspects; piping and instrument designs (or plant model) are analyzed critically during this process for effects of potential problems that could arise in each vessel or pipeline in the process; potential causes for failure are evaluated, consequences of failure are reviewed, and safeguards against failure are assessed for adequacy; can be used for continuous or batch processes; useful for evaluation of pilot plant operation, engineering, normal operation, expansion/modification, and incident investigation.
Failure mode and effects analysis	A methodical study of component failure; includes all effect analysis system diagrams and reviews all components in the system that could fail, such as pumps, instrument transmitters, seals, valves, temperature controllers, and other components; potential mode of failure, consequence of failure, hazard class, probability of failure, and compensating provisions are evaluated; multiple concurrent failures are assessed; data are analyzed for each component or multiple component failure, and a series of recommendations appropriate to risk management are developed; applicable to individual system components; useful for evaluation of pilot plant operation, engineering, normal operation, expansion/modification, and incident investigation.

TABLE 12.3 *Continued*

Methodology	Description
Fault tree analysis	A qualitative or quantitative model of all undesirable outcomes from a specific event; includes catastrophic occurrences such as explosion, toxic gas release, and rupture; a deductive technique that uses logic symbols to graphically represent possible sequences; the resulting diagram looks like tree with many branches, each branch listing the sequential events (failures) for different independent paths to the top event; assessment of probability using failure rate data is used to calculate probability of occurrence of the undesired event; useful in evaluating the effect of alternative actions on reducing the probability of occurrence of the undesired event; applicable to equipment and system design features and operational procedures that are interrelated; applicable to highly redundant systems; useful for evaluation of pilot plant operation, engineering, normal operation, expansion/modification, and incident investigation.
Other	Event tree, cause-consequence analysis, human methodologies, error analysis, cause-consequence analysis, relative ranking, and others.

Note: Each type of process hazard analysis technique or combination of techniques should be evaluated carefully by a trained expert for appropriateness and compliance with 29 CFR 1910.119(e).

Source: Information from 55 FR 29150, July 17, 1990, EPA (1987), and AIChE (1992).

The team should have expertise in the various areas of process technology, process design, operating procedures and practices, instrumentation, routine and nonroutine tasks, safety and health, and other relevant areas, as the need dictates. At least every five years after the completion of the initial process hazard analysis, the analysis must be updated and revalidated by a team that meets the standard's requirements. Documentation pertaining to the PHA and updates/revalidation must be kept on file for the life of the facility.

Operating Procedures. Employers covered under the PSM standard must develop and implement written operating procedures, consistent with the process safety information, that provide clear instructions for safely conducting activities involved in each covered process. Elements to be addressed in the procedures are presented in Table 12.4.

Training. The PSM standard requires that each employee presently involved in operating a process or a newly assigned process covered under the standard,

TABLE 12.4   Elements to Include in Operating Procedures for Facilities Covered Under the Process Safety Standard

Steps for Each Operating Phase

- Initial start-up
- Normal operations
- Temporary operations
- Emergency shutdown, including the conditions under which emergency shutdown is required, and the assignment of shutdown responsibility to qualified operators to ensure that emergency shutdown is executed in a safe and timely manner
- Emergency operations
- Normal shutdown
- Start-up during turnaround or after an emergency shutdown

Operating Limits

- Consequences of deviation
- Steps required to correct or avoid deviation

Safety and Health Considerations

- Properties of, and hazards presented by, the chemicals used in the process
- Precautions necessary to prevent exposure, including engineering controls, administrative controls, and personal protective equipment
- Control measures to be taken if physical contact or airborne exposure occurs
- Quality control for raw materials and control of hazardous chemical inventory levels
- Any special or unique hazards
- Safety systems (e.g., interlocks, detection or suppression systems) and their functions

including maintenance and contractor employees, must be trained in an overview of the process and in its operating procedures. Training must include emphasis on the specific safety and health hazards; emergency operations, including shutdown; and safe work practices applicable to the employee job tasks. If an employee is already operating the process on the effective date of the standards, the employer has the option, in lieu of initial training, of certifying in writing that the employee has the required knowledge, skills, and abilities to safely carry out the duties and responsibilities specified in the operating procedure.

Refresher training must be provided at least every three years, or more often if necessary, to ensure that the employee understands and adheres to the current operating procedures of the process. The employer, in consultation with the employees involved in operating the process, must determine the appropriate frequency of refresher training. Documentation of the training must be kept, including employee's identity, the date of training, and means used by the employer to verify that the employee understood the training.

Contractors. PSM applies to contractors performing maintenance or repair, turnaround, major renovation, or specialty work on or adjacent to a covered process. It does not apply to contractors providing incidental services that do not influence process safety, such as janitorial, food and drink, laundry, delivery, or other supply services.

Employers who use contractors to perform work in and around processes that involve highly hazardous chemicals have to establish a screening process so that they hire and use only contractors who accomplish the desired job tasks without compromising the safety and health of any employees at the facility. Employer responsibilities include:

- Obtaining and evaluating information regarding the contract employer's safety performance and programs
- Informing contract employers of the known potential fire, explosion, or toxic release hazards related to the contractor's work and the process
- Explaining to contract employers the applicable provisions of the emergency action plan
- Developing and implementing safe work practices consistent with PSM standards to control the entrance, presence, and exit of contract employers and contract employees in covered process areas
- Evaluating periodically the performance of contract employers in fulfilling their obligations under the PSM standard
- Maintaining a contract employee injury and illness log related to the contractor's work in the process areas

The contractor employer also has responsibilities, which includes the following:

- Ensuring that the contract employees are trained in the work practices necessary to perform their jobs safely
- Ensuring that the contract employees are instructed in the known potential fire, explosion, or toxic release hazards related to their jobs and the process, and in the applicable provisions of the emergency action plan
- Documenting that each contract employee has received and understood the training required by the standard by preparing a record that contains the identity of the contract employee, the date of training, and the means used to verify that the employee understood the training
- Ensuring that each contract employee follows the safety rules of the facility, including the required safe work practices required in the operating procedures section of the standard

- Advising the employer of any unique hazards presented by the contract employer's work

Pre-Start-Up Safety Review. A pre-start-up safety review is required before any covered new process facilities or modified facilities—when the modification is significant enough to require a change in the process safety information—can be brought on line. The pre-start-up safety review includes confirmation that construction and equipment are in accordance with design specifications, as well as assurance that safety, operating, maintenance, and emergency procedures are in place and are adequate. For new facilities, the pre-start-up review ensures that a PHA has been performed and that recommendations have been resolved or implemented before start-up; for modified facilities, any changes, other than "in-kind replacements," to the facilities must go through management of change procedures. These include updates to P&IDs, operating procedures and instructions, training, and other aspects of the PSM standard. (See "Management of Change," presented later in this section.)

Mechanical Integrity of Equipment. Employers must establish and implement written procedures to maintain the ongoing integrity of process equipment, including pressure vessels and storage tanks, piping systems, relief and vent systems and devices, emergency shutdown systems, pumps, and controls such as monitoring devices and sensors, alarms, and interlocks. Each employee involved in maintaining the mechanical integrity of the equipment must be trained in the process hazards and the procedures applicable to his or her job tasks. Inspection and testing must be performed on the process equipment, using the following criteria:

- Procedures must follow recognized and generally accepted good engineering practices.
- Frequency of inspections and tests of process equipment must be consistent with applicable manufacturer's recommendations, state and federal regulations, and good engineering practices, and must be performed more frequently if determined to be necessary by prior operating experience.
- Testing documentation must include the date of the inspection or test, the name of the person who performed the inspection or test, the serial number or other identifier of the equipment on which the inspection or test was performed, a description of the inspection or test performed, and the results of the inspection or test.

Equipment deficiencies outside the acceptable limits defined by the process safety information must be corrected before further use, unless other steps are taken to ensure a safe operation. In these cases, the deficiencies must be corrected in a safe and timely manner. In constructing new plants and equipment, the employer must ensure that equipment, as it is fabricated, is suitable for the process application for which it will be used. Appropriate field checks and inspections should be performed to ensure that equipment is installed properly and is consistent with design specifications, engineering drawings, and the manufacturer's instructions.

Hot Work Permit. The employer is responsible for issuing a hot work permit for hot work operations conducted on or near a covered process. The permit must document that the fire prevention and protection requirements specified under 29 CFR 1910.252(a) are met. Further information, such as date of authorization for hot work and identity of the object on which the hot work is to be performed, must be included. The permit should be kept on file until the work is complete or, as a good management practice, filed at the facility for a period of time designated by the safety professional.

Management of Change. Employers covered under the PSM standard must establish and implement written procedures to manage changes to process chemicals, technology, equipment, and procedures, as well as changes to facilities that affect a covered process. Considerations to be addressed in the procedures include the technical basis for the proposed change, the impact of the change on employee safety and health, modifications to operating procedures, necessary time period for the change to be effected, and authorization requirement for the change.

As with other aspects of the PSM standard, employees who operate a process and maintenance and contract employees whose job tasks will be affected by a change in the process must be informed of, and trained in, the change prior to start-up of the process or start-up of the affected part of the process. Updates to information, procedures, and other documents also should be made before start-up.

Incident Investigation. Any incident that resulted in, or reasonably could have resulted in, a catastrophic release of a highly hazardous chemical in the workplace must be investigated by the employer. The incident investigation must be initiated as promptly as possible, but within no more than 48 hours. An incident investigation team must be established and must include at least one person knowledgeable of the process; if a contractor is involved, a contract employee must also participate. A report must be prepared at the conclusion of the investigation that includes the date and time of the incident, the date the investigation began, a description of the incident, factors contributing to the incident, and recommendations resulting from the investigation.

The report must be reviewed by all affected personnel whose job tasks are relevant to the incident findings, including contract employees where applicable. The incident investigation must be retained for five years. In some cases, if the release is also regulated by other agencies or authorities, investigation and reporting must meet their requirements, which may be different and possibly more stringent.

Emergency Planning and Response. Emergency planning and response are required under the PSM standard for an entire facility, with planning being in accordance with requirements under 29 CFR 1910.38(a). In addition, the plan must include procedures for handling small releases. In some cases, the hazardous waste and emergency response provisions under 29 CFR 1910.120(a), (p), and (q) also apply. Emergency planning at the community level is discussed later in this chapter.

Compliance Audits. To ensure that PSM is effective, employers must certify that they have evaluated compliance with the provisions of PSM every three years. The compliance audit must be conducted by at least one person knowledgeable in the process. A findings report must be written, and the employer must determine and document an appropriate response to each of the findings of the compliance audit. Corrected deficiencies must be documented, and the two most recent compliance reports should be kept on file.

Trade Secrets. Employers must make available all information necessary to comply with PSM to those persons responsible for compiling the process safety information, those developing the PHA, those responsible for developing the operating procedures, and those performing incident investigation, emergency planning and response, and compliance audits, without regard to the possible trade secret status of such information. However, nothing in PSM precludes the employer from requiring those persons to enter into confidentiality agreements not to disclose such information.

Release Mitigation Techniques

Release mitigation techniques can be utilized by facility owners and operators to prevent disaster. Process hazard analysis is one of the best methods of pinpointing factors that can cause (or avert) an accident or incident. Release mitigation techniques can be used, such as prerelease control, utilization of prerelease protection equipment, management activities, and safety systems and procedures. Other factors include facility siting and sizing, facility security, chemical selection and storage practices, and emergency preparedness. Table 12.5 provides information about these release mitigation techniques.

SYSTEM RELIABILITY

Series and Parallel Systems

Series Systems. A series system is one that has components that are interrelated to the degree that if one component fails, the system fails. All components function independently of one another; they may have different reliability

TABLE 12.5   Examples of Release Mitigation Factors and Activities

Factors Pertaining to the Facility

- Buffer zones to protect population areas
- Plant layout to minimize zone impact
- Drainage control
- Sizing of systems to minimize zone of impact and to reduce inventories
- Routine security patrols
- Adequate boundary security such as fencing and controlled accesses
- Locks on drain valves and pump starters
- Adequate lighting

Factors Pertaining to Chemical Selection and Storage Practices

- Use of low-toxicity chemicals when possible
- Good inventory control
- Adequate segregation practices
- Use of low-pressure vessels when possible
- Formalized load/unload practices, including equipment inspection and proper containment

Factors Pertaining to Emergency Preparedness

- Adequate local emergency groups to assist during an incident
- Adequate emergency systems such as fire suppression systems and isolation capability
- Contracts with emergency cleanup vendors
- Adequate emergency training

Prerelease Control Factors

- Adequate preventative maintenance of equipment
- Adequate inspection and testing of equipment
- Scheduled comprehensive safety audits
- Adequate process controls for operations monitoring
- Alarm capability for early warning of problems
- Adequate release prevention equipment
- Other control mechanisms specified under the PSM standard

Prerelease Protection Equipment

- Adequate containment for tanks, process equipment, and overhead piping
- Neutralization capabilities
- Adequate equipment such as flares or incinerators, scrubbers or adsorbers, spray curtains, and other equipment
- Adequate spill cleanup supplies, fire-fighting equipment, air monitors, and other necessary equipment

TABLE 12.5 *Continued*

Management Activities

- Adequate training, including employee safety training, PSM training, hazard communication training, emergency response training, certification of operators on equipment or systems, and other training as necessary
- Membership in community emergency planning and response groups
- Development of an adequate release control program and accident/incident investigation program as defined in the PSM standard
- Participation in research/technical conferences
- Development of safety loss prevention program
- Development of formal procedures for notification of accidental releases
- Development of other safety procedures and programs, as necessary

Safety Systems and Procedures

- Adequate backup systems and redundant systems
- Use of formal valve lockout program
- Use of automatic shutoffs, bypass systems, and surge systems
- Use of manual overrides and interlocks
- Use of alarms
- Use of formalized testing program for all safety equipment
- Use of formalized color coding/labeling program

Source: Information from AIChE, *Guidelines for Safe Storage and Handling of High Toxic Hazard Materials* (1988a), AIChE, *Guidelines for Vapor Release Mitigation* (1988b), and EPA (1991).

rates for separate components, or the rates may be the same. The reliability of this type of system can be calculated by multiplying the reliabilities of the individual components, as shown in the following equation:

$$R_{\text{system}} = (R_1)(R_2)(R_3)(R_4) \ldots \tag{12.1}$$

EXAMPLE 12.1

A system consists of four components, each with a reliability rate of 0.98. Calculate the system reliability.

Solution. To calculate the system reliability, multiply the reliabilities of the individual components. In this case, because the component reliabilities are equal, the system reliability can be calculated as follows:

$$(0.98)^4 = 0.92$$

If the system consists of 10 components, each with a reliability rate of 0.98, the system reliability becomes:

$$(0.98)^{10} = 0.82$$

As can be seen from the products, the more components in series, the less reliable the system, because all it takes is one component to fail for the system to fail.

EXAMPLE 12.2

Another system consists of four components that are not the same. The reliabilities of the components are:

Component A—0.98
Component B—0.95
Component C—0.99
Component D—0.76

Calculate the system reliability.

Solution. The system reliablity is calculated as follows:

$$(0.98)(0.95)(0.99)(0.76) = 0.70$$

As can be seen from this case, one component with poorer reliability, such as Component D, can significantly affect the reliability of the entire system. Note also that the system reliability can be no higher than the lowest reliability among all components.

Parallel Systems. To increase the reliability of a system, components can be operated in parallel. In a parallel system, the system will fail only if all parrallel components or subsystems fail. To calculate reliability of a parallel system, use the products of unreliability (or failure) to determine reliability, as follows:

$$P_s = 1 - P_f \tag{12.2}$$

where P_s = Probability of success

and P_f = Probability of failure

Likewise:

$$P_f = 1 - P_s \tag{12.3}$$

EXAMPLE 12.3

Two components are operated in parallel, each with a reliability rate of 0.9. Calculate the system's reliability.

Solution. Use the product of the unreliabilities or failures:

$$P_f = (1-0.9)^2 \text{ or } 0.01$$

thus:

$$P_s = (1-0.01) = 0.99$$

For systems that are combinations of components in series and in parallel, it is necessary to combine the reliability equiations for both types of systems. This is illustrated in the following example.

EXAMPLE 12.4

The preceding parallel system is added to three other independent components in series, each with a reliability rate of 0.95. Calculate the system reliability.

Solution. To calculate the reliability of the entire system, multiply the calculated reliability of the parallel system with the product reliability of the independent components:

$$(0.99)(0.95)(0.95)(0.95) = 0.85$$

Failures

Typical manufactured items will experience failures over time in a manner similar to that presented in the graph presented in Figure 12.1, which is typically termed the "bathtub curve." The initial failures during the burn-in or break-in peroid occur at a higher rate than during the normal operating period for a variety of reasons. These include design or test inadequacies, material inadequacies, improper setup, and other reasons. The next period consists of normal or chance failures, which are typically caused by overstress, predictable design margins, improper use, and other unknown factors. The third period—end of life wear-out—also has a higher than normal failure rate, which can be caused by aging, material corrosion or wear, inadequate or improper maintenance, and other factors.

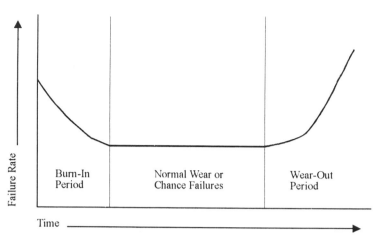

Figure 12.1   Bathtub curve.

Mean Time Between Failures

Another method of assessing the reliability of a component or system is by using mean time between failures (MTBF). The reliability of a component in terms of its service time can be obtained by use of the following equation:

$$R(t) = e^{-\lambda t}$$

where $R(t)$ = reliability as a function of time, λ = failure rate, and t = time

Further,

$$\lambda = 1/\theta$$

where θ = mean time between failures.

EXAMPLE 12.5

A component has a mean time between failures of 1 failure per 100 hours. What is the probability that it will operate for 200 hours? Calculate the probability of failure after 150 hours.

Solution. To calculate the probability of operation at 200 hours, use the reliability equation as follows:

$$R(t) = e^{-\lambda t}; \lambda = 1/\theta = 1/100; t = 200$$

Therefore, $R(t) = e^{-200/100} = 0.135$

To calculate the probability of failure after 150 hours, first determine the reliability; that is, the probability that the component will work for 150 hours:

$$R(t) = e^{-\lambda t}; \lambda = 1/\theta = 1/100; t = 150$$

Therefore, $R(t) = e^{-150/100} = 0.223$

Because $P_f = 1 - R(t)$, the probability of failure is:

$$1 - 0.223 = 0.777$$

The failure rate of a series system is the sum of the failure rates of its components. For instance, three components in series, each with a failure rate of 0.05, would give the system an overall failure rate of 0.15.

EXAMPLE 12.6

Calculate the mean time between failure for a four-component series system that has the following component reliabilities, assuming the reliabilities are given for 100 hours of operation.

Component A—0.90
Component B—0.95
Component C—0.87
Component D—0.99

Solution. Solve for λ for component A:

$$R(t) = 0.90 = e^{-\lambda(100)}$$
$$\lambda = (1.05) 10^{-3} \text{ failures per hour}$$

Solve for λ for component B:

$$R(t) = 0.95 = e^{-\lambda(100)}$$
$$\lambda = (0.51) 10^{-3} \text{ failures per hour}$$

Solve for λ for component C:

$$R(t) = 0.87 = e^{-\lambda(100)}$$
$$\lambda = (1.39) 10^{-3} \text{ failures per hour}$$

Solve for λ for component D:

$$R(t) = 0.99 = e^{-\lambda(100)}$$

$$\lambda = (0.10)\,10^{-3}\text{ failures per hour}$$

The system failure rate is determined by summing the individual component failure rates:

$$(1.05)\,10^{-3} + (0.51)\,10^{-3} + (1.39)\,10^{-3} + (0.10)\,10^{-3} = (3.05)\,10^{-3}$$

Since the MTBF (θ) is the reciprocal of λ, the failure rate,

$$\theta = 1/(3.05)\,10^{-3}\text{, or 328 hours}$$

Fault Tree Analysis

Fault tree analysis is used to predict the probability of a catastrophic event. An example of a fault tree is presented in Figure 12.2. Boolean algebra—a mathematical system devised for the analysis of symbolic logic, in which all variables have the value of either zero or one—can be used to determine the most probable means for creating the catastrophic event. From the figure, the symbol used for elements B_1, B_2, and A_2 represents an "and" gate; that is, both events must happen or both components must fail. The symbol used for element A_1 represents an "or" gate; that is, either event must happen or either component must fail.

EXAMPLE 12.7

Determine the simplified expression for the fault tree in Figure 12.2 (*Source:* BCSP 1993.)

Solution. From the fault tree:

$$(X_1 \bullet X_2) \bullet [X_2 + (X_1 \bullet X_4) + X_3]$$

where: $\bullet$ represents the "and gate" and $+$ represents the "or gate."

Using Distributive Law: $A(B + C) = (A \bullet B) + (A \bullet C)$:

$$(X_1 \bullet X_2 \bullet X_2) + [(X_1 \bullet X_1) \bullet (X_2 \bullet X_4)] + (X_1 \bullet X_2 \bullet X_3)$$

From the Law of Tautology: $A \bullet A = A$:

$$(X_1 \bullet X_2) + (X_1 \bullet X_2 \bullet X_4) + (X_1 \bullet X_2 \bullet X_3)$$

From Distributive Law:

$$X_1 \bullet X_2 \bullet (1 + X_4 + X_3)$$

From the Laws of Sets: $A + 1 = 1$ and $A \bullet 1 = A$

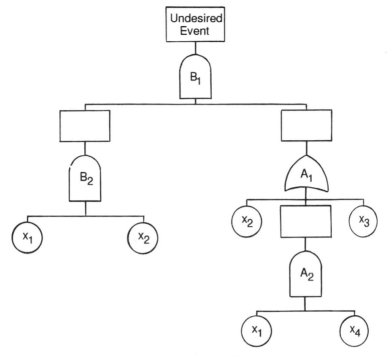

Figure 12.2 Example of fault tree.

$$X_1 \bullet X_2 \bullet 1 = X_1 \bullet X_2$$

All other expressions for the catastrophic event are more complicated (i.e., involve additional terms) than $X_1 \bullet X_2$; thus, this expression gives the highest probability of the event is happening.

EXAMPLE 12.8

Using Figure 12.2, calculate the probability of occurrence of a catastrophic event if the probabilities of occurrence are:

$$X_1 = 0.0600$$

$$X_2 = 0.0025$$

$$X_3 = 0.0060$$

$$X_4 = 0.0003$$

Solution.   Using $X_1 \bullet X_2$ as the highest probability of occurrence:

$$(0.0600) \bullet (0.0025) = 1.5 \times 10^{-4}$$

EMERGENCY PLANNING AND COMMUNITY RIGHT-TO-KNOW

Overview

EPA regulations that address the issue of hazard assessment and emergency response planning were developed as part of the Emergency Planning and Community Response Act (EPCRA), incorporated into Title III of the Superfund Amendments and Reauthorization Act of 1986 (known as SARA Title III). These regulations required each community to set up a formal local emergency planning committee (LEPC), which has the mission of assessing hazards in the area and developing an emergency response plan.

The LEPC includes representation of local elected officials, the fire department, the police, the civil defense unit, hospitals, health and first aid groups, local environmental entities, broadcast and print media, industry, and citizens from the community. This extensive representation allows input into the planning process from all affected parties in the local area. In addition, a state emergency response commission (SERC) has been formed in every state to aid in emergency response coordination.

Developing a Plan

The EPCRA, or SARA Title III, regulations required all LEPCs to develop a plan for emergency response to incidents that could occur in the designated local area, such as a county or a township. The deadline for the plans to be submitted to the EPA and the SERC was October 1988. Updates to the original plan are required if changes in local hazards or in the emergency response procedures occur.

Several guidance documents are available on the subject of local emergency planning, including:

- *Hazardous Materials Emergency Planning Guide*, published by the National Response Team (NRT) of the National Oil and Hazardous Substances Contingency Plan
- *Criteria for Review of Hazardous Materials Emergency Plans*, also published by the NRT
- *Guide for Development of State and Local Emergency Operations Plans*, published by the Federal Emergency Management Agency

Developing a thorough, workable emergency response plan can be a large task, depending on the number of facilities and amounts of chemicals stored in the area. Elements to be included in a local emergency response plan, as required by SARA Title III and recommended by NRT (NRT 1988), are discussed in the following paragraphs.

Planning Factors. Planning factors include the identification and description of all facilities in the local areas that possess extremely hazardous substances; identification of other facilities that may contribute to risk in the district or may be subject to risks as a result of being within close proximity of facilities with extremely hazardous substances; documentation of methods for determining that a release of extremely hazardous substances has occurred, and the area of population likely to be affected by a release; other information, such as findings from the hazard analysis, geographical features, demographic features, and other planning data.

Concept of Operations. Included are the designations of community emergency coordinator and facility emergency coordinators who will make determinations necessary to implement the plan.

Emergency Notification Procedures. This part of the emergency plan includes a description of procedures for providing reliable, effective, and timely notification by the facility coordinators and community emergency coordinator, to persons designated in the plan and to the affected public, that a release has occurred; other information provided may include emergency hotline numbers and lists of names and numbers of organizations and agencies that are to be notified in the event of a release; optional information may include a description of methods to be used by facility emergency coordinators to notify community and state emergency coordinators of a release.

Direction and Control. Included are descriptions of methods and procedures to be followed by facility owners and operators and local emergency and medical personnel in response to a release of extremely hazardous substances; and information identifying organizations and persons who provide direction and control during an incident, including a chain of command.

Resource Management. This section of the plan includes a description of emergency equipment and facilities in the community, identification of persons responsible for such equipment and facilities, and a list of all personnel resources available for emergency response.

Health and Medical. Information provided in this section includes descriptions of methods and procedures to be followed by facility owners and operators, and by local emergency and medical personnel to respond to a release of extremely hazardous substances, as well as information on major types of emergency medical services, first aid, triage, ambulance service, and emergency medical care.

Personal Protection of Citizens. This section of the plan includes descriptions of methods in place in the community and at each affected facility for determining areas likely to be affected by a release, as well as a description of methods for indoor protection of the public.

Procedures for Testing and Updating the Plan. Provided here are descriptions of methods and schedules for exercising the emergency response plan and procedures for evaluating and updating the plan.

Training. This part of the plan gives descriptions of the training programs, including schedules for training of local emergency response and medical personnel.

Other Information Regarded as Essential by NRT. Other types of information that are regarded by NRT as necessary to the overall completeness of the emergency plan include general information, including a description of essential information to be recorded in an actual incident and signatures of the LEPC chairperson and other official/industry representatives endorsing the plan; instructions for plan use and record of amendments, including listings of organizations and persons receiving the plan or plan amendments, and other data about the dissemination of the plan; a description of communication methods among responders; the identification of warning systems and emergency public notification; a description of the methods used for public information and community relations prior to any emergency; descriptions of procedures for responders to enter and leave the incident area, including safety precautions, medical monitoring, sampling procedures, and designation of personal protective equipment; the identification of major tasks to be performed by fire fighters, including a listing of fire response and HAZMAT personnel; a description of the command structure for multiagency/multijurisdictional incident management systems; descriptions of major law enforcement tasks related to responding to releases, including security-related tasks; descriptions of methods to assess areas likely to be affected by an ongoing release; a description of agencies responsible for providing emergency human services; a description of the chain of command for public works actions and a listing of major tasks; a description of major containment and mitigation activities for major types of HAZMAT incidents; descriptions of major methods for cleanup; a listing of reports required following an incident and methods of evaluating response activities; and other information outlined by the National Response Team as appropriate for inclusion in an emergency plan.

REFERENCES

American Institute of Chemical Engineers, *Guidelines for Hazard Evaluation Procedures*, 2d ed., Center for Chemical Process Safety, New York, 1992.

American Institute of Chemical Engineers, *Guidelines for Safe Storage and Handling of High Toxic Hazard Materials*, prepared by Arthur D. Little, Inc., and Richard LeVine for Center for Chemical Process Safety, New York, 1988a.

American Institute of Chemical Engineers, *Guidelines for Vapor Release Mitigation,* prepared by Richard W. Prugh and Robert W. Johnson for Center for Chemical Process Safety, New York, 1988b.

Federal Register, 55 FR 29150, July 17, 1990.

National Response Team, "Criteria for Review of Hazardous Materials Emergency Plans," National Response Team of the National Oil and Hazardous Substances Contingency Plan, Washington, DC, 1988.

Occupational Safety and Health Administration, OSHA 313, "Process Safety Management Guidelines for Compliance," U.S. Department of Labor, Occupational Safety and Health Administration, 1993.

Occupational Safety and Health Administration, "Standard for Process Safety Management of Highly Hazardous Chemicals," 29 CFR 1910.119, 1992.

U.S. Environmental Protection Agency, "Environmental Protection Agency Accidental Release Questionnaire," OMB #2050-0065, Washington, DC, 1991.

U.S. Environmental Protection Agency, "Prevention Reference Manual: User's Guide Overview for Controlling Accidental Releases of Air Toxics," EPA/600/8-87/028, prepared by Radian Corporation for Office of Research and Development, Air and Energy Engineering Research Laboratory, Research Triangle Park, NC, 1987.

BIBLIOGRAPHY

AICHE–CCPS (1994), *Guidelines for Chemical Process Documentation*, American Institute of Chemical Engineers—Center for Chemical Process Safety, New York.

_____ (1994), *Guidelines for Evaluating the Characteristics of Vapor Cloud Explosions, Flash Fires, and Bleves*, American Institute of Chemical Engineers—Center for Chemical Process Safety, New York.

_____ (1994), *Guidelines for Implementing Process Safety Management Systems*, American Institute of Chemical Engineers—Center for Chemical Process Safety, New York.

_____ (1994), *Guidelines for Process Safety Fundamentals for General Plant Operations*, American Institute of Chemical Engineers—Center for Chemical Process Safety, New York.

_____ (1993), *Guidelines for Auditing Process Safety Management Systems*, American Institute of Chemical Engineers—Center for Chemical Process Safety, New York.

_____ (1993), *Guidelines for Engineering Design for Process Safety*, American Institute of Chemical Engineers—Center for Chemical Process Safety, New York.

CMA (1989), "Evaluation Process Safety in the Chemical Industry: A Manager's Guide to Quantitative Risk Assessment," Chemical Manufacturers Association, Washington, DC.

_____ (1987), "An Analysis of Risk Assessment Methodologies for Process Emissions from Chemical Plants," Chemical Manufacturers Association, Washington, DC.

_____ (1986), "Site Emergency Response Planning Handbook," Chemical Manufacturers Association, Washington, DC.

_____ (1985), "Community Awareness and Emergency Response Program Handbook," Chemical Manufacturers Association, Washington, DC.

_____ (1985), "Process Safety Management (Control of Acute Hazards)," Chemical Manufacturers Association, Washington, DC.

Davis, Daniel S., et al. (1989), *Accidental Releases of Air Toxics: Prevention, Control and Mitigation*, Noyes Data Corporation, Park Ridge, NJ.

Kelly, Robert B. (1989), *Industrial Emergency Preparedness*, Van Nostrand Reinhold, New York.

NRT (1987), "Hazardous Materials Emergency Planning Guide," National Response Team of the National Oil and Hazardous Substances Contingency Plan, Washington, DC.

OMB (1988), "Using Community Right to Know: A Guide to a New Federal Law," OMB Watch, Office of Management and Budget, Washington, DC.

U.S. EPA (1993), "Guiding Principles for Chemical Accident Prevention, Preparedness, and Response," Office of Solid Waste and Emergency Response, U.S. Environmental Protection Agency.

13

CONFINED SPACE SAFETY

Many workplaces have spaces that are considered to be "confined" because their configurations hinder work activities and because, in many instances, employees who work in them face increased risk of exposure to serious physical injury from existing hazards. OSHA defines the term "confined space" as one that has limited or restricted means of entry or exit, is large enough for an employee to enter and perform assigned work, and is not designed for continuous occupancy by the employee. Examples of such spaces include underground vaults, tanks, storage bins, pits and diked areas, vessels, and silos. A "permit-required confined space" (also termed "permit space") is one that meets the definition of a confined space and has one or more of the following characteristics:

- The confined space contains or has potential to contain a hazardous atmosphere;
- The confined space contains a material that has the potential for engulfing an entrant;
- The confined space has an internal configuration that might cause an entrant to be trapped or asphyxiated by inwardly converging walls or by a floor that slopes downward and tapers to a smaller cross section; and/or
- The confined space contains any other recognized serious safety or health hazards.

Examples of permit-required confined spaces are shown in Figures 13.1 and 13.2. A complete set of OSHA definitions that pertain to this standard are presented at the end of this chapter. A flowchart for evaluating applicability of the standard to a specific confined space is presented in Figure 13.3.

OSHA's standard for confined spaces is delineated in 29 CFR 1910.146. This standard, effective on April 15, 1993, contains the requirements for practices and procedures to protect employees who work in general industry from hazards associated with entry into permit-required confined spaces (permit spaces). OSHA estimates that approximately 224,000 establishments have permit spaces and that approximately 1.2 million workers enter permit spaces annually (OSHA 1993). The standard includes the requirement to develop and implement a written program and a permit system, training and education requirements, and requirements for emergencies.

WRITTEN PROGRAM

The employer who allows employee entry of permit spaces must develop and implement a written program. Examples of items to include in a written program for permit spaces are presented in Table 13.1.

PERMIT SYSTEM

The employer must provide a method of permitting entry into permit spaces. The permit serves as a document of verification that required pre-entry preparations have been completed and that the space is safe to enter. Examples of types of information to include on an entry permit, recommended by OSHA in 1910.146, Appendix D (sample permit), are summarized and presented in Table 13.2. The permit is signed by the entry supervisor and must be posted at entrances or otherwise made available to entrants before they enter a permit space.

The duration of an entry permit cannot exceed the time required to complete an assignment. The entry supervisor must terminate entry and cancel permits when an assignment has been completed or when new conditions arise. The new conditions must be documented on the canceled permit and are to be used in revising the permit-space program. All canceled entry permits must be kept by the employer for at least one year.

TRAINING AND EDUCATION

Proper training must be provided to all workers who are required to work in permit spaces. Training must take place before the initial work assignment begins, and the employer must ensure that the employees have acquired the

Figure 13.1 Example of permit-required confined space.

Figure 13.2 Example of permit-required confined space.

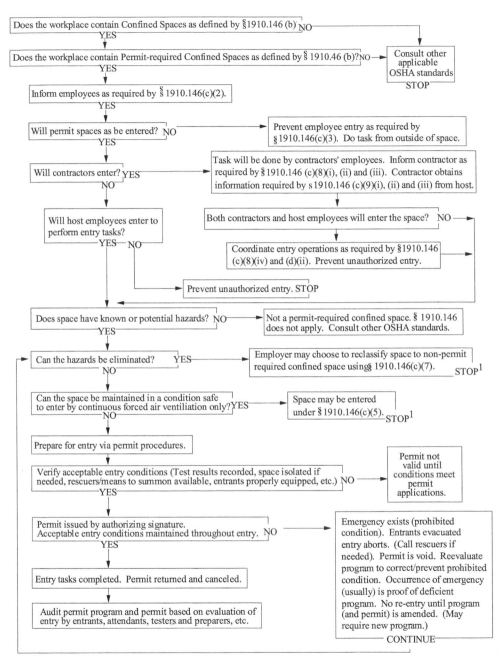

Does the workplace contain Confined Spaces as defined by §1910.146 (b)? NO ⟶ Consult other applicable OSHA standards. STOP

YES ↓

Does the workplace contain Permit-required Confined Spaces as defined by § 1910.46 (b)? NO ⟶ Consult other applicable OSHA standards. STOP

YES ↓

Inform employees as required by § 1910.146(c)(2).

YES ↓

Will permit spaces as be entered? NO ⟶ Prevent employee entry as required by §1910.146(c)(3). Do task from outside of space.

YES ↓

Will contractors enter? YES ⟶ Task will be done by contractors' employees. Inform contractor as required by §1910.146 (c)(8)(i), (ii) and (iii). Contractor obtains information required by s 1910.146 (c)(9)(i), (ii) and (iii) from host.

NO ↓

Will host employees enter to perform entry tasks?

Both contractors and host employees will enter the space? NO ⟶

Coordinate entry operations as required by §1910.146 (c)(8)(iv) and (d)(ii). Prevent unauthorized entry. ⟶

YES NO ⟶ Prevent unauthorized entry. STOP

Does space have known or potential hazards? NO ⟶ Not a permit-required confined space. § 1910.146 does not apply. Consult other OSHA standards.

YES ↓

Can the hazards be eliminated? YES ⟶ Employer may choose to reclassify space to non-permit required confined space using§ 1910.146(c)(7). STOP[1]

NO ↓

Can the space be maintained in a condition safe to enter by continuous forced air ventilation only? YES ⟶ Space may be entered under § 1910.146(c)(5). STOP[1]

NO ↓

Prepare for entry via permit procedures.

↓

Verify acceptable entry conditions (Test results recorded, space isolated if needed, rescuers/means to summon available, entrants properly equipped, etc.) NO ⟶ Permit not valid until conditions meet permit applications.

YES ↓

Permit issued by authorizing signature. Acceptable entry conditions maintained throughout entry. NO ⟶ Emergency exists (prohibited condition). Entrants evacuated entry aborts. (Call rescuers if needed). Permit is void. Reevaluate program to correct/prevent prohibited condition. Occurrence of emergency (usually) is proof of deficient program. No re-entry until program (and permit) is amended. (May require new program.) CONTINUE

YES ↓

Entry tasks completed. Permit returned and canceled.

↓

Audit permit program and permit based on evaluation of entry by entrants, attendants, testers and preparers, etc.

[1] Spaces may be evacuated and re-evaluated if hazards arise during entry

Source: 29CFR 1910.146, Appendix A.

Figure 13.3 Confined space flowchart.

TABLE 13.1   Items to Be Included in a Written Confined Space Program

Identification, Evaluation, and Control of Hazards

- Methods for identifying and evaluating hazards before employee entry
- Methods for testing conditions in the permit space before entry operations and for monitoring the space during entry
- Use of appropriate testing methods for atmospheric hazards of oxygen, combustible gases or vapors and toxic gases or vapors (in that order)
- Evaluation and implementation of means, procedures, and practices, such as specifying acceptable entry conditions; isolating the permit space; providing barriers; verifying acceptable entry conditions; purging, making inert, flushing, or ventilating the permit space—to eliminate or control hazards as necessary for safe permit-space entry operations

Items Related to Personnel and Personnel Safety

- Identification of employee job duties
- Identification of use of protective equipment and other equipment necessary for safe entry, and methods of ensuring employee is properly provided with this equipment, including testing and monitoring equipment, ventilating and lighting equipment, communications equipment, barriers, shields, and ladders; specification that employer must provide, maintain, and require all equipment at no cost to the employee
- Identification of necessary measures to prevent unauthorized entry of permit spaces
- Specification and ensure that at least one attendant is stationed outside the permit space for the duration of entry operations
- Specification of the coordination of entry operations when employees of more than one employer are to be working in the permit space

Items Related to Emergencies

- Procedures for summoning rescue and emergency services
- Procedures to be followed during an emergency in one or more permit spaces being monitored by an attendant who is required to monitor multiple spaces

Other Items to Be Included

- A system for preparation, issuance, use, and cancellation of entry permits
- Specification of the review of established entry operations and the annual revision of the permit-space entry program
- Specification of other requirements listed under 29 CFR 1910.146(d)

Source: Information from 29 CFR 1910.146 and OSHA (1993).

understanding, knowledge, and skills necessary for the safe performance of their duties.

Additional training is required if the employee's job duties change, if there is a change in the permit-system program or the permit-space operation presents a new hazard, and when an employee's job performance shows that additional training is necessary. Once training is completed, employees must receive a certificate of training that must be made available for inspection by the employees and their authorized representatives.

Employees must be trained in their assigned duties, which may include those of an authorized entrant, an attendant, or an entry supervisor. The assigned duties for employees working in permit spaces in these roles are summarized in Table 13.3.

There are numerous training aids available to the safety professional who designs the confined-space training class. Classroom training may include a basic regulatory overview of the standard, as well as information about the types and number of confined spaces at the facility, confined-space testing methods, proper documentation, the permit itself, and other aspects of the program. Field training may include "mock" confined space testing, barricading and signing, confined space entry, rescue, and other aspects of the program.

TABLE 13.2   Examples of Types of Information to Include on an Entry Permit

Testing and Equipment Information

- Test results, including initial atmospheric checks for oxygen, explosives, and toxics; atmospheric checks after isolation and ventilation to ensure acceptable limits; additional periodic atmospheric tests; results of continuous monitoring for parameters such as percent oxygen, lower flammable limit, carbon monoxide, aromatic hydrocarbons, hydrogen cyanide, hydrogen sulfide, sulfur dioxide, or ammonia, as necessary; results of tests and calibrations for monitoring equipment
- Tester's initials or signature and time of test
- Special equipment and procedures, including personal protective equipment such as self-contained breathing apparatus (SCBA) and other personal protective equipment (PPE); safety equipment such as safety harnesses and lifelines for entry and standby persons, hoisting equipment, equipment with proper electrical ratings, nonsparking tools, explosion proof lighting, fire extinguishers, and other equipment
- Communications equipment and procedures to be used for maintaining contract during entry

Entry Information

- Name of permit space and job site
- Purpose of entry and known space hazards
- Type of work to be performed

TABLE 13.2 *Continued*

- Measures to be taken to isolate permit spaces and to eliminate or control space hazards, such as locking out or tagging out of equipment and procedures for purging, making inert, ventilating, and flushing permit spaces
- Date and authorized duration of entry
- Acceptable entry conditions documented and met
- Additional permit(s), such as hot work permits, that have been issued to authorize work in the permit space
- Other information needed to ensure safety of employees, including secured area and adequate procedures

Personnel Information

- Name and signature of supervisor who authorizes entry
- Name(s) of authorized entrant(s), eligible attendants, and individual(s) authorized to be entry supervisors
- Names of standby personnel
- Personnel training information
- Names and telephone numbers of rescue and emergency services, including ambulance, fire, safety, and others

Additional Information

- Additional site-specific or job-specific information, included at the discretion of the employer
- Additional information required under 29 CFR 1910.146(f)

Source: Information from 29 CFR 1910.146, Appendix D, and 29 CFR 1910.146(f).

TABLE 13.3   Assigned Duties of Employees Working in Permit Spaces

Duties of the Authorized Entrant

- Know the confined space hazard, including information on the mode of exposure, such as inhalation or dermal absorption; know signs or symptoms, and consequences of exposure
- Use appropriate personal protective equipment properly, including face and eye protection and other forms of barrier protection such as gloves, aprons, and coveralls
- Maintain communication—by telephone, radio, visual observation, or other means—with attendants, as necessary, to enable attendants to monitor the entrant's status as well as to alert the entrant to evacuate
- Exit from the permit space as soon as possible when ordered by an authorized person, when the entrant recognizes that the warning signs or symptoms of exposure exist, when a prohibited condition exists, or when an automatic alarm is activated
- Alert the attendant when a prohibited condition exists or when warning signs or symptoms of exposure exist

TABLE 13.3   *Continued*

Duties of an Attendant

- Remain outside permit space during entry operations unless relieved by another authorized attendant
- Perform non-entry rescues when specified by employer's rescue procedure
- Know existing and potential hazards, including information on the mode of exposure, signs or symptoms, consequences of exposure, and their physiological effects
- Maintain communication with and keep an accurate account of those workers entering the permit space
- Order evacuation of the permit space when a prohibited condition exists, when a worker shows signs of physiological effects of hazard exposure, when an emergency outside the confined space exists, and when the attendant cannot effectively and safely perform required duties
- Summon rescue and other services during an emergency
- Ensure that unauthorized persons stay away from permit spaces or exit immediately if they have entered the permit space
- Inform authorized entrants and entry supervisor of entry by unauthorized persons
- Perform no other duties that interfere with the attendant's primary duties

Duties of the Entry Supervisor

- Know the confined space hazards, including information on the mode of exposure, signs, or symptoms, and consequences of exposure
- Verify emergency plans and specified entry conditions such as permits, tests, procedures, and equipment before allowing entry
- Terminate entry and cancel permits when entry operations are completed or if a new condition arises
- Take appropriate measures to remove unauthorized entrants
- Ensure that entry operations remain consistent with the entry permit and that acceptable entry conditions are maintained

Source: Information from 29 CFR 1910.146(h)-(j) and OSHA (1993).

OTHER ELEMENTS OF THE PROGRAM

Safety Requirements

There are several safety requirements that must be met if an employer has confined spaces onsite. For instance, confined spaces onsite should be evaluated to determine whether any are spaces that are permit-required confined spaces. If so, then the employer must inform exposed employees of their existence and location and the danger posed by the spaces. Danger signs or an equivalent means can be used to accomplish this end. If employees or contractors are not to enter and work in permit spaces, employers must take effective measures to prevent their employees from entering the permit spaces, as well as comply with other requirements.

If hazardous conditions are detected during entry of a permit space, employees must immediately leave the space and the employer must evaluate the space to determine the cause of the hazardous atmospheres. In addition, when entry to permit spaces is prohibited, the employer must take effective measures to prevent unauthorized entry.

Any non-permit confined spaces must be reevaluated when there are changes in their use or configuration, and, as applicable, they must be reclassified. Conversely, if testing and inspection data prove that a permit-required confined space no longer poses hazards, then that space can be reclassified as a non-permit space.

Entry Requirements

If entry is required into a permit space to eliminate hazards and to obtain data, the employer must follow procedures set forth under 29 CFR 1910.146(d)–1910.146(k):

- Establish a permit-required confined space program (see 29 CFR 1910.146(d))
- Establish a permit system (see 29 CFR 1910.146(e))
- Prepare an entry permit (see 29 CFR 1910.146(f))
- Provide training (see 29 CFR 1910.146(g))
- Ensure that authorized entrants, attendants, and supervisors understand their duties (see 29 CFR 1910.146(h)–29 CFR 1910.146(j))
- Ensure that rescue and emergency services training and appropriate information are provided (see 29 CFR 1910.146(k))

If testing and inspection data prove that hazards in a permit space have been eliminated, the space can be declassified to a non-permit space. The employer must document, with a certificate, that all hazards have been eliminated. The certificate should include all data collected and must be made available to employees entering the space. Further, the certificate must include the date, location of the space, and the signature of the person making the certification.

Contractors

When an employer arranges to have contractors perform work that involves permit space entry, several requirements must be met. These include informing contractors of permit spaces and permit-space entry requirements, any identified hazards, the employer's experience with the space, such as knowledge of hazardous conditions, and precautions or procedures to be followed when in or near permit spaces.

When employees of more than one employer are conducting entry operations, the affected employers must coordinate entry operations to ensure that affected employees are appropriately protected from permit-space hazards. Contractors must also be given any other pertinent information regarding hazards and operations in permit spaces and must be debriefed at the conclusion of entry operations.

Contractors themselves have additional responsibilities. These include obtaining any available information regarding permit-space hazards and entry operations from the host employer, coordinating entry operations with the host employer when both host employer personnel and contractor personnel are working in or near permit spaces, and informing the host employer of the permit-space program that the contractor will follow and of any hazards confronted or created in permit spaces, either in a debriefing or during the entry operation.

Emergencies

The employer must ensure that rescue service personnel are provided with and trained in the proper use of personal protective and rescue equipment, including respirators. Rescue service personnel must be trained in their assigned duties and must have had authorized entrants' training. Rescuers are required to be trained in first aid and CPR, and at least one rescuer must be certified in first aid and CPR.

Practice rescue exercises must be performed yearly, and rescue services must be provided access to permit spaces so that they can practice rescue operations. Moreover, rescuers must be informed of the hazards of the permit space.

Where appropriate, an authorized entrant who enters a permit space must wear a chest or full body harness with a retrieval line attached to the center of the person's back near shoulder level, or above the head. Wristlets may be used if the employer can demonstrate that the use of a chest or full-body harness is infeasible or creates a greater hazard. The other end of the retrieval line must be attached to a mechanical device or to a fixed point outside the permit space. A mechanical device must be available to retrieve a person from a vertical-type permit space more than 5 feet deep.

In the case of chemical emergencies, if the entrant is injured through exposure to a hazardous substance for which there is a material safety data sheet (MSDS) at the worksite, the MSDS must be made available to the medical facility treating the exposed entrant.

Alternate Procedures

In some cases, the employer may use alternate procedures for worker entry into a permit space that has as its only hazard an actual or potentially hazardous atmosphere, as indicated in 29 CFR 1910.146(c)(5)(i) and

1920.146(c)(5)(ii). For example, if an employer can demonstrate with monitoring and inspection data that the atmosphere can be made safe for entry by the use of continuous forced-air ventilation alone, exemptions from some requirements, such as permits and attendants, may be allowed. A safety professional should be called upon to evaluate carefully (perhaps with advice from local OSHA representatives) the potential for alternate procedures and to develop and implement work plans for these spaces.

DEFINITIONS

The following OSHA definitions (see 29 CFR 1910.146(b)) relate to the topic of confined space safety.

Acceptable entry conditions. The conditions that must exist in a permit space to allow entry and to ensure that employees involved in a permit-required confined space entry can safely enter and work within the space.

Attendant. An individual stationed outside one or more permit spaces who monitors the authorized entrants and who performs all attendant's duties assigned in the employer's permit-space program.

Authorized entrant. An employee who is authorized by the employer to enter a permit space.

Blanking or blinding. The absolute closure of a pipe, line, or duct by the fastening of a solid plate (such as a spectacle blind or a skillet blind) that completely covers the bore and that is capable of withstanding the maximum pressure of the pipe, line, or duct with no leakage beyond the plate.

Confined space. A space that (1) is large enough and so configured that an employee can bodily enter and perform assigned work, and (2) has limited or restricted means for entry or exit (for example, tanks, vessels, silos, storage bins, hoppers, vaults, and pits are spaces that may have limited means of entry), and (3) is not designed for continuous employee occupancy.

Double block and bleed. The closure of a line, duct, or pipe by closing and locking or tagging two in-line valves and by opening and locking or tagging a drain or vent valve in the line between the two closed valves.

Emergency. Any occurrence (including any failure of hazard control or monitoring equipment) or event, internal or external to the permit space, that could endanger entrants.

Engulfment. The surrounding and effective capture of a person by a liquid or finely divided (flowable) solid substance that can be aspirated to cause death by filling or plugging the respiratory system or that can exert enough force on the body to cause death by strangulation, constriction, or crushing.

Entry. The action by which a person passes through an opening into a permit-required confined space. Entry includes ensuing work activities in that space and is considered to have occurred as soon as any part of the entrant's body breaks the plane of an opening into the space.

Entry supervisor. The person (such as the employer, foreman, or crew chief) responsible for determining whether acceptable entry conditions are present at a permit space where entry is planned, for authorizing entry and overseeing the entry operations, and for terminating entry as required by the standard. *Note:* An entry supervisor also may serve as an attendant or as an authorized entrant, as long as that person is trained and equipped as required by the standard for each role he or she fills. The duties of entry supervisor may be passed from one individual to another during the course of an entry operation.

Hazardous atmosphere. An atmosphere that may expose employees to the risk of death, incapacitation, impairment of ability to self-rescue (that is, escape unaided from a permit space), injury, or acute illness from one or more of the following causes: (1) flammable gas, vapor, or mist in excess of 10% of its lower flammable limit (LFL); (2) airborne combustible dust at a concentration that meets or exceeds it LFL (*note:* this concentration may be approximated as a condition in which the dust obscures vision at a distance of 5 feet (1.52 m) or less); (3) atmospheric oxygen concentration below 19.5% or above 23.5%; (4) atmospheric concentration of any substance for which a dose or a permissible exposure limit is published in 1910 Subpart G—Occupational Health and Environmental Control, or in 1910 Subpart Z—Toxic and Hazardous Substances, and which could result in employee exposure in excess of its dose or permissible exposure limit (*note:* an atmospheric concentration of any substance that is not capable of causing death, incapacitation, impairment of ability to self-rescue, injury, or acute illness due to its health effects is not covered under this provision); and (5) any other atmospheric condition that is immediately dangerous to life or health (*note:* for air contaminants for which OSHA has not determined a dose or permissible exposure limit, other sources of information, such as Material Safety Data Sheets that comply with the Hazard Communication Standard, 29 CFR 1910.1200, published information, and internal documents, can provide guidance in establishing acceptable atmospheric conditions).

Hot work permit. An employer's written authorization to perform operations (for example, riveting, welding, cutting, burning, and heating) capable of providing a source of ignition.

Immediately dangerous to life or health (IDLH). Any condition that poses an immediate or delayed threat to life or that would cause irreversible adverse health effects or that would interfere with an individual's ability to escape unaided from a permit space. *Note:* Some materials—hydrogen fluoride gas and cadmium vapor, for example—may produce immediate transient effects that, even if severe, may pass without medical attention, but are followed by sudden, possibly fatal collapse 12 to 72 hours after exposure. The victim "feels normal" upon recovery from transient effects until collapse. Such materials in hazardous quantities are considered to be "immediately" dangerous to life or health.

Inerting. The displacement of the atmosphere in a permit space by a noncombustible gas (such as nitrogen) to such an extent that the resulting atmosphere is noncombustible. *Note:* This procedure produces an immediately dangerous to life and health (IDLH) oxygen-deficient atmosphere.

Isolation. The process by which a permit space is removed from service and completely protected against the release of energy and material into the space by such means as blanking or blinding; misaligning or removing sections of lines, pipes, or ducts; a double block and bleed system; lockout or tagout of all sources of energy; or blocking or disconnecting all mechanical linkages.

Line breaking. The intentional opening of a pipe, line, or duct that is or has been carrying flammable, corrosive, or toxic material, an inert gas, or any fluid at a volume, pressure, or temperature capable of causing injury.

Non-permit confined space. A confined space that does not contain or, in respect to atmospheric hazards, have the potential to contain any hazard capable of causing death or serious physical harm.

Oxygen-deficient atmosphere. An atmosphere containing less than 19.5% oxygen by volume.

Oxygen enriched atmosphere. An atmosphere containing more than 23.5% oxygen by volume.

Permit-required confined space (permit space). A confined space that has one or more of the following characteristics: (1) contains or has a potential to contain a hazardous atmosphere; (2) contains a material that has the potential for engulfing an entrant; (3) has an internal configuration such that an entrant could be trapped or asphyxiated by inwardly converging walls or by a floor that slopes downward and tapers to a smaller cross-section; or (4) contains any other recognized serious safety or health hazard.

Permit-required confined space program (permit space program). An employer's overall program for controlling, and, where appropriate, for protecting employees from, permit space hazards and for regulating employee entry into permit spaces.

Permit system. An employer's written procedure for preparing and issuing permits for entry and for returning the permit space to service following termination of entry.

Prohibited condition. Any condition in a permit space that is not allowed by the permit during the period when entry is authorized.

Rescue service. The personnel designated to rescue employees from permit spaces.

Retrieval system. The equipment (including a retrieval line, chest or full-body harness, wristlets, if appropriate, and a lifting device or anchor) used for non-entry rescue of persons from permit spaces.

Testing. The process by which the hazards that may confront entrants of a permit space are identified and evaluated. Testing includes specifying the tests that are to be performed in the permit space. *Note:* Testing enables employers both to devise and implement adequate control measures for the protection of authorized entrants and to determine whether acceptable entry conditions are present immediately prior to, and during, entry.

REFERENCES

U.S. Department of Labor, 29 CFR 1910.146, "Permit-required Confined Spaces," Occupational Safety and Health Administration, Washington, DC., 1994.

U.S. Department of Labor, "Permit-Required Confined Spaces (Permit Spaces)," OSHA 3138, Occupational Safety and Health Administration, Washington, DC., 1993.

14

CONSTRUCTION SAFETY

OSHA derives its authority to regulate construction activities from Section 107 of the Contract Work Hours and Safety Standards Act, also known as the Construction Safety Act, which is codified under 40 U.S.C. 333. In general, the law requires a contractor of federal, federally financed, or federally assisted construction to ensure that laborers are not required to work in surroundings or under working conditions that are unsanitary, hazardous, or dangerous to health and safety. Regulations pertaining to these requirements are promulgated under 29 CFR 1926. Topics addressed in the regulations are presented in Table 14.1. Requirements for the ship-repairing and shipbuilding industries are delineated under 29 CFR 1915 and 1916, respectively.

STANDARDS INCORPORATED INTO 29 CFR 1926 AND
MATERIALS APPROVED FOR INCORPORATION
BY REFERENCE

Some requirements set forth in 29 CFR 1926 are also defined in general industry standards under 29 CFR 1910. A list of general industry standards that are incorporated into the body of construction standards is presented in Appendix A to Part 1926. In addition, materials from other agencies and technical organizations have been incorporated into the regulations by reference. These are presented in the Finding Aids at the back of 29 CFR 1926.

TABLE 14.1 Topics Addressed in 29 CFR 1926, Subparts A–X

Subpart	Topics Included
Subpart A	General—including purpose and scope; variances from safety and health standards; inspections and right of entry; rules of practice for administrative adjudications for enforcement of safety and health standards
Subpart B	General Interpretations—including interpretation of statutory terms; federal contracts for "mixed" types of performance; relationship to the Service Contract Act and Walsh-Healey Public Contracts Act; rules of construction; other information
Subpart C	General Safety and Health Provisions—including safety training and education; recording and reporting of injuries (reserved); first aid medical attention; fire protection and prevention; housekeeping; illumination; sanitation; personal protective equipment; acceptable certification; shipbuilding and ship repairing; incorporation by reference; and definitions
Subpart D	Occupational Health and Environmental Controls—including medical services and first aid; sanitation; occupational noise exposure; ionizing radiation; nonionizing radiation; gases, vapors, fumes, dusts, and mists; illumination; ventilation; asbestos, tremolite, anthophyllitye, and actinolite; hazard communication
Subpart E	Personal Protection and Life-Saving Equipment—including head protection; hearing protection; eye and face protection; respiratory protection; safety belts, lifelines, and lanyards; safety nets; working over or near water; definitions
Subpart F	Fire Protection and Prevention—including general protection and prevention standards; flammable and combustible liquids; liquefied petroleum gas (LP-Gas); temporary heating devices; definitions
Subpart G	Sign, Signals, and Barricades—including accident prevention signs and tags; signaling requirements; barricade requirements; definitions
Subpart H	Materials Handling, Storage, Use and Disposal—including general requirements of storage; rigging equipment for material handling; and disposal of waste materials
Subpart I	Tools—Hand and Power—including general requirements; requirements for hand tools; power operated hand tools; abrasive wheels and tools; woodworking tools; lever and ratchet jacks, and screw and hydraulic jacks
Subpart J	Welding and Cutting—including gas welding and cutting; arc welding; fire prevention; ventilation and protection in welding, cutting, and heating; welding, cutting, and heating in way of preservative coatings
Subpart K	Electrical—including installation requirements such as wiring design and protection, wiring methods, components, and

TABLE 14.1 *Continued*

Subpart	Topics Included
	equipment for general use, hazardous (classified) locations, and other information; safety-related work practices such as lockout and tagging of circuits and other requirements; safety-related maintenance and environmental considerations such as maintenance of equipment and environmental deterioration of equipment; safety requirements for special equipment such as battery locations and battery charging; definitions
Subpart L	Scaffolding—including scaffolding requirements; requirements for guardrails, handrails, and covers
Subpart M	Floor and Wall Openings—including definitions and roof widths
Subpart N	Cranes, Derricks, Hoists, Elevators, and Conveyors—including equipment requirements for cranes and derricks; helicopters; material hoists, personnel hoists, and elevators; base-mounted drum hoists; overhead hoists; conveyors; aerial lifts
Subpart O	Motor Vehicles, Mechanized Equipment, and Marine Operations—including equipment; requirements for motor vehicles; material-handling equipment; pile-driving equipment; site clearing; marine operations and equipment; definitions
Subpart P	Excavation—including definitions; general requirements; requirements for protective systems; soil classification; sloping and benching; and shoring for trenches
Subpart Q	Concrete and Masonry Construction—definitions; general requirements; requirements for equipment and tools; requirements for cast-in-place and precast concrete; requirements for lift-slab construction operations; requirement for masonry construction
Subpart R	Steel Erection—including flooring requirements; structural steel assembly; bolting, riveting, fitting-up, and plumbing-up
Subpart S	Underground Construction, Caissons, Cofferdams—including requirements for underground construction and caissons; cofferdams; compressed air; definitions
Subpart T	Demolition—including preparatory operations; stairs, passageways, and ladders; chutes; removal of materials through floor openings; removal of walls, masonry, sections, and chimneys; manual removal of floors; removal of walls, floors, and material with equipment; storage; removal of steel construction; mechanical demolition; selective demolition by explosives
Subpart U	Blasting and Use of Explosives—including general provisions; blaster qualification; surface and underground

(Continued)

TABLE 14.1 *Continued*

Subpart	Topics Included
	transportation of explosives; storage of explosives and blasting agents; loading of explosives or blasting agents; initiation of explosive charges—electric blasting
Subpart V	Power Transmission and Distribution—including general requirements; tools and protective equipment; mechanical equipment; material handling; grounding for protection of employees; overhead and underground lines; construction in energized substations; external load helicopters; lineman's body belts; safety straps and lanyards; definitions
Subpart W	Rollover Protective Structures; Overhead Protection— including rollover protective structures for material handling equipment; minimum performance criteria for rollover protective structures for designated scrapers, loaders, dozers, graders, and crawler tractors; protective frame test procedures and performance requirements for wheel-type agricultural and industrial tractors used in construction; overhead protection for operators of agricultural and industrial tractors
Subpart X	Stairways and Ladders—including scope, application and definitions; general requirements; stairways; ladders; training requirements

Source: 29 CFR 1926.

TYPES OF CONSTRUCTION PROJECTS

The types of construction projects covered under the OSHA construction standards are varied. Examples include, but are not limited to, civil projects, such as highways, bridges, and airports; housing projects, including single-family, multifamily, and other housing; buildings such as libraries, government buildings, art museums, theaters, and other buildings; educational facilities; health care facilities such as nursing homes, hospitals, community mental health facilities; manufacturing facilities; and marine facilities.

WRITTEN SAFETY PROGRAMS AND OTHER
GENERAL PROVISIONS OF SAFETY

It is the responsibility of the employer to initiate and maintain safety programs that ensure accident prevention. These programs include such aspects as safety education and training, first aid and medical care, fire protection and prevention, housekeeping, illumination, sanitation, personal protective equipment, life safety, medical surveillance, and other programs, as applicable. General safety and health provisions are set forth under 29 CFR 1926 Subpart C.

General Safety Education and Training

One of the most important programs for the construction industry is safety education and training. It is the employer's responsibility to provide safety training and education for all employees. The safety officer or other specialist should instruct employees in the recognition and avoidance of unsafe conditions and the regulations applicable to the work environment to control or eliminate any hazards or other exposure to illness or injury. Examples of items to be included in construction safety training are presented in Table 14.2.

TABLE 14.2 Examples of Items to Be Included in Construction Safety Training

Pre-Start-up Training—Supervisors

- Review overall safety plan
- Review emergency plan
- Review arrangements with hospitals and other emergency medical personnel
- Review list of other agency personnel that may be needed during an emergency, such as fire marshall, police, state and local emergency response coordinators
- Review list of first aid and other medical supplies
- Review list of fire protection equipment
- Review personal protective equipment needed for project tasks, including hard hats, foot protection, eye protection, hearing protection, protective clothing, respiratory protection, lifelines, safety belts, and other protective equipment
- Review schedule for testing fire protection equipment, personal protective equipment, and other safety equipment
- Review need for safety signs and barricades
- Review accident investigation techniques
- Review employee training for safe work practices
- Review required qualifications/training of employees
- Review contractor's safety record and safety training
- Review recordkeeping/reporting requirements
- Review rights of employees

Pre-Start-up Training—Employees

- Review safety plan and emergency plan, including location of first aid facility, emergency procedures, and location of fire protection equipment
- Complete hazard communication training, basic training, and training about toxic substances on the job site
- Complete training on personal protective equipment
- Complete confined-space training
- Review requirements for scaffolding
- Review requirements for perimeter guarding
- Review housekeeping requirements

(Continued)

TABLE 14.2 *Continued*

- Review material-handling procedures, rigging procedures, and crane safety
- Review electrical safety
- Review trenching and excavation requirements
- Review any other special project requirements and procedures

Weekly Safety Meetings

- Review topics related to safe working conditions and practices that are part of the job during the coming week, such as safe use of tools, use of personal protective equipment, safe lifting techniques, and other pertinent safety information
- Inventory emergency equipment, personal protective equipment, signs and barricades, and other safety equipment
- Address any questions employees may have about specific job tasks

Safety meetings should be held before employees begin work at a project site and should be held periodically thereafter. Weekly safety meetings are typical and are normally held at the job location. All training should be documented. New employees who join the work force after the construction project has begun should receive training before they are allowed to begin work. In addition to employee training, supervisors should receive safety orientation so that they are familiar with the supervisor's role and responsibilities in safety. This orientation should include the written safety programs, the project site safety plan, and specific safety aspects pertaining to job conditions. The supervisor should be able to recognize unsafe conditions and take corrective actions before work begins. The management chain of escalation for problems that cannot be resolved immediately should be specified. As with other employees, supervisors should receive training before they begin their assignment, and additional supervisor safety meetings should be held periodically. Attendance at orientation classes for supervisors and subsequent safety meetings should be documented.

Access to Employee Exposure and Medical Records

Construction employees, like employees of general industry and maritime industry, are allowed to request and receive access to their records of exposure and medical records. The employer must provide access in a reasonable time, place, and manner. When the employee first enters employment, and at least annually thereafter, the employer must inform the employee of the existence, location, and availability of the employee's records, information about the employee's rights to access to the records, and other pertinent information.

Housekeeping

Good housekeeping is required under the general safety provisions, and the need for practicing good housekeeping on a daily basis cannot be overstated.

Because housekeeping practices are highly visible, maintaining good house-keeping sets the tone for all other safety practices on the construction site. Clearing up scrap lumber with protruding nails, keeping passageways clear, segregating and properly disposing of waste, properly disposing of oily, flammable, or hazardous wastes, and other good management practices should be strictly enforced throughout the construction project.

Competent Person

Numerous subparts of the construction standards require a "competent person" to develop and implement safety programs or to perform hazard evaluations. This term refers to a safety professional or other trained specialist who is capable of identifying existing and predictable hazards in the workplace and is authorized to take corrective action to eliminate them.

Other General Safety and Health Provisions

Other aspects of the general safety and health provisions outlined in Subpart C are detailed further in other parts of the construction safety standard. These include first aid and medical attention, illumination, and sanitation, which are fully outlined in Subpart D; fire protection and prevention, specified in Subpart F; and personal protective equipment, described in Subpart E.

OCCUPATIONAL HEALTH AND ENVIRONMENTAL CONTROLS

Occupational health and environmental controls are outlined in Subpart D. These controls cover medical services and first aid, sanitation, illumination, hazard communication, exposure to hazardous chemicals, and other related topics.

Medical Services and First Aid

The construction contractor (employer) is responsible for ensuring that medical personnel are available for advice and consultation on matters of occupational health. If there is not an easily accessible infirmary, clinic, hospital, or physician, there must be a person certified in first-aid training at the worksite.

Sanitation

At construction sites, sanitation is usually provided by temporary facilities, since many sites are remote or undeveloped. Potable water that meets U.S. Public Health Service Drinking Water Standards must be provided at all construction sites. Portable containers used to dispense drinking water must be capable of being tightly closed and must be equipped with a tap. The water container must be well marked and must not be used for any purpose other than dispensing drinking water. The use of a common drinking cup is prohibited, and single-use cups must be provided in a sanitary container. All cups must be disposed of in a receptacle after a single use.

Outlets for nonpotable water that may be used for fire fighting or industrial uses must be properly identified by signs. There can be no cross connection or potential for cross connection between potable and nonpotable water supplies. Where job sites are not provided with a sanitary sewer, toilet facilities must be provided, as defined in 29 CFR 1926.51. If food service facilities are located at the job site, they must meet applicable local/state laws and regulations.

Washing facilities must be available for employees engaged in the application of paints, coatings, herbicides, or insecticides, or in other operations where contaminants may be harmful to the employees. Showers may be required or recommended under certain conditions, such as when employees are performing construction activities involving asbestos-containing materials or when airborne exposure to cadmium, lead, or other regulated substances is above the permissible exposure limits. Whenever protective clothing is required, the employer must provide change rooms equipped with storage facilities for street clothes and separate storage facilities for the protective clothing.

Eating and drinking are prohibited in a toilet room or in any area exposed to a toxic material. Enclosed workplaces must be constructed so as to prevent the entrance or harborage of rodents, insects, and other vermin. An extermination program must be initiated when the presence of vermin is detected.

Illumination

OSHA specifies lighting requirements for construction areas, ramps, runways, corridors, offices, shops, and storage areas in 29 CFR 1926.56, Table D-3. These requirements are summarized as follows:

- 3 footcandles—Acceptable for general construction areas, concrete placement, excavation and waste areas, accessways, active storage areas, loading platforms, refueling, and field maintenance areas.
- 5 footcandles—Acceptable for general construction area lighting; indoors for warehouses, corridors, hallways, and exitways; for tunnels, shafts, and general underground work areas. An exception provides that a minimum of 10 footcandles is required at tunnel and shaft headings during drilling, mucking, and scaling. Bureau of Mines-approved cap lights are acceptable for use in the tunnel headings.
- 10 footcandles—Acceptable for general construction plants and shops (e.g., batch plants, screening plants, mechanical/electrical equipment rooms, carpenter shops, rigging lofts, barracks or living quarters, locker or dressing rooms, mess halls, and indoor toilets and workrooms).

- 30 footcandles—Acceptable for first aid stations, infirmaries, and offices.

The standard refers to the American National Standard A11.1-65, "Practice for Industrial Lighting," for areas not covered specifically in 1926.56.

Hazard Communication

The hazard communication standard applies to all hazardous chemicals used in the construction industry. Table 14.3 provides examples of the types of hazardous chemicals that may be found at construction job sites.

Under the hazard communication standard, employers must ensure that employees are provided with information and training about hazardous chemicals in the workplace. This requirement for the construction industry is

TABLE 14.3 Examples of Potentially Hazardous Chemicals Commonly Found on Construction Projects

Health Hazards
- Carcinogen or potential carcinogen, such as coal tar pitch volatiles
- Toxic or highly toxic chemical, such as pentachlorophenol
- Reproductive toxin, such as lead
- Corrosive, such as etching agents, potassium hydroxide, muriatic acid
- Irritant, such as bleaching agents, ammonium solutions, magnesium dust
- Sensitizer, such as fiberglass dusts
- Hepatotoxin, such as arsenic compounds, some solvents
- Neurotoxin, such as xylene
- Nephrotoxin, such as turpentine, some solvents
- Agent that damages blood, such as benzene, carbon monoxide
- Agent that damages lungs, such as asbestos, tars, dusts
- Agent that damages eyes or skin, such as sodium hydroxide, methanol

Physical Hazards
- Combustible liquid, such as diesel oil, form oils
- Water reactive agent, such as potassium*
- Flammable agent, such as acetone, adhesives and glues, gasoline, kerosene
- Explosive, such as dynamite
- Oxidizer, such as calcium hypochlorite
- Pyrophoric, such as yellow phosphorus*
- Compressed gas, such as acetylene gas, oxygen, hydrogen, carbon dioxide

*Not as commonly found on construction sites as other materials may be.
Source: Information from NIOSH (1987), NFPA (1991), DOE (1993), and 29 CFR 1910.1200, Appendix A.

delineated under 29 CFR 1926.59, and it mirrors the general industry standard for hazard communication under 29 CFR 1910.1200.

Although promulgated at an earlier date, this standard was not enforced in the construction industry until 1989 because of legal proceedings. Since then it has been enforceable, and its violation has been one of the most frequently cited during OSHA inspections (OSHA 1993 and DOE 1993). Absence of a written program, lack of employee hazard communication training, and non-availability of material safety data sheets (MSDSs) at the worksite are all common violations found at construction sites.

Exposure to Hazardous Chemicals

Asbestos. The standard for asbestos was updated in 1994 under 29 CFR 1926.1101. The new standard set the permissible exposure limit (PEL) as an 8-hour time-weighted average at 0.1 feet/cubic centimeter (f/cc). In addition, OSHA established four classes of activities that trigger different provisions in the standard. All insulation, surfacing materials, and floor tile, if installed before 1980, is considered to contain asbestos, unless proven to contain less than 1%.

Lead. The OSHA standard for lead in the construction industry is promulgated under 29 CFR 1926.62. The standard applies to all construction work where an employee may be occupationally exposed to lead. The standard sets the PEL for airborne concentrations of lead at 50 $\mu g/m^3$ as an 8-hour time-weighted average. The action level—the level at which an employer must begin certain compliance activities outlined in the standard—is an airborne concentration of 30 $\mu g/m^3$ as an 8-hour time-weighted average. Requirements include exposure monitoring; medical surveillance; use of engineering, work practice, and administrative controls to reduce and maintain employee lead exposure; training; and related requirements.

Cadmium. Cadmium exposure in the construction industry is regulated under 29 CFR 1926.63. The PEL is 5 $\mu g/m^3$ as an 8-hour time-weighted average for all cadmium compounds, including dusts and fumes. Regulated areas must have warning signs, protective work clothing and equipment (including respirators) must be provided, employers must perform medical surveillance, cadmium hazards must be communicated, air monitoring must take place during work activities, and other pertinent requirements apply.

4,4′ Methylenedianiline (MDA). The OSHA standard for exposure to MDA in the construction industry is outlined in 29 CFR 1926.60. The 8-hour time-weighted average and the 15-minute short-term exposure are set at 10 parts per billion (ppb) and 100 ppb, respectively. In addition, there is an action level set at 5 ppb as a time-weighted average. Numerous requirements apply, including exposure monitoring, medical surveillance, use of protective clothing and equipment (including respirators), training, housekeeping, and others.

Other Aspects of Subpart D

Other aspects included under Subpart D of the standard are requirements for noise exposure; ionizing and nonionizing radiation; gases, vapors, fumes, dusts, and mists; and ventilation. Other topics covered are process safety management of highly hazardous chemicals; hazardous waste operations and emergency response; and criteria for design and construction of spray booths.

PERSONAL PROTECTIVE AND LIFESAVING EQUIPMENT

Requirements for personal protective and lifesaving equipment are delineated in Subpart E of 29 CFR 1926. Employees can provide their own personal protective equipment, but the employer is responsible for ensuring its adequacy.

Foot and Head Protection

ANSI standard Z41.1 for foot protection is incorporated by reference in OSHA's standard. OSHA also refers to ANSI Z89.1 and Z89.2 for standards for head protection. ANSI Z89.1 specifies helmets for the protection against impact and penetration of falling and flying objects; ANSI Z89.2 specifies helmets for employees exposed to high-voltage electrical shock and burns.

Eye and Face Protection

For eye and face protection, OSHA specifies that equipment must meet the requirements of ANSI Z87.1. In addition, OSHA provides a guide for selecting eye and face protection, according to operation and hazards. This guide, presented in 1926.102, Table E, is summarized as follows:

- Acetylene burning, cutting, and welding—Operation hazards include sparks, harmful rays, molten metal, and flying particles. Recommended protectors include welding goggles with tinted lenses (eyecup- or coverspec-type) lenses and coverspec-type welding goggles with tinted plate lenses. Table E-2 in 1926.102 provides a guide for the selection of the proper shade numbers of filter lenses or plates used in welding.
- Chemical handling—Operation hazards include splash, acid burns, and fumes. Recommended protectors include flexible fitting goggles with hooded ventilation, a face shield with plastic or mesh window, and (for severe exposure) a face shield over goggles.
- Chipping—Operation hazards include flying particles. Recommended protectors include flexible-fitting goggles with regular ventilation, cushioned-fitting goggles with rigid body, spectacles with sideshields (metal frame, plastic frame, or metal-plastic frame), and eyecup-type chipping goggles with clear safety lenses (eyecup- or coverspec-type).

- Electric (arc) welding—Operation hazards include sparks, intense rays, and molten metal. Recommended protectors include coverspec-type welding goggles with tinted plate lenses, welding helmets, and spectacles with sideshields (metal frame, plastic frame, or metal-plastic frame) and tinted lenses in combination with welding helmets. Table E-2 in 1926.102 provides a guide for the selection of the proper shade numbers of filter lenses or plates used in welding.

- Furnace operations—Operation hazards include glare, heat, and molten metal. Recommended protectors include eyecup-type welding goggles with tinted lenses (eyecup- or coverspec-type), coverspec-type welding goggles with tinted plate lenses, and (for severe exposure) a plastic or mesh window face shield. As for welding operations, see Table E-2 in 1926.102 for filter lens shade numbers for protection against radiant energy.

- Light grinding—Operation hazards include flying particles. Recommended protectors include flexible fitting goggles with regular ventilation, cushioned fitting goggles with a rigid body, spectacles with sideshields (metal frame, plastic frame, or metal-plastic frame), and a plastic or mesh window face shield.

- Heavy grinding—Operation hazards include flying particles. Recommended protectors include flexible fitting goggles with regular ventilation, cushioned fitting goggles with rigid body, chipping goggles with clear safety lenses (eyecup- or coverspec-type), and (for severe exposure) a plastic or mesh window face shield.

- Laboratory—Operation hazards include chemical splash and glass breakage. Recommended protectors include flexible fitting goggles with hooded ventilation and a plastic or mesh window face shield in combination with spectacles with sideshields (metal frame, plastic frame, or metal-plastic frame).

- Machining—Operation hazards include flying particles. Recommended protectors include flexible fitting goggles with regular ventilation, cushioned fitting goggles with rigid body, spectacles with sideshields (metal frame, plastic frame, or metal-plastic frame), and a plastic or mesh window face shield.

- Molten metals—Operation hazards include heat, glare, sparks, and splash. Recommended protectors include welding goggles with tinted lenses (eyecup- or coverspec-type) and a plastic or mesh window face shield in combination with spectacles with sideshields (metal frame, plastic frame, or metal-plastic frame).

- Spot welding—Operation hazards include flying particles and sparks. Recommended protectors include flexible fitting goggles with regular ventilation, cushioned fitting goggles with rigid body, spectacles with sideshields (metal frame, plastic frame, or metal-plastic frame), and a plastic or mesh window face shield.

Table E-3 in 1926.102 lists the maximum power or energy density for which adequate protection is afforded by glasses of optical densities from 5 through 8.

Hearing Protection

Hearing protection is addressed in 1926.101. Protective devices are required when it is not feasible to reduce the noise levels or duration of exposure to acceptable levels as defined in Table D-2, "Permissible Noise Exposure," in 1926.52. Exposures to sound levels of 90 dBA on the A scale are allowed for eight hours per day without the use of hearing protection. If the sound levels are greater than 90 dBA, a safety professional should review the time and frequency of exposure to determine the need for hearing protection. If hearing protection is required, the protective devices must be determined adequate by a competent person. Plain cotton is not an acceptable protective device.

Respiratory Protection

Respiratory protection for the construction industry is defined under 1926.103. Selection of respirator type by hazard is addressed in the standard, and is presented in Table 14.4. Other requirements are addressed such as use, maintenance, and care of respirators, permissible practice, minimal acceptable program, air quality for air-supplied respirators, and identification of gas mask canisters. This latter topic is discussed in Chapter 16.

Safety Belts, Lifelines, Lanyards, Safety Nets, and Other Safety Apparatus

Safety belts, lifelines, and lanyards are addressed in 1926.104. Lifelines, safety belts, and lanyards can be used only for employee safeguarding and are not to be used for in-service loading. Lifelines must be secured above the point of operation to an anchor or structural member capable of supporting a minimum dead weight of 5,400 pounds. If a lifeline is used in areas where it may be subject to cutting or abrasion (such as rock scaling), the lifeline must be a minimum of ⅞-inch wire core manila rope. For all other applications, a minimum of ¾-inch manila rope or equivalent, with a minimum breaking strength of 5,400 pounds, is allowed.

Safety belts must be a minimum of one-half inch nylon or equivalent, with a maximum length to provide for a fall of no greater than 6 feet. As with lifelines, the rope must have a nominal breaking strength of 5,400 pounds. There are also requirements for safety belt and lanyard hardware, such as withstanding a tensile loading of 4,000 pounds without cracking, breaking, or permanently deforming.

Safety nets are addressed in 1926.105. Safety nets must be provided when the workplaces are more than 25 feet above ground or water surfaces, or other surfaces, where the use of ladders, scaffolds, catch platforms, temporary

TABLE 14.4   Selection of Respirators for the Construction Industry

Hazard	Respirator Type
Oxygen deficiency	Self-contained breathing apparatus. Hose mask with blower. Combination air-line respirator with auxiliary self-contained air supply or an air-storage receiver with alarm.
Gas and vapor contaminants immediately dangerous to life and health	Self-contained breathing apparatus. Hose mask with blower. Air-purifying, full facepiece respirator with chemical canister. Self-rescue mouthpiece respirator (for escape only). Combination air-line respirator with auxiliary self-contained air supply or an air-storage receiver with alarm.
Gas and vapor contaminants not immediately dangerous to life and health	Air-line respirator. Hose mask without blower. Air purifying half-mask or mouthpiece respirator with chemical cartridge.
Particulate contaminants immediately dangerous to life and health	Self-contained breathing apparatus. Hose mask with blower. Air-purifying full facepiece respirator with appropriate filter. Self-rescue mouthpiece respirator (for rescue only). Combination air-line respirator with auxiliary self-contained air supply or an air-storage receiver with alarm.
Particulate contaminants not immediately dangerous to life and health	Air-purifying half-mask or mouthpiece respirator with filter pad or cartridge. Air-line respirator. Air-line abrasive-blasting respirator. Hose mask without blower.
Combination gas, vapor, and particulate contaminants immediately dangerous to life and health	Self-contained breathing apparatus. Air-purifying full facepiece respirator with chemical canister and appropriate filter (gas mask and filter). Self-rescue mouthpiece respirator (for escape only). Combination air-line respirator with auxiliary self-contained air supply or an air-storage receiver with alarm.
Combination gases, vapor, and particulate contaminants not immediately dangerous to life and health	Air-line respirator. Hose mask without blower. Air-purifying half-mask or mouthpiece respirator with chemical cartridge and appropriate filter.

Note: In this table, "immediately dangerous to life and health" is defined as a condition that either poses an immediate threat to life and health, or an immediate threat of severe exposure to contaminants, such as radioactive materials, that are likely to have adverse delayed effects on health.

floors, safety lines, or safety belts is impractical. Work should not begin until the safety net is in place and tested. The net should be extended 8 feet beyond the edge of the work surface where employees are exposed. The net should be placed as close under the work surface as practical, but no greater than 25 feet below the work surface, and it must have sufficient clearance to prevent the user's contact with surfaces or structures below.

The mesh size of the net should not be greater than 6 inches by 6 inches, and all new nets must meet accepted performance standards of 17,500 pounds minimum impact resistance, certified by the manufacturer. Edge ropes must provide a minimum breaking strength of 5,000 pounds. Forged steel safety hooks or shackles must be used to fasten the net.

When employees are working over or near water, where danger of drowning exists, the employees must be provided with a U.S. Coast Guard-approved life jacket or buoyant work vest, as specified in 1926.106. The life jackets and/or buoyant vests must be inspected for defects that would alter their strength or buoyancy, both prior to and after each use. Ring buoys with at least 90 feet of line must also be provided, and the distance between ring buoys cannot exceed 200 feet. In addition at least one lifesaving skiff must be provided.

Other Requirements

Other requirements in Subpart E include protective clothing for fire brigades and respiratory protection for fire brigades. These are addressed in 1926.97 and 1926.98, respectively.

FIRE PROTECTION AND PREVENTION

Fire protection and prevention for the construction industry is delineated under Subpart F of 1926. This subpart defines requirements for fire protection, fire prevention, flammable and combustible liquids, liquified petroleum gas (LP-Gas), temporary heating devices, fixed extinguishing systems, fire detection systems, and employee alarm systems.

General Requirements

The nature of construction work often mandates that portable fire-fighting equipment be used at the job site. OSHA allows a variety of portable fire-fighting equipment in the standards under 1926.150. Included are (with some restrictions) fire extinguishers rated at 2A and above, a 55-gallon open drum of water with two fire pails, and a half-inch garden-type hose capable of discharging at least 5 gallons per minute. Maintenance, training, and other requirements also apply.

General fire prevention provisions, contained in 1926.151, include locating combustion engine exhausts away from combustible materials, prohibition of smoking in the vicinity of operations that pose a fire hazard, appropriate signage in "no smoking" areas, and other safeguards pertaining to ignition hazards. Other fire prevention topics addressed include temporary buildings, open yard storage—including management of piles of combustible materials—and indoor storage.

Flammable and Combustible Liquids

General Requirements. Because storage of flammable and combustible liquids can pose a fire hazard in certain instances, OSHA regulates such storage under 1926.152. General requirements include the use of approved containers and portable tanks for storage and handling of flammable and combustible materials. Flammable materials stored or used in quantities greater than one gallon must be contained in metal cans, unless the material is highly viscous. Flammable and combustible liquids must not be stored in areas used for safe passage of people, such as stairways and exits.

Indoor Storage. Approved flammable storage cabinets that are labeled "Flammable—Keep Fire Away" should be used for storage of flammable and combustible liquids in quantities of more than 25 gallons. No more than 60 gallons of flammable liquids or 120 gallons of combustible liquids should be stored in an single cabinet. Inside storage rooms must be constructed to meet the required the fire-resistive rating for their use, as referenced in NFPA 251, "Standard Methods of Fire Test of Building Construction and Material." Electrical wiring and equipment located in flammable and combustible liquid storage rooms should be approved for Class I, Division 1, Hazardous Locations. Other requirements apply as well.

Outside Storage. When flammable or combustible liquids are stored outside or in outside buildings, the quantity per container cannot be greater than 60 gallons and the maximum quantity in any one area or pile cannot exceed 1,100 gallons. Piles or groups of containers must be placed at least 20 feet from any building, and there must be a 5-foot clearance between the piles. The area must be graded to divert possible spills away from buildings, or the area must have a 12-inch high curb or dike. Portable tanks must have emergency venting and other fire safety devices as specified in NFPA 30, "The Flammable and Combustible Liquids Code."

Fire Control. OSHA requires construction sites that store 60 gallons or more of flammable or combustible liquids to have at least one portable fire extinguisher of a rating of not less than 20-B units. If the flammable or combustible materials are stored inside, the portable extinguisher must be located outside, but within 10 feet of the room used for storage. For outside storage areas, a portable fire extinguisher of not less than 20-B units must be located between 25 and 75 feet of the storage area.

Other Requirements. The standard also addresses dispensing activities related to flammable or combustible materials, handling of the materials at the point of final use, service and refueling areas, and tank storage. The requirements for tank storage include design and construction standards for atmospheric tanks, low-pressure vessels, and pressure vessels, and aboveground and below-ground tanks are both addressed. Other topics covered include tank spacing, venting, drainage and diking, corrosion protection, piping, testing, and other aspects pertaining to tanks.

Liquified Petroleum Gas (LP-Gas)

LP-Gas requirements are defined in 1926.153. Containers, valves, connectors, manifold valve assemblies, and regulators must be of an approved type. The container valves and accessories must have a rated working pressure of at least 250 pounds per square inch (psig). Safety relief valves or other safety devices are required and must allow free vent to the outer air, with discharge at least 5 feet horizontally away from any opening into a building below the discharge.

Dispensing operations involving the filling of fuel containers for trucks or motor vehicles from bulk storage containers must be performed at a distance of at least 10 feet from the nearest masonry-walled building, and at least 25 feet from other buildings or other construction. If there is a building opening, dispensing must be at least 25 feet from the opening. Filling of portable containers, or of containers mounted on skids, from storage containers must be performed at least 50 feet away from the nearest building.

OSHA specifies safety requirements for LP-Gas containers and equipment that cannot be located practically outside buildings and must be used inside buildings or structures. Included are requirements for the use of regulators, hose design, portable heaters, and other equipment. Other requirements related to LP-Gas storage include storage outside buildings, fire protection, design pressure and classification of containers not constructed in accordance with Department of Transportation (DOT) specifications, and marking of containers.

Temporary Heating Devices

Temporary heating devices are addressed in 1926.154. These devices require fresh air in sufficient quantities to maintain the health and safety of workers. If they are used in confined spaces, special care must be taken to ensure that there is proper ventilation for adequate combustion, for the health and safety of the workers, and that the temperature rise in the area is limited. In addition, heaters must be placed at least 10 feet away from combustible tarpaulins, canvas, or similar coverings. There are additional requirements for flammable liquid-fired heaters, the prohibition of solid fuel salamanders in buildings or on scaffolds, and the stability of heaters.

Fire Suppression Equipment and Other Requirements

Requirements for fixed fire suppression equipment are addressed in 1926.156 and 1926.157. These requirements apply to all fixed extinguishing systems installed to meet OSHA standards, except for automatic sprinkler systems, which are covered by general industry standards outlined in 1910.159. Fire detection systems are addressed in 1926.158, and employee alarm systems are addressed in 1926.159.

All fixed systems must be operable, and employees must be warned if the extinguishing agents are hazardous to employee safety and health. Designated employees must be trained to inspect, maintain, operate, and repair fixed extinguishing systems. The employer must review the training annually to ensure that employees remain up-to-date in the functions that they are to perform.

Fire detection systems must be installed properly, maintained, and tested to ensure proper reliability and operability. Fire detection systems installed outdoors or in the presence of corrosive atmospheres must be protected from corrosion. The detection system must operate in time to control or extinguish a fire.

Employee alarm systems must provide warning for emergency action as called for in the employer's emergency action plan, or for reaction time to ensure safe escape of employees from the workplace or the immediate work area. The systems must be installed properly, maintained, and tested to ensure their reliability and operability. Spare alarm devices and components should be available for those parts of the system that are subject to wear or destruction. A backup means of alarm must be in place when the employee alarm system is out of service.

**NFPA 241—Safeguarding Construction, Alteration,
and Demolition Operations**

The National Fire Protection Association has published a standard for safeguarding construction, alteration, and demolition operations, NFPA 241, approved by ANSI and effective 1993. The standard addresses the following topics:

- Temporary construction, equipment, and storage—including temporary offices and sheds, temporary enclosures, and equipment
- Processes and hazards—including hot work operations, temporary heating equipment, smoking, waste disposal, flammable and combustible liquids and flammable gases, and explosive materials
- Utilities—in particular, electrical utilities, including temporary wiring, branch circuits, and lighting
- Fire protection—including the owner's responsibility for fire protection, site security, fire alarm reporting, access for fire fighting, standpipes, and first-aid fire-fighting equipment

- Safeguarding construction and alteration operations—including scaffolding, shoring, and forms; construction material and equipment storage; permanent heating equipment; utilities; fire cutoffs; and fire protection during construction
- Safeguarding roofing operations—including asphalt and tar kettles, single-ply and torch-applied roofing systems, fire extinguishers for roofing operations, and fuel for roofing operations
- Safeguarding demolition operations—including special precautions, temporary heating equipment, smoking, demolition using explosives, utilities, fire cutoffs, and fire protection during demolition
- Safeguarding underground operations—including special precautions, emergency procedures and systems, underground equipment and storage, and electrical safeguarding.

The standard stresses early planning, scheduling, and implementation of fire prevention measures, fire protection systems, rapid communications, and on-site security. In addition, items to be emphasized at the work site include good housekeeping, installation of new fire protection systems as construction progresses, preservation of existing systems during demolition, organization and training of an on-site fire brigade, a pre-fire plan developed with the local fire department, and consideration of special hazards resulting from previous occupancies.

SIGNS, SIGNALS, AND BARRICADES

Signs, signals, and barricades are addressed in Subpart G of 1926. Included are requirements for accident prevention signs and tags, signaling, and barricades. Accident prevention signs are required when a hazard exists where work is being performed. These signs can take the form of danger signs, caution signs, exit signs, safety instruction signs, directional signs, traffic signs, and accident prevention signs. Examples of danger signs, caution signs, and accident prevention signs are presented in Figures 14.1 through 14.3.

Flagmen are to use standard signaling, as specified in ANSI D6.1, "Manual on Uniform Traffic Control Devices for Streets and Highways," to provide traffic control when operations warrant protection on or adjacent to a highway or street. Barricades must also meet these same standards. Likewise, crane and hoist signaling must be in accordance with applicable ANSI standards.

MATERIAL HANDLING, STORAGE, USE, AND DISPOSAL

Requirements for material handling, storage, use, and disposal are detailed in Subpart H of 1926. Included are general requirements for material storage, rigging equipment for material handling, and disposal of waste materials. Examples of general requirements for material storage are outlined in Table 14.5.

Figure 14.1   Example of danger sign that might be found on a construction site. (With permission from EMED Co., Buffalo, N.Y.)

Figure 14.2   Example of caution sign that might be found on a construction site. (With permission from EMED Co., Buffalo, N.Y.)

Rigging Equipment

General Information. Rigging equipment for material handling is regulated under 1926.251. Recommended safe working loads for rigging equipment are defined in Tables H-1 through H-20 of the standard. These tables specify the following:

- Table H-1—defines rated capacities for alloy steel chain slings, sizes one-fourth inch to one and three-fourths inch, including the chain size

Figure 14.3 Example of accident prevention sign that might be found on a construction site.

working load limit for a single branch sling at a 90-degree loading, a double sling at varying vertical angles, and triple and quadruple slings at varying vertical angles.

- Table H-2—defines the maximum allowable wear at any point of the chain link for chain sizes of one-fourth inch to one and three-fourths inch.
- Tables H-3–H-6—define the rated capacities for single-leg slings of different rope types and diameters, and different applications such as vertical, choker, and vertical basket.
- Tables H-7–H-10—define the rated capacities for two-leg and three-leg bridle slings of different rope types and diameters, and different applications.
- Tables H-11 and H-12—define the rated capacities for strand laid grommet (hand tucked) and cable laid grommet (hand tucked) cables of varying diameters and applications.
- Tables H-13 and H-14—define the rated capacities for strand laid endless slings (mechanical joint) and cable laid endless slings (mechanical joint), for different rope types, diameters, and applications.
- Tables H-15–H-18—define rated capacities for manila rope slings, nylon rope slings, polyester rope slings, and polypropylene rope slings. Included are applications for eye and eye slings and endless slings for various applications.

TABLE 14.5   Examples of General Requirements for Material Storage

General Requirements

- All materials stored in tiers must be stacked, racked, blocked, interlocked, or otherwise secured to prevent sliding, falling, or collapse.
- Maximum safe load limits of floors within buildings and structures, in pounds per square foot, must be conspicuously posted in all storage areas, except for floors or slabs on grade.
- Maximum safe loads must not be exceeded.
- Aisles and passageways must be kept clear to provide for the free and safe movement of material handling equipment and employees.
- Aisles and passageways must be kept in good repair.
- When a difference in road or working levels exists, means such as ramps, blocking, or grading must be used to ensure the safe movement of vehicles between the two levels.

Material Storage Requirements

- Material stored inside buildings under construction must not be placed within 6 feet of any hoistway or inside floor openings, nor within 10 feet of an exterior wall that does not extend above the top of the material stored.
- Employees required to work on stored material in silos, hoppers, tanks, and similar storage areas must be equipped with lifelines and safety belts.
- Noncombustible materials must be segregated in storage.
- Bagged materials must be stacked by stepping back the layers and cross-keying the bags at least every 10 bags high.
- Materials must not be stored on scaffolds or runways in excess of supplies needed for immediate operations.
- Brick stacks must not be more than 7 feet in height. When a loose brick stack reaches a height of 4 feet, it must be tapered back 2 inches for every foot above the 4-foot level.
- When masonry blocks are stacked higher than 6 feet, the stack must be tapered back one-half block per tier above the 6-foot level.
- Used lumber must have all nails withdrawn before stacking.
- Lumber must be stacked on level and solid support sills, and it must be stacked so as to be stable and self-supporting.
- Lumber piles cannot exceed 20 feet in height, provided that lumber to be handled manually is not stacked more than 16 feet high.
- Cylindrical materials such as structural steel, poles, pipe, and bar stock must be stacked and blocked so that they do not spread or tilt.

Housekeeping

- Storage areas must be kept free from accumulation of materials that constitute hazards resulting from tripping, fire, explosion, or pest harborage.
- Vegetation control should be exercised when necessary.

TABLE 14.5 *Continued*

Dockboards (Bridge Plates)

* Portable and powered dockboards must be strong enough to carry the load imposed on them.
* Portable dockboards must be secured in position, either by being anchored or by being equipped with devices that will prevent their slipping.
* Handholds, or other effective means, must be provided on portable dockboards to permit safe handling.
* Positive protection must be provided to prevent railroad cars from being moved while dockboards or bridge plates are in position.

Source: Information from 29 CFR 1926.250.

* Table H-19—defines safe working loads for shackles of varying sizes.
* Table H-20—defines the number and spacing of U-bolt wire rope clips required, including drop-forged clips and clips of other material.

Rigging equipment must not be loaded in excess of its recommended safe working load.

Sling Inspections. A key requirement for rigging equipment is inspection of the equipment prior to use on each shift. Inspections must be performed by a competent person, designated by the employer, and must include a thorough inspection of the sling, all fastenings, and attachments. Because slings are often the final means of attachment and support in the rigging system used to hoist, lift, and move materials and equipment, the daily inspection provides a means of finding defects before the equipment is put in use. Other inspections should occur as conditions warrant. Examples of types of damage that can occur to wire rope include the following:

* Randomly distributed broken wires in one rope lay
* Wear or scraping of outside individual wires
* Kinking of the rope
* Crushing of the rope
* "Bird caging" or the twisting or distending of a strand or strands in a multistrand rope owing to torsion
* "High stranding," or the raising of one strand of the rope
* Bulges in the rope
* Gaps or excessive clearance between strands
* Core protrusion
* Heat damage, torch burns, or electric arc strikes

- Unbalanced, severely worn area
- Cracks, deformities, wearing, or corrosion of the end attachments
- Stretched or twisted hook throat opening

Examples of inspection items for chain slings include:

- Localized stretch or wear
- Stretch throughout the entire length of chain
- Grooving
- Twisted or bent links
- Cracks
- Gouges or nicks
- Corrosion pits
- Burns
- Integrity of master links and hooks
- Stretched or twisted hook throat opening

Any damaged or defective sling must be removed from service immediately. Additional inspection requirements apply to alloy steel chains.

Safe operating temperatures apply to wire rope, natural rope, and synthetic fibers. If a sling is exposed to temperatures outside the safe operating range, the sling must be removed from service permanently. When rigging equipment is no longer needed, it should be removed from the immediate work area.

Disposal of Waste Materials

OSHA regulates the dropping of materials and the disposal of waste material under 1926.252. Materials cannot be dropped more than 20 feet outside the exterior of a building without the use of chutes that are closed on all sides. Likewise, if debris is dropped through holes in the floor without the use of chutes, the drop area must be enclosed with barricades at least 42 inches high and at least 6 feet back from the projected edge of the opening above. Warning signs indicating a falling hazard must be posted at each level.

As work progresses, all waste material, lumber, and rubbish must be removed from the work area. Disposal of all waste material must be in accordance with federal, state, and local laws.

HAND AND POWER TOOLS

General Requirements

Hand and power tools are addressed in Subpart I of 1926, and the general requirements for their use are outlined in 1926.300. General requirements specify that it is the employer's responsibility to ensure that all hand and power tools—whether furnished by employer or employee—are maintained in a safe condition.

When power operated tools are designed to accommodate guards, they must be equipped with guards. Examples of machines that normally require point of operation guarding include guillotine cutters, shears, alligator shears, power presses, milling machines, power saws, jointers, portable power rolls, and forming rolls and calenders. Guarding methods include barrier guards, two-handed tripping devices, electronic safety devices, and others. In *addition* to guarding (not *in lieu* of guarding), there are special hand tools that can be used for placing and removing material without the operator's having to place a hand in danger.

Other general requirements also apply; these pertain to exposure of blades, guarding of abrasive wheel machinery, personal protective equipment, and switches.

Hand Tools

Hand tools are addressed in 1926.301. This section of the standard outlines requirements associated with electric power-operated tools, pneumatic power tools, fuel-powered tools, hydraulic power tools, and powder-actuated tools.

Other Tools

Requirements are also detailed for other tools, including abrasive wheels and tools, woodworking tools, jacks, air receivers, and mechanical power transmission apparatus. These requirements can be found in 1926.303–1926.307, respectively.

WELDING AND CUTTING

Welding and cutting are addressed in Subpart J of the construction standards. Safe transport and handling of compressed gas cylinders is required under 1926.350 and includes tilting and rolling the cylinders to be moved; securing the cylinders in an upright position during transport on powered vehicles; ensuring that cylinders are kept far enough away from welding or cutting operations so that sparks, hot slag, or flame will not reach them (or providing a flame-resistant shield); opening the cylinder valve slowly to prevent damage to the regulator; ensuring that leaks at the valve stem are repaired immediately or that the cylinder is taken out of service; and other safety requirements.

Requirements for arc welding and cutting are outlined in 1926.351. The use of manual electrode holders specifically designed for arc welding and cutting is required, as are grounding devices, instruction to employees on safe means of arc welding and cutting, and the shielding of all arc welding and cutting operations with noncombustible or flameproof screens, whenever practicable.

Other topics addressed include fire prevention; ventilation and protection; and cutting, welding, and heating of surfaces that are covered by a preservative coating. These topics are regulated under 1926.352–1926.354, respectively.

ELECTRICAL

Electrical safety for the construction industry is outlined under Subpart K of the standards. This set of regulations is similar to that for general industry, which is covered in Chapter 10.

SCAFFOLDING

Scaffolding is regulated under Subpart L of the construction standards. Examples of general requirements for scaffolds and scaffold types are presented in Table 14.6.

TABLE 14.6   Examples of Scaffolding Requirements and Scaffold Types

---

General Requirements

- Scaffolding must have sound footing/anchorage, capable of carrying maximum load capabilities.
- Erection, movement, alteration, or dismantlement of scaffolding must be performed under the supervision of competent persons.
- Guardrails and toeboards are required on open ends and sides of scaffolds more than 10 feet above the ground; scaffolds that are 4 to 10 feet in height and have a minimum horizontal dimension in either direction of less than 45 inches must have standard guardrails on all open sides and ends of the platform.
- Guardrails must be 2 by 4 inches, or the equivalent, and approximately 42 inches high, with a midrail when required; supports must be at intervals not to exceed 8 feet; toeboards must be a minimum of 4 inches in height.
- If persons can pass or work under the scaffold, screened guardrails are required between the midrail and the toeboard.
- Any damaged parts of the scaffold must be immediately repaired or replaced.
- All load-carrying timber members of the scaffold framing must be a minimum of 1,500 fiber (Stress Grade) construction grade lumber; all planking must be scaffold grades, or equivalent; permissible spans for 2- by 10-inch or wider planks vary, depending on workload and lumber thickness (see Table L-3 in the standards); the maximum permissible span for 1¼ by 9 inches or wider of full thickness must be 4 feet with a medium duty loading of 50 pounds per square foot.

TABLE 14.6 *Continued*

- Scaffold planks must extend over their end supports not less than 6 inches or more than 12 inches.
- An access ladder or equivalent safe access must be provided.
- Scaffold poles, legs, or uprights must be plumb, secured, and rigid.
- Overhead hazards require hard hats and overhead protection.
- All slippery hazards must be eliminated immediately.
- No welding, burning, riveting, or open flame work can be performed on any staging suspended by fiber or synthetic rope.
- Wire, synthetic, or fiber rope used for scaffold support must be capable of supporting at least six times the rated load.
- Use of shore or lean-to-scaffolds is prohibited.
- Materials being hoisted onto a scaffold must have a tag line.
- Employees must not work on scaffolds during storms or high winds.
- Tools, materials, and debris must not be allowed to accumulate at the worksite so as to become a hazard.

Scaffold Types

- Wood pole scaffold—double pole or independent pole—a wood scaffold supported from the base by a double row of uprights, independent of support from the walls and constructed of uprights, ledgers, horizontal platform bearers, and diagonal bracing.
- Tube and coupler scaffold—an assembly consisting of tubing that serves as posts, bearers, braces, ties, and runners; a base supporting the posts; and special couplers that serve to connect the uprights and to join the various members.
- Tubular welded frame scaffold—a sectional panel or frame metal scaffold substantially built up of prefabricated welded sections consisting of posts and horizontal bearers with intermediate members.
- Manually propelled mobile scaffold—a portable rolling scaffold supported by casters.
- Elevating and rotating work platform—a vehicle-mounted platform that can be adjusted to varying heights and to provide rotational capability.
- Outrigger scaffold—a scaffold supported by outriggers or thrustouts projecting beyond the wall or face of the building or structure, the inboard ends of which are secured inside the building or structure.
- Masons' adjustable multipoint suspension scaffold—a scaffold having a continuous platform supported by bearers suspended by wire rope from overhead supports, so arranged and operated as to permit the raising or lowering of the platform desired working positions.
- Two-point suspension scaffold (swinging scaffold)—a scaffold, the platform of which is supported by hangers (stirrups) at two points, suspended from overhead supports so as to permit the raising or lowering of the platform to the desired working position by tackle or hoisting machines.

(Continued)

TABLE 14.6 *Continued*

- Stone setters' adjustable multipoint suspension scaffold—a swinging-type scaffold having a platform supported by hangers suspended at four points so as to permit the raising or lowering of the platform to the desired working position by the use of hoisting machines.
- Single-point adjustable suspension scaffold—a manually or power-operated unit designed for light duty use, supported by a single wire rope from an overhead support so arranged and operated as to permit the raising or lowering of platform desired working positions.
- Boatswain's chair—a seat supported by slings attached to a suspended rope, designed to accommodate one workman in a sitting position.
- Carpenters' bracket scaffold—a scaffold consisting of wood or metal brackets supporting a platform.
- Bricklayers' square scaffold—a scaffold composed of framed wood squares that support a platform, limited to light and medium duty.
- Horse scaffold—a scaffold for light or medium duty, composed of horses supporting a platform.
- Needle beam scaffold—a light duty scaffold consisting of needle beams supporting a platform.
- Plasters', decorators', and large area scaffold—a light duty scaffold with a large platform area.
- Interior hung scaffold—a scaffold suspended from a ceiling or roof structure.
- Ladder jack scaffold—a light duty scaffold supported by brackets attached to ladders.
- Window jack scaffold—a scaffold, the platform of which is supported by a bracket or jack that projects through a window opening.
- Roofing or "bearer bracket"—a bracket used in slope roof construction, having provisions for fastening to the roof or supported by ropes fastened over the ridge and secured to some suitable object.
- Crawling boards or "chicken ladder"—a plank with cleats spaced and secured at equal intervals, for use by a worker on roofs; not designed to carry any material.
- Float or ship scaffold—a scaffold hung from overhead supports by means of ropes and consisting of a substantial platform having diagonal bracing underneath, resting on and securely fastened to two parallel plank bearers at right angles to the span.
- Form scaffold—a figure-four form scaffold or metal-bracket-form scaffold used for light to medium duty.
- Pump jack scaffold—a scaffold consisting of poles secured to the work wall, with the platform being held by brackets.

Note: Specific requirements for each type of scaffolding are outlined in 1926.451, and in 1926.453 for manually propelled mobile ladder stands and scaffolds, and should be reviewed carefully by a safety professional or other competent person before a scaffolding is erected, moved, dismantled, or altered.

FLOOR AND WALL OPENINGS

Floor and wall openings are addressed in the construction standards under 1926 Subpart M. Floor openings and floor holes must be guarded by standard railing and toeboards or cover. Wall openings, open-sided floors, platforms, and runways also have guarding and other requirements. Work on low-pitched roofs at heights greater than 16 feet have fall protection/warning line requirements.

CRANES, DERRICKS, HOISTS, ELEVATORS, AND CONVEYORS

Cranes and Derricks

Cranes, derricks, hoists, elevators, and conveyors for the construction industry are regulated under Subpart N of the standards. General equipment and operational requirements for cranes and derricks are found in 1926.550; examples are presented in Table 14.7. Specific requirements are included in 1926.550 for crawlers, locomotive and truck cranes, hammerhead tower cranes, overhead and gantry cranes, derricks, floating cranes and derricks, and crane- or derrick-suspended personnel platforms.

TABLE 14.7   Examples of General Requirements for Cranes and Derricks

General Requirements—Capacity and Inspection

- Manufacturer's specifications and limitations must be complied with.
- Rated load capacities and recommended operating speeds, special hazard warnings, and instructions must be conspicuously posted on all equipment; instructions and warnings must be visible to the operator while the operator is at the control station.
- The employer must designate a competent person to inspect all machinery and equipment prior to each use, and during use, to make sure it is in safe operating condition.
- A thorough annual inspection of the hoisting machinery must be made by a competent person or by a government or private agency recognized by the U.S. Department of Labor; the employer must maintain a record of the dates and results of inspections for each hoisting machine and piece of equipment.
- If defective, wire rope must be taken out of service.
- Modifications to the equipment must not be made without the manufacturer's approval, and the factor of safety must not be reduced.

(Continued)

TABLE 14.7 *Continued*

General Requirements—Equipment

- Belts, gears, shafts, pulleys, sprockets, spindles, drums, fly wheels, chains, and other reciprocating, rotating, and otherwise moving parts or equipment must be guarded if the parts are exposed to contact by employees or otherwise create a hazard; guarding must meet requirements of ANSI B15.1.
- All exhaust pipes must be guarded or insulated in areas where contact by employees is possible during the performance of normal duties.
- All windows in cabs must be safety glass, or equivalent, and cannot introduce any visible distortion that will interfere with the safe operation of the machine.
- Where necessary for rigging or service requirements, a ladder, or steps, must be provided to give access to a cab roof.
- Guardrails, handholds, and steps must be provided on cranes for easy access to the car and cab, and must conform to ANSI B30.5.
- Fuel tank filler pipe must be protected or positioned so as to not allow spills or overflow to run onto the engine, exhaust, or electrical equipment of any machine being fueled.
- Equipment and machines must meet certain requirements specified in 1926.550 when operated near power lines, unless electrical distribution and transmission lines have been deenergized and visibly grounded at the point of work or unless insulating barriers have been erected to prevent physical contact with the lines.
- Power Crane and Shovel Association Mobile Hydraulic Crane Standard No. 2 must be complied with.
- Sideboom cranes mounted on wheel or crawler tractors must meet the requirements of SAE J743a.

General Requirements—Other

- Hand signals to cane and derrick operators must be those prescribed by the applicable ANSI standard.
- Accessible areas within the swing radius of the rear of the rotating superstructure of the crane must be barricaded to prevent an employee from being struck or crushed by the crane.
- Whenever internal combustion engine-powered equipment exhausts in enclosed spaces, tests must be made and recorded to verify that employees are not exposed to unsafe concentrations of toxic gases or oxygen-deficient atmospheres.
- Platforms and walkways must have antiskid surfaces.

Operational/Personnel Platform Requirements

- The use of a crane or derrick to hoist employees on a personnel platform is prohibited, except when the erection, use, and dismantling of conventional means of reaching the worksite, such as a personnel hoist, ladder, stairway, aerial lift, elevating work platform, or scaffold, would be more hazardous or is not possible because of structural design or worksite conditions.
- Hoisting of personnel on a personnel platform is prohibited during dangerous weather conditions.

TABLE 14.7 *Continued*

- Hoisting of a personnel platform must be performed in a slow, controlled, and cautious manner.
- A crane must be uniformly level within 1% of level grade and located on firm footing; hoisting of employees while a crane is traveling is prohibited.
- Load lines must be capable of supporting at least seven times the maximum intended load; where rotation resistant rope is used, the lines must be capable of supporting at least ten times the maximum intended load.
- The total weight of a loaded personnel platform and related rigging must not exceed 50% of the rated capacity for the radius and configuration of the crane or derrick.
- Use of machines that have live booms (boom in which lowering is controlled by a brake without aid from other devices that slow the lowering speeds) is prohibited.
- When an occupied personnel platform is in a stationary working position, load and boom hoist drum brakes, swings, brakes, and locking devices must be engaged.
- A trial lift with an unoccupied personnel platform loaded to at least the anticipated lift weight must be made from the ground level, or any other level where employees will enter the platform, to each location at which the personnel platform is to be hoisted and positioned; a trial lift must be performed immediately prior to placing personnel on the platform.
- If a crane or derrick is moved and set up in a new location or returned to a previously used location, the trial lift must be repeated.
- A visual inspection of the crane or derrick system, including crane or derrick, rigging, personnel platform, and the crane or derrick base support, must be conducted by a competent person immediately after the trial lift to determine whether the testing has exposed any defect or produced any adverse effect on any component or structure.
- A prelift meeting is required and must be attended by the crane or derrick operator, signal persons (as required), employees to be lifted, and the person responsible for the task to be performed.
- The crane or derrick operator must remain at the controls at all times when the crane engine is running and the personnel platform is occupied; employees being hoisted must remain in continuous sight of and in direct communication with the operator or signal person.
- Before employees exit or enter a hoisted personnel platform that is not landed, the platform must be secured to the structure where the work is to be performed.
- Employees being hoisted must use a body belt/harness system with a lanyard appropriately attached to the lower load block or overhaul ball, or to a structural member within the personnel platform capable of supporting a fall impact for employees using the anchorage.
- Other requirements apply, including those pertaining to instruments and components, personnel platform design criteria and specification, and others.

Field Applications for Cranes and Derricks

An example of an operational checklist that is part of the Department of Energy's construction safety training guide is presented in Table 14.8. This checklist provides general requirements for work platforms suspended from lattice or hydraulic crane booms, as well as some DOE-specific requirements. A similar checklist based on specific equipment and operations can be developed by a trained safety professional for use at the construction worksite.

An example of the types of inspection items that might be on a daily visual inspection report for a crane or derrick is presented in Table 14.9. Like that in Table 14.8, this checklist is from the Department of Energy construction safety training guide. In addition to daily inspections, other, more in-depth inspections of the equipment are necessary on a periodic basis.

Helicopter Cranes

Helicopter cranes are regulated under 1926.551. The regulations cover briefings prior to operation, slings and tag lines, personal protective equipment, loose gear and other objects, housekeeping, operator responsibility, hooking and unhooking loads, static charge, weight limitation, ground lines, visibility, approach, signal systems, communications, personnel, and other aspects related to helicopter cranes.

TABLE 14.8 Example of an Operational Checklist

Operational Checklist for Suspended Work Platforms:
Lattice or Hydraulic Crane Booms

GENERAL
( ) No less-hazardous means to reach area.
( ) Evidence that other means are not available or practical.
( ) Crane meets ANSI B30.5-1968.
( ) 50% of load chart for greatest radius.
( ) Full-cycle operation test of intended load.
( ) Prelift plans showing boom angle and max intended load.
( ) Lift in accordance with manufacturer's recommendations.
( ) Level deck within 1% and proper use of outriggers.
( ) Verify stability of footing during full-cycle test.
( ) Power load lowering on load line.
( ) Lifting bridle shackled to closed hook.
( ) Platform design four times intended load.
( ) Standard guardrails 42″ height—200 lb. load any direction.
( ) No more than four employees.
( ) Each employee considered 250 lb.
( ) Not to be used during adverse weather.
( ) Proper communications.
( ) Crane must be equipped with positive-acting (anti-two-block) device, not just an audible warning system, before employees can be hoisted.

TABLE 14.8 *Continued*

CRANE: SELECT THE BEST-SUITED CRANE

( ) Power load lowering boom and hoist system: no free fall on either system.
( ) Use main hoist system—lifts on jibs should be discouraged.
( ) Refer to manufacturers' load and range charts.
( ) Plan the lift—determine maximum radius and height, allow 10 feet or more between load block and boom point, a maximum height of platform; allow for rigging length.
( ) Position crane for maximum stability.

PLATFORM

( ) Securely attached to hook.
( ) Determine actual height of platform, rigging materials, and personnel.
( ) Provide safety belt attachment—above the hook as fastened to load block anchor, ¾-inch wire rote sling or equivalent—don't fabricate with wire rope clips.
( ) Inspect platform and rigging.

PERSONNEL

( ) Conduct briefing of operation with all personnel involved, including crane operator and signalmen.
( ) Establish firm communication procedures—radio communications recommended.
( ) Assign responsibility for one person in basket to watch distance between load block and boom—don't rely on ATB.
( ) Provide tag lines as required.
( ) Platform personnel equipped with safety belts and lanyards and tied off to approved anchor.

OPERATION

( ) Ensure no power lines or other obstruction in area—check out during dry run required by OSHA.
( ) Keep line movements and swing at slow speeds to preclude dynamic and shock loads and resultant movement.
( ) Keep other cranes, equipment, and personnel out of the lift area.
( ) Dog off boom and hoist drums when elevation or work station is reached.
( ) Tie off platform to structure if personnel are to leave platform—remove tie-off when complete and prior to crane operation.
( ) Crane operator to remain in cab at controls at all times.
( ) Do not travel crane while personnel are aloft.
( ) Tag line shall be used whenever possible to control platform.

Source: DOE (1993).

Other Hoists, Elevators, Conveyors, and Aerial Lifts

Material hoists, personnel hoists, and elevators are regulated under 1926.552. Minimum safety factors for personnel hoists are provided. Requirements for base-mounted drum hoists are outlined in 1926.553. Overhead hoists, conveyors, and aerial lifts are addressed in 1926.554–1926.556, respectively.

TABLE 14.9   Example of a Daily Visual Inspection Report

Daily Visual Inspection Report

GENERAL

Facility Name: _____

Crane No.: _____   Manufacturer: _____

Capacity: Main Hoist: _____   Auxiliary Hoist: _____

Date: _____   Time: _____

OK	Needs Attention	
		FUNCTIONAL MECHANISMS
( )	( )	Master switch
( )	( )	Push-button pendant and stations
( )	( )	Controllers
( )	( )	Return springs
( )	( )	Directional controls
( )	( )	Speed steps
( )	( )	Upper limit switch
( )	( )	Bridge brakes
( )	( )	Trolley brake
( )	( )	Hoist brake
( )	( )	AIR/HYDRAULIC SYSTEMS
		(Deterioration or leakage.)
( )	( )	HOISTING CABLE AND REEVING
		(If there are any signs of EXCESSIVE WEAR, BROKEN STRANDS, KINKS, STRETCHES, TWISTS, evidence of being CRUSHED, or if cable is not following drum grooves, check *Needs Attention*.)
( )	( )	ROPE SLINGS AND END CONNECTIONS
		(If there are any signs of EXCESSIVE WEAR, BROKEN WIRES, STRETCHES, KINKS, or TWISTS, check *Needs Attention*.)
( )	( )	HOIST OR LOAD ATTACHMENT CHAINS AND END CONNECTIONS
		(If there are any signs of EXCESSIVE WEAR, TWISTS, DISTORTED LINKS, or STRETCHES, check *Needs Attention*.)
( )	( )	HOOKS
		(If there is any evidence of CRACKS, INCREASE IN THROAT OPENING, or TWIST from plane or unbent hook, check *Needs Attention*.)
( )	( )	LOAD BLOCKS
		(Check for loose fasteners.)

TABLE 14.9 *Continued*

Comments: _____

_____

_____

Signed: _____

Source: DOE (1993).

MOTOR VEHICLES, MECHANIZED EQUIPMENT, AND MARINE OPERATIONS

Motor vehicles, mechanized equipment, and marine operations are covered under Subpart O of the construction standards. Topics addressed include equipment safety, motor vehicles, material handling equipment, pile-driving equipment, general requirements for site clearing, and marine operations and equipment. Safety guidelines include the requirements for seat belts, brakes, fenders, audible alarms, rollover guards, and other protective devices.

Marine operations and equipment requirements address material handling operations, access to barges, working surfaces of barges, first aid and lifesaving equipment, and commercial diving operations. Other, more detailed requirements for marine operations are defined in 29 CFR 1918, "Safety and Health Regulations for Longshoring."

EXCAVATIONS

All open excavations, including trenches, are regulated under Subpart P of 1926. Underground construction (tunneling), regulated under Subpart S, is discussed later in this chapter.

General Requirements

General requirements for safe management of excavations include the following:

- Locate, remove, or support all surface encumbrances that would create a hazard to employees.
- Determine the location of utility installations, such as sewer, telephone, fuel, electric, water lines, and other underground installations that may reasonably be expected to be encountered.
- Structural ramps that are used solely by employees as a means of access or egress from excavations must be designed by a competent person and must adhere to design criteria. A stairway, ladder, ramp, or other safe

means of egress must be located in trench excavations that are 4 feet or more in depth so as to require no more than 25 feet of lateral travel for employees.

- Employees exposed to public vehicular traffic must be provided with, and must wear, warning vests or other suitable garments marked with or made of reflectorized or high-visibility material.
- No employee is permitted underneath loads handled by lifting or digging equipment.
- There must be a warning system—such as barricades, hand or mechanical signals, or stop logs—for mobile equipment operated adjacent to an excavation or near the edge of an excavation.
- In oxygen-deficient or hazardous atmospheres, or expected oxygen deficient or hazardous atmospheres, the atmospheres in the excavation must be tested before employees enter an excavation greater than 4 feet in depth. Adequate precautions must be taken to prevent employee exposure to oxygen-deficient or hazardous atmospheres, including proper respiratory protection or ventilation. When controls are used to reduce the level of atmospheric contaminants to acceptable levels, testing must be conducted as often as necessary to ensure that the atmosphere remains safe. (Other requirements are set forth in Subparts D and E of 1926).
- Employees must be protected from hazards associated with water accumulation in an excavation.
- If an excavation endangers the stability of an adjoining building, wall, or other structure, support systems such as shoring, bracing, or underpinning must be provided to ensure the stability of such structures for the protection of employees.
- Safeguards, including scaling and installation of protective barricades, must be provided to protect employees from loose rock or soil that could pose a hazard by falling or rolling from an excavated face.
- Daily inspections of the excavation, the adjacent areas, and protective systems must be made by a competent person.
- When employees or equipment are required or permitted to cross over excavations, walkways, or bridges, standard guardrails must be provided. Physical barriers must be provided at remotely located excavations. Wells, pits, shafts, and other similar excavations must be barricaded or covered. When these excavations are temporary, once the exploration or other operations are completed, they must be backfilled.

Requirements for Protective Systems

Protective systems are required in excavations to protect employees from cave-ins. The only times that protective systems are not required are when excavations are made entirely in stable rock or when excavations are less than 5 feet in depth and examination of the ground by a safety specialist or other competent person provides no indication of a potential cave-in. Protective systems can take the form of sloping, shoring, or shielding.

OSHA allows four options for the use of sloping and benching systems. These include making all slopes at an angle not steeper than one and one half horizontal to one vertical (34 degrees measured from the horizontal); use of design configurations, based on soil type, specified in Appendices A and B of the subpart; use of designs based on other tabulated data approved by a registered professional engineer; and design by a registered professional engineer.

Likewise, OSHA allows several options for support systems such as timber shoring, shield systems, and other protective systems. These include design configurations provided in Appendices A, C, and D of the subpart, manufacturers' tabulated data, other tabulated data approved by a registered professional engineer, and design by a registered professional engineer.

Appendix A provides soil classification information, and Appendix B provides allowable sloping and benching systems, based on soil classification. Specifically, soil classification definitions and maximum allowable slopes for excavations less than 20 feet are as follows:

- *Stable rock* means natural solid mineral matter that can be excavated with vertical sides and remain intact while exposed. A vertical or 90-degree slope is allowed.

- *Type A soil* means cohesive soils with an unconfined compressive strength of 1.5 tons per square foot (tsf) or greater; examples include clay, silty clay, sandy clay, clay loam, and, in some cases, silty clay loam and sandy clay loam; cemented soils such as caliche and hardpan are also considered to be Type A; soil is *not* Type A if it is fissured, is subject to vibration from heavy traffic or other causes, or has been previously disturbed. It is also *not* Type A if the material is part of a sloped, layered system where the layers dip into the excavation on a slope of 4 to 1 or greater, or if other factors require it to be classified as a less stable material. A maximum allowable slope is ¾ to 1, or a 53-degree angle. Excavations in Type A soil that are 12 feet or less in depth are allowed a short-term (24 hours or less) maximum slope of ½ to 1, or a 63-degree angle.

- *Type B soil* means cohesive soil with an unconfined compressive strength greater than 0.5 tsf but less than 1.5 tsf; or granular cohesionless soils, including angular gravel (similar to crushed rock), silt, silt loam, sandy loam, and, in some cases, silty clay loam and sandy clay loam. Type B soil is also previously disturbed soil, unless it is classified as Type C, as well as soil that meets the unconfined compressive strength or cementation requirements for Type A, but which is fissured or subject to vibration. Also included is dry rock that is not stable and material that is part of a sloped, layered system where the layers dip into the excavation on a slope less steep than 4 to 1, but only if the material would otherwise be classified as Type B. Maximum allowable slope for this type of soil is 1 to 1, or a 45-degree angle.

- *Type C soil* means cohesive soil with an unconfined compressive strength of 0.5 tsf or less, or granular soils including gravel, sand, and loamy sand. Type C also includes submerged soil or soil from which water is freely seeping, rock that is not stable, and material in a sloped, layered system where the layers dip into the excavation or a slope of 4 to 1 or steeper. Maximum allowable slopes are 1½ to 1, or a 34-degree angle.

Examples of maximum allowable simple slopes for Type A, (including short-term maximum allowable), Type B, and Type C are presented in Figures 14.4 through 14.7. Other sloping/benching configurations are allowable, including the simple bench and multiple bench. In layered soils where the more stable soil is on the bottom, the design incorporates allowable slopes of both soil types. For layered soils having the more unstable soil on bottom, the protective system should incorporate the required slope for the less stable soil throughout.

Requirements for timber shoring are fully defined by OSHA in Appendix C of the Subpart, including arrangements and timber sizes. Requirements for aluminum hydraulic shoring are presented in Appendix D, and alternatives to

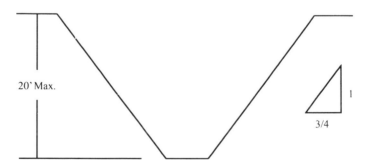

Figure 14.4   Maximum allowable slope: Type A soil 20 ft. or less deep.

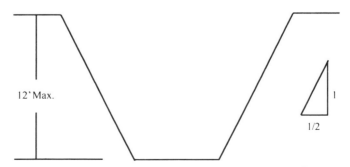

Figure 14.5   Maximum allowable slope: Type A soil 12 ft. or less deep.

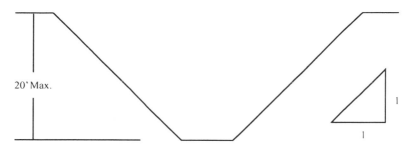

Figure 14.6   Maximum allowable slope: Type B soil 20 ft. or less deep.

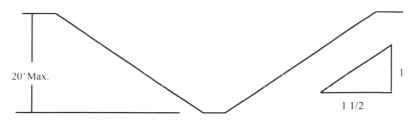

Figure 14.7   Maximum allowable slope: Type C soil 20 ft. or less deep.

timber shoring, including pneumatic/hydraulic shoring, trench jacks, trench shields, and other systems, are presented in Appendix E.

CONCRETE AND MASONRY CONSTRUCTION

Requirements for concrete and masonry construction are outlined in Subpart Q of 1926. These include general requirements and more specific requirements for equipment and tools, cast-in-place concrete, precast concrete, lift-slab construction operations, and masonry construction. Safe work requirements include guarding all reinforcing steel to eliminate the hazard of impalement, ensuring equipment safety through guarding and automatic shutoff devices on concrete mixers, limiting access to areas where a masonry wall is being constructed, ensuring formwork meets ANSI A10.9, and numerous other requirements.

STEEL ERECTION

Requirements for steel erection are presented in Subpart R of the construction standards. They include requirements for flooring, structural steel assembly, and bolting, riveting, fitting-up, and plumbing-up. These requirements are outlined in 1926.750–1926.752.

UNDERGROUND CONSTRUCTION, CAISSONS, COFFERDAMS, AND COMPRESSED AIR

Underground construction, including tunneling and related operations such as work in a compressed air environment, are defined in Subpart S of 1926. The standard, updated in 1989, applies to the construction of underground tunnels, shafts, chambers, and passageways. It also applies to cut-and-cover excavations, both those physically connected to ongoing underground construction tunnels and those that create conditions characteristic of underground construction. Hazards include reduced natural ventilation and light, difficult and limited access and egress, exposure to air contaminants, fire, and explosion. In the past, employees performing tasks in this sector of the construction industry have experienced a higher accident and injury rate than those in other sectors of the industry (OSHA 1991).

Competent Person

As required in most of the construction standards, a "competent person," (see page 367) has numerous duties and responsibilities. These include determining whether air contaminants are present at sufficient levels to be hazardous to life and health, testing the atmosphere for flammable limits, inspecting all drilling equipment prior to each use, inspecting hauling equipment before each shift, and visually checking all hoisting machinery, equipment, anchorages, and rope at the beginning of each shift and during hoisting, as necessary.

Safety Instruction

Employees must be taught to recognize and avoid hazards associated with underground construction. Topics, as appropriate for the worksite, must include air monitoring, ventilation and illumination, communication, flood control, mechanical and personal protective equipment, explosives and fire prevention and protection, and emergency procedures, including evacuation plans and check-in/check-out procedures. Check-in/check-out procedures are discussed in the next paragraph.

Access and Egress and Check-In/Check-Out

Employers must provide safe access to and egress from all workstations, including protection from being struck by excavators, hauling machines, and other mobile equipment. In addition, employers must prevent any unauthorized entry underground. Completed or unused sections of an underground work area must be barricaded. Unused openings must be covered, fenced off, or posted with warning signs that state "Keep Out" or a similar message.

In addition to access and egress requirements, the employer must maintain a check-in/check-out procedure to ensure that aboveground personnel have an accurate count of the number of persons underground in an emergency. At least one designated person must be on duty aboveground whenever anyone is working underground. This person is responsible for securing immediate aid for and keeping an accurate count of employees underground in case of an emergency. This check-in/check-out procedure is not required, however, when the underground construction is sufficiently completed so that permanent environmental controls are effective and when remaining construction activity will not cause an environmental hazard or structural failure of the construction.

Hazardous Classifications

Underground construction operations must be classified as potentially gassy operations or gassy operations under the following circumstances:

- Potentially gassy operations—This classification applies when air monitoring shows, for more than a 24-hour period, 10% or more of the lower explosive limit (LEL) for methane or other flammable gases measured at 12 inches ±0.25 inch from the roof, face, floor, or walls in any underground work area, or when the geological formation or history of the area shows that 10% or more of the LEL for methane or other flammable gases is likely to be encountered in the underground operation.
- Gassy operations—This classification applies when air monitoring shows, for three consecutive days, 10% or more of the LEL for methane or other flammable gases measured at 12 inches ±0.25 inch from the roof, face, floor, or walls in any underground work area, or when methane or other flammable gases emanating from the strata have ignited, indicating the presence of such gases, or when the underground work operation is connected to a currently gassy underground work area and is subject to a continuous course of air containing a flammable gas concentration.

Underground construction gassy operations can be declassified to potentially gassy when air monitoring results remain below 10% of the LEL for methane or other flammable gases for three consecutive days. When a gassy operation exists, additional safety precautions are required. These include ensuring that ventilation requirements are met; using diesel equipment only if it is approved for use in gassy operations; posting each entrance with warning signs; prohibiting smoking and personal sources of ignition and keeping a fire watch; and suspending all operations in the affected area until all special requirements are met or the operation is declassified. Additional monitoring requirements also apply. These are described in the following paragraphs.

Air Monitoring

General Requirements. Air monitoring is required to determine proper ventilation and quantitative measurements of potentially hazardous gases. In instances where monitoring of airborne contaminants is required "as often as necessary," a competent person is responsible for determining which substances to monitor and the frequency, taking into consideration factors such as worksite, location, geology, history, work practices, and operating conditions.

The atmosphere in all underground areas must be tested quantitatively for carbon monoxide, nitrogen dioxide, hydrogen sulfide, and other toxic gases, dusts, vapors, mists, and fumes to ensure that limits set forth in 1926.55 (threshold limit values for airborne contaminants for construction) are met. Quantitative tests for methane must also be performed to determine whether an operation is gassy or potentially gassy.

Oxygen and Hydrogen Sulfide. Specific testing must include oxygen and hydrogen sulfide. Oxygen concenterations must be at least 19.5% but no more than 22%. When air monitoring indicates the presence of 5 parts per million (ppm) or more of hydrogen sulfide, testing must be conducted in the affected areas at the beginning and midpoint of each shift until the concentration of hydrogen sulfide has been less than 5 ppm for three consecutive days. If hydrogen sulfide is present above 10 ppm, employees must be notified and continuous monitoring must be performed. At a concentration of 20 ppm, a visual and aural alarm must signal that additional measures may be required (e.g., respirators, increased ventilation, or evacuation) to maintain allowable exposure levels.

Other Precautions. When a competent person determines that there are contaminants present that are dangerous to life or health, the employer must post notices of the condition at all entrances to the underground job site to inform employees of hazardous conditions and must ensure that necessary precautions are taken. Whenever 5% or more of the LEL for methane or other flammable gases is detected in any underground work area, steps must be taken to increase ventilation air volume or otherwise control the gas concentration, unless the employer is operating in accordance with the potentially gassy or gassy operation requirements. When 10% or more of the LEL for methane or other flammable gases is detected where welding, cutting, or other "hot" work is performed, work must be suspended until the concentration is reduced to less than 10% of the LEL. When there is a concentration of 20% or more of the LEL, all employees must be immediately withdrawn to a safe location aboveground, except those necessary to eliminate the hazard. In addition, electrical power, except for pumping and ventilating equipment, must be cut off from the area endangered by the flammable gas until the concentration of the gas is reduced to less than 20% LEL.

Monitoring Potentially Gassy and Gassy Operations. Air monitoring of potentially gassy and gassy operations must include testing for oxygen in the affected work areas. Flammable gas monitoring is also required, and continuous automatic equipment should be used when rapid excavation machines are being used; manual monitoring is adequate for other applications. Other requirements include performing local gas tests prior to, and continuously during, any hot work, as well as testing continuously for flammable gas when employees are working underground using drill and blast methods and prior to reentry after blasting.

Records. Records of all air quality tests, including location, date, item, substances tested, and results, must be kept at the worksite for inspection by OSHA upon request.

Ventilation

There are a number of requirements for ventilation in underground construction activities. First, fresh air must be supplied to all underground work areas in sufficient amounts to prevent any dangerous or harmful accumulation of dusts, fumes, mists, vapors, or gases. A minimum of 200 cubic feet of fresh air per minute is to be supplied for each employee underground. Mechanical ventilation, with reversible airflow, must be provided in all work areas, except where natural ventilation is demonstrated to be sufficient. Where blasting or drilling is performed, or other types of work operations that may cause harmful amounts of dust, fumes, or vapors, the velocity of the airflow must be at least 30 feet per minute.

For potentially gassy or gassy operations, ventilation systems must meet additional requirements. Ventilation systems used during these operations must have controls located aboveground for reversing airflow.

Cranes and Hoists for Underground Construction

There are some aspects of hoisting that are unique to underground construction, and these are covered in Subpart S. In general, the standards for cranes and hoists delineated in Subpart N are incorporated by reference in the Subpart pertaining to underground construction. Provisions for hoisting in underground construction include the following:

- Securing or stacking materials, tools, and other objects being raised or lowered into a shaft to prevent the load from shifting. snagging, or falling into the shaft
- Using a flashing warning light for employees at the bottom of the shaft and at subsurface shaft entrances whenever a load is above these locations or is being moved in the shaft

- Following procedures for the proper lowering of loads when a hoistway is not fully enclosed and employees are at the bottom of the shaft
- Informing and instructing employees of maintenance and repair work that is to commence in a shaft served by a cage, skip, or bucket
- Providing a warning sign at the shaft collar, at the operator's station, and at each underground landing for work being performed in the shaft
- Using connections between the hoisting rope and cage or skip that are compatible with the wire rope used for hoisting
- Using cage, skip, and load connections that will not disengage as a result of the force of the hoist pull, vibration, misalignment, release of lift force, or impact
- Ensuring that all spin-type connections are maintained in a clean condition, and assuring that wire rope wedge sockets, when used, are properly seated

In addition, there are requirements for cranes, which include the use of limit switches or anti-two-block devices. These operational aids are necessary for limiting the travel of loads if operational controls malfunction. They are not to be used as a substitute for other operation controls. Further requirements also apply to hoists, including assigning load and speed rates for hoists, use of "deadman type" control levers, and others.

Emergencies

At worksites that have more than 25 employees who work underground at one time, employers are required to provide rescue teams or rescue services that include at least two five-person teams. One must be on the job site or within one-half hour travel time, and the other must be within two hours travel time. Where there are fewer than 25 employees underground at one time, the employer must provide or make available in advance one five-person rescue team onsite or within one-half hour travel time.

The members of a rescue team must be qualified in rescue procedures as well as in the use of fire-fighting equipment and use of respiratory protection, including SCBA. Rescue team members should also be familiar with job site conditions. As part of the emergency procedures, the employer must provide self-rescuers, to be available to all employees at underground work stations who might be trapped by smoke or gas. These respirators must be approved by the National Institute for Occupational Safety and Health (NIOSH) and by the Mine Safety and Health Administration (MSHA).

In addition, a person must be designated to be responsible for securing immediate aid for workers and for keeping an accurate count of employees underground. Emergency lighting, in the form of a portable hand or cap lamp, must be provided to all underground workers in their work areas to provide adequate light for escape.

Illumination

As in all construction operations, the standard requires that proper illumination be provided during underground construction operations, as defined in Subpart D. When explosives are used or handled at the job site, only acceptable portable lighting equipment is allowed within 50 feet of any underground heading.

Fire Prevention and Control

There are requirements in addition to those delineated in Subpart F for underground construction. Open flames and fires are prohibited in all underground construction activities, except for hot work operations. During hot work operations such as welding, noncombustible barriers must be installed below work being performed in or over a shaft or raise. The number of fuel gas and oxygen cylinders kept underground must be limited to those amounts necessary to perform welding, cutting, or other hot work over the next 24-hour period. When work is completed, the gas and oxygen cylinders must be removed.

Smoking is restricted except in areas free of fire and explosion hazards, and the employer is required to post signs prohibiting smoking and open flames where these hazards exist. Numerous other restrictions apply, including limitations on the piping of diesel fuel from the surface to an underground location and the prohibition of gasoline underground. Leaks and spills of flammable or combustible fluids must be cleaned up immediately. Other required fire prevention measures include the use of fire-resistant barriers, fire-resistant hydraulic fluids, proper location and storage of combustible materials, and other requirements in respect to electrical installations underground, lighting fixtures, and fire extinguishers.

Other Requirements for Underground Construction

Other requirements apply for subsidence areas, shafts, and other aspects of underground construction. These requirements are delineated in 1926.800.

Caissons and Cofferdams

Requirements for caissons—large watertight chambers within which work is performed—are detailed in 1926.801. Requirements for cofferdams—temporary enclosures built in the water and pumped dry to permit work on bridge piers and similar structures—are delineated in 1926.802.

Compressed Air

Requirements for working in compressed air atmospheres are defined in 1926.803 and should be carefully reviewed and understood by a safety specialist or other competent person who will represent the employer while work activities are being performed. Aspects of working in a compressed air atmosphere that are regulated include medical aspects, telephone and signal communications, signs and records, compression and decompression, manlocks and special decompression chambers, the compressor plant, ventilation and air quality, sanitation, fire prevention and protection, and bulkheads and safety screens. Appendix A to Subpart S provides decompression tables that are computed for working chamber pressures from 0 to 14 pounds and from 14 to 80 pounds per square inch gauge, inclusive, by 2-pound increments, and for exposure times for each pressure extending from one-half to over eight hours, inclusive.

DEMOLITION

Demolition activities are regulated under Subpart T of the construction standards. These standards do not cover removal of asbestos or selective demolition by explosives, which are covered under Subpart D (specifically 1926.58) and Subpart U, respectively.

Prior to permitting employees to start demolition operations, an engineering survey of the structure must be made to determine the condition of the framing, floors, and walls and the possibility of any unplanned collapse of any portion of the structure. The survey must be made by a competent person and documented. Shoring and bracing of walls and floors are required when a building to be demolished has been damaged by fire, flood, explosion, or other cause. Electricity, gas, water, steam, and other utilities must be shut off, capped, or otherwise controlled outside the building before the work is to commence. Other hazards, such as hazardous chemicals, gases, explosives, flammable materials, or other dangerous substances in pipes, tanks, or other equipment, must be identified and purged before work can begin. Other hazards, such as wall openings, holes in the floor, and the like, must be identified and controlled before work commences.

Stairs, passageways, and ladders that are adequate for employee use must be designated, and all other access ways must be closed at all times. Proper illumination must be provided for these access areas.

Dropping material outside the exterior walls of a structure is prohibited unless the area is effectively protected. Material chutes at an angle of more than 45 degrees must be entirely enclosed, except for openings equipped with closures at or about floor level for the insertion of materials. The chute must have a guardrail of 42 inches, and other pertinent requirements apply.

Walls, masonry sections, and chimneys must be removed carefully so that demolition activities do not threaten safe carrying capacities of the floors. Employees are not permitted to work on the top of a wall when weather conditions constitute a hazard. Structural or load-supporting members on any floor must not be cut or removed until all stories above such a floor have been demolished and removed. In buildings of "skeleton-steel" construction, the steel framing may be left in place during the demolition of masonry. Where this is done, all steel beams, girders, and similar structural supports must be cleared of all loose material as the masonry demolition progresses downward. Storage of waste material must not exceed allowable floor loads.

Safe supports, as defined in 1926.855, must be provided to workers who are removing floors. When steel construction is dismantled, it should be dismantled column length by column length, tier by tier.

Mechanical demolition requirements are presented in 1926.859. In general, no workers can be permitted in any area that can be adversely affected by demolition operations, when balling or clamming is being performed. The weight of the demolition ball cannot exceed 50% of the crane's rated load, and other pertinent requirements apply.

BLASTING AND USE OF EXPLOSIVES

Blasting and the use of explosives are covered under Subpart U of 1926. Only authorized and qualified persons can handle and use explosives, and these persons must understand the numerous requirements pertaining to explosives. The standard is comprehensive in its scope and covers general requirements, blaster qualifications, transportation of explosives, storage; handling, and other requirements.

POWER TRANSMISSION AND DISTRIBUTION

Subpart V of the construction standards defines requirements that apply to the construction of electric transmission and distribution lines and equipment. Areas covered under Subpart V include general requirements for testing and deenergizing lines and equipment, tools and protective equipment, mechanical equipment, material handling, grounding for protection of employees, overhead and underground lines, construction in energized substations, external-load helicopter cranes, and lineman's body belts, safety straps, and lanyards.

ROLLOVER PROTECTIVE STRUCTURES—
OVERHEAD PROTECTION

Rollover protective structures (ROPS) are required for construction material-handling equipment such as rubber-tired self-propelled scrapers, rubber-tired front-end loaders, rubber-tired dozers, wheel-type agricultural and industrial tractors, crawler tractors, crawler-type loaders, and motor graders.

Requirements for ROPS are detailed in Subpart W of the construction standards.

There are performance requirements for ROPS that include extensive testing for vehicle overturn, side and rear load application, dynamic and static loadings, crushing, and other impacts. Other engineering requirements apply as well.

STAIRWAYS AND LADDERS

Requirements for stairways and ladders are promulgated under Subpart X. In general, a stairway or ladder is required when there is a break in elevation of 19 inches or more, and when there is no other means of access provided. Employees must be trained to recognize hazards associated with stairways and ladders. Training should include the nature of fall hazards in the work area; the correct procedures for erecting, maintaining, and disassembling the fall protection systems to be used; the proper construction, use, placement, and care in handling of all stairways and ladders; the maximum intended load-carrying capacities of ladders used; and the standards contained in Subpart X. Specific requirements for stairways and ladders are presented in 1926.1052 and 1925.1053, respectively.

DIVING

Requirements for diving and related support operations for employees who work within the waters of the United States and U.S. territories are defined under Subpart Y of the construction standards. These requirements also apply to general industry, ship repairing, shipbuilding, ship breaking, and longshoring.

In addition to requirements for predive, dive, and postdive procedures, there are numerous recordkeeping requirements. These include medical records, a current copy of the safe practices manual, recordings of dives, decompression procedure assessment evaluations, equipment inspections and testing records, and records of hospitalizations.

TOXIC AND HAZARDOUS SUBSTANCES

Toxic and hazardous substances are covered under Subpart Z and Subpart D (see page 367 for Subpart D requirements). Most of the substances covered in Subpart Z and Subpart D of the construction standards are covered in the general industry standards under Subpart Z of 1910. Substances that are regulated under Subpart Z of 1926 are presented in Table 14.10.

TABLE 14.10 Substances Regulated under Subpart Z of 1926

• 4-Nitrobiphenyl	• Formaldehyde
• alpha-Naphthylamine	• Benzidine
• Methyl chloromethyl ether	• Benzene
• 3,3'-Dichlorobenzidine	• Ethylene oxide
• bis-Chloromethyl ether	• Coke oven emissions
• beta-Naphthylamine	• Acrylonitrile
• 4-Aminodiphenyl	• Vinyl chloride
• 2-Acetylaminofluorene	• Inorganic Arsenic
• 4-Dimethylaminoazobenzene	• Ethyleneamine
• N-Nitrosodimethylamine	• beta-Propiolactone
• 1,2-Dibromo-3-chloropropane	

Requirements outlined for specific toxic and hazardous substances include permissible exposure limits, personal protective equipment requirements, restrictions to a work area, warning signs and other labeling, training, hygiene facilities and practices, contamination control, medical surveillance, and others.

FIELD VIOLATIONS

Types of violations of construction standards most commonly found at worksites include violations of electrical standards, scaffolding standards, occupational health standards, and standards for ladders and stairways. The most frequently cited OSHA construction standards in 1991, by category type, are presented in Figure 14.8 (OSHA 1993).

OSHA has reported that 90% of the fatalities occuring at construction sites are caused by: falls from elevations (33%); being struck by vehicles or other objects (22%); trench cave-ins or being crushed by equipment (18%); and electric shock (17%) (*Professional Safety*, 1994). In 1994, OSHA issued new enforcement guidelines, in which these areas are to be given priority during site inspections. After reviewing these areas, the compliance inspector may conclude that the inspection of other conditions is not needed.

REFERENCES

Code of Federal Regulations, 29 CFR 1910.1200 (including Appendices), U.S. Department of Labor, Occupational Safety and Health Administration, Washington, DC, 1995.

"Inspectors Will Focus on Major Construction Hazards," *Professional Safety*, Vol. 39/No. 12, December 1994, p. 54.

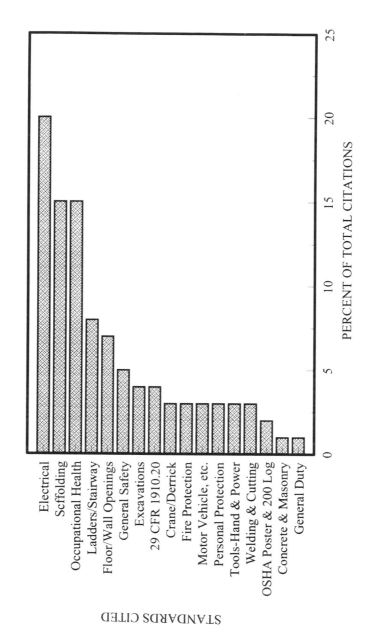

Source: Information from OSHA (1993).

Figure 14.8 Frequently cited OSHA construction standards in 1991.

National Fire Protection Association, *Fire Protection Guide for Hazardous Materials,* 10th ed., NFPA, Quincy, MA, 1991.

National Institute for Occupational Safety and Health, *Pocket Guide to Chemical Hazards,* DHHS (NIOSH) Publication No. 85-114, U. S. Department of Health and Human Services, Public Health Service, Centers for Disease Control, National Institute for Occupational Safety and Health, Washington, DC, 1987.

Occupational Safety and Health Administration, *The 100 Most Frequently Cited OSHA Construction Standards in 1991: A Guide for the Abatement of the Top 25 Associated Physical Hazards,* U.S. Department of Labor, Occupational Safety and Health Administration, Washington, DC, 1993.

Occupational Safety and Health Administration, *Underground Construction (Tunneling),* OSHA 3115, U.S. Department of Labor, Occupational Safety and Health Administration, Washington, DC, 1991.

U.S. Department of Energy, *Construction Safety in DOE,* DE93 019172, Pacific Northwest Laboratory, U.S. Department of Energy, Richland, WA, 1993.

BIBLIOGRAPHY

NIOSH (1989), "Fatal Accident Circumstances and Epidemiology (FACE) Report: Distribution Line Technician Dies After Contacting Energized Conductor, January 20, 1989," National Institute for Occupational Safety and Health, Morgantown, WV.

NIOSH (1992), "Fatal Accident Circumstances and Epidemiology (FACE) Report: Roofer's Helper Electrocuted When Ladder Platform Hoist Contacts a Powerline, South Carolina, June 24, 1992," National Institute for Occupational Safety and Health, Morgantown, WV.

———— (1992), "NIOSH Alert: Request for Assistance in Preventing Lead Poisoning in Construction Workers," U.S. Department of Health and Human Services, Public Health Service, Center for Disease Control, National Institute for Occupational Safety and Health, Washington, DC.

———— (1991), "Fatal Accident Circumstances and Epidemiology (FACE) Report: Electrical Lineman Electrocuted After Contacting Energized Trailer-Mounted Line Tensioner in South Carolina, August 5, 1991," National Institute for Occupational Safety and Health, Morgantown, WV.

———— (1991), "Fatal Accident Circumstances and Epidemiology (FACE) Report: Ironworker Dies Following an 18-Foot Fall from Structural Steel Framework, Alaska, October 20, 1991," National Institute for Occupational Safety and Health, Morgantown, WV.

———— (1990), "Fatal Accident Circumstances and Epidemiology (FACE) Report: Construction Laborer Electrocuted After Handling Damaged Energized Extension Cord in Virginia, October 11, 1990," National Institute for Occupational Safety and Health, Morgantown, WV.

———— (1990), "Fatal Accident Circumstances and Epidemiology (FACE) Report: Laborer Touching Suspended Cement Bucket Electrocuted When Crane Cable Contacts 7200-Volt Powerline in North Carolina, March 1, 1990," National Institute for Occupational Safety and Health, Morgantown, WV.

OSHA (1993), "Ground-Fault Protection on Construction Sites," OSHA 3007, U.S. Department of Labor, Occupational Safety and Health Administration, Washington, DC.

_____ (1993), "Lead in Construction," OSHA 3142, U.S. Department of Labor, Occupational Safety and Health Administration, Washington, DC.

_____ (1993), "Occupational Exposure to Cadmium in the Construction Industry," OSHA 3139, U.S. Department of Labor, Occupational Safety and Health Administration, Washington, DC.

_____ (1993), "4,4-Methylenedianiline (MDA) in the Construction Industry," OSHA 3137, U.S. Department of Labor, Occupational Safety and Health Administration, Washington, DC.

_____ (1992), "Concrete and Masonry Construction," OSHA 3106, U.S. Department of Labor, Occupational Safety and Health Administration, Washington, DC.

_____ (1992), "Construction Accidents: The Workers' Compensation Data Base 1985–1988," U.S. Department of Labor, Occupational Safety and Health Administration, Washington, DC.

_____ (1991), _"Excavations,"_ OSHA 2226, U.S. Department of Labor, Occupational Safety and Health Administration, Washington, DC.

PART IV

PRINCIPLES OF INDUSTRIAL HYGIENE AND OCCUPATIONAL HEALTH ENGINEERING

15

CHEMICAL HAZARD ASSESSMENT AND COMMUNICATION

Exposures from hazardous substances in industrial applications can be evaluated in several ways. This chapter discusses methods typically used by industry to provide workers with information or real-time data about exposures to workplace hazardous materials. These methods include the use of material safety data sheets, documented exposure limits, medical surveillance programs, and workplace monitoring. In addition, the Hazard Communication Standard—which sets the requirements for communication of hazards to employees—is discussed, and the requirements in case of exposure to hazardous materials in laboratories are presented.

MATERIAL SAFETY DATA SHEETS

All chemicals that are manufactured, imported, sold, or used in a manufacturing process must be accompanied by a material safety data sheet (MSDS), as defined under 29 CFR 1910.1200. Chemical manufacturers and importers must provide an MSDS to the distributor and employers with their initial shipment of chemicals and with the first shipment after data on the MSDS has been updated. Distributors who sell to other distributors or employers, likewise must provide an MSDS and updates to their customers. An MSDS or equivalent must be available for review by employees in the workplace and must contain, at a minimum, the information listed in Table 15.1. The American National Standards Institute (ANSI) has published a guideline on MSDS preparation—ANSI Z400—that is commonly used by manufacturers and other MSDS preparers, although not required by OSHA. The ANSI MSDS guideline includes 16 sections and an explanation of what to include in each section.

TABLE 15.1   Information To Be Included on an MSDS

Chemical and Common Names

- Single substances—Chemical and common name(s) must be identified.
- Mixture, tested as a whole—Chemical and common names of the ingredients that contribute to the known hazards and the common name(s) of the mixture itself must be identified.
- Mixture, not tested as a whole—Chemical and common names of all ingredients that have been determined to be health hazards and that comprise 1% or greater, or that are carcinogens and comprise 0.1% or greater, must be identified; the chemical or common name(s) of chemicals in the mixture that could be released in amounts greater than the permissible exposure limit must be identified; the chemical or common name(s) of all ingredients determined to present a physical hazard in the mixture must be identified.

Physical and Chemical Characteristics

- Pertinent physical characteristics such as vapor pressure, flash point, boiling point, melting point, and specific gravity must be included.
- Pertinent chemical characteristics such as solubility and molecular weight and other such information must be included.

Physical Hazards

- Potential for fire, explosion, and reactivity must be included.
- Incompatibilities and segregation practices may be included.

Health Hazards

- The health hazards of the chemical must be included.
- Signs and symptoms of exposure, and any medical conditions that are generally recognized as being aggravated by exposure to the chemical, must be delineated.

Primary Routes of Entry

- Information pertaining to primary routes of entry, such as inhalation, ingestion, skin absorption, and eye or skin contact, must be delineated.

Permissible Exposure Limit

- The OSHA permissible exposure limit must be included.
- Other information on exposure limits, such as information from the American Conference of Governmental Industrial Hygienists, must be included.

TABLE 15.1 *Continued*

Carcinogen or Potential Carcinogen

- If the chemical is listed in the National Toxicology Program's *Annual Report on Carcinogens* (latest edition), this information should be included.
- If the chemical has been published as a carcinogen or potential carcinogen in the International Agency for Research on Cancer (IARC) *Monographs* (latest edition), this information should be included.
- If OSHA has listed the chemical as a carcinogen or potential carcinogen, this information should be included.

Safe Handling Procedures

- Procedures such as appropriate industrial hygiene practices, protective measures for handling the chemical during repair and maintenance of contaminated equipment, and procedures for cleanup of spills and leaks should be delineated.

Control Measures

- Control measures such as appropriate engineering controls, protective clothing and equipment, and work practices are to be detailed.

Emergency and First Aid Procedures

- Procedures pertaining to emergency actions to be taken for exposure, through each of the primary routes of entry, should be delineated.

Manufacturer's and Other Information

- The name, address, and telephone number of the chemical manufacturer or importer, the employer, or other responsible party who prepares or distributes the MSDS, should be included.
- The date of preparation of the MSDS or of the last change to the MSDS should be included.

EXPOSURE LIMITS

Exposure Limits Defined by OSHA and ACGIH

OSHA has published exposure limits for hundreds of hazardous airborne contaminants in terms of permissible exposure limits (PELs) and time-weighted averages (TWAs). These limits, which are enforceable standards, define the maximum time-weighted exposure over an 8-hour work shift of a 40-hour workweek that should not be exceeded. These limits are expressed in parts per million (ppm) and/or milligrams per cubic meter (mg/m^3) and are published in 29 CFR §1910.1000.

The American Conference of Governmental Industrial Hygienists (ACGIH), which defines airborne contaminant exposure similarly, terms this exposure concentration the threshold limit value–time weighted average (TLV–TWA). TLV–TWAs are defined as a concentration to which nearly all workers can be repeatedly exposed over a normal 8-hour workday in a 40-hour workweek without adverse affects. Like OSHA, ACGIH has documented TLV–TWAs for hundreds of chemicals in the annual publication "Threshold Limit Values for Chemical Substances and Physical Agents and Biological Exposure Indices" (ACGIH 1995). Although not legally enforceable, ACGIH standards are considered industry standards.

In some instances, industry standards established by ACGIH may differ from OSHA standards. This is typically because OSHA standards must be adjusted through regulatory amendments, which is a lengthy and time-consuming process, whereas ACGIH standards are updated annually, based on the latest scientific data. Thus, to ensure adequate protection of workers, it is considered good management practice for employers to abide by the ACGIH standard for a particular chemical if it is lower than the OSHA standard. If the OSHA standard is lower, the OSHA standard must be met.

For some chemicals, ACGIH and OSHA have documented additional exposure concentrations. These include a short-term exposure limit (STEL) and a ceiling limit (C). The STEL is a 15-minute time-weighted average exposure limit that should not be exceeded.[1] This limit is not a stand-alone limit, and it must be included in the 8-hour time-weighted average. The STEL generally applies to a substance whose toxic effects are considered to be chronic, but for which there may be recognized acute effects at a particular limit. ACGIH recommends that exposure at this limit should not be repeated more than four times daily and that at least a 60-minute rest period between exposures should be allowed.

A ceiling limit is a maximum concentration that should not be exceeded for any period of time. Ceiling limits are established for chemicals that, at certain concentrations, may produce acute poisoning during very short exposures.

Because OSHA and ACGIH exposure limits are based on an 8-hour work-day, 40-hour workweek time-weighted average, worker exposure may safely exceed published exposure limits for periods of time during the workday, as long as they are compensated with periods of time when exposures are below the limit, so that the 8-hour time-weighted average is not exceeded. The time-weighted average is calculated as follows:

$$TWA = \sum C_i t_i / \sum t_i \qquad (15.1)$$

---

[1] The 15-minute STEL applies to almost all chemicals; however, asbestos has a 30-minute excursion limit established by OSHA.

EXAMPLE 15.1

The TWA for methyl ethyl ketone is 200 ppm, with an STEL of 300 ppm. If a worker is exposed to this chemical at a level of 250 ppm for 4 hours, has the TWA or STEL been exceeded?

Solution. Using the Equation 15.1:

$$TWA = 250(4)/8 = 125 \text{ ppm}$$

Thus, the TWA is not exceeded. Nor is the STEL of 300 exceeded; therefore, this exposure level meets ACGIH guidelines.

In addition to the effects of single substances, factors related to a chemical mixture exposure—mainly additive effects, if the same organ system is affected—should be considered. The additive effects of mixtures for the TLV may be calculated by the following equation:

$$C_1/T_1 + C_2 T_2 + \ldots C_n T_n = 1 \qquad (15.2)$$

If the sum of the mixture exceeds 1, then the TLV of the mixture is considered to be exceeded.

EXAMPLE 15.2

Air contains a mixture of 450 parts acetone (TLV = 750 ppm) and 150 ppm of methyl ethyl ketone (TLV = 200 ppm). Is the TLV exceeded?

Solution: $450/750 + 150/200 = 1.35$

Yes, the TLV is considered to be exceeded.

The STEL and ceiling limits should not be exceeded at any time during the workday. For many chemicals, there is not enough toxicological data available for ACGIH or OSHA to publish an STEL or ceiling limit. When an STEL has not been published, ACGIH recommends that worker exposure at levels of three times the TLV–TWA should last for no more than 30 minutes during a workday. In addition, if there is no ceiling limit, a concentration level of five times the TLV–TWA should never be exceeded.

Odor threshold is a physical property of a chemical and has no relation to acceptable exposure concentrations. There are numerous chemicals that are toxic at levels below the odor threshold, such as methylene chloride and carbon monoxide. Thus, odor—or the absence of odor— should never be used in determining acceptable workplace concentrations. In cases where workers are using air purifying (cartridge) respirators, odor (and taste) is considered a warning property indicating breakthrough of the filter material.[2] When noticed, the work area should be evacuated immediately.

Exposure Limits for Asbestos

Asbestos is regulated under 29 CFR 1910.1001 (general industry) and 29 CFR 1926.58 (construction industry). Asbestos is the name of a class of magnesium-silicate minerals that occur in fibrous form. Minerals included in this group are chrysotile, crocidolite, amosite, anthophylite asbestos, tremolite asbestos, and actinolite asbestos. Revised in 1994, asbestos exposure limits include an 8-hour time weighted average of 0.1 feet/cubic centimeter (f/cc) and a 30-minute short-term exposure limit (excursion limit) of 1f/cc.

Exposure Limits for Radioactive Materials

The Nuclear Regulatory Commission (NRC) has codified the practice of maintaining all radiation exposures to workers and the general public as low as reasonably achievable (known as the ALARA concept). Public and worker protection standards for specific radioactive materials and wastes have been established by OSHA, EPA, NRC, the Mine Safety and Health Administration (MSHA), and other agencies. Worker-limiting requirements are categorized by body part and organ and defined in numerous regulations, including 10 CFR 20 Subpart C, 29 CFR §1910.96, 40 CFR Part 191, 30 CFR Part 57, and other citations. See Chapter 18 for more information on radiation safety.

MEDICAL SURVEILLANCE PROGRAMS

Medical surveillance programs offer information about the biological effects chemicals may be having on an individual worker. Medical surveillance is also required by OSHA for workers who are routinely exposed to certain concentrations of regulated chemicals or dusts, such as lead, benzene, formaldehyde, acrylonitrile, asbestos, and coke oven emissions. Medical surveillance is required for workers who routinely wear a respirator or who take part in hazardous waste operations and emergency response activities. The employer is responsible for administering and paying for the medical examinations and other medical surveillance activities.

---

[2]In general, cartridge respirators are not allowed for chemicals that have permissible exposure limits below the odor threshold.

Tests included in a typical medical surveillance program can vary, depending on specific chemical exposures. An initial questionnaire is mandatory for workers who are exposed to asbestos and is generally included in some form in most medical surveillance programs. The OSHA asbestos questionnaire includes occupational history, past medical history, chest and other illnesses, family history, smoking habits, and other pertinent information.

Other aspects of a medical surveillance program may include a physical examination, general health survey, blood count, chemistry screen, urinalysis, spirometry test, chest X ray, audiogram, and electrocardiogram. Breath analysis may be included for workers exposed to chemicals that can be detected through this type of screening.

All examinations should be performed on a regularly scheduled basis, with the time between examinations dependent on types of exposures. The time of testing, such as at the end of a shift or the end of the week, can often be critical to the accurate observation of effects on blood or urine, so this should be taken into account in establishing an examination schedule.

Biological effects of chemical overexposure are documented by ACGIH, and can be seen in blood, urine, and respiratory functions. Effects on blood can be seen directly or indirectly through blood counts and blood screening. Indirect indicators are red and white blood cells, plasma, and other blood components. Examples of chemicals that affect blood composition include parathion, carbon monoxide, aniline, and nitrobenzene.

Overexposure can be ascertained directly by the presence of a specific chemical in the bloodstream at elevated levels. Chemicals that can be tested in this manner include cadmium, lead, toluene, and perchlorethylene.

In addition to blood screening, chemistry screening can provide information on cholesterol, triglycerides, total protein, blood urea nitrogen, calcium, phosphorus, and other chemical components. Urinalysis is also a very useful screening test. This test can provide direct detection of numerous chemicals in the urine. Examples include mercury, cadmium, lead, methyl ethyl ketone, and phenol.

Breath analysis can be used to detect the presence of some chemicals. Standards for acceptable concentrations of chemicals in the breath have been developed by ACGIH for chemicals such as benzene, carbon monoxide, toluene, and trichloroethylene.

Decreases in lung capacity can be determined with the spirometry test. When necessary, this test can be coupled with chest X rays. In some cases, respiratory capacity tests may be performed more often than other parts of the medical surveillance tests if exposure warrants.

WORKPLACE MONITORING

Workplace monitoring can be performed using a wide range of analyzers. These include single-chemical or total-concentration analyzers, which are relatively easy to use and typically provide quantitative information about a single chemical or chemical species. For evaluating a multichemical environment, a more sophisticated instrument is needed. Examples of monitors for these applications, including information about instrument operation, chemicals detected, and limitations, are presented in Table 15.2.

TABLE 15.2   Examples of Analyzers for Workplace Monitoring

Single-Chemical or Total-Concentration Analyzers

CHEMILUMINESCENCE ANALYZER FOR MONITORING OXIDES OF NITROGEN

- Instrument operation—Oxides of nitrogen (NO_x) are converted to nitric oxide (NO) and reacted with ozone to generate light emissions that are monitored by a photomultiplier tube. Nitrogen dioxide (NO_2) concentrations are determined by intermittent direct sampling of the stream (without No_x conversion) and by subtracting the NO concentration from the NO_x concentration.
- Chemicals detected—NO_x or NO_2 can be detected in the parts per million (ppm) range.
- Limitations—Negative interferences may occur at high humidities for instruments calibrated with dry span gas. Olefins and organic sulfur compounds, if present, will positively interfere with NO_x monitoring.

OXYGEN METER WITH AN ELECTROCHEMICAL SENSOR

- Instrument operation—An oxygen meter typically uses an electrochemical sensor. The electrochemical sensor has a semipermeable membrane to allow air into the cell through diffusion and uses an electrolytic, current-conducting solution to register current changes directly proportional to the amount of oxygen in the atmosphere. The current is amplified and displayed on a meter. Alarms can be set to sound if the oxygen concentration drops below a preset percentage.
- Chemicals detected—This instrument is programmed to read the percentage of oxygen in the atmosphere. Other applications for an electrochemical sensor include a sulfur dioxide sensor and a hydrogen sulfide analyzer.
- Limitations—The sensor responds to the partial pressure of oxygen and is therefore altitude sensitive, giving reduced readings at higher altitudes. The sensor in an oxygen meter may be affected by oxidants such as ozone. Carbon dioxide will poison the detector cell.

COMBUSTIBLE GAS INDICATOR (CGI)

- Instrument operation—A CGI utilizes a sensor to measure the relative resistance changes produced by gases burning on hot filaments, one of which is coated with a catalyst. All readings on the combustible gas meter are relative to the calibrant gas, usually hexane, methane, or pentane. A CGI is used to determine whether flammable/combustible material is present in a concentration that could be dangerous. Concentrations between the lower and upper explosive limits (LEL and UEL, respectively), are considered immediately dangerous.

TABLE 15.2 *Continued*

- Chemicals detected—Flammable gases can be measured with this instrument. The concentrations are measured in percent LEL and can be converted to ppm using vendor response curves and conversion factors.
- Limitations—Oxygen variations will affect combustion and, consequently, proper operation of the instrument. Lean and enriched mixtures will give inaccurately low and high readings, respectively. Temperature differences between calibration and use can also affect the instruments accuracy. In addition, silicones, halides, and lead compounds will coat the detector unit and render it inoperable. Thus, its use in atmospheres containing unknown vapors can be limited. Corrosive atmospheres, over time, can damage internal components of the instrument and limit its use. Under oxygen-deficient conditions, the instrument will not provide an accurate reading.

LENGTH-OF-STAIN DETECTOR TUBE

- Instrument operation—The length-of-stain detector tube is a colorimetric visual indicator. It is operated by drawing a fixed volume of sample air through a tube with a squeeze bulb or small hand pump. Diffusional models (dosimeters) are also available. A length-of-stain dosimeter operates by allowing an air sample to diffuse through the tube over a 1- to 8-hour period. Both types of tubes, contain a length of granulated resin or gel impregnated with a reactive chemical. The granules change color when specific types of air contaminants are introduced. Generally, length-of-stain detector tubes that use a pump allow the chemical concentration to be read directly on the tube, based on the length of color change. In the diffusional models, a color chart is provided by the manufacturer.
- Chemicals detected—The detector tube can be used for determining the presence of most hydrocarbons, acids, bases, organic amines, and alcohols. Accuracy is typically better than ±25% of the actual concentration in the ppm range.
- Limitations—Temperature and humidity can affect the length-of-stain color change, and calibration charts must be used properly to make corrections. Moreover, similar chemicals can interfere positively with the detector tube's reading.

PHOTOIONIZATION DETECTOR (PID)

- Instrument operation—A PID uses high-energy ultraviolet (UV) light to ionize volatile organic compounds in an air sample. Ionization of the sample produces a current that is proportional to the number of ions measured.
- Chemicals detected—The PID can measure total concentrations of many organic and some inorganic gases. Major air components such as oxygen, nitrogen, and carbon dioxide are not ionized in the process. The detector can quantify chemicals in the parts per billion (ppb) to ppm range, depending on the compound.

(*Continued*)

TABLE 15.2 *Continued*

- Limitations—The instrument cannot detect some compounds if the probe used has a lower energy level than the compound's ionization potential. In addition, high humidity can dampen the detector's response. If the instrument does not have a filter, charged particles can damage the internal parts. Likewise, atmospheres containing corrosive gases can cause damage unless a corrosion resistant instrument is used.

FLAME IONIZATION DETECTOR (FID)

- Instrument operation—The FID mixes an air sample with hydrogen and combusts the sample in a detector cell, ionizing the gases and vapors. The instrument electronically measures the current flow through the flame as the sample is burned and amplifies this measurement on an analog display.
- Chemicals detected—Total hydrocarbons are measured in the ppm range.
- Limitations—Ultrahigh-purity hydrogen fuel and high purity air, free from hydrocarbons, must be used to ensure accurate readings. Moreover, the combustion chamber is very sensitive and can be damaged by corrosive or reactive gases. The portable version of the instrument is limited in atmospheres that are oxygen deficient, because lack of oxygen can cause flameouts.

INFRARED (IR) ANALYZER

- Instrument operation—An IR analyzer uses a spectrometer to read chemical "fingerprints," or concentration intensities, created by passing infrared frequencies through a heated air sample. The basic instrument has a fixed cell pathlength and is calibrated for one or a few predetermined chemicals.
- Chemicals detected—The instrument can detect volatile organic compounds in the spectral range of 2.5 to 15 microns (μm). In the most sophisticated instruments, numerous compounds can be detected in the ppb to ppm ranges.
- Limitations—This instrument cannot be used in flammable or explosive atmospheres. In addition, excessively humid or corrosive atmospheres can cause damage to the instrument. Water vapors and carbon dioxide can interfere with the instrument's readings.

PORTABLE GAS CHROMOTOGRAPH (GC)

- Instrument operation—A portable GC can be used with other detectors, such as a PID or FID, to identify and measure specific compounds. The portable model uses a packed or capillary column to concentrate and separate the compounds according to their vapor pressures. After separation, a detector quantifies the individual compounds (peaks). The identity of a compound or peak is qualitatively or quantitatively determined by its retention time in the GC.
- Chemicals detected—The GC can separate organic gases and vapors for quantitation with a detector. Quantitation limits are dependent on the detector used.
- Limitations—For selective quantitative results, the instrument must be calibrated with the specific analyte. The instrument may not be consistently sensitive to all organic compounds. Moreover, mixtures of polar and nonpolar compounds can cause peak
superimpositions, which may require changes in column type, length of column, or operating conditions.

TABLE 15.2 *Continued*

Multichemical Analyzers

ON-LINE MASS SPECTROMETER (MS)

- Instrument operation—This on-line system utilizes a central MS unit and a vacuum pump system to sequentially draw samples and analyze chemicals from remote locations as far away as 1,500 feet. As many as 50 different workplace locations can be connected with chemical-resistant tubing. Up to 25 chemicals can be analyzed per location. Depending on the number of chemicals analyzed and number of ports attached, the time elapsing between the drawing of a sample for a particular location can range from a few minutes to several hours.
- Chemicals detected—An on-line MS can quantify most chemicals that normally are read on a lab MS, at ppm levels. Metals cannot be quantified. The lower the detection limit required, the longer the dwell time in the analyzer. Lengthened dwell times can substantially increase the time between samples.
- Limitations—In certain cases, high chemical concentrations may not be monitored optimally by this system. Corrosive chemicals, dust, and humidity may cause damage to the system. In addition, the system is operationally complex and may require extensive calibration and maintenance.

ON-LINE GC/FID

- Instrument operation—The GC/FID is an automatic system designed for continuous monitoring of a wide variety of volatile organic compounds. The automated GC separates chemicals by vapor pressure, and the FID quantifies the chemicals. The system can be configured to sequentially sample up to 24 lines from 50 to 100 feet away.
- Chemicals detected—Organic vapors can be detected in the low ppm ranges.
- Limitations—The system is operationally complex and may require extensive calibration and maintenance.

ON-LINE FOURIER TRANSFORM INFRARED (FTIR) SPECTROMETER

- Instrument operation—This instrument is commonly used for quantitative spectroscopic applications in the mid-infrared range. The instrument operates similarly to a standard IR, but uses variable pathlengths and wavelengths to allow for quantitation of numerous chemicals.
- Chemicals detected—Chemicals detected will vary, depending on the setup of the instrument, but the equipment is capable of detecting the same chemicals detected by any IR spectrometer.
- Limitations—Sample conditioning, which can potentially remove chemicals of interest, is required for sample streams containing moisture or corrosive chemicals. In addition, the system's complexity requires extensive calibration and may require extensive maintenance.

PERSONAL MONITORING

Personal monitoring is performed to determine the actual chemical concentrations to which workers are exposed during the workday. This type of monitoring is performed periodically under typical, representative working conditions. The samples are ordinarily taken over an 8-hour shift if chemical-use operations are continuous. In some cases, sampling may be performed over a short period of time within a shift to quantify peak exposures.

Personal sampling usually is accomplished through two methods—sampling with a personal sampling pump and passive or diffusional sampling. Sampling with a personal sampling pump is generally accepted by OSHA for documenting chemical exposures. Passive sampling is typically used for supplemental sampling and screenings. Sampling verification using OSHA methods, when specified, is required. The following paragraphs provide information about both of these sampling methods.

Sampling with a Personal Sampling Pump

Sampling with a personal sampling pump is standard across industry and is documented in ASTM standards and by ACGIH. For this type of sampling, a battery-powered personal sampling pump is typically used in conjunction with an air metering device and an adsorbent tub sampler such as a charcoal or silica gel tube sampler. Sampling with a personal sampling pump yields accurate results because the volume of air sampled is metered and can be accurately quantified. Factors that must be considered in calculating sampled air volume include time, flow rate, pressure, and workplace temperature and humidity.

Charcoal Tube Adsorption Sampler.   This sampler uses a charcoal sampling tube containing two sections of activated charcoal and a sampling pump that draws a sample at a stable flow rate. The carbon tube is taken to a lab and desorbed into a GC using carbon disulfide (or another recommended desorber) to determine the identify and concentration of the chemical adsorbed. Based on air flow rate and time, a workplace concentration is determined.

With the use of this sampling device, numerous organic chemicals can be detected at limits in the ppm range. The method is useful for determining air-borne time-weighted average concentrations of many of the organic chemicals listed by OSHA in 29 CFR §1910.1000.

There are some limitations to sampling with charcoal tube samplers. High humidity can reduce the adsorptive capacity of activated charcoal for some chemicals. Further, mixtures of polar and nonpolar compounds are difficult to recover (desorb) from activated charcoal.

Silica Gel Tube Sampler.   This sampler is very similar to the charcoal tube sampler except that silica gel, which is an amorphous form of silica derived from sodium silicate and sulfuric acid, is used instead of charcoal. Unlike charcoal, silica gel is well suited for sampling polar contaminants because they are removed easily from the adsorbent with common solvents. In addition, amines and some inorganic substances not suitable for charcoal sampling can be sampled with a silica gel tube sampler. The major disadvantage of this type of sampler is that it will adsorb water.

Passive or Diffusional Sampling

Passive or diffusional sampling is performed without the use of a personal sampling pump. Sampling devices consisting of badges or dosimeters are attached to the worker near the breathing zone. The dosimeters collect vapors, by diffusion onto a medium such as charcoal or other adsorbent, to indicate chemical concentrations in the breathing zone. The vapors diffuse through the medium over time at a rate dependent on the cross-sectional area of the diffusion cavity, the diffusion coefficient, and the length of the diffusion path. Diffusion factors can be determined by using the manufacturer's supplied calibration factors or by applying diffusion laws. In some samplers, the adsorbent will have to be desorbed and analyzed. Other types of passive samplers are colorimetric and use length of stain or color change as the concentration indicator.

Like the charcoal tube sampler described previously, passive charcoal adsorbent tube samplers can detect numerous organic chemicals in the ppm range. Its limitations are also similar.

For colorimetric badges or dosimeters, the manufacturer's color graph must be used. The main limitation of these types of samplers is that they can give only a gross assessment of chemical concentration. In addition, humidity and temperature must be taken into account in reading the calibration curves.

OTHER MONITORING

Sampling for Asbestos Exposure

Sampling for airborne asbestos is required during any demolition or renovation project that involves any of the OSHA-listed asbestos minerals—chrysotile, crocidolite, amosite, anthophyllite, tremolite, and actinolite. Procedures are defined by OSHA in 29 CFR 1910.1001, Appendix A. ASTM in D 4240, NIOSH in Publication No. 79-127, and ACGIH-AIHA also provide guidance for asbestos exposure sampling (see the Bibliography).

A sample is collected by pumping air through an open-faced filter membrane. OSHA specifies a mixed cellulose ester filter membrane designated by the manufacturer as suitable for asbestos counting. A section of the membrane is converted to an optically transparent homogenous gel, and the asbestos particles are sized and counted by phase-contrast microscopy at a magnification of 400 to 500 times.

All sampling must be conducted so as to be representative of typical working conditions. Because there are OSHA worker exposure limits for times as short as 30 minutes and as long as 8 hours, selection of an appropriate sampling time is an important consideration. For the most reliable results, several samples should be taken over an 8-hour shift to allow for quantitation of peaks as well as an 8-hour time weighted average. These samples must include personal samples taken in the workers breathing zone and can also include static samples at fixed locations if the dust is uniformly distributed over a large area.

Sampling for Radiation Exposure

Most sampling for worker radiation exposure is performed using a personal dosimeter. Personal sampling must be performed on all workers who can potentially receive a dose in any calendar quarter excess of 25% of the allowable radiation dose limits. In addition, all workers who enter a high radiation area (an area where there is potential for a major portion of the body to receive a dose of greater than 100 millirems in one hour) must be sampled. Some states may have more stringent or additional requirements.

The personal dosimeter must be processed by processors accredited through the National Voluntary Laboratory Accreditation Program (NVLAP). Personal dosimeter performance standards are outlined in ANSI Standard N13.11.

Fixed and portable samplers are also available for sampling indoor and outdoor atmospheres that are potentially radioactive. These samplers include alpha particle analyzers, beta particle analyzers, gamma particle analyzers, scintillation counting systems, radon monitors, and other systems.

Monitoring Indoor Air Quality

The indoor air quality of nonmanufacturing areas, such as office space or low-chemical-use labs, is a key part of environmental health. As a result of negative publicity about Legionnaires' disease, excessive radon exposure in homes and workplaces, and the phenomenon known as "sick building syndrome," indoor air quality has gained national focus. Periodic monitoring of indoor air quality can help ensure that the air in a building is fresh and free from contamination. If occupants complain of specific symptoms such as headaches, dizziness, or general malaise, the complaints should be investigated immediately.

Sampling for indoor air quality microbial parameters involves the collection of microbial particulate. Collection methods used for any particulate aerosol can be used in collecting samples for indoor air quality analysis. The use of culture plate impactors, including multiple- and single-stage sieve devices, is common for office environments, as is the use of slit-to-agar samplers. Filter canisters are used for microorganisms that are resistant to desiccation. Sampling for volatile organic compounds (VOCs) in nonindustrial indoor air environments is typically performed using sorbents or activated charcoal combined with analysis involving GC and MS.

HAZARD COMMUNICATION

Employers have a duty to communicate the hazards of chemicals to employees, as defined in the Hazard Communication Standard under 29 CFR 1910.1200. This standard requires information to be prepared and transmitted about hazardous chemicals stored, used, or handled in the workplace.

Chemical Manufacturers, Importers, and Distributors

Chemical manufacturers, importers, and distributors of hazardous chemicals are required to provide MSDSs and appropriate labels for the hazardous materials they produce or import. Information to be included on the MSDS is discussed earlier in this chapter. Container labels should include the following information:

- Identity of the hazardous chemical(s)
- Appropriate hazard warnings, such as health hazard, physical hazard, and other warning information
- Name and address of chemical manufacturer, importer, or other responsible party

Employers

Employers must prepare and implement a written hazard communication program. The key aspect of this program is employee training on the hazards of the chemicals in the workplace (OSHA 1991). All employees must be trained on the hazards of the chemicals they will be exposed to in the workplace before work begins. If a new hazard is introduced, employees should receive training on that hazard before working with the hazardous chemical. Training should include how to read an MSDS and a chemical label; the location of the MSDS; the hazards of the chemical; methods and observations that may be used to detect the presence or release of a hazardous chemical in the work area; protective equipment required when working with the hazardous chemical; safe work practices; disposal information; emergency and first aid information; signs and symptoms of overexposure; and other relevant information.

In addition to training, employers are responsible for labeling containers of chemicals that have been transferred to point-of-use containers. Some exceptions apply, as defined in the standard. Examples of chemical labels that may be found in the workplace are shown in Figure 15.1.

Figure 15.1 Examples of chemical labels. With permission from EMED Co., Buffalo, N.Y.

EXPOSURE TO HAZARDOUS CHEMICALS IN LABORATORIES

Laboratory workers are routinely exposed to numerous hazardous chemicals. The use of chemicals in laboratories, however, is quite different from the use of chemicals in an industrial setting. Exposure is typically limited to small quantities of chemicals, on a short-term basis, in operations where chemicals and procedures change frequently. As a result, OSHA regulates laboratory activities in a separate standard under 29 CFR 1910.1450. At the core of the standard is a written chemical hygiene plan and an adequate training program.

Chemical Hygiene Plan

The chemical hygiene plan is a written plan that is developed by the employer. It must specify a training and information program for employees exposed to

Figure 15.1 *Continued*

hazardous chemicals in the laboratory. In addition, it must establish the following:

- Appropriate work practices
- Standard operating procedures
- Methods of control
- Measures for appropriate maintenance and use of protective equipment
- Medical examinations
- Special precautions for work with particularly hazardous substances

A sample chemical hygiene plan (nonmandatory) is presented in Appendix A of 29 CFR 1910.1450. A list of references that may be consulted in developing a chemical hygiene plan is presented in Appendix B of the same section of the standard.

Employee Information and Training

Another part of the standard requires an employer to establish a training and information program that meets the Hazard Communication Standard. This includes training on the physical and health hazards of chemicals to which the worker is exposed. Moreover, the training program must include information about the laboratory standard; the existence and content of the chemical hygiene plan; permissible exposure limits for regulated substances and recommended exposure limits for nonregulated substances; signs and symptoms associated with exposures to hazardous chemicals; and the location and availability of known reference materials, including MSDSs, on the hazards, safe handling, storage, and disposal of hazardous chemicals in the workplace (OSHA 1994). The training plan must consist of four major elements:

- The components of the chemical hygiene plan and how it is implemented in the workplace
- The hazards of the chemicals in the work area and the protective measures employees can take
- Specific procedures put into effect by the employer to provide protection, including engineering controls, work practices, and personal protective equipment
- Methods and observations (e.g., continuous monitoring procedures, visual appearance, and smell) that workers can use to detect the presence of hazardous chemicals

Medical Examinations and Consultation

Medical surveillance is not required in the standard; however, a company may elect to include its laboratory workers in their medical surveillance program, particularly if the workers routinely work with chemicals that are carcinogens or potential carcinogens. If an employee exhibits signs or experiences symptoms associated with exposure to a hazardous chemical used in the laboratory, the employer must offer a medical examination and follow-up treatment and examinations. Further, if any employee is exposed routinely to a chemical or chemicals above the action level (or in cases where there is no action level, above the PEL), and if the substance has exposure monitoring or medical surveillance requirements, then that employee must be offered periodic medical examinations. If there is a spill, leak, explosion, or other accident in the area that results in a potentially significant exposure to a hazardous chemical, the exposed employee must be offered medical consultation to determine whether there is a need for a medical examination.

The employer is required to give the physician specific information on the identity of the hazardous chemical, the conditions under which exposure occurred, and a description of the signs and symptoms experienced by the worker. The employer must also obtain from the physician any written opinion for a recommended medical examination, follow-up examination, and the attendant test results; any detected medical conditions of the employee that might pose increased risk; and a statement that the employee was informed of the medical examination/consultation results.

Hazard Identification

Employers must make certain that labels on containers of hazardous chemicals are not removed or defaced. They must also maintain and make available to employees MSDSs received with incoming shipments of chemicals; however, an employer is not required to prepare an MSDS for laboratory mixtures, except in cases where a chemical is produced in the laboratory for another user outside the laboratory.

Recordkeeping

Employers must establish and maintain for each employee an accurate record of exposure monitoring results and any medical consultation and examinations, including tests and physicians' medical opinions. These records must be kept, transferred, and made available to employees as other employee exposure and medical records are, as defined in 29 CFR 1910.20.

REFERENCES

American Conference of Governmental Industrial Hygenists, "Threshold Limit Values for Chemical Substances and Physical Agents and Biological Exposure Indices," Cincinnati, OH, 1996.

American National Standards Institute, "Hazardous Industrial Chemicals—Material Safety Data Sheets—Preparation," ANSI Z400.1, American National Standards Institute, New York, 1993.

Occupational Safety and Health Administration, "Exposure to Hazardous Chemicals in Laboratories," OSHA 3119, U.S. Department of Labor, Washington, DC, 1994.

Occupational Safety and Health Administration, "Hazard Communication Guidelines for Compliance," OSHA 3111, U.S. Department of Labor, Washington, DC, 1991.

BIBLIOGRAPHY

ACGIH (1989), *Air Sampling Instruments for Evaluation of Atmospheric Contaminants*, 7th ed., American Conference of Governmental Industrial Hygienists, Cincinnati, OH.

ACS (1985), *Safety in Academic Chemistry Laboratories*, 4th ed., American Chemical Society, Washington, DC.

American Conference of Government Industrial Hygienists—American Industrial Hygiene Association (1975), "Aerosol Hazards Evaluation Committee: Recommended Procedures for Sampling and Counting Asbestos Fibers," *American Industrial Hygiene Journal*, Vol. 36.

AIHA (1989), *Odor Thresholds for Chemicals with Established Occupations Health Standards*, American Industrial Hygiene Association, Akron, OH.

ANSI (1983), "Personnel Dosimetry Performance, Criteria for Testing," ANSI N13.11, American National Standards Institute, New York.

Ashford, N.A., and C.S. Miller (1991), *Chemical Exposures—Low Levels and High Stakes*, Van Nostrand Reinhold, New York.

ASTM (1995), *Annual Book of ASTM Standards*, Vol. 11.03, "Atmospheric Analysis; Occupational Health and Safety," American Society of Testing and Materials, Philadelphia.

———— (1995), *Annual Book of ASTM Standards*, vol. 11.04, "Pesticides; Resources Recovery; Hazardous Substances and Oil Spill Responses; Waste Disposal; Biological Effects," American Society of Testing and Materials, Philadelphia.

Clayton, George D., and Florence E. Clayton (1991), *Patty's Industrial Hygiene and Toxicology*, Volume 1, Parts A and B, 4th ed., John Wiley & Sons, New York.

Coleman, Daniel R., et al. (1990), "Automatic Continuous Air Monitoring at Fixed Sites with Minicams (TM)," pp. 114–117, *Proceedings of the Third Annual Hazardous Materials Management Conference/Central*, Chicago.

Daisey, Joan M. (1987), "Real Time Portable Organic Vapor Sampling Systems: Status and Needs," prepared for the American Conference of Governmental Industrial Hygienists, Cincinnati, OH, Lewis Publishers, Boca Raton, FL.

Department of Energy (1991), "The Status of ANSI N13.11—The Dosimeter Performance Test Standard," DE91 017874, prepared by C.S. Sims, Oak Ridge National Laboratory, Oak Ridge, TN.

DHHS (1995), *Annual Report on Carcinogens*, National Toxicology Program, U.S. Department of Health and Human Services, Government Printing Office, Washington, DC.

Dillon, H.K., and M.H. Ho, eds. (1991). *Biological Monitoring of Exposure to Chemicals: Metals*, John Wiley & Sons, New York.

Erb, Jeff, Evelyn Ortiz, and Gayle Woodside (1990), "On-line Characterization of Stack Emissions," pp. 40–45, *Chemical Engineering Progress*, Vol. 86/No.5.

Ho, M.H., ed. (1987), *Biological Monitoring of Exposure to Chemicals: Organic Compounds*, John Wiley & Sons, New York.

Hodgson, Ernest (1988), *Dictionary of Toxicology*, Van Nostrand Reinhold, New York.

Kusnetz, S., and M.K. Hutchinson (1979), *A Guide to the Work-Relatedness of Disease*, National Institute for Occupational Safety and Health, Cincinnati, OH.

Lewis, Richard J., Sr. (1990), *Rapid Guide to Hazardous Chemicals in the Workplace*, 2d ed., Van Nostrand Reinhold, New York.

Lipton, Sidney, and Jeremiah Lynch (1987), *Health Hazard Control in the Chemical Process Industry*, John Wiley & Sons, New York.

NIOSH (1987), *NIOSH Manual of Analytical Methods*, DHHS (NIOSH) Publication No. 84-100, 3rd ed., edited by P.M. Eller for U.S. Department of Health and Human Services, National Institute of Occupational Safety and Health, Cincinnati, OH.

_____ (1987), "Pocket Guide to Chemical Hazards," DHHA (NIOSH) Publish No. 85-114, U.S. Department of Health and Human Services, National Institute for Occupational Safety and Health, Washington, DC.

_____ (1979), "Membrane Filter Method for Evaluation Airborne Asbestos Fibers," DHEW (NIOSH) Publication No. 79-127, prepared by N.A. Leidel et al., National Institute of Occupational Safety and Health, Department of Health Education and Welfare, Rockville, MD.

NIOSH, OSHA, and U.S. EPA (1985), "Occupational Safety and Health Guidance Manual for Hazardous Waste Site Activities," DHHS (NIOSH) Publication No. 85-115, National Institute of Occupational Safety and Health, Occupational Safety and Health Administration, U.S. Coast Guard, and U.S. Environmental Protection Agency, U.S. Department of Health and Human Service, U.S. Government Printing Office, Washington, DC.

NRC (1992), "Occupational Radiation Exposure at Commercial Nuclear Power Reactors and Other Facilities," 22nd Report, NUREG-0713-V11/XAB, Nuclear Regulatory Commission, Washington, DC.

Oak Ridge Associated Universities (1988), "A Compendium of Major U.S. Radiation Protection Standards and Guides: Legal and Technical Facts," prepared by W.A. Mills et al., Oak Ridge. TN.

OSHA (1991), *OSHA Analytical Methods Manual*, Part 2, "Inorganics," Occupational Safety and Health Administration, OSHA Analytical Laboratories, Salt Lake City, UT.

_____ (1990), *OSHA Analytical Methods Manual*, Part 1, "Organics," Occupational Safety and Health Administration, OSHA Analytical Laboratories, Salt Lake City, UT.

Proctor, Nick H., James P. Hughes, and Michael L. Fischman (1990), *Chemical Hazards of the Workplace*, 2d ed., Van Nostrand Reinhold, New York.

Reichert, Richard K. (1990), "Pitfalls and Protocols for Medical Surveillance of Hazardous Waste Workers," *Proceedings of the Third Annual Hazardous Materials Management Conference/Central*, Chicago, 119–129.

Sherman, Janette (1998), *Chemical Exposure and Disease: Diagnostic and Investigative Techniques*, Van Nostrand Reinhold, New York.

Technical Assessment Task Force (1990), *Reproductive Health Hazards in the Workplace*, Van Nostrand Reinhold, New York.

U.S. EPA (1990), *Compendium of Methods for the Determination of Toxic Organic Compounds in Indoor Air*, EPA/600/4-90/010, prepared by Engineering Science, Inc., for U.S. Environmental Protection Agency, Atmospheric Research and Exposure Assessment Laboratory, Office of Research and Development and Quality Assurance Division, Environmental Monitoring Systems Laboratory, Research Triangle Park, NC.

_____ (1989), "Fourier Transform Infrared Spectroscopy as a Continuous Monitoring Method: A Survey of Applications and Prospects," EPA/600/D-90/003, prepared by Entropy Environmentalist, Inc., for U.S. Environmental Protection Agency, Research Triangle Park, NC.

_____ (1984), "Biological Effects of Radiofrequency Radiation," EPA/600/8-83/026F, prepared by D.F. Cahill and J.A. Elder, eds., Health Effects Research Laboratory, Office of Research and Development, U.S. Environmental Protection Agency, Research Triangle Park, NC.

WOH (latest editions), *IARC Monographs on the Evaluation of the Carcinogenic Risk of Chemicals to Man*, World Health Organization Publication Center, Albany, NY.

Williams, Phillip L., and James L. Burson (1989), *Industrial Toxicology—Safety and Health Applications in the Workplace*, Van Nostrand Reinhold, New York.

16

PERSONAL PROTECTIVE EQUIPMENT

Personal protective equipment (PPE) for general industry is regulated under 29 CFR 1910 Subpart I, with specific aspects of PPE being addressed as follows:

- General requirements— §1910.132
- Eye and face protection— §1910.133
- Respiratory protection— §1910.134
- Head protection— §1910.135
- Foot protection— §1910.136
- Electrical protective equipment— §1910.137
- Hand protection— §1910.138

In addition, Appendix B to §1910.120 provides a general description and discussion of the levels of protection and protective gear for hazardous waste operations and emergency response. PPE for the construction industry, regulated under 29 CFR 1926, is discussed in Chapter 14 of this book.

GENERAL REQUIREMENTS

It is the employer's responsibility to provide PPE, as necessary, to protect employees from hazards in the workplace. Types of PPE can include protective equipment for eyes, face, head, and extremities; protective clothing, respiratory devices, and protective shields and barriers. Hazards that warrant such protective equipment include process hazards, chemical hazards, radiological hazards, and mechanical irritants that could cause injury through absorption, inhalation, or physical contact. Examples of PPE that might be used during a chemical spill cleanup operation (and an associated warning sign and spill diking equipment) as shown in Figure 16.1. Examples of a chemical-resistant glove, an armguard, and an apron that might be worn during chemical transfer or other chemical operations are shown in Figure 16.2.

If employees provide their own PPE, it is the employer's responsibility to ensure that it is adequate, properly maintained, and sanitary. In respect to PPE, the employer must also:

- Assess the workplace to determine whether hazards are present, or likely to be present, that necessitate the use of PPE and prepare written certification of the assessment
- Select the types of PPE that will protect the affected employees from identified hazards
- Communicate selection decisions to employees
- Ensure proper use of PPE by employees
- Ensure proper fit of equipment for each employee

Figure 16.1   Examples of PPE for spill cleanup.

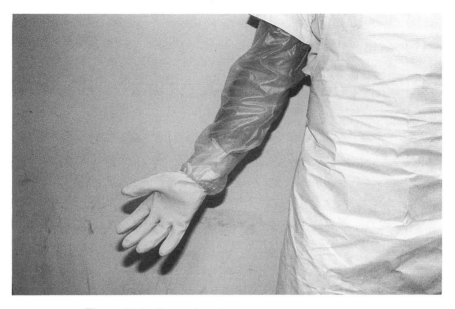

Figure 16.2 Examples of PPE for chemical activities.

- Provide training to each employee required to use PPE, including when PPE is necessary, what PPE is necessary, proper wearing of PPE, the limitations of PPE, the proper care and maintenance of PPE, and the useful life and disposal of PPE
- Ensure that employees demonstrate an understanding of the training before being allowed to perform work requiring the use of PPE
- Provide written certification of the employee's training

Hazard assessment is an important part of the PPE program and should be performed by a trained specialist. Examples of hazards that may be found in an industrial workplace are presented in Table 16.1.

EYE AND FACE PROTECTION

Eye and face protection is required when employees are exposed to potential hazards such as flying particles, molten metal, liquid chemicals, acid or caustic liquids, chemical gases or vapors, or potentially injurious light radiation. All eye and face devices must comply with ANSI Z87.1-1989, "American National Standard for Occupational and Educational Eye and Face Protection."

TABLE 16.1   Examples of Hazards That May Be Found in an Industrial Workplace

Basic Hazard Categories	Possible Sources of Hazards
Impact	Movement of machinery or process parts or elements; movement of personnel that could result in a collision with stationary objects
Penetration	Sharp objects such as cutting blades, saws, picks, and pointed protrusions that might pierce the feet or cut the hands
Compression (rollover)	Objects that might roll over and crush the feet, such as barrels or other containers; moving equipment that might become unbalanced and roll over; pinching objects and pinch points of processes
Chemical	Chemical handling operations, chemical processes, and chemical storage units
Heat	Sources of high temperatures such as boilers, steam distribution piping, ovens, heating coils, and thermal test chambers that could result in burns, eye injury, or ignition of protective equipment
Harmful dust	Operations such as grinding, milling, and other abrasive operations; asbestos removal projects; mining operations
Light radiation	Welding, cutting, and brazing operations; heat treating operations; equipment such as furnaces and high-intensity lights

Source: Information from 29 CFR Subpart I, Appendix B.

RESPIRATORY PROTECTION

OSHA Requirements

Whenever feasible, engineering controls must be used by employers to prevent atmospheric contamination from dusts, fogs, fumes, mists, gases, smokes, sprays, or vapors. When effective engineering controls are not feasible, employers are required to provide respirators when such equipment is needed to protect an employee's health. The respirators must be applicable and suitable for the purpose intended. Further, the employer must establish and maintain a respiratory protection program that includes elements defined in Table 16.2.

In addition to establishing and maintaining a respiratory protection program, the employer must ensure the safe and correct use of respirators. A qualified individual must specify the respirator type, ensure that procedures for normal operation and emergencies are established and maintained, and ensure that personnel are familiar with safe practices and respirator maintenance.

Testing for proper fit is required for users of negative-pressure respirators. Fit testing protocols can be either qualitative or quantitative. Qualitative fit test protocols include isoamyl acetate (banana oil) odor threshold screening; saccharin solution aerosol taste threshold screening; and the irritant fume protocol. Essentially, these protocols require the respirator user to wear the respirator for 5 to 10 minutes before beginning test exercises in a fit test chamber. Exercises include normal and deep breathing, turning the head from side to side, moving the head up and down, talking, grimacing, bending over, and others. Depending on the protocol used, if the respirator user smells banana oil, tastes saccharin, or senses the irritant fume, the respirator does not fit adequately to afford proper protection. Depending on the chemical exposure, a combination of qualitative fit tests may be required.

Quantitative fit tests are more precise in measuring the effectiveness of the respirator seal and are often conducted when respirators are used to protect against particularly hazardous chemicals that could be immediately dangerous to life or health (IDLH). Sodium chloride or another appropriate aerosol (termed the "challenge agent") is generated, diluted, and measured in the test chamber and inside the respirator face piece. A high-efficiency particulate filter is used during the test to exclude introduction of the challenge agent through the cartridge. A fit factor can be calculated simply by dividing the average concentration of the test chamber at the beginning and at the end of the test by the average concentration inside the respirator face piece. Both

TABLE 16.2 Minimum Requirements of a Respiratory Protection Program

- Establish written operating procedures for selection and use of respirators.
- Select respirators on the basis of hazards to which the worker is exposed.
- Instruct and train the user in the proper use of respirators and their limitations.
- Provide for regular cleaning and disinfecting of respirators; if respirators are used by more than one employee, provide for thorough cleaning and disinfecting after each use.
- Provide a storage location for respirators that is convenient, clean, and sanitary.
- Provide for the regular inspection of respirators during cleaning; replace any worn or deteriorated parts; provide for monthly inspections and for inspections after each use of respirators designated for emergency use, such as self-contained devices.
- Maintain appropriate surveillance of work area conditions and the degree of employee exposure or stress.
- Provide regular inspection and evaluation to determine the continued effectiveness of the program.
- Ensure that persons assigned to tasks requiring the use of respirators are determined to be physically able to perform the work and use the equipment.
- Select respirators from among those jointly approved by the Mine Safety and Health Administration (MSHA) and the National Institute for Occupational Safety and Health (NIOSH), as specified under the provisions of 30 CFR Part 11.

Source: 29 CFR 1910.134.

qualitative and quantitative fit tests should be conducted under the surveillance of a trained specialist. This person must know how to perform the tests properly, calibrate the needed equipment, and make the necessary calculations. In addition, fit test requirements are prescribed by OSHA for certain chemicals, so regulations should be reviewed thoroughly in establishing fit test protocols.

Proper maintenance of respirators includes inspection, cleaning, and proper storage. Respirators used routinely should be inspected, before and after each use, for items presented in Table 16.3.

Respirators that are used for emergency escape and other nonroutine use must be inspected after each use and at least monthly, and must be cleaned after each use. Other respirators must be cleaned and disinfected as frequently as necessary to ensure that proper protection is provided. Maintenance on respirators must be performed by an experienced person, with parts designed specifically for the particular respirator.

Finally, respirators should be stored so that any damage to them is minimized in the storage process. The storage area should be protected from dust, sunlight, excessive cold, heat, or moisture, and chemicals that might cause damage.

Respirator Types

Typical respiratory protective devices include the chemical cartridge respirator, particulate respirator, gas mask, supplied air respirator, and self-contained breathing apparatus (SCBA). The selection of a respirator is based on the chemicals used in the area, concentrations, expected exposure time, and oxygen present in the atmosphere.

Chemical Cartridge Respirators. Chemical cartridge respirators are air-purifying respirators and are used for respiratory protection against specific chemicals of relatively low concentrations. Because they have no air-supplying capability, the cartridges cannot be used in atmospheres containing less than 19.5% oxygen or in atmospheres that are IDLH. The respirators come in half-face (orinasal) or full facepiece models, with cartridge holders attached to the facepiece or belt. Typical mask materials include natural rubber, neoprene, silicon rubber, and thermoplastic elastomer. Powered models are also available for situations involving a high work crate. A powered air-purifying respirator uses a blower to force ambient air through the chemical cartridges.

Particulate Respirators. Particulate respirators are designed for protection against dusts, mists, fumes, and other particulates. Before a worker uses this type of respirator, the atmosphere should be tested; it must contain at least 19.5% oxygen and cannot be IDLH. Designs of particulate respirators include:

- Single use particulate respirators

TABLE 16.3   Respirator Elements to Inspect Before and After Each Use

Facepiece

- Examine for excessive dirt.
- Check for cracks, tears, and holes.
- Inspect for distortion caused by improper storage.
- Look for brittleness in material.
- Inspect for cracks or deep scratches on lens of full facepieces.
- Inspect for incorrectly mounted lenses on full facepieces.
- Examine for broken mounting clips.
- Check for broken canister holders.
- Look for badly worn threads.

Head Straps and Head Harness

- Examine for breaks and tears.
- Inspect for proper elasticity.
- Check for broken or malfunctioning buckles.
- Examine for excessive wear that might allow slippage.

Exhalation Valve

- Examine for cracks, tears, and holes in valve material.
- Inspect for proper valve seating.
- Look for any distortion in valve.
- Inspect for dirt.

Air-Purifying Apparatus

- Inspect for proper air-purifying cartridge or canister.
- Check shelf life of air-purifying cartridge or canister.
- Check for proper seating of cartridge or canister in holder.
- Inspect for dents and cracks in cartridge or canister.
- Inspect holder for cracks, chips, and stripped threads.

Breathing Tube

- Examine for cracks, tears, and holes.
- Inspect for damaged end connectors, gaskets, and o-rings.
- Check for material brittleness.
- Check for missing hose clamps.

Air Supply System

- Inspect for appropriate attachments and end fittings.
- Check operation of airflow regulators and valves.
- In self-contained breathing apparatus, check cylinder to determine whether it is fully charged.

- Particulate respirators for protection against dusts and mists
- Particulate respirators for protection against dust, fumes, and mists
- Particulate respirators with high-efficiency filters

The first three designs listed here are approved for use in atmospheres where the permissible exposure limit as a time-weighted average is equal to or greater than 0.05 mg/m^3 or 2 × 10^6 particles/ft^3. The high-efficiency particulate respirator has the greatest versatility and can be used for protection against dusts, mists, asbestos-containing dusts and mists (if approved by NIOSH/MSHA), and radionuclides. This type of respirator provides protection against dusts, fumes, and mists having permissible exposure limits as a time weighted average of less than 0.05 mg/m^3.

Gas Masks. Gas masks are similar to chemical cartridge respirators. They can be powered or nonpowered, but are all full-face masks that are to be used in an emergency. Like chemical cartridge respirators, they are not effective in atmospheres of less than 19.5% oxygen or in IDLH atmospheres (NIOSH 1990).

Gas masks afford protection against a specific chemical or chemical type such as ammonia, chlorine, acid gases, organic vapors, carbon monoxide, sulfur dioxide, and pesticides. Many models have approval for protection against additional vapors, gases, and dusts or mists. Gas mask canisters have been assigned colors for use against varying contaminants in order to make identification of the proper canister easy and consistent. The matrix of colors defined under 29 CFR 1910.134 is presented in Table 16.4.

Supplied Air Respirators. Supplied air respirators are also known as air-line respirators or atmosphere-supplying respirators. The air supply can be either continuous flow or pressure demand. These respirators are used for the same applications as those of cartridge respirators; similarly, they are not approved for use in IDLH atmospheres or atmospheres containing less than 19.5% oxygen. This limitation is based on safety considerations, since the air supply or air line could fail. However, there is an exception to this standard if an auxiliary tank of air, permitting escape, is incorporated into the respirator system (AIHA 1991). Moreover, when supplied air respirators are used in an IDLH atmosphere, standby personnel must be present with suitable rescue equipment. Approved types of supplied air respirators include continuous flow, pressure demand, and continuous flow abrasive blasting types.

In using a supplied air respirator, a line is attached to the user from a supply of respirable air. Full facepieces or hoods must be worn with these devices. Typical mask materials include natural rubber, neoprene, silicon rubber, thermoplastic, and polyvinyl chloride. Hose lengths and air supply pressure ranges specified by the manufacturer (and approved by NIOSH/MSHA) must be used. The maximum length of hose allowed is 300 feet, and the maximum inlet pressure is 125 pounds per square inch gauge (psig).

TABLE 16.4   Defined Canister Colors

Atmospheric Contaminants to Be Protected Against	Colors Assigned*
Acid gases	White
Hydrocyanic acid gas	White, with ½-inch green stripe completely around the canister near the bottom
Chlorine gas	White, with ½-inch yellow stripe completely around the canister near the bottom
Organic vapors	Black
Ammonia gas	Green
Acid gases and ammonia gas	Green with ½-inch white stripe completely around the canister near the bottom
Carbon monoxide	Blue
Acid gases and organic vapors	Yellow
Hydrocyanic acid gas and chloropicrin vapor	Yellow with ½-inch blue stripe completely around the canister near the bottom
Acid gases, organic vapors, and ammonia gases	Brown
Radioactive materials, excepting tritium and noble gases	Purple (Magenta)
Particulates (dusts, fumes, mists, fogs, or smokes) in any combination with any of the aforementioned gases or vapors	Canister color for contaminant, as designated previously, with ½-inch gray stripe completely around the canister near the top
All of the aforementioned atmospheric contaminants	Red, with ½-inch gray stripe completely around the canister near the top

*Gray shall not be assigned as the main color for a canister designed to remove acids or vapors.

Note: Orange shall be used as a complete body color, or stripe color, to represent gases not included in this table. The user must refer to the canister label to determine the degree of protection that canister will afford.

Source: 29 CFR 1910.134, Table 1-1.

SCBA. The SCBA is the only respirator approved by NIOSH/MSHA for use in oxygen-deficient and IDLH atmospheres. SCBAs are also worn in emergencies and when unknown atmospheres must be entered. Examples of SCBAs stored for emergency use are shown in Figure 16.3.

A widely used SCBA is the open-circuit pressure demand system. This apparatus maintains a positive pressure inside the facepiece at all times, which provides the highest level of respiratory protection. The system has a full facepiece and a facepiece- or belt-mounted regulator that regulates air flow from a bottle worn on the worker's back. Service life is generally 30 to 60 minutes.

Other types of SCBAs include the open-circuit demand system and the closed-circuit demand and pressure demand systems. The open-circuit de-

Figure 16.3 Examples of SCBAs.

mand system maintains a positive pressure during inhalation. Closed-circuit systems allow the worker's exhalation to be filtered (chemically scrubbed) and rebreathed, while providing supplemental oxygen from a supply source. This type of system has a longer service life but does not provide protection as adequately as the open-circuit systems.

Open-circuit pressure demand systems for "escape only" typically have a service life of 5 to 15 minutes. These devices have full facepieces or hoods, but do not have a large air bottle associated with them.

Closed-circuit "escape only" systems usually have only a mouthpiece, rather than a facepiece or hood. These systems have a service life of up to 60 minutes.

HEAD PROTECTION

Employees must wear protective helmets when working in areas where there is a potential for injury to the head from falling objects. Protective helmets designed to reduce electrical shock hazard must be worn by employees who work near exposed electrical conductors that could contact the head. Protective helmets must comply with ANSI Z89.1-1986, "American National Standard for Personnel Protection—Protective Headwear for Industrial Workers."

FOOT PROTECTION

Foot protection is required for employees who work in areas where there is a potential for foot injuries as a result of falling or rolling objects, or of objects piercing the sole, and where there is exposure of the employee's feet to electrical hazards. Protective footwear must comply with ANSI Z41-1991, "American National Standard for Personal Protection—Protective Footwear."

ELECTRICAL PROTECTIVE EQUIPMENT

Electrical protective equipment typically consists of rubber insulating gloves and sleeves, matting and blankets, covers, and line hoses. Rubber insulting equipment must be in compliance with American Society for Testing and Materials (ASTM) standards, including:

- ASTM D 120-87—Specification for Rubber Insulating Gloves
- ASTM D 178-93—Specification for Rubber Insulating Matting
- ASTM D 1048-93—Specification for Rubber Insulating Blankets
- ASTM D 1049-93—Specification for Rubber Insulating Covers
- ASTM D 1050-90—Specification for Rubber Insulating Line Hose
- ASTM D 1051-87—Specification for Rubber Insulating Sleeves

Like other protective equipment, electrical protective equipment must be inspected, cleaned, and properly stored and maintained. In addition, OSHA specifies AC and DC proof-test requirements for each class of equipment (Classes 0–4) and requires marking of the equipment according to its class.

HAND PROTECTION

Employers must select and require employees to use appropriate hand protection when there is potential exposure to hazards, including absorption of harmful substances, severe cuts or lacerations, severe abrasions, punctures, chemical burns, thermal burns, and harmful temperature extremes. A trained specialist should select the proper hand protection based on the performance characteristics of the hand protection relative to the tasks to be performed, conditions present, duration of use, and the hazards and potential hazards identified.

EYEWASHES AND EMERGENCY SHOWERS

ANSI Standard Z358.1 provides recommended practices for eyewashes and emergency showers. Eyewashes and emergency showers must be available in all chemical work areas and must be inspected periodically to ensure proper flow and, in the case of eyewashes, flow direction. In addition, the area around eyewashes and emergency showers must be kept clear for easy access. Easily readable signs that mark the location of the chemical safety devices should be posted.

DEFINED LEVELS OF PPE FOR EMERGENCY RESPONSE AND WASTE CLEANUP

Types of PPE to be worn during emergency response and waste cleanup are defined by OSHA under 29 CFR 1910.120, Appendix B. The Environmental Protection Agency (EPA) has also defined these same levels in guidance documents. The four recognized levels of PPE and their corresponding uses are presented in Table 16.5.

CHEMICAL PROTECTIVE CLOTHING

Selection of chemical protective clothing (CPC) is one of the more difficult tasks encountered by the safety or industrial hygiene specialist. Typically, protective clothing must afford protection against varying types of chemical hazards; a single material is not resistant to all chemicals. Thus, engineering judgments must be made as to the best materials for face shields, gloves, aprons, boots, and protective suits to be used during a specific work task.

The professional who selects CPC for a given task must consider multiple variables. These include potential and actual employee chemical exposures, chemical toxicity, and CPC performance information. Most manufacturers provide permeation data about chemical protective clothing materials. Permeation is the diffusion of a chemical through a material. The movement of a chemical through a material may not be noticeable inasmuch as it occurs on a molecular basis. Thus, the data provided give the safety specialist an indication of the expected life of the material, based on concentrations of chemicals encountered. In addition, protective clothing should be inspected often for signs of degradation or weakening of the material, as well as for tears, holes, and punctures.

Physical resistance factors that should also be considered in selecting chemical protective clothing include abrasion resistance, cut and puncture resistance, tensile strength, tear strength, flammability, and resistance to varying temperature extremes. These data may also be available from the manufacturer or from reference sources. Other physical properties to consider

TABLE 16.5   Levels of PPE and Their Corresponding Uses

Level A

- PPE includes a positive pressure SCBA, a fully encapsulating chemical-resistant suit, chemical-resistant inner and outer gloves, and chemical-resistant safety boots; a two-way communications device such as a two-way radio is also necessary; other optional clothing includes a hard hat, coveralls, long cotton underwear, and a cooling unit.
- Level A protection is worn when the highest level of protection is needed for respiratory, skin, and eye protection; applications include unknown atmospheres and known high-hazard atmospheres.

Level B

- PPE includes a positive-pressure SCBA, chemical-resistant clothing (hooded one- or two-piece chemical splash suit, disposable chemical-resistant coveralls, overalls and long-sleeved jacket, or equivalent clothing), inner and outer chemical-resistant gloves, chemical-resistant safety boots, and a two-way communication device; optional clothing could include a hard hat, long cotton underwear, or disposable boot covers.
- Level B protection is worn when the highest level of respiratory protection is needed, but a lesser level of skin and eye protection is required; applications include oxygen-deficient and IDLH atmospheres where substances do not represent a severe skin and eye hazard.

Level C

- PPE includes a full facepiece air-purifying respirator, chemical-resistant clothing, chemical-resistant inner and outer gloves, chemical-resistant safety boots, and a two-way communication device; optional clothing includes a hard hat, disposable boot covers, long cotton underwear, a faceshield, and an escape mask.
- Level C protection is worn when the atmosphere contaminants are known and the concentrations have been measured and found to meet the criteria for use of an air-purifying respirator, skin and eye exposure must be unlikely.

Level D

- PPE includes coveralls or other work uniform, safety glasses, and safety boots.
- Level D protection is worn when there is no respiratory hazard and when minimum skin and eye protection is required.

Source: 29 CFR 1910.120, Appendix B.

include the material and strength of closures, seam strength, material flexibility and weight, and ease of movement within the material.

Typical materials used for chemical protective clothing include butyl, polycarbonates, polyvinyl chloride, neoprene, chlorinated polyethylene, nitrile, and viton. General chemical compatibilities for these protective materials are presented in Table 16.6.

TABLE 16.6 Chemical Compatibilities for Selected Protective Materials

Butyl

- Compatible with moderate to strong acids, ammonia solutions, alcohols, inorganic salts, ketones, phenols, and aldehydes
- Incompatible with petroleum distillates, solvents, alkanes; limited use with esters and ethers

Polycarbonates

- Compatible with weak acids, ammonium solutions, alcohols, inorganic salts, phenols, some bases and aldehydes
- Incompatible with aggressive petroleum distillates; limited use with ketones

Polyvinyl Chloride

- Compatible with moderate to strong acids, bases, ammonium solutions, inorganic salts, petroleum distillates, alcohols, alkanes, and some aldehydes
- Incompatible with petroleum distillates, ketones, concentrated solvents, and phenols

Neoprene

- Compatible with moderate acids, bases, ammonium solutions, alcohols, inorganic salts, some solvents, some phenols, some aldehydes, and ethers
- Incompatible with petroleum distillates, esters; limited use with ketones

Chlorinated Polyethylene

- Compatible with moderate to strong acids, bases, ammonium solutions, some petroleum distillates, alcohols, inorganic salts, phenols, alkanes, and aldehydes
- Limited use with ketones and solvents

Nitrile

- Compatible with moderate to strong acids, bases, most ammonium solutions, some petroleum distillates, some solvents, and some alcohols
- Incompatible with ketones; limited use with inorganic salts and aldehydes

Viton

- Compatible with acids, bases, some ammonium solutions, petroleum distillates, alcohols, inorganic salts, most solvents, and phenols
- Incompatible with ketones; limited use with aldehydes

Note: The specific chemical or chemical mixture used in the workplace should be evaluated by a trained specialist for compatibility with any material used for protective clothing.
Source: Information from NFPA (1989) and Woodside (1993).

HEAT STRESS AS IT RELATES TO PPE

Heat-stress violations are enforceable (and fineable) under the general duty clause of 29 CFR 1910; thus, the wearing of PPE in hot and humid environments is a common concern among safety and health professionals, particularly for those workers wearing emergency response PPE of Levels A and B and fire-fighting ensembles. Examples of fire-fighters' bunker gear and boots are presented in Figures 16.4 and 16.5.

Manifestations of heat stress include:

- Behavioral changes
- Elevated core temperature of the body, termed hyperthermia
- Failure of the temperature regulatory mechanism
- Circulatory fatigue
- Depletion of water and body salts
- Inflammation of sweat glands

Types of heat illnesses that result from wearing PPE that does not allow proper skin ventilation and body temperature control include heat stroke, heat hyperpyrexia, fainting, heat exhaustion or prostration, heat cramps, heat rash, and heat fatigue.

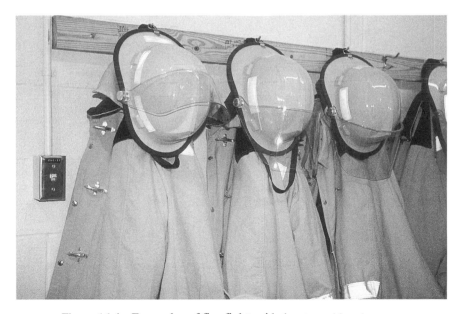

Figure 16.4   Examples of fire-fighters' helmets and bunker coats.

Figure 16.5   Examples of fire fighters' boots.

Factors that influence the bodys ability to tolerate heat include:

- Age—Older workers are more susceptible to heat-related illness than younger workers.
- Sex—Women are more susceptible to heat-related illness than men.
- General health—Workers with heart disease, high blood pressure, and other pulmonary and heart-related conditions are more susceptible to heat-related illness than workers in excellent health.
- Medication—Workers taking medications for high blood pressure or other conditions that cause loss of water are more susceptible to heat-related illness than those who have normal, established water loss patterns.
- Body fat—Workers who are obese are more susceptible to heat-related illnesses than workers who are within the normal weight range.

Controls for heat stress include the use of cool vests worn under PPE, cool-off rooms, limiting the time of exposure in PPE, ensuring adequate rest periods, ensuring adequate replacement of fluids, and other engineering and adminstrative control measures.

REFERENCES

American Conference of Governmental Industrial Hygienists, *Guidelines for Selection of Chemical Protective Clothing*, 3d ed., A.S. Schwope, P.P. Costas, J.O. Jackson, and D.J. Weitzman, eds., ACGIH, Cincinnati, OH, 1987.

American Industrial Hygiene Association, *Respiratory Protection: A Manual and Guideline*, 2d ed., Craig E. Colton, Lawrence R. Birkner, and Lisa M. Brosseau, eds., ACGIH, Akron, OH, 1991.

Forsberg, Krister, and S.Z. Mansdorf, *Quick Selection Guide to Chemical Protective Clothing*, 2d ed., Van Nostrand Reinhold, New York, 1993.

National Fire Protection Association, *Hazardous Materials Response Handbook*, Henry F. Martin, ed., Quincy, MA, 1989.

National Institute for Occupational Safety and Health, "Certified Equipment List," DHHS (NIOSH) Publication No. 90-102, U.S. Department of Health and Human Services, U.S. Government Printing Office, Washington, DC, 1990.

Woodside, Gayle, *Hazardous Materials and Hazardous Waste Management: A Technical Guide*, John Wiley & Sons, New York, 1993.

BIBLIOGRAPHY

ASTM (1989), *Chemical Protective Clothing Performance in Chemical Emergency Response*, Special Technical Publication 1037, Perkins and Stull, eds., American Society of Testing and Materials, Philadelphia.

DREO (1991), "Heat Stress Caused by Wearing Different Types of CW Protective Garments," DREO Technical Note 91-14, prepared by S.D. Livingstone and R.W. Nolan, Defence Research Establishment Ottawa, National Defence, Ottawa, ONT.

Naval Health Research Center (1995), "Cool Vests Worn Under Firefighting Ensemble Increases Tolerance to Heat," Report No. 94-6, prepared by R.D. Hagan, K.A. Huey, and B.L. Bennett, Naval Medical Research and Development Command, Bethesda, MD.

U.S. Army (1995), "General Procedure for Clothing Evaluations Relative to Heat Stress," Technical Note No. TN 95-5, prepared by Leslie Levine, Michael N. Sawke, and Richard R. Gonzalez, U.S. Army Research Institute of Environmental Medicine, Natick, MA.

U.S. EPA (1992), "Limited-Use Chemical Protective Clothing for EPA Superfund Activities," EPA/600/R-92/014, prepared by Arthur D. Little, Inc, for Risk Reduction Engineering Laboratory, Office of Research and Development, U.S. Environmental Protection Agency.

Waxman, Michael F. (1996), *Hazardous Waste Site Operations: A Training Manual for Site Professionals*, John Wiley & Sons, Inc. New York.

17

INDUSTRIAL VENTILATION

Air contaminants in industrial workplaces pose hazards, because they may be toxic, flammable, or irritating; they may produce unsafe (wet or oily) conditions; or they may be merely a nuisance. This chapter first describes the physical forms of different types of air contaminants, then discusses general ventilation and local exhaust systems. OSHA regulations regarding ventilation are found at 29 CFR 1910.94. These regulations cover abrasive blasting; grinding, polishing, and buffing operations; spray finishing operations; and open surface tanks.

AIR CONTAMINANTS

Air contaminants can be hazardous to health or life safety or may cause nuisance conditions. Air contaminants can be in the form of gas, vapor, dust, fume, smoke, or mist. These terms have distinct meanings in the environmental, health, and safety field. Definitions published by the American Society of Safety Engineers (1988) are presented here:

- Gas—A state of matter in which a material has a very low density and viscosity, can expand and contract greatly in response to changes in temperature and pressure, easily diffuses into other gases, and readily and uniformly distributes itself throughout any container. A gas can be changed to the liquid or solid state only by the combined effect of increased pressure and decreased temperature (below the critical temperature).

- Vapor—The gaseous phase of a substance that is a liquid at normal temperature and pressure.
- Dust—Suspended particles of solid matter (such as pollen or soot) in such a fine state of subdivision that they may be inhaled, swallowed, or absorbed into the body. Dusts do not diffuse in air, but settle under the influence of gravity. *Dust* is a descriptive term for airborne solid particles that range in size from 0.1 to 25 microns. Dusts above 5 microns in size do not usually remain airborne long enough to present an inhalation problem.
- Fume—A gaslike emanation containing minute solid particles arising from the heating of a solid body, such as lead, as distinct from a gas or vapor. This physical change is often accompanied by a chemical reaction, such as oxidation. Fumes flocculate and sometimes coalesce. Odorous gases and vapors should not be called fumes. As distinguished from dusts, fumes are finely divided solids produced by other methods of subdividing, such as chemical processing, combustion, explosion, or distillation. Some solids when heated to a liquid produce a vapor that, while it arises from the molten mass, immediately condenses to a solid without returning to its liquid state. Fumes are much finer than dusts, containing particles from 0.1 to 1 micron in size.
- Smoke—Carbon or soot particles less than 0.1 micron in size that result from the incomplete combustion of carbonaceous materials such as coal or oil; an air suspension (aerosol) of particles, often originating from combustion or sublimation.
- Mist—Fine liquid droplets or particles, measuring from 40 to 500 microns, suspended in or falling through the air, or a thin film of moisture condensed on a surface in droplets. Mist is generated by condensation from the gaseous to the liquid state, or by breaking up a liquid into a dispersed state by splashing, foaming, or atomizing. Examples in industrial operations are the oil mist produced during cutting and grinding, acid mists from electroplating, acid or alkali mists from pickling operations, and paint spray mists.

GENERAL VENTILATION

General ventilation, as opposed to local ventilation, involves the ventilation of relatively large spaces where air contaminants are diffuse or overall climate control is needed. As in pollution prevention, it is often best to minimize problems near their sources before contaminants become dispersed or mixed with other materials. Under certain conditions, however, general ventilation may be the better choice, such as when the source is relatively small and difficult to capture locally.

General ventilation can be divided into two broad categories, dilution ventilation and heat control ventilation. Dilution ventilation is used to bring in fresh air to keep the concentration of hazardous contaminants below threshold limit values (TLV) or to minimize nuisance odors. Heat control is needed in process areas generating large amounts of heat, such as at foundries.

The American Conference of Governmental Industrial Hygienists (ACGIH) suggests the following principles for dilution ventilation design (1992):

- Determine the amount of air needed to provide enough dilution to maintain the contaminant below its TLV. Air volumes needed for many common chemicals and solvents have been compiled by ACGIH, but may also be determined independently based on the TLV and evaporation rate for a particular material.
- Increase the quantity of dilution air to account for incomplete mixing of the air. This multiplying factor, usually called K, typically ranges from 1 to 10.
- Choose locations for the air supply and exhaust outlets so that the air passes through the contaminated zone and does not transport contaminants into the breathing zone of workers.
- Replace exhausted air by using a replacement air system.
- Avoid reintroducing exhausted air by discharging the exhaust high above the roof line or by making sure that no window, outside air intakes, or other building openings are located near the exhaust outlet.

Many chemicals pose both health and fire or explosion hazards. When workers are involved, the TLV is the controlling factor because it is lower than the concentration needed to protect against fires or explosions.

To calculate the required dilution ventilation rate for steady state conditions, the following equation is used:

$$Q = \frac{(403)(10^6)(SG)(ER)(K)}{MW(C)} \tag{17.1}$$

where Q is the ventilation rate in cubic feet per minute (cfm), SG is the specific gravity of the contaminant as a liquid, ER is the evaporation rate of the liquid in pints per minute, MW is the molecular weight of the liquid, K is the adjustment factor for incomplete mixing, and C is the concentration of the gas or vapor, usually the TLV, expressed in parts per million (ppm).

EXAMPLE 17.1
Dilution Ventilation Rate

Calculate the fresh air flow rate needed at steady state to maintain the concentration of isopropyl alcohol below its TLV of 400 ppm if the evaporation rate of the alcohol is 1 pint per hour. To adjust for incomplete mixing, use a value of 5 for K. The molecular weight of isopropyl alcohol is 60.09, and the specific gravity is 0.785.

Solution. Use Equation 17.1 to calculate the air flow rate.

$$Q = \frac{(403)(10^6)(0.785)(1/60)(5)}{60.09(400)}$$

$$= 1,097 \text{ cfm}$$

When dilution ventilation must be provided for a mixture of chemicals, Equation 17.1 is used in a slightly different manner to calculate the air flow rate. If the chemicals in the mixture have the same effects on the human body, their effects are considered additive and the air flow rate is the sum of the individual flow rates required for each chemical. For example, if the required air flow rate for one chemical is 1,000 cfm and the other 2,000 cfm, the air flow rate for the mixture of the two chemicals would be 3,000 cfm. If the chemicals in the mixture are expected to have different health effects, they are considered to act independently and the required air flow is the highest individual air flow. Thus, if one chemical requires 1,500 cfm and the other 1,200 cfm, the required air flow would be 1,500 cfm.

LOCAL EXHAUST SYSTEMS

Local exhaust systems are used to capture air contaminants at the source so that they do not become widely dispersed into the surrounding air space. Local exhaust systems are used more often than general ventilation in industrial workplaces for many reasons. One of the major reasons is that many chemicals and compounds pose fire hazards or are toxic and, consequently, are regulated very stringently. It would be difficult to meet these strict regulations with only general ventilation. Other situations in which a local exhaust system would be the method of choice are those in which the contaminant source is near an employee work area or where the source or sources vary widely in time, location, or size.

Components of a local exhaust system typically include hoods, ducts, fans, and air cleaners. Hoods are used to initially capture the air contaminants at the source. Fans set up the necessary air flow that draw the contaminants into the hood and transport them through the duct work to air cleaners. Air cleaners remove, treat, or destroy contaminants before the air stream is released. Each of these components is discussed in the following sections.

Hoods

The shape, size, and placement of an air exhaust hood are very important design elements in a local exhaust system. Several commonly used hood types are presented in Figure 17.1, and applications for these types are summarized in Table 17.1. Hoods may be placed above the source of air contaminants, such as with a canopy hood; to the side, such as with a side-open hood or side-slotted hood; or below, such as with a grilled or slotted downdraft hood. Hoods can also be designed to enclose the contaminant source either partially or totally, as in laboratory exhaust hoods and paint spray booths.

Because air velocities decrease rapidly away from the hood, it is important that velocities in the capture zone be powerful enough to pull air contaminants through the hood. Guidelines for capture velocities suggested by the ACGIH for various work conditions are shown in Table 17.2. Velocities in the lower end of the ranges shown are used when air currents in the work area are minimal or help to move contaminants into the hood, when contaminants have low toxicity or present only a nuisance, when production is light and intermittent, or when the air mass moving through the hood is large. Higher capture velocities are used when air currents in the work area can disrupt the flow into the hood, contaminants are highly toxic, production is heavy and continuous, or small local hoods are used.

Capture velocities around a hood will vary with distance from the hood and the shape of the hood. For unflanged hoods having round or essentially square cross sections, the centerline velocity in feet per minute is calculated by the following equation, where Q is the air flow through the hood in cubic feet per minute (cfm), x is the distance in feet from the face of the hood along the centerline, and A is the hood cross-sectional area in square feet:

$$V = \frac{Q}{10x^2 + A} \qquad (17.2)$$

The preceding equation is valid only when x is within 1.5 times the hood diameter (round hoods) or side length of the hood (square hoods).

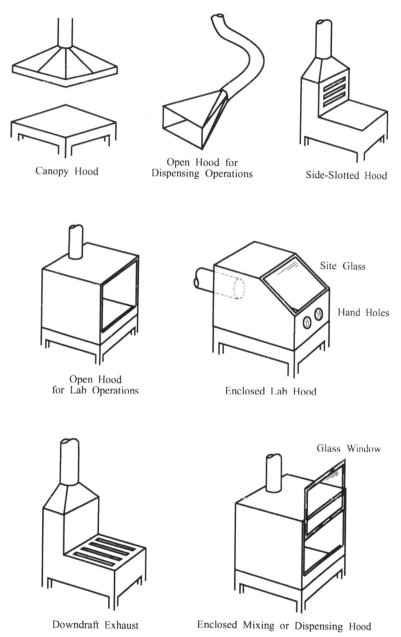

Canopy Hood

Open Hood for
Dispensing Operations

Side-Slotted Hood

Open Hood
for Lab Operations

Enclosed Lab Hood

Site Glass

Hand Holes

Downdraft Exhaust

Enclosed Mixing or Dispensing Hood

Glass Window

Figure 17.1 Selected examples of hood types.

TABLE 17.1   Examples of Selected Hood Types and Industrial Applications

Hood Type	Typical Industrial Application
Canopy hood	Automated plating operations, degreasing operations, and hot processes
Open hood	Low- to medium-toxicity chemical dispensing operations, flammable chemical dispensing operations, lab operations
Slotted hood (side mounted)	Floor applications for operations involving fumes heavier than air, low-hazard bench operations
Slotted or open hood (downdraft)	Felting or brushing operations, grinding operations, low-level dust-producing operations
Enclosed hood or booth	Highly toxic chemical dispensing, lab operations involving infectious materials, radioactive material applications, extremely dusty operations, furnace or oven applications, spraying operations, high-agitation mixing operations

Note: Some applications may require respiratory protection.

TABLE 17.2   ACGIH Guidelines for Hood Capture Velocities

Dispersal Conditions	Capture Velocity, fpm	Example Applications
Contaminants released with practically no velocity into quiet air	50–100	Evaporation from tanks and open vessels
Contaminants released at low velocity into moderately still air	100–200	Spray booths, intermittent container filling, low-speed conveyor transfers, welding, plating, pickling
Contaminants released at considerable velocity or into zone of rapid air movement	200–500	Spray painting in shallow booths, barrel filling, conveyor loading, crushers
Contaminants released at high initial velocity into zone of very rapid air movement	500–2,000	Grinding, abrasive blasting, tumbling

Source: Information from ACGIH (1992).

EXAMPLE 17.2
Unflanged Hood Capture Velocity

A 6-inch diameter unflanged hood has a centerline velocity at the face of the hood of 4,000 feet per minute (fpm). At what distance will the centerline velocity have decreased to half its face value? What is the centerline velocity at a distance 1.5 diameters from the face of the hood?

Solution. First calculate the cross-sectional area of the hood and the air flow into the hood.

$$A = \left(\frac{3}{12} \text{ft}\right)^2 \pi$$

$$= 0.1963 \text{ ft}^2$$

$$Q = VA$$

$$= 4,000 \frac{\text{ft}}{\text{min}} \times 0.1963 \text{ ft}^2$$

$$= 785.2 \text{ cfm}$$

If the centerline velocity has decreased to 2,000 fpm (half its face value), the distance from the hood face at which this reduction occurs is:

$$2,000 \frac{\text{ft}}{\text{min}} = \frac{785.2 \text{ ft}^3/\text{min}}{10x^2 + 0.1963 \text{ ft}^2}$$

$$x^2 = \frac{\left(\dfrac{785.2}{2,000} - 0.1963\right)}{10} \text{ft}^2$$

$$x = 0.14 \text{ ft}$$

$$= 1.7 \text{ in}$$

At a distance equal to 1.5 times the hood diameter (the limit of Equation 17.2), the centerline velocity decreases to:

$$V = \frac{785.2 \text{ ft}^3/\text{min}}{10(1.5 \times 0.5 \text{ ft})^2 + 0.1963 \text{ ft}^2}$$

$$= 135 \text{ fpm}$$

So, at a distance of less than 9 inches from the face of the hood, the centerline velocity has decreased to less than 4% ($135 \div 4,000$) of its maximum at the face.

Hoods can be designed with flanges around their faces to prevent the collection of uncontaminated air from behind the hood. As a result, the capture velocities in front of the hood increase and thus are more effective in removing contaminants. Alternatively, flanges can be used to decrease the flow rate through the hood needed to achieve a desired capture velocity. For a flanged hood, Equation 17.2 is modified as shown here (ACGIH 1992):

$$V = \frac{Q}{0.75(10x^2 + A)} \tag{17.3}$$

EXAMPLE 17.3
Flanged Hood Capture Velocity

Assume that the hood in Example 17.2 has a flanged design. At what distance from the face will the centerline velocity be half of its maximum value?

Solution. Using Equation 17.3:

$$2,000\frac{\text{ft}}{\text{min}} = \frac{785.2 \ \text{ft}^3/\text{min}}{0.75(10x^2 + 0.1963 \ \text{ft}^2)}$$

$$x^2 = \frac{\dfrac{785.2}{2,000} - 0.1963}{0.75(10)} \ \text{ft}^2$$

$$x = 0.162 \ \text{ft}$$

$$= 1.9 \ \text{in}$$

Velocity equations for various hood types, including the round and square hoods just described, are summarized in Table 17.3.

Besides capture velocity, there are other factors to consider in hood design. Capture velocities drop off quickly away from the hood, so it is important to place the hood as close as possible to the source of air contaminants. However, the hood must be placed so that the air movement does not carry the contaminants into the breathing zone of employees. Likewise, vents must not be exhausted into work areas. Hoods should not interfere with the way the worker performs the job, because the worker may modify or remove the exhaust equipment to make the work easier.

Ducts

A properly designed duct system is one that has enough air flow to pull contaminants through exhaust hoods and keep solid particulates from settling

TABLE 17.3   Centerline Velocities for Various Hood Types

Hood Type	Ratio of Width to Length of Hood Opening (W/L)	Centerline Velocity, V (fpm)
Unflanged		
Simple	$\dfrac{W}{L} \geqslant 0.2$ or round opening	$V = \dfrac{Q}{(10x^2 + A)}$ Q = air flow through hood face (cfm) x = distance in front of hood along hood centerline A = cross-sectional area of hood opening (ft^2)
Slot, single	$\dfrac{W}{L} \leqslant 0.2$	$V = \dfrac{Q}{(3.7Lx)}$
Slots, multiple	$\dfrac{W}{L} \geqslant 0.2$	$V = \dfrac{Q}{(10x^2 + A)}$
Flanged		
Simple	$\dfrac{W}{L} \geqslant 0.2$ or round opening	$V = \dfrac{Q}{[0.75(10x^2 + A)]}$
Slot, single	$\dfrac{W}{L} \leqslant 0.2$	$V = \dfrac{Q}{(2.6Lx)}$
Slots, multiple	$\dfrac{W}{L} \geqslant 0.2$	$V = \dfrac{Q}{[0.75(10x^2 + A)]}$
Canopy		
	To suit work	$V = \dfrac{Q}{(1.4PD)}$ P = perimeter of canopy opening (ft) D = height above work surface (ft)
Booth		
	To suit work	$V = \dfrac{Q}{A}$ A = flow cross-sectional area

Note: For a multiple slot opening, the width, W, is the entire width of the slotted area, including the spaces between slots.

Source: Information from ACGIH (1992).

out. Ducts should be constructed of materials that are sufficiently durable for handling a particular air contaminant, such as a solid abrasive material or corrosive chemical. Both the shape of the duct cross-section and the interior finish are important in minimizing friction losses.

For noncorrosive air contaminants, there are four classes of exhaust systems, as shown in Table 17.4 (ACGIH 1992). Industrial duct systems are usually constructed of black iron or galvanized steel. Galvanized steel is not recommended where temperatures exceed 400°F. Corrosive contaminants require the use of coatings or special plastics, nonmetallic materials, or alloys.

To minimize frictional losses, round duct is preferable to rectangular duct because, for a given cross-sectional area, the surface area of a cylinder is smaller. If round duct cannot be used, then rectangular duct should be as nearly square as possible, again, to minimize surface area and frictional losses.

Different air contaminants require different minimum velocities to transport them through a duct without settling or plugging. Minimum duct velocities recommended by the ACGIH for various applications are shown in Table 17.5.

Besides the straight piping sections of a duct system, other components include elbows, tapers, tees or branch connectors, dampers, equalizers, blast gates, and cleanouts. Elbows change the direction of flow in a duct. Consequently, they are subject to greater wear and abrasion and must be made of thicker-gauge material than the straight sections. Tapers are transitions between piping of different cross sections. Tee or branch connectors are used to merge a line of ducting into another. Dampers are used to shut off the air flow in a duct to prevent fires from spreading to other areas through the ducting sys-

TABLE 17.4   Exhaust System Classes for Noncorrosive Applications

Class		Description
1	Light Duty	Nonabrasive applications, as with replacement air, general ventilation, and gaseous emissions control
2	Medium Duty	Moderately abrasive particulate in light concentration, as with buffing and polishing, woodworking, and grain dust
3	Heavy Duty	Highly abrasive particulate in low concentration, as with abrasive cleaning operations, dryers and kilns, boiler breeching, and sand handling
4	Extra Heavy Duty	Highly abrasive particulate in high concentration, as with materials conveying high concentrations of particulates in all examples listed under Class 3 (usually used in heavy industrial plants such as steel mills, foundries, mining, and smelting)

Source: Information from ACGIH (1992).

tem and to protect against back pressure and contaminant loss from air cleaners. Equalizers are permanent dampers welded in place to balance the air flow throughout the system. Blast gates are like equalizers, but they are adjustable. Blast gates are used to adjust the air flow after the ducting system is installed. Because they are adjustable, the system can be fine-tuned to obtain necessary air flows in all parts of the system. However, they are not the preferred method for design for a number of reasons, one of which is that they *are adjustable* and workers may change them indiscriminately and disrupt flows in other areas of the system. Cleanouts may be placed at regular intervals in ducts that carry dusts in order to remove accumulated material.

TABLE 17.5 Minimum Duct Design Velocities

Type of Contaminant	Design Velocity (fpm)	Example Applications
Vapors, gases, smoke	Any desired velocity 1,000–2,000 fpm is usually most economic	All vapors, gases, and smoke
Fumes	2,000–2,500	Welding
Very fine light dust	2,500–3,000	Cotton lint, wood flour, litho powder
Dry dusts and powders	3,000–4,000	Fine rubber dust, jute lint, cotton dust, light shavings, soap dust, leather shavings
Average industrial dust	3,500–4,000	Grinding dust, dry buffing lint, wool jute dust (shaker waste), coffee beans, shoe dust, granite dust, silica flour, brick cutting, clay dust, limestone dust
Heavy dust	4,000–4,500	Heavy and wet sawdust, metal turnings, foundry tumbling barrels and shake-out, sandblast dust, wood blocks, hog waste, brass turnings, cast iron boring dust, and lead dust
Heavy or moist materials	≥4,500	Lead dusts with small chips, moist cement dust, sticky buffing lint, quicklime dust

Source: Information from ACGIH (1992).

Air Flow and Pressure Losses

A key element of exhaust system design is balancing the air flow throughout the system so that every hood has the necessary air flow to remove contaminants. The basic steps in this procedure, outlined in the ACGIH's *Industrial Ventilation* manual (ACGIH 1992), are summarized as follows:

- Determine the type, size, and design flow rate of each exhaust hood in the system, according to the operation or contaminant source being handled.
- Determine the minimum duct velocity needed for transport of the contaminants.
- Calculate the size of each branch duct by dividing the design flow rate by the minimum duct velocity. If particulate matter is being transported, choose a commercially available duct size with a smaller area to ensure that the actual duct velocity is greater than the required minimum.
- Determine the design length of each segment and the number and type of fittings and elbows.
- Calculate the pressure losses for the system resulting from friction and fittings.
- Check for flow balance at entries. Adjust volumetric flow rate, duct size, or hood design to balance the system.
- Select the collector and fan based on the final flow rate and calculated system resistance.

Pressure losses can be calculated by either the velocity pressure method or the equivalent foot method. Both methods are described in the ACGIH manual. The velocity pressure method is often preferred because it is generally quicker, all pressure losses are on the same basis, and recalculation of branch duct size is quicker for a balanced duct design. The following paragraphs discuss the velocity pressure method, giving a simple example problem to illustrate some of the basic calculations.

There are three basic pressure terms used extensively in the design of a ventilation system: static pressure, velocity pressure, and total pressure (measured in inches of water gage, or wg). These terms are defined as follows:

- Static pressure (SP)—The pressure in a duct, measured perpendicular to the flow. Static pressure can be either positive (blowing) or negative (sucking) pressure, in respect to local atmospheric pressure. Static pressure represents potential energy in the system and may be converted to kinetic energy (velocity pressure) and heat resulting from friction and other losses.
- Velocity pressure (VP)—The pressure required to accelerate air from zero velocity to some velocity (V). Velocity pressure is always positive and is exerted in the direction of air flow.

- Total pressure (TP)—The sum of static and velocity pressures. Total pressure can be either positive or negative. Total pressure is negative only when static pressure is both negative and larger than velocity pressure because velocity pressure, is always positive.

Thus, the following equation relates the three pressure terms:

$$TP = SP + VP \qquad (17.4)$$

Air flow in an exhaust system is governed by conservation of mass and energy. Conservation of mass means that mass flow rates must balance. Conservation of energy means that all energy changes must balance. Energy losses resulting from friction and other losses are also part of the energy balance. Energy losses that result as the air flows from one point in the system (A) to another (B) are represented by h_L in the following equation:

$$TP_A = TP_B + h_L \qquad (17.5)$$

This equation can also be written in terms of static and velocity pressures, using the relationship in Equation 17.4:

$$SP_A + VP_A = SP_B + VP_B + h_L \qquad (17.6)$$

Energy losses can occur at entry into the hood, along the duct, and at elbows and fittings.

Another equation used in air flow calculations relates the air velocity (V) to the velocity pressure. For standard air with a density of 0.075 pounds mass per cubic foot:

$$V(\text{fpm}) = 4{,}005 \sqrt{VP} \qquad (17.7)$$

The velocity pressure in Equation 17.7 is measured in inches of water.

To obtain the right amount of air flow through each hood and duct, it is important to balance the resistance or static pressure in each duct branch. In designing a branched duct system such as that shown in Figure 17.2, the static pressure is first calculated for each branch based on the minimum duct velocity needed, expected duct size (cross section and length), elbows, and fittings. Then, at the point where each branch joins the other, the static pressures are compared. If they are not reasonably close, either the air flow or the duct size of the branches is changed to balance the pressure. General rules used in making these adjustments are as follows:

- If the larger static pressure, SP_L, is no more than 5% greater than the smaller static pressure, SP_S, the system is reasonably balanced and no changes are made. When static pressure is negative, the absolute values

Figure 17.2   Branched duct system in industrial setting.

of the static pressures are used for this comparison. For example, if two static pressures are −6 in wg and −4 in wg, SP_L is −6 in wg.

- If the difference is between 5% and 20%, the air flow for the duct with the smaller static pressure, Q_S, is recalculated by the following equation:

$$Q_R = Q_S \sqrt{\frac{SP_L}{SP_S}} \qquad (17.8)$$

where Q_R is the revised air flow.

- If the difference is greater than 20%, the duct size of the branch with the smaller static pressure is decreased.

The following example is an illustration of a branched duct system showing how the preceding equations in this section are used. There are many other considerations in the design of an exhaust system that are not included here, and the reader is referred to the ACGIH's *Industrial Ventilation* manual for additional detail.

EXAMPLE 17.4
Simple Branched Duct System

Figure 17.3 shows two branches of a duct system for a woodshop. One branch handles the exhaust for a table saw and the other for a planer. Check the system for balance and make adjustments as necessary. What will be the combined air flow in the duct downstream of the connection?

Solution. Calculate the cross section area (A), flow rate (Q), velocity pressure (VP), and initial static pressure (SP) for each branch. For branch A:

$$A_A = \left(\frac{3}{12}\,\text{ft}\right)^2 \pi = 0.1963 \text{ ft}^2$$

$$Q_A = 0.1963 \text{ ft}^2 \left(4000\frac{\text{ft}}{\text{min}}\right) = 785.2 \text{ cfm}$$

$$VP_A = \left(\frac{4000}{4005}\right)^2 - 0.9975 \text{ in wg}$$

The initial static pressure is equal to the negative of the velocity pressure because total pressure is zero outside the hood (atmospheric pressure) (refer to Equation 17.4). Therefore, the initial static pressure of branch A is:

$$SP_A = 0.9975 \text{ in wg}$$

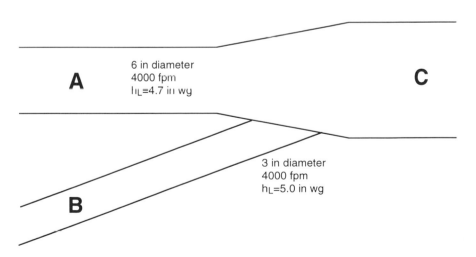

Figure 17.3 Branched duct system in Example 17.4.

For branch B, the calculations are as follows:

$$A_B = \left(\frac{1.5}{12} \text{ft}\right)^2 \pi = 0.04909 \text{ ft}^2$$

$$Q_B = 0.04909 \text{ ft}^2 \left(4000 \frac{\text{ft}}{\text{min}}\right) = 196.3 \text{ cfm}$$

$$VP_B = \left(\frac{4000}{4005}\right)^2 = 0.9975 \text{ in wg}$$

$$SP_B = -VP_B = -0.9975 \text{ in wg}$$

The static pressure at the point where the branches join is the initial pressure plus head or energy losses. The static pressures for the two branches are therefore:

$$SP_A = -0.9975 + -4.7 = -5.6975 \text{ in wg}$$

$$SP_B = -0.9975 + -5.0 = -5.9975 \text{ in wg}$$

Checking the static pressures for balance:

$$\frac{SP_B}{SP_A} = \frac{5.9975}{5.6975} = 1.053$$

Because the difference is between 5% and 20%, keep the duct sizes the same, but recalculate a new air flow rate for the duct with the smaller static pressure.

$$Q_A = 785.2 \text{ cfm} \sqrt{\frac{5.9975}{5.6975}} = 806 \text{ cfm}$$

The total air flow in duct C is the sum of the air flows in the two branches:

$$Q_C = 196 \text{ cfm} + 806 \text{ cfm} = 1002 \text{ cfm}$$

Fans

Fans are used to move air in an exhaust system. There are two major categories of fans, axial flow and centrifugal. A summary of various types of fans in these two categories is provided in Table 17.6, including a description of each fan's basic design and general applications.

Axial fans move air in the same direction as it enters the fan, in the direction of the axis of rotation. Centrifugal fans move air perpendicular to the direction it enters the fan, or perpendicular to the axis of rotation. In general, axial fans are used for low-pressure, high-volume air in general ventilation and heating, ventilation, and air-conditioning (HVAC) systems. Centrifugal fans are generally used for higher-pressure, lower-volume systems. The centrifugal, radial blade fan is commonly used in industrial local exhaust systems because of its high mechanical strength and resistance to abrasion. An example of an

TABLE 17.6   Types of Fans Used in Ventilation Systems

Type of Fan	Basic Design	General Applications
Axial (flow in the direction of the axis of rotation)		
Propeller	Disc blade (no duct); narrow or propeller blade; 2 or more blades	Low pressure, high volume; general ventilation; ventilation through wall without duct; replacement air
Tube-axial	Mounted inside duct; narrow or propeller blade; 4 to 8 blades	Low- to medium-pressure system with ducts; heating, ventilation, and air-conditioning; drying ovens, paint spray booths, fume exhaust
Vane-axial	Propeller within a cylinder; short vanes, large hub; guide vanes straighten flow	Low-, medium-, and high-pressure systems; heating, ventilation, and air-conditioning; clean air
Power exhausters	Varied propeller design	Low pressure, high volume; roof ventilation for general factory, kitchen, and warehouse environments
Centrifugal (flow perpendicular to the axis of rotation)		
Forward curved (squirrel cage)	Blades curved toward direction of rotation; 24 to 64 shallow blades	Low (primarily) to medium pressure; heating, ventilation, and air-conditioning; domestic furnaces, central station units, packaged air-conditioning equipment from room to rooftop units, clean air or after air cleaner
Radial	Blades straight or in radial direction from hub; 6 to 10 blades	High-pressure industrial systems; material handling
Backward inclined Backward curved	Blades inclined opposite to direction of rotation; 9 to 16 blades	Heating, ventilation, and air-conditioning; clean air industrial systems; corrosive or abrasive materials
Airfoil	Blades inclined opposite to direction of rotation; 9 to 16 blades	Heating, ventilation, and air-conditioning; clean air industrial systems (large size for power savings)

(Continued)

TABLE 17.6 *Continued*

Type of Fan	Basic Design	General Applications
Tubular	Similar to airfoil backward inclined or backward curved blade; mixed flow impellers	Low pressure return air in heating, ventilation, and air-conditioning
Power exhausters	Varied propeller design; airfoil or backward inclined impeller	Low pressure, high volume; roof ventilation for general factory, kitchen, and warehouse environments

Source. Information from ACGIH (1992).

industrial size centrifugal fan is shown in Figure 17.4; the air inlet faces forward and the discharge outlet is at the top. Examples of centrifugal fans with attached ducting are shown in Figures 17.5 and 17.6.

An exhaust fan is selected based on a number of factors, including capacity, type of exhaust stream, physical limitations, motor drives, safety features, and accessories (ACGIH 1992). Capacity factors include required air flow rate and pressure. Considerations for the type of airstream include what type of material will be handled *through* the fan (smoke, dust, fume, moisture, particulates) and whether the material is explosive, flammable, corrosive, or high temperature. Accessory equipment that should be considered include drains, cleanout doors, split housing, and shaft seals.

Air Cleaners

Air cleaners are used to remove contaminants from an exhaust stream. The type of air cleaner is selected based on the type and quantity of contaminant (dust, mist, fume, gas, vapor), required cleaning efficiency, and other characteristics of the exhaust stream.

Dust and Particle Collectors. Air-cleaning devices for particulate or dust removal are either air filters or dust collectors. Air filters are used to remove low concentrations of dust in air. Typical applications are heating, ventilation, and air-conditioning systems. Dust collectors are classified in four major groups: electrostatic precipitators, fabric collectors, wet collectors, and dry centrifugal collectors.

In an electrostatic precipitator, the gas in an exhaust stream is ionized by passing through an electric field. Particulates in the exhaust become charged when the gas ions collide with and attach to them. These charged particles move toward an oppositely charged collector plate where they collect, lose their charge, and are then removed. Low-voltage precipitators are often used

Figure 17.4 Centrifugal fan.

Figure 17.5 Local exhaust system with centrifugal fan.

Figure 17.6   Local exhaust system with centrifugal fan.

with roasters, kilns, and coolers in chemical manufacturing. High-voltage precipitators are often used with fly ash and with dryers and kilns in metal mining and rock operations.

Fabric collectors work by filtering the particulates through "cloth." Filtering may be accomplished by straining, impingement, interception, diffusion, or electrostatic charge. Fabrics may be woven or nonwoven and made of natural materials such as cotton, wool, or synthetic materials such as Dacron or Orlon. The fabric collector may take the form of bags or tubes, envelopes or flat bags, or pleated cartridges. Fabric filters are "conditioned" by the residual dust cake that remains after cleaning. This residual acts as a filter aid and increases removal efficiency. Fabric collectors are frequently used in many different industries.

Wet collectors use water to trap particulates. Types of collectors include scrubbers (chamber or spray tower and packed tower), wet centrifugal collectors, wet dynamic precipitators, orifice-type collectors, and venturi collectors.

Dry centrifugal collectors include gravity separators, inertial separators, dynamic precipitators, cyclone collectors, and high-efficiency centrifugal collectors. Gravity separators are limited to removal of extremely coarse, heavy particulates. Cyclones are commonly used to remove coarse particulates, serving as precleaners to more efficient air-cleaning devices. Centrifugal collectors, even high-efficiency designs, are not as efficient as other types (electrostatic, fabric, wet collectors) for small-size particles.

Gaseous Contaminants. Air-cleaning devices for gaseous contaminants are used to remove and/or treat the contaminants. Water, chemical solutions, activated carbon, and molecular sieves are typically used to remove contaminants by physical/chemical means. Thermal oxidizers, direct combustors, and catalytic oxidizers are commonly used to treat or destroy combustible gaseous contaminants.

REFERENCES

American Conference of Governmental Industrial Hygienists, *Industrial Ventilation: A Manual of Recommended Practice*, ACGIH, Cincinnati, OH, 1992.

American Society of Safety Engineers, *Dictionary of Terms Used in the Safety Profession*, ASSE, Des Plaines, IL, 1988.

BIBLIOGRAPHY

Alden, John L., and John M. Kane (1982), *Design of Industrial Ventilation Systems*, Industrial Press, New York.

Dinardi, Salvatore R. (1995), *Calculation Methods for Industrial Hygiene*, Van Nostrand Reinhold, New York.

Fundamentals of Industrial Hygiene, NSC (1988), Chicago.

Talty, John T., ed. (1988), *Industrial Hygiene Engineering: Recognition, Measurement, Evaluation and Control*, 2d ed., Noyes Data Corporation, Park Ridge, NJ.

18

RADIATION SAFETY

As a result of the proliferation of electronic devices, there has been increased public attention on the effects of ionizing and nonionizing radiation. The Radiation Control for Health and Safety Act of 1968 covers both ionizing and nonionizing radiations emitted from any electrical product. The Occupational Safety and Health Administrations requirements pertaining to radiation are promulgated under 29 CFR 1910.96 and 1910.97 for ionizing and nonionizing radiation, respectively. In addition, radioactive materials and wastes are regulated by the Nuclear Regulatory Commission and the U.S. Environmental Protection Agency.

Nonionizing radiation is radiation within the electromagnetic spectrum, which includes ultraviolet and infrared radiation, radio frequencies, microwaves, and lasers. Ionizing radiation includes alpha, beta, and gamma rays, X rays, neutrons, high-speed electrons and protons, and other atomic particles. This chapter discusses radiation hazards and methods of protection against these hazards, as well as radioactive material/waste disposal.

NONIONIZING RADIATION

Nonionizing radiation—radiation within the electromagnetic spectrum—can be defined as radiation with sufficient energy to excite atoms or electrons, but with insufficient energy to remove electrons from an atom or to cause the formation of ions. Even though nonionizing radiation does not cause the formation of ions, there is still a potential health hazard to employees who are exposed to this type of radiation in a work area. Types of nonionizing radiation, shown in Figure 18.1, that are typically found in the workplace include ultraviolet and infrared radiation, high-intensity visible light, radio frequencies, microwaves, and lasers.

Ultraviolet Radiation

Although the primary source of ultraviolet (UV) radiation is the sun, this type of radiation is also found in the workplace. Sources include fluorescent lighting, welding operations, plasma torches, and laser operations. In water treatment operations and other sterilization applications, UV light is used as a germicide for bacteria and molds. It is also used in wastewater treatment operations—typically in combination with other treatment methods such as ozonation—for the destruction of chlorinated compounds and pesticides.

Hazards. The most critical range of UV radiation is from 240 to 320 nanometers. This is the range at which UV radiation has the highest biological impact, the primary biological effect being on the skin and eyes. Symptoms are dependent on skin type, dose, and time of exposure; they include the reddening of the skin, the formation of blisters, and peeling of the skin. Once the source is removed, the symptoms subside. Exposure to continuous ultraviolet radiation results in increased pigmentation in the upper layer of the skin. The American Conference of Governmental Industrial Hygienists has published threshold limit values (TLVs) for ultraviolet radiation in the spectral region between 180 and 400 nanometers (ACGIH 1996). It is presumed that continuous exposure over a long period of time (i.e., many years) has the potential to cause skin cancer. In addition, skin aging may be a consequence of continuous exposure to UV radiation over a long period of time.

Exposure of the eyes to ultraviolet radiation can occur during welding operations. If the exposure is above the TLV, the worker may experience inflammation of the conjunctiva or of the cornea if proper protective equipment is not used. The eye does not build up a protective layer, so symptoms may recur with each overexposure. In addition to inflammation of the eye, exposure to ultraviolet radiation may cause blurred or decreased visual acuity. These symptoms are temporary and will disappear when exposure ceases.

Safety Mechanisms. In protecting against ultraviolet radiation, exposure time and shielding are the main considerations. The shorter the exposure, the less effect the ultraviolet radiation will have on the worker. When exposure is such that shielding is required, personal protective clothing for skin protection may include gloves, a long-sleeved shirt, coveralls, and a face shield. Eye protection may include filter lenses of varying shade numbers, depending on the operation. For example, industry standards recommend that filter lenses of shade No. 3 or 4 be used for torch brazing and soldering; filter lenses of shade No. 5 or 6 be used for oxygen cutting, medium gas welding, and arc welding up to 30

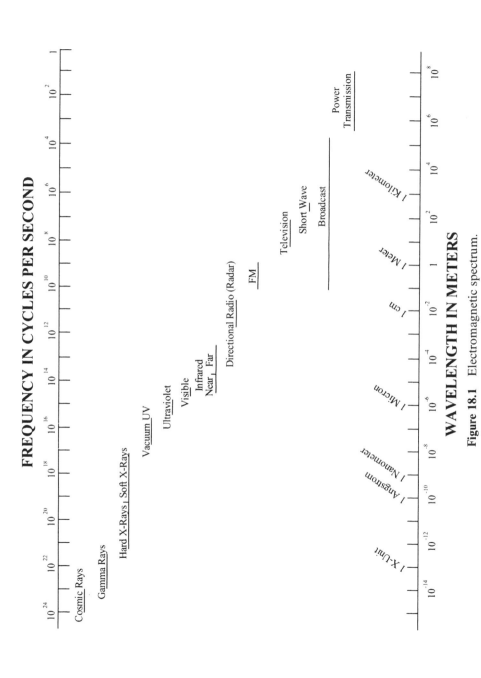

Figure 18.1 Electromagnetic spectrum.

amps; and filter lenses of shade No.r 6 or 8 be used for heavy gas welding and arc welding and cutting from 30 to 75 amps. In addition, flash goggles should be worn under all arc welding helmets (Talty 1988).

Infrared Radiation

Exposure to infrared (IR) radiation can result from any surface that is of a higher temperature than the receiver. Sources of IR radiation include heat-producing operations such as drying/baking operations, heating of metal parts, dehydration of materials, and other heating operations that use a furnace or oven.

Hazards. IR radiation is perceptible to the worker in the form of heat to the skin. Exposure in the spectral range of 750 nanometers to 1.5 microns can cause acute skin burns. Moreover, IR radiation in the shorter wavelength region can cause damage to the eye—particularly to the cornea, iris, retina, and lens. In particular, IR radiation in the range of 400 to 700 nanometers emits a bluish light, and chronic exposure to this blue light can cause retinal injury. ACGIH has published TLVs for IR radiation in the range of 760 nanometers to 1.4 microns.

Safety Mechanisms. IR radiation is best controlled by administrative controls and personal protective equipment. Exposure to IR radiation should be limited, and eye protection should be worn to reduce IR radiation levels.

High-Intensity Visible Light

High-intensity visible light can cause injury to the eye if energy levels associated with the light are high enough. Sources include welding, carbon arc lamps, and some lasers. Typical controls for this hazard include filtering the source, shielding the eye from it, and enclosure of the source.

Radio Frequencies

Radio frequencies can induce electrical currents in conductors, and they can induce the displacement of current in semiconductors. The first application—which results in the transfer of patterned energy—is used for radio and television. The second application allows radio frequency to be used as a heat source. Radio frequency heating is used in industry for the hardening of metals, bonding and laminating, rubber vulcanization, plastics molding, and other applications.

The most common hazard associated with exposure to radio frequencies is the thermal effect on the body. A temperature increase can harm the eye lens, male reproductive capability, and central nervous system. Nonthermal effects from exposure to radio frequencies include the potential for demodulation of heart and central nervous system activity and the potential for enhanced re-actions to certain types of chemicals. TLVs for this type of radiation are established, and exposure should be monitored in areas where the TLV could be reached or exceeded. Warning signs may also be needed. An example of the warning sign defined by OSHA is presented in Figure 18.2.

Extremely Low Frequency Radiation

Extremely low frequency (ELF) radiation and its effects on individuals are presently under study. Increased interest in this type of radiation is primarily focused on the concern that ELF radiation from household appliances, wiring in the home, and power transmission lines may have detrimental health effects. Studies have indicated that exposure to ELF radiation can cause changes in heartbeat and respiration, and that chronic exposure to ELF magnetic fields may increase the risk of cancers such as leukemia and central nervous system tumors (Clayton and Clayton 1991). Certainly, more research on the effects of ELF radiation is warranted. There are no national standards pertaining to exposure to ELF radiation, but guidelines have been generated for medical exposures to magnetic fields (including exposure to radiation from MRI devices) and for exposures to static magnetic fields.

Lasers

The word *laser* is an acronym for "light amplification by stimulated emission of radiation." Lasers are used for numerous purposes, including alignment, welding, trimming, fiberoptic communication systems, surgical applications, and other purposes. A laser can use ultraviolet, infrared, visible, or microwave radiation and produces a concentrated light beam that is coherent, mon-ochromatic, and typically of high power density. The three elements of a laser include:

- An optical cavity consisting of at least two mirrors, one of which is par-tially transmissive;
- An active laser medium that can be excited from an unenergized ground state to a relatively long-lived excited state; and
- A means of "pumping" or supplying the excitation energy to the active laser medium.

Examples of types of lasers are presented in Table 18.1.

Hazards. The primary hazard from laser radiation is exposure to the eye and, secondarily, the skin. If radiation levels are kept below those that damage the eye, there will be no harm to other body tissues. The hazards associated with a laser are dependent on several factors, including wavelength and intensity of beam, duration of exposure, body part exposed, and means of exposure.

Note: For size requirements, refer to 29 CFR 1910.97.

Figure 18.2 Radiation hazard warning.

NIOSH has provided information about laser class, potential hazard, and requirements by class, as detailed in Table 18.2.

Safety Mechanisms. Safety requirements, by class, are presented in Table 18.2. In addition, there are several safety controls—both engineering and administrative—that can further protect an employee from the hazards of lasers:

- Provide interlocks on equipment, as appropriate.
- Ensure proper employee training.
- Ensure proper signing of area and equipment.
- Ensure that surfaces surrounding lasers are nonreflective.
- Ensure proper shielding from beam, as appropriate.
- Ensure that beam direction is controlled.
- Ensure that flammable solvents and combustible materials are stored away from lasers.

TABLE 18.1   Types of Lasers

Solid Host Lasers

- Ruby (chromium)—pulsed operation, with up to 1,000 megawatts per pulse and 1 watt of power (continuous wave)
- Neodymium YAG—rapid-pulsed operation, with 10 megawatts per pulse and 100 watts of power (continuous wave)
- Neodymium glass—rapid-pulse operation, with 10 megawatts per pulse and 100 watts of power (continuous wave)

Gas Lasers

- Helium neon (neutral atom)—continuous wave of up to 100 milliwatts
- Argon (ion gas)—continuous wave of 1 to 20 watts
- Carbon dioxide (molecular gas)—continuous wave of 10 to 5,000 watts
- Nitrogen (molecular gas)—pulsed or rapid-pulsed operation with 250 milliwatts of power (continuous wave)

Other Lasers

- Gallium (semiconductor diode)—pulsed or continuous wave operation with 1 to 20 watts power per pulse
- Hydrogen chloride (chemical laser)—continuous wave or pulsed operation
- Organic dye laser (liquid laser)—continuous wave or pulsed operation with power per pulse of 1 megawatt

Source: Information from Talty (1988).

- Ensure that protective eyewear is prescribed and used.
- Perform periodic medical examinations—in particular, eye examinations—on personnel working on or near lasers, as appropriate.

Exposure limits developed by the American National Standards Institute (ANSI) are published in the ANSI standard Z136. They include exposure limits for direct ocular exposure intrabeam viewing; viewing a diffuse reflection of a laser beam or an extended-source laser; and exposure of the skin to a laser beam.

Microwave Radiation

Microwaves are produced by the deceleration of electrons in an electrical field. Microwave radiation has applications in communications, navigational technology, and commercial uses. Sources of microwave energy include klystrons, magnetrons, backward wave oscillators, and semiconductor transit time devices, and these sources may operate in a continuous wave, in an intermittent mode, or in a pulsed mode (Clayton and Clayton 1991).

TABLE 18.2 Laser Classifications and Requirements

Class I

- *Potential hazard*—Incapable of creating biological damage.
- *Special requirements*—None.

Class II

- *Potential hazard*—Low power; beam may be viewed directly under carefully controlled conditions.
- *Special requirements*—Posting of signs in area; control of beam direction.

Class III

- *Potential hazard*—Medium power; beam cannot be viewed.
- *Special requirements*—Well-controlled area; no specular surfaces; terminate beam with diffuse material and minimum reflection; eye protection for direct beam viewing.

Class IV

- *Potential hazard*—High power; direct and diffusely reflected beam cannot be viewed or touch the skin.
- *Special requirements*—Restricted entry to facility; interlock; fail-safe system; alarm system; panic button; good illumination of at least 150 footcandles; light-colored diffuse room surfaces; operate by remote control; design to reduce fire hazard, buildup of fumes, etc.

Class V

- Classes II, III, and IV that are completely enclosed so that no radiation can leak out.

Source: Information from Talty (1988).

There are four microwave band designations (Talty 1988):

- Very high frequency (VHF)—VHF has a wavelength of 10 to 1 meters and a frequency of 30 to 300 megahertz; it is used for FM broadcasting, television, air traffic control, and radionavigation.
- Ultrahigh frequency (UHF)—UHF has a wavelength of 1 to 0.1 meters and a frequency of 0.3 to 3 gigahertz; it is used for television, citizens band radio, microwave point-to-point, microwave ovens, telemetry, and meteorological radar.
- Superhigh frequency (SHF)—SHF has a wavelength of 10 to 1 centimeters and a frequency of 3 to 30 gigahertz; it is used for satellite communication, airborne weather radar, altimeters, ship-borne navigation radar, and microwave point-to-point.

- Extrahigh frequency (EHF)—EHF has a wavelength of 1 to 0.1 centimeters and a frequency of 30 to 300 gigahertz; it is used for radio astronomy, cloud detection radar, and space research.

The biological hazards associated with microwaves are similar to those caused by radio frequency and include primarily hazards resulting from thermal heating. Secondary concerns include nonthermal effects such as the potential for demodulation of heart and central nervous system activity and the potential for enhanced reactions to certain types of chemicals. In addition, because people may come into contact with microwave equipment, grounding is essential in order to prevent electric shock or burns. Hazards are best controlled through good engineering designs, safe operating practices, and adequate maintenance of equipment.

IONIZING RADIATION

Ionizing radiation can be defined as any electromagnetic or particulate radiation with enough energy to produce ions when it interacts with atoms and molecules. The main types of ionizing radiation include X rays, gamma rays, alpha particles, beta particles, neutrons, high-speed electrons and protons, and other atomic particles. Their uses include radiography, tracing the flow of materials in pipes, sterilization of food and medical supplies, utilization in nuclear reactors, and as cross-linking agents to improve the properties of plastics.

Alpha Particles

Alpha particles originate in the nuclei of radioactive atoms during the process of disintegration. An alpha particle is a high-energy particle composed of two protons and two neutrons. It has a mass of four atomic mass units (amu) and a charge of +2. The energy level of an alpha particle varies, but can be as high as 10 million electron volts (MeV).

The alpha particle causes more ionization than a does a beta particle or gamma radiation in the absorbing material; however, because of its large mass and positive charge, the alpha particle can travel only short distances. Alpha particles can be easily shielded with paper-thin materials. As a result, the danger of exposure from alpha particle radiation is through respiration and ingestion, rather than through penetration of the skin. Once in the body, alpha particles can concentrate in bone, lungs, kidney, and liver and can cause damage to tissue, and other internal structures.

Beta Particles

A beta particle is an electron emitted by the nucleus of an atom during radioactive decay. Electrons can be positively or negatively charged. The mass of these particles is insignificant. The energy level of a beta particle typically ranges between 0.017 and 4 MeV.

As a result of its smaller mass, the penetration capability of the beta particle is greater than that of the alpha particle, but less than that of gamma- and X-rays. Shielding materials for beta particles must be more substantial than those used for alpha particles, and typically include aluminum and other materials of low atomic weight. In addition, these particles have the potential to form secondary gamma and X-radiation, which make them more hazardous than the alpha particle.

Gamma Radiation

Gamma radiation is not a particle; instead, it is defined as energy waves in the electromagnetic spectrum. Gamma rays are emitted by the nucleus of certain radionuclides during their decay scheme. The energy levels of gamma radiation can range between 0.008 and 10 MeV.

The process of ionization in the tissue may cause irreparable damage to living cells. Because gamma radiation has no mass or charge, but is a wave, it has the capability for deep penetration and presents a greater hazard in the workplace than either alpha or beta particles.

There are approximately 240 radioactive isotopes, or radionuclides, that can undergo radioactive decay to reach a more stable energy level. The rate of this decay is measured in terms of a "half-life." A half-life is the amount of time needed for half of the active atoms of any given quantity to decay. Half-lives of several radionuclides and the associated gamma-radiation energies are presented in Table 18.3. A pictorial view of the decay of a radioactive material at 100 millirems per hour (mR/hr) is shown in Figure 18.3. As can be seen in the figure, after seven half-lives, the material has decayed to less than 1 percent.

Shielding to protect against gamma radiation requires dense material such as lead or iron. Shield thickness is calculated based on shield material, radiographic source (e.g., cobalt-60 or radium-226), and distance. Standard tables provide information about half-value thicknesses for typical materials and sources of radiation.

X-Radiation

X-radiation is also electromagnetic radiation. It is similar in properties to gamma radiation, but does not have as high an energy level as gamma radiation. X-radiation is formed when high-speed electrons are slowed down or stopped. In the process of slowing down, the electrons give up energy in the form of X-radiation. The quantity of energy released is dependent on the speed of the electrons and the characteristics of the medium or striking target. Typically, X-radiation is produced by a machine, although some X-radiation can result from radioactive decay. As in shielding against gamma radiation,

TABLE 18.3 Examples of the Half-Life and Gamma-Radiation Energy of Selected Radionuclides

Radionuclide	Half-Life	Gamma-Radiation Energy (MeV)
Sulfur	88 days	none
Selenium	120 days	0.12, 0.14, 0.26, 0.28, 0.40
Zinc	245 days	1.12
Manganese	303 days	0.84
Sodium	2.6 years	1.277
Iron	2.6 years	none
Cobalt	5.27 years	1.3, 1.12
Lead	21 years	0.047
Strontium	28.1 years	none
Nickel	92 years	none
Radium	1602 years	0.186
Carbon	5730 years	none

Source: Information from National Safety Council (1988).

shielding to protect against X-radiation requires dense material and should be carefully calculated by an expert.

Biological Effects of Ionizing Radiation

The human body can tolerate some amount of exposure to ionizing radiation without adverse effect, for we live in a world with background radiation caused by cosmic radiation and radiation in the earth. However, there are exposure levels that are considered dangerous to human health. Typically, the effects of

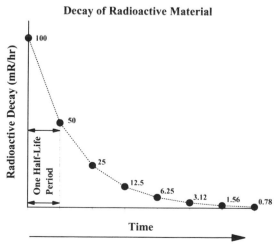

Figure 18.3 Example of decay of radioactive material.

radiation are classified as somatic effects and genetic effects. The somatic effects include damage to bone, tissue, and organs; cancer; cataracts; and other diseases. Genetic effects include birth defects, the production of mutant genes, and aberrations in chromosomes.

Exposure limits to ionizing radiation are provided in a number of standards and sources, which include the following:

- OSHA—29 CFR 1920.96 (general industry) and 29 CFR 1926.53 (construction industry)
- NRC—10 CFR 20.101, 10 CFR 20.104, and 10 CFR 20.105
- EPA—40 CFR 141.15, 40 CFR 141.16, and 40 CFR 61.102

OSHA and NRC have both published exposure limits for the body:

- Whole body; head and trunk; active blood-forming organs; lenses of eyes; and gonads—1¼ rems per calendar quarter
- Hand and forearms; feet and ankles—18¾ rems per calendar quarter
- Skin of whole body—7½ rems per calendar quarter.

Radiation Protection Program

The elements to be included in a radiation protection program vary from company to company, depending on the size, number, and type of radiation sources. The following elements are typically included in such a program:

- Program documentation and administration
- Training on safe work practices
- Training on emergency procedures
- Training on warning symbols and signs
- Control measures
- Surveys and monitoring
- Medical surveillance
- Recordkeeping
- Incident reporting

An excellent source of information useful to the safety engineer when setting up a radiation protection program is the National Council on Radiation Protection and Measurements (NCRP). The organization has published more than 100 reports on all topics of radiation, including radiation exposure, radiation measurement and assessment, radiation protection, engineering and administrative controls, and radiation in the medical/health fields. A listing of selected publications from NCRP is presented in Appendix C.

Warning Symbol. OSHA requires a standard symbol to depict radiation hazards. That symbol is shown in Figure 18.4.

Recordkeeping. The need for accurate records related to radiation safety cannot be overstated. Examples of records that may be maintained as part of a radiation protection program are presented in Table 18.4.

Surveys and Monitoring. An effective radiation protection program must include assessment of radiation exposure in the area. Typical monitoring instruments include film badges, thermoluminescence detectors, dosimeters, Geiger-Mueller counters, radon monitors, and ionization chambers. Monitoring should be performed on a specified schedule, and equipment should likewise be calibrated according to a schedule recommended by the manufacturer.

Incident Reporting. Employers must notify the proper authorities when there is any incident involving radiation exposure that exceeds specified guidelines. An effective way to ensure that notifications are made promptly is for the safety engineer to develop a matrix that provides information about:

- Exposure levels that require notification to proper authorities
- Statutory time requirements for reporting an incident
- Name of agency, contact person, phone number, and any other information needed in order to make the incident notification

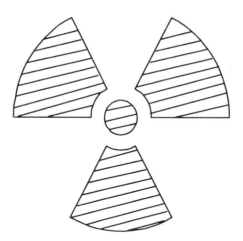

Note: Cross-hatched area is to be magenta or purple;
background is to be yellow. Size and angle requirements
are found in 29 CFR 1910.96.

Figure 18.4   OSHA symbol for radiation hazards.

TABLE 18.4   Examples of Records Related to Radiation Safety That May Be Maintained as Part of a Radiation Protection Program

Radiation Protection Program Records

- Authorizing documents
- Accreditations and certifications, such as accreditation of dosimetry, hospital accreditations, laboratory accreditation, and accreditation of nuclear plant training program
- Guidance documents
- Records detailing qualifications, positions, and training of employees, in the radiation protection program
- Written documents for radiation protection programs, such as a radiological training program, an instrument calibration program, a personal protective equipment program, exposure and assessment programs, ALARA programs, and engineering control programs

Individual Records

- Exposure categories for individuals, such as occupational exposure, occasional exposure, management visitors, and other categories
- Personal data
- External dosimetry assessment records
- Internal dosimetry records
- Occupational exposure history
- Abnormal exposures
- Training records

Workplace Records

- Facility description, including design features relevant to radiation protection, potentially affected populations, emission/release point, and other information
- Records of controlled areas
- Documents detailing ventilation and exhaust systems
- Access control points
- Radiation work permits or other work authorizations
- Area radiation and contamination records
- List and description of personal protective equipment
- ALARA documentation
- Records of radioactive material shipments
- Inventory of radioactive material
- Records of accidents and incidents

Environmental Records

- Preoperating monitoring program, including climatic, topographic, and land use data, demographic studies, and radiological surveillance records

(Continued)

TABLE 18.4 *Continued*

- Operation of environmental monitoring program and records
- Reports of radioactive releases and dose assessments
- Reports from offsite investigations and special studies

Radiation Protection Instrumentation Records

- Equipment specifications, such as type of detector, energy, range, sensitivity, and other information
- Records pertaining to calibration facility and source certification
- Calibration records for each instrument
- Maintenance records
- Instrument inventory

Source: NCRP (1992).

- Name and phone number of facility security contact
- Name and phone number of communications contact
- Sample information to report, such as date, time, facility name and location, description of incident, threat to surrounding areas, incident response, and other information
- Other pertinent information

Once an incident is reported, there is usually a requirement for a follow-up report, which then becomes part of the public record.

DISPOSAL OF RADIOACTIVE WASTES

Definitions for Radioactive Wastes

Definitions for radioactive wastes are published by the NRC. There are basically three major types of radioactive wastes: high-level radioactive wastes (HLW), transuric radioactive wastes (TRU), and low-level radioactive wastes (LLW).

HLW are those processing wastes derived from nuclear reactors including irradiated reactor fuel, liquid wastes resulting from the operation of the first-cycle solvent extraction system or the equivalent, and solids derived from the conversion of high-level radioactive liquids. Weapons by-products can also contain high-level radioactivity.

TRU wastes include wastes containing elements with atomic numbers greater than uranium (92), such as plutonium and curium, and that contain more than 100 nanocuries per gram (nCi/g) alpha-emitting transuranic isotopes with half-lives greater than 20 years. Some TRU materials are exceptionally long-lived, such as plutonium-239, which has a half-life of 24,400 years. TRU waste is generated from the reprocessing of plutonium-bearing fuel and irradiated targets and from operations required to prepare the recovered plutonium for weapons use. The wastes include TRU metal scrap, glassware, process equipment, soil, laboratory wastes, filters, and wastes contaminated with TRU materials.

LLW wastes are defined as wastes containing radioactivity that is neither HLW nor TRU. In general, LLW is divided into three classes: A, B, and C, with Class C containing the highest concentration of radionuclides. Examples of LLW include solidified liquids, filters and resins, and lab trash from nuclear reactors, hospitals, research institutions, and industry (NRC 1989).

Disposal Requirements

Low-Level Radioactive Wastes. Disposal of LLW is regulated under 10 CFR Part 61. This regulation pertains to near-surface disposal and includes requirements for site suitability, site design, and other elements. Suitability of a site for LLW disposal is determined by the site's natural characteristics, such as hydrogeology, climate, soil chemistry, and other factors.

After selection criteria are met, the LLW disposal facility must be designed to meet the following basic requirements:

- The facility should be designed to minimize water contact with waste during storage, disposal, and after disposal.
- The design should minimize the need for any active maintenance after closure to ensure long-term isolation of the wastes.
- The design should complement the site's natural characteristics and provide for erosion control and for run-on and run-off management.

LLW disposal facilities must be monitored continuously throughout the life of the facility and after closure to ensure that the facility poses no threat to human health or to the environment. Typically, upgradient and downgradient groundwater monitoring wells are used for this purpose. Other management practices include intruder protection for all Class C facilities and some Class B facilities. For facilities accepting Class C wastes, waste stabilization and disposal barriers are also required.

Transuric Wastes. NRC must approve the disposal of these wastes. Requirements include disposal in a geologic repository or equivalent in most cases.

High-Level Radioactive Wastes. Disposal of HLW is regulated under 10 CFR Part 60. This type of waste requires permanent isolation in a geologic repository. Technical performance criteria for HLW repositories are defined in the regulations. These requirements include the following:

- All geologic repository operations areas must be designed so that until permanent closure has been completed, radiation exposure, radiation levels, and release of radioactive materials must meet standards set by NRC and EPA.

- Waste must be retrievable for up to 50 years.
- Upon permanent closure, engineered barriers must be designed to effect substantially complete containment.
- Any release of radionuclides must be small and occur over long periods of time.
- Containment of HLW within waste packages must be substantially complete for a period of 300 to 1,000 years after permanent closure of the facility.

In addition to technical performance requirements, there are also design criteria for the geologic repository operations area that must be met before a license is issued to a facility.

REFERENCES

American Conference of Governmental Industrial Hygienists, *1996–1997 Threshold Limit Values for Chemical Substances and Physical Agents and Biological Exposure Indices*, ACGIH, Cincinnati, OH, 1996.

Clayton, George D., and Frances E. Clayton, eds., *Patty's Industrial Hygiene and Toxicology: General Principles*, Volume 1, Part B, John Wiley & Sons, New York, 1991.

National Council on Radiation Protection and Measurements, *Maintaining Radiation Protection Records*, NCRP Report No. 114, NCRP, Bethesda, MD, 1992.

National Safety Council, *Fundamentals of Industrial Hygiene*, 3d ed., NSC, Chicago, IL, 1988.

Nuclear Regulatory Commission, "Regulating the Disposal of Low-Level Radioactive Waste: A Guide to the Nuclear Regulatory Commission's 10 CFR Part 61," NUREG/BR-0121, Office of Nuclear Material Safety and Safeguards, U.S. NRC, Washington, DC, 1989.

Talty, John T., ed., *Industrial Hygiene Engineering: Recognition, Measurement, Evaluation and Control*, prepared for the National Institute for Occupational Safety and Health, Noyes Data Corporation, Park Ridge, NJ, 1988.

BIBLIOGRAPHY

ANSI (1983), "Personnel Dosimetry Performance Criteria for Testing," ANSI N13.11, American National Standards Institute, New York.

API (1988), "Naturally Occurring Radioactive Material (NORM) in Oil and Gas Production Operations: Video Presentation," American Petroleum Institute, Washington, DC.

DOE (1993), "Electric Power Lines: Questions and Answers on Research into Health Effects," DE94 005637, Bonneville Power Administration, Portland, OR.

———— (1991), "The Status of ANSI N13.11—The Dosimeter Performance Test Standard," DE91 017874, prepared by C.S. Sims, Oak Ridge National Laboratory, Oak Ridge, TN.

———— (1988), "Review of EPA, DOE, and NRC Regulations on Establishing Solid Waste Performance Criteria," DE88 015331, ORNL/TM-9322, prepared by A.J. Mattus, T.M. Gilliam, and L.R. Dole, Oak Ridge National Laboratory, Department of Energy, Oak Ridge, TN.

Knoll, G.F. (1989), *Radiation Detection and Measurement*, 2d ed., John Wiley & Sons, New York.

ORAU (1988), "A Compendium of Major U.S. Radiation Protection Standards and Guides: Legal and Technical Files," prepared by W.A. Mills et al., Oak Ridge Associated Universities, Oak Ridge, TN.

19

NOISE AND HEARING CONSERVATION

Excessive noise in the workplace can result in permanent hearing loss. For this reason, there are federal regulations controlling occupational noise exposure, enforced by the U.S. Occupational Safety and Health Administration (OSHA) (40 CFR 1910.95). These regulations set out requirements for permissible exposure levels, monitoring of noise levels, employee hearing tests, hearing protectors, training, and recordkeeping.

In dealing with any type of "pollution," one tries to prevent it by eliminating it at the source. For noise pollution, engineering and administrative controls should be considered first. When these types of controls are not effective or are not feasible, personal hearing protection must be provided and *used*.

This chapter begins with an introduction to some fundamental concepts of sound waves, sound pressure and energy levels, and sound measurement. Following the introduction is a discussion of hearing loss resulting from excessive noise. The OSHA standard for noise exposure is outlined next, followed by a discussion of techniques for noise abatement and control.

SOUND AND NOISE

Sound is produced by any pressure variation caused by vibrations. A sound wave is a longitudinal vibration of a conducting medium such as air or water. Sound waves can be represented as sinusoidal patterns with given amplitudes and frequencies. The intensity of sound, which is related to its loudness, is represented by the amplitude of the sound wave. The frequency of a sound wave is the number of times the vibrating object completes its cycle of motion in a period of one second. Frequency is measured in Hertz (Hz), which is cycles per

494

second. Pitch is related to frequency. A higher pitched sound is one that is higher in frequency. *Noise* means essentially, unwanted sound and, in that sense, is a subjective term.

EXAMPLE 19.1
Sound Waves

Which of the following sound wave curves in Figure 19.1 have the same amplitude? Which represents the highest pitched sound and why?

Solution. Sound waves B and C have the same height, or amplitude. Sound wave A represents the highest-pitched sound, because it has the highest frequency.

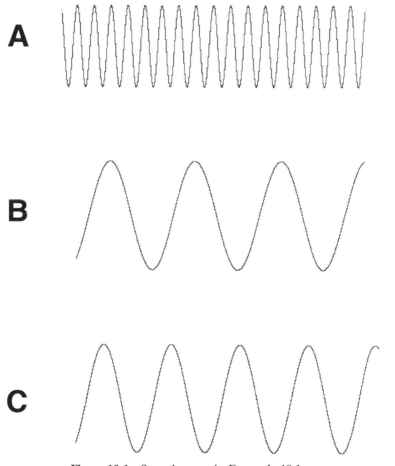

Figure 19.1   Sound waves in Example 19.1.

Sound pressure is measured in units of pressure: micropascals (µPa); newtons per square meter (N/m²), microbars (µbar), and dynes per square centimeter (d/cm²). One microbar equals 1 d/cm², which equals 0.1 N/m², which equals 0.1 Pa. The range of audible sound pressure is very large; the sound pressure at the top of the range is more than 10 million times as strong as the weakest sound. Like earthquakes, sound pressure is translated to a more practical scale which makes it easier to use and apply. For sound, the scale is logarithmic and the unit is the decibel (dB). Sound pressure level, in decibels, is measured against a reference level. At a reference tone of 1,000 Hz, a sound pressure of 20 µPa represents the hearing threshold, p_o, of an average person. The pressure level, L_p, of an audible sound, measured in decibels, is calculated by the following equation:

$$L_p = 20 \log \frac{p}{p_o} \qquad (19.1)$$

Sound pressure is usually proportional to the square root of the sound power. Therefore, sound power or energy, L, which is also measured in decibels, is calculated as:

$$L = 20 \log \frac{q^{0.5}}{q_o^{0.5}} = 10 \log \frac{q}{q_o} \qquad (19.2)$$

where q is the sound power being measured and q_o is the reference power level.

EXAMPLE 19.2
Differences in Sound Pressure and Energy

If the sound energy is doubled, how great is the increase in decibels? If the sound energy is increased by a factor of 100, what is the increase in decibels? If sound *pressure* is doubled or increased 100-fold, what are the corresponding increases in decibels for sound pressure?

Solution. Set up the equation to solve the difference in two levels of sound energy, using Equation 19.2.

$$L_2 - L_1 = 10 \log \frac{q_2}{q_o} - 10 \log \frac{q_1}{q_o}$$

Simplifying the equation

$$L_2 - L_1 = 10 \left(\log \frac{q_2}{q_o} - \log \frac{q_1}{q_o} \right)$$

$$= 10(\log q_2 - \log q_o - \log q_o - \log q_1 + \log q_o)$$

$$= 10(\log q_2 - \log q_1)$$

$$= 10 \log \frac{q_2}{q_1}$$

Substituting a value of 2 for the doubling of sound power, the difference in decibels is

$$L_2 - L_1 = 10 \log 2$$

$$= 10 \times 0.30103$$

$$= 3 \text{ dB}$$

Therefore, a doubling of sound energy represents an increase of 3 dB. If the sound energy is increased by a factor of 100, the increase in dB is

$$L_2 - L_1 = 10 \log 100$$

$$= 10 \times 2$$

$$= 20 \text{ dB}$$

If sound pressure is doubled, the same type of equation can be derived using Equation 19.1. The increase in dB is

$$L_{p_2} - L_{p_1} = 20 \log 2$$

$$= 6 \text{ dB}$$

If sound pressure is increased 100 times, the increase in sound pressure level is

$$L_{p_2} - L_{p_1} = 20 \log 100$$

$$= 40 \text{ dB}$$

As seen in the preceding example, decibels cannot be added directly. For example, if one source of sound energy of 60 dB is combined with another source of equal measure, the resulting sound energy is 63 dB, not 120 dB. The decibel level of multiple sound sources can be estimated on an energy basis. To calculate the effect of combined sources, you must first convert each source to its relative power level, add these levels together, then convert the total back to decibels.

EXAMPLE 19.3
Combining Decibels from Multiple Sources

Estimate the decibel level from two sources, one having a power level of 80 dB and the other 90 dB.

Solution. The relative power level of the 80 dB source, q_r, is calculated from Equation 19.2.

$$L = 10 \log \frac{q}{q_o} = 10 \log q_r$$

$$80 \text{ dB} = 10 \log q_r$$

$$q_r = 10^8$$

The relative power level of the 90 dB source is, therefore, 10^9. The decibel level of the two sources combined is

$$10 \log(10^8 + 10^9) = 90.4 \text{ dB}$$

Show how adding a third source of 85 dB source increases the sound power level to 91.5 dB.

EXAMPLE 19.4
Subtracting Decibels by Removing Sound Sources

If there are 3 sources of sound in one area that have individual sound energy readings of 80 dB, 85 dB, and 90 dB, what would be the decibel level if the 90 dB source is removed?

Solution. There are several ways to solve this problem. Since the individual decibel levels are known for each source, you can simply calculate the combined effect of the 80 dB and 85 dB sources.

$$10 \log(10^8 + 10^{8.5}) = 86.2 \text{ dB}$$

Sometimes, however, only the combined decibel level will be known. In that case, you subtract the relative power values, that is, the combined value minus the source you are removing. From Example 19.3, you have a 91.5 dB sound energy level for the three sources combined. Converting this energy level to its relative power value and subtracting the relative power value of the 90 dB source, you have the remaining sound energy level for the 80 dB and 85 dB sources.

$$10 \log(10^{9.15} - 10^9) = 86.2 \text{ dB}$$

Another way of calculating the decibel level of combined sounds is to look at the incremental increase (or decrease) in levels. These are the same types of calculations as those in Example 19.2.

EXAMPLE 19.5
Another Way of Combining Decibels

Calculate the change in decibel level for Examples 19.3 and 19.4, using the methods shown in Example 19.2.

Solution. When the 80 dB and 90 dB sources are combined, the increase in decibels is

$$10 \log \frac{(10^8 + 10^9)}{10^9} = 10 \log(10^{-1} + 1)$$

$$= 0.4 \text{ dB}$$

When the 90 dB source is removed from the 80 dB and 85 dB sources, the change in decibels is

$$10 \log \frac{(10^{9.15} - 10^9)}{10^{9.15}} = 10 \log(1 - 10^{-0.15})$$

$$= -5.35 \text{ dB}$$

Therefore, the resulting decibel level is $91.5 - 5.35 = 86.15$ dB or 86.2 dB.

SOUND LEVEL MEASUREMENTS

There are three internationally standardized weighting networks for sound pressure weighting: A, B, and C. Very low frequencies are largely filtered out by the A-network, moderately filtered by the B-network, and filtered hardly at all by the C-network. The A-network is often used to assess noise exposure because it is thought to best relate to injuries that such noise can cause to the human ear. The American Conference of Governmental Industrial Hygienists (ACGIH) has adopted the A-weighted sound level measurement for noise exposure assessment. Noise limits in the OSHA regulations are stated in decibels on the A-scale (dBA).

Sound levels are measured with a sound level meter or dosimeter. The sound level meter measures the sound intensity at a particular moment. The sound level meter is hand held and can be used to measure sound levels at various locations throughout the workplace and at different times of the day in order to estimate overall exposure levels. The dosimeter measures the overall exposure directly, because it is worn by the employee and measures sound levels wherever the employee travels. The dosimeter is like a sound level meter that automatically integrates measurements over time. Standards for both sound level meters and personal noise dosimeters are set by the American National Standards Institute (ANSI) (see Bibliography).

HEARING LOSS

Natural hearing loss as a result of aging is called presbycusis.[1] Hearing loss caused by continuous exposure to noise over a long period of time is referred to as noise-induced hearing loss. Noise-induced hearing loss does not refer to sudden hearing loss resulting from acoustic trauma or immediate physical injury to the ear. Noise-induced hearing loss commonly involves cumulative injury to the hair cells of the inner ear.

Hearing loss may be temporary or permanent. A standard threshold shift or loss in hearing is a change of 10 dB or more at 2,000, 3,000, and 4,000 Hz. A temporary threshold shift (TTS) can result from brief, but intense, sounds. Hearing eventually recovers, usually overnight or within 14 hours (NSC 1988). A permanent threshold shift (PTS) can result from sounds that are prolonged or recurring. Factors that contribute to PTS include (NSC 1988):

- Sound levels in excess of 60–80 dB,
- Sound levels with most of their energy in the speech frequencies,
- Longer periods of exposure,
- Shorter breaks (quiet periods) with intermittent sounds,
- Increasing rise time and/or burst duration of the sound, and
- Lower individual tolerance to sound.

STANDARDS OVERVIEW

OSHA regulations covering noise control and hearing conservation are found at 40 CFR 1910.95, 1926.52, and 1926.101. The Part 1910 standards cover general industry, whereas the Part 1926 standards cover only the construction industry. The construction industry standards are not nearly as extensive as those for general industry; however, they do have certain important elements in common, such as limiting exposure time at different noise levels.

The general industry standard, in addition to limiting exposure time, sets out requirements for exposure monitoring, audiometric testing, hearing protection, employee training, and recordkeeping. Because it is more extensive, this section focuses on the detailed requirements of the general industry standard.

Permissible Noise Exposure

OSHA limits exposure to noise to a time-weighted average (TWA) of 90 dB over an 8 hour period, when measured on the A-network or A-scale of a standard sound level meter at slow response. Using the A-scale, noise levels in the OSHA standard are written as dBA. The amount of exposure time is reduced

---

[1] The root word, *presbus*, originates from New Latin: Greek and means "elder."

as the noise level increases. These time limits are shown in Table 19.1 for noise levels ranging from 90 dBA to 115 dBA. As shown in the table, the allowable exposure time is cut in half each time the noise level increases by 5 dBA. For example, at 90 dBA, the allowable exposure is 8 hours, but at 95 dBA, the allowable exposure is 4 hours, and at 100 dBA it is reduced to 2 hours.

When a person is exposed to different noise levels throughout an 8-hour period, the combined effect must be considered. This is done by calculating a combined exposure fraction, using the following equation:

$$\sum_{i=1}^{n} \frac{C_i}{T_i} \qquad (19.3)$$

where n is the number of different noise levels, C_i is the amount of time a person is exposed to noise level i, and T_i is the total amount of time permitted at that noise level (from Table 19.1). Any amount of time the person is exposed to noise levels of less than 90 dBA is not included in the sum. If the sum of the individual C/T fractions is greater than 1, the allowable exposure is exceeded.

TABLE 19.1   Permissible Noise Exposure

Duration per Day (Hours)	Sound Level dBA Slow Response
8	90
6	92
4	95
3	97
2	100
1.5	102
1	105
0.5	110
0.25 or less	115

Notes:
1. When the daily noise exposure is composed of two or more periods of noise exposure of different levels, their combined effect should be considered, rather than the individual effect of each. If the sum of the combined exposure fractions exceeds unity, then, the mixed exposure should be considered to exceed the limit value.
2. Exposure to impulsive or impact noise should not exceed 140 dBA peak sound pressure level.

Source: Table G-16, 40 CFR 1910.95(a).

EXAMPLE 19.6
Calculating Allowable Exposure Time to Mixed Noise Levels

Determine whether the allowable exposure is exceeded for the following noise levels and exposure times.

Noise Level	Actual Exposure Time
85 dBA	2 hours
90 dBA	4 hours
95 dBA	1 hour
105 dBA	15 minutes
90 dBA	45 minutes

Solution. Calculate the exposure fraction for each noise level and then add them together.

Noise Level	Exposure Fraction
85 dBA	0 (there is no limit for 85 dBA)
90 dBA	$4 \div 8 = 0.5$
95 dBA	$1 \div 4 = 0.25$
105 dBA	$(15 \div 60) \div 1 = 0.25$
90 dBA	$(45 \div 60) \div 8 = 0.09$

The sum of the exposure fractions is

$$0 + 0.5 + 0.25 + 0.25 + 0.09 = 1.09$$

Because the total exposure fraction, 1.09, is greater than 1, the allowable exposure is exceeded.

If the allowable exposure is exceeded, the employer must use administrative and engineering controls to reduce the exposure to allowable levels. If such controls are not feasible, then hearing protection must be provided to the employee and used to reduce exposure to allowable levels.

The limits in Table 19.1 are based on continuous noise, not intermittent or impact-type noise. However, noise is considered continuous if it involves maxima at intervals of 1 second or less. An example is the noise produced by a power tool that switches on and off, with the off periods being 1 second or less.

Exposure to impulsive or impact-type noise should not exceed 140 dBA. This type of noise, like a short, high-level burst of compressed air, is characterized by a rapid rise, high peak value, and rapid decay. Impulsive or impact noise generally occurs in bursts lasting less than half a second and does not repeat more than once per second (NSC 1988).

Monitoring

A time-weighted average of 85 dBA for an 8-hour period is defined as the OSHA action level. When information indicates that any employee is exposed to noise equal to or greater than the action level, the employer must implement monitoring of noise exposure levels. Monitoring must be designed to identify those employees who should be included in a hearing conservation program and to allow the proper hearing protectors to be selected. Employees are entitled to observe the monitoring. The employer must relate the results of monitoring to each employee who is exposed to noise levels at or above the 85 dBA action level.

In measuring noise levels, all continuous, intermittent, and impulsive sound levels from 80 to 130 dBA must be included. Measuring instruments must be accurately calibrated. When exposure conditions make area sampling inappropriate (such as when workers move around a lot), there are significant variations in sound level, or impulsive sound makes up a significant part of the noise, personal sampling must be used instead. Monitoring must be repeated whenever there is a change in production, process, equipment, or controls that increases noise levels to the extent that additional employees are exposed to 85 dBA or more or that hearing protectors no longer provide adequate protection.

Audiometric Testing

The employer must establish audiometric testing for any employee who is exposed to an 8-hour TWA noise level of 85 dBA or more. A baseline audiogram is first obtained, and annual testing is repeated as long as the employee remains exposed to these noise levels. Baseline testing must be conducted within six months of the employees exposure to these noise levels. However, if mobile test vans are relied on for the testing, the employer has up to a year to obtain the baseline audiogram. During this time extension, the employee must wear hearing protection until the testing is completed.

Audiometric testing must be performed by a qualified person such as a licensed or certified audiologist, otolaryngologist[2] or other physician, or by a technician. A technician must be certified by the Council of Accreditation in Occupational Hearing Conservation or must have satisfactorily demonstrated competence in audiometric examinations, obtaining valid audiograms, and properly using, maintaining, and checking calibration and proper functioning of the audiometers. A technician who uses microprocessor audiometers, however, does not need to be certified. Audiometers must meet the specifications in the American National Standard Institute (ANSI) Specification for Audiometers, S3.6-1969. This standard also applies to their use and maintenance.

[2] Physician who specializes in the diagnosis and treatment of disorders of the ear, nose, and throat.

The annual audiogram is compared with the baseline to determine whether there has been a change in hearing, more specifically, a standard threshold shift. If a standard threshold shift is indicated beyond that owing to natural hearing loss with age, the employer may obtain a retest and consider the results as the annual audiogram. Baseline comparisons may be made by a qualified technician; however, problem audiograms must be evaluated by the audiologist, otolaryngologist, or physician to determine whether further evaluation is needed.

If a standard threshold shift occurs, the employee must be notified within 21 days. Unless the shift is found to be not related to or not aggravated by occupational noise exposure, the employer must:

- Fit employees not previously using hearing protectors with hearing protectors, train them in their use, and require them to use the protectors;
- Refit employees using hearing protectors, retrain them, and provide them with protectors that offer greater attenuation, if necessary;
- Refer employees for a clinical audiological evaluation or otological examination if additional testing is needed or if a medical pathology of the ear is suspected to have been caused or aggravated by the hearing protectors;
- Inform employees of the need for an otological examination if a medical pathology is indicated that is unrelated to the use of hearing protectors.

If subsequent testing shows that the threshold shift does not persist, the employee does not have to wear hearing protection if the 8-hour TWA noise level is less than 90 dBA although hearing protection must be made available to employees at levels above 85 dBA, as described in the next subsection.

Baseline audiograms may be updated if the annual audiogram shows significant improvement in hearing, or conversely, if the standard threshold shift is shown to be persistent. In the latter case, the purpose of changing the baseline when there is a persistent hearing loss is to avoid identifying the same threshold shift repeatedly in the annual audiogram. In any case, the original baseline audiogram is always retained as long as the employee remains at the facility.

Hearing Protectors

The employer must make hearing protectors available to any employee who is exposed to an 8-hour TWA noise level of 85 dBA or greater. The employer must also make sure that hearing protectors are worn (as opposed to just being available) by employees who are exposed to noise greater than the permissible levels stipulated in Table 19.1. If an employee has already experienced a standard threshold shift in hearing or has not yet had a baseline audiogram conducted within the first six months of exposure, hearing protectors must be worn any time the noise level exceeds the 8-hour TWA 85 dBA. Hearing protectors must attenuate or reduce the noise level to at least an 8-hour TWA of 85 dBA if the employee has experienced a threshold shift, and to at least 90 dBA for all other employees.

Training

Employees who understand the importance of hearing protection and are given the tools to preserve their own hearing will be better motivated to participate in a hearing conservation program. The employer is responsible for training any employees who are exposed to noise levels of 85 dBA or greater as a time-weighted 8-hour average, in the hazards of noise exposure and the need for hearing protection. Training must be repeated annually, and the information covered must be updated to reflect any changes in work activities and protective equipment. Training must address:

- The effects of noise on hearing;
- The purpose of hearing protectors, the advantages, disadvantages, and attenuation of the different types of hearing protectors, and their selection, fitting, use, and care; and
- The purpose of audiometric testing and how the tests are conducted.

The employer must make available to these employees (or their representatives) copies of the OSHA noise exposure standard, as well as any information regarding this standard that is supplied to the employer by OSHA. The employer must provide to OSHA, upon request, all materials related to the training program.

Recordkeeping

Employers must maintain records of both noise exposure monitoring and audiometric testing. Upon request, records must be provided to employees, former employees, authorized employee representatives, and the assistant secretary for Occupational Safety and Health. Monitoring records must be kept for at least two years. Audiometric test records must be kept as long as the person remains employed at the facility. If ownership of the facility changes, the previous employer must transfer these records to the new employer, who must continue to keep them for the required periods.

Audiometric records must include the following information:

- Name and job classification of the employee;
- Date of the audiogram;
- Name of the examiner;
- Date of the last acoustic or exhaustive calibration of the audiometer;
- Employee's most recent noise exposure assessment; and

- Measurements of the background sound pressure levels in the audiometric test rooms.

NOISE CONTROL AND ABATEMENT

Noise exposure can be limited in the workplace through engineering controls, administrative controls, and personal hearing protection devices. As in pollution prevention, the most effective means of controlling noise is to reduce it at the source through engineering design. When engineering and administrative controls are not feasible, personal hearing protectors are needed.

Engineering controls reduce sound energy either at the source or at the point where it reaches the individual. Sound can be attenuated or reduced by either enclosing the source or isolating the individual in a soundproof booth or room. Other engineering controls include equipment and process substitution or modification, preventative maintenance, operational changes, and the use of sound-absorbing materials. Table 19.2 includes some suggestions from the National Safety Council (1988) for reducing sound levels through engineering controls.

Administrative controls involve balancing production and employee shifts to maintain exposures within safe levels. Employee work schedules may be structured so that an employee's time in a high-noise area does not exceed the time-weighted 8-hour average of 90 dBA. Where feasible, production schedules may be adjusted so that high-noise machines operate for shorter periods during the day. Administrative controls also include providing quiet areas for break and lunch periods where employees can relax and obtain relief from production noise.

TABLE 19.2 Engineering Controls for Noise Suggested by the National Safety Council

Preventative Maintenance

- Replace or adjust worn, loose, or unbalanced machine parts.
- Lubricate machine parts and use cutting oils.
- Use properly shaped and sharpened cutting tools.

Equipment Substitution

- Use larger, slower machines in place of smaller, faster ones.
- Use step dies instead of single-operation dies.
- Use presses instead of hammers.
- Use rotating shears instead of square shears.
- Use hydraulic instead of mechanical presses.
- Use belt drives instead of gears.

TABLE 19.2 *Continued*

Process Substitution

- Use compression instead of impact riveting.
- Use welding instead of riveting.
- Use hot instead of cold working.
- Use pressing for rolling or forging.

Vibrating Surfaces

- Reduce driving force of vibrating surfaces.
- Minimize rotational speed.
- Isolate driving force.
- Dampen vibration.
- Provide additional support for machine.
- Increase material stiffness.
- Increase material mass of vibrating members.
- Change size to change resonance frequency.
- Reduce the area for radiating sound.
- Reduce the overall size of vibrating surface.
- Perforate surface.

Sound Transmission

THROUGH SOLIDS

- Use flexible mountings, flexible sections in pipe runs, and flexible shaft couplings.
- Use fabric sections in ducts.
- Use resilient flooring.

THROUGH AIR

- Use sound-absorptive material on walls and ceilings.
- Use sound barriers and sound absorption along sound transmission path.
- Place individual machines in enclosures.
- Place high-noise machines in insulated rooms.
- Isolate operator in soundproof booth.
- Use baffles.

Sound Produced by Gas Flow

- Use intake and exhaust mufflers.
- Use fan blades designed to reduce turbulence.
- Use large, low-speed fans instead of smaller, high speed fans.
- Reduce flow velocities.
- Increase cross-sectional area.
- Reduce pressure.
- Reduce air turbulence.

Source: National Safety Council (1988).

507

After it is reduced by engineering and administrative controls, noise exposure can be reduced further by the use of personal hearing protectors. Hearing protectors come in a variety of styles: aural inserts, superaural protectors, circumaural protectors, and helmets. Aural inserts are earplugs, made from soft materials such as rubber, plastic, and foam. Earplugs that are designed for one-time, throwaway use are often the formable type that are squeezed or rolled before placement in the ear; they quickly reexpand to fill the ear canal and block noise. Premolded earplugs are made of soft materials such as silicone, rubber, or plastic. The premolded type come in various shapes and sizes to accommodate differences in ear canals, which can occur even in the same individual. Superaural protectors block the external opening of the ear. They are made of soft, rubberlike materials that are held against the ear by a light band or head suspension. Circumaural protectors are earmuffs that fit over the ears and rest against the sides of the head. The space between the outer and inner liners of the muffs is filled with liquid, air, or, most commonly, foam. The seal that rests against the head is made of a soft, pliable material such as vinyl. Helmet protectors enclose the entire head and may support earmuffs. Helmets are used in extremely high-noise areas and where protection is needed against bumps and impacts.

The OSHA regulations include methods for estimating how well hearing protectors reduce noise (Appendix B, 40 CFR 1910.95). This appendix describes four methods using EPA's Noise Reduction Rating (NRR), which must be shown on hearing protector packages. The NRR is a measure of how much sound is reduced; an employee's TWA exposure is calculated by subtracting the NRR from the noise exposure level. The method of calculation differs, based on how noise is measured in the work area: by instrument type (dosimeter or sound level meter) or by weighting network (C-weighted or A-weighted sound measurements).

REFERENCE

National Safety Council, *Fundamentals of Industrial Hygiene*, Barbara A. Plog, ed. NSC, Chicago, IL, 1988.

BIBLIOGRAPHY

Abercrombie, Stanley A. (1988), "Dictionary of Terms Used in the Safety Profession," American Society of Safety Engineers, Des Plaines, IL.

ACGIH (1995), "Threshold Limit Values (TLVs) for Chemical Substances and Physical Agents and Biological Exposure Indices (BEIs)," American Conference of Governmental Industrial Hygienists, Cincinnati, OH.

ANSI (1989), ANSI S3.6-1989, "Specification for Audiometers," American National Standards Institute, New York.

_____ (1991), ANSI S1.25-1991, "Specification for Personal Noise Dosimeters," American National Standards Institute, New York.

_____ (1983), ANSI S1.4-1983, "Specification for Sound Level Meters," American National Standards Institute New York.

National Institute for Occupational Safety and Health (1991), "NIOSH Publications on Noise and Hearing," PB92-155852, U.S. Department of Health and Human Services, Cincinnati, OH.

Suter, A.H., and J.R. Franks, eds. (1990), "A Practical Guide to Effective Hearing Conservation Programs in the Workplace," U.S. Department of Health and Human Services, National Institute for Occupational Safety and Health, Cincinnati, OH.

U.S. DOL (1992), "Hearing Conservation," OSHA 3074 (booklet), U.S. Department of Labor, Washington, DC.

20

ERGONOMICS

Ergonomics refers to the integration of worker, task, tool, and workstation to achieve a safe and comfortable working environment. In such an environment, a worker can perform with greater job satisfaction and efficiency. The term, *ergonomics*, is derived from *ergon* relating to work and strength, and *nomos*, referring to law or rule. In the United States the study of *human factors* may be used synonymously with *ergonomics*. Ergonomics is the integration of many disciplines, including anatomy, physiology, medicine, orthopedics, psychology, and sociology with the goal of making work easy and efficient (NSC 1988).

The focus of ergonomics should be to fit the task to the worker, not the other way around. In their jobs, workers can be exposed to heavy lifting, repetitive motion, vibration, noise, and eyestrain, which can result in injuries or related problems involving the tendons, muscles, or nerves. Many of these problems develop from light-duty or seemingly harmless activities whose ill effects accumulate over time because of continuous or repetitive exposure; hence, the term *cumulative trauma disorders* (CTDs). Consequently, ergonomics is concerned with the design and function of tools, workstations, controls, displays, safety devices, as well as with lighting and noise.

CUMULATIVE TRAUMA DISORDERS

Cumulative trauma disorders (CTDs) describe a broad category of ergonomics-related problems that result from repetitive motion or exposure involving forceful exertion, vibration, mechanical compression, awkward body positions or posture, noise, lighting, and temperature. A common element of many CTDs is the use of force combined with repetitive motion (OSHA 1991). The following sections describe common CTDs.

Tendonitis

Tendonitis is an inflammation of the tendon resulting from overuse or unaccustomed use of the wrist and shoulder, a type of injury common among power press operators, welders, painters, and assembly line workers. Additional damage to the tendon can result from continued exposure. Tendon fibers can fray or tear, the tendon can become thick, bumpy, or irregular, or the injured area may calcify (OSHA 1991).

Tenosynovitis

Tenosynovitis results when there is an excessive secretion of synovial fluid by the synovial sheath surrounding the tendon. Normally, synovial fluid lubricates the tendon to reduce friction during movement; however, repetitive movements can result in overproduction of the fluid, which causes swelling and pain (OSHA 1991). Tenosynovitis is most common on the back of the hand and wrist (IBM undated). Tenosynovitis can cause carpal tunnel syndrome, as discussed in the next subsection.

Carpal Tunnel Syndrome

Carpal tunnel syndrome results from pressure placed on the median nerve, which passes through the carpal tunnel of the wrist. The median nerve provides the sense of touch to the thumb, index and middle fingers, and half of the fourth, or ring, finger. Pressure on the median nerve can cause tingling, numbness, or severe pain in the wrist and hand. Damage to the median nerve can result in permanent loss of sensation and even partial paralysis (OSHA 1991).

In addition to the median nerve, many tendons, nerves, and blood vessels also pass through the carpal tunnel of the wrist. Any swelling of the tendons, through inflammation or overproduction of synovial fluid (tenosynovitis), can cause pressure on the median nerve. Palm-held tools can press on the median nerve, causing carpal tunnel syndrome. Carpal tunnel syndrome is common in many areas of work requiring repetitive motion, including computer data entry, meat processing (deboning), assembly line work, machine operation control, letter sorting, and cashiering (OSHA 1991).

Trigger Finger

Trigger finger results when the tendon controlling the finger develops a bump or groove that prevents the tendon from sliding smoothly and locks the finger in a bent position. Straightening the finger is difficult, and movement can be jerky and painful. Trigger finger, which results from using tools with hard or sharp edges, is common among meat processors, assembly line workers, and carpenters (OSHA 1991).

De Quervain's Disease

In De Quervain's disease, the tendon sheath of the thumb becomes inflamed because of excessive friction between two tendons and their common sheath. This excessive friction can be caused by hand motions used in twisting and forceful gripping, similar to clothes wringing. Butchers, cutters, packers, and seamstresses are susceptible to this type of CTD (OSHA 1991).

Hand-Arm Vibration Syndrome

Hand-arm vibration syndrome (HAVS) is a form of CTD involving damage to nerves and blood vessels from vibrating tools. High-frequency vibrations can cause vascular tissues to thicken, reducing or eliminating blood flow (Robinson and Lyon 1994). Nerve and vascular damage can result in numbness, pain, blanching of the fingers, and loss of dexterity (NSC 1992). Factors affecting the development of HAVS include vibration level and spectrum, type and design of tool, hours and method of tool use, environmental conditions, vibration tolerance of the worker, and tobacco or drug use by the worker (NSC 1992). Raynaud's syndrome is a common vibration-induced disorder.

Raynaud's Syndrome

Raynaud's syndrome, a vibration-induced CTD, results when damaged blood vessels cannot transport enough oxygen to skin and muscles in the hands. Symptoms can include numbness and tingling in the fingers; skin that turns pale, ashen, and cold; and even loss of sensation and control. Extremely cold temperatures can aggravate symptoms and turn fingers white, hence the term "white finger" used in reference to this syndrome (OSHA 1991).

Back Pain

Although back problems are caused by both sudden and cumulative injuries, the majority are caused by cumulative injuries, or CTDs. Back problems develop from injuries to muscles, ligaments, tendons, and spinal discs. Activities that contribute to back problems include excessive twisting, bending, and reaching and lifting or moving large or heavy loads. Poor physical condition, poor posture, and staying in the same position too long also contribute to back problems. Workers who sit for long periods without adequate support of the lower back are susceptible to back pain (OSHA 1991).

Hearing Loss

Excessive noise can cause hearing loss, the severity depending on worker tolerance and the intensity, frequency, and duration of the noise. Hearing loss can be temporary, and the individual can recover when removed from the source of noise. Repetitive exposure, however, can cause irreversible damage to the inner ear, resulting in permanent hearing loss. Even when noise is not likely to cause hearing problems, it may be annoying and affect worker satisfaction and performance. For a detailed discussion of noise-induced problems and control, see Chapter 19.

GENERAL GUIDELINES

Much of good ergonomic design is just common sense. Here are a few general guidelines:

- Eliminate the hazard instead of treating the symptom.
- Design to accommodate more than just the average user. Make tools and the work environment adjustable or designed to fit a majority (for example, 90%) of the users.
- Design for the tallest workers where working height of the hands cannot be adjusted.
- Limit the amount of reaching and twisting required in handling materials.
- Design the *bend* in the tool to prevent twisting and bending of the wrist and use of excessive force.
- Avoid sharp or hard edges where hands contact tools.
- Incorporate changes in position or short breaks into the job or workstation to avoid static work situations.
- Use controls and displays that respond in the way most people expect them to, such as a knob that turns clockwise to increase volume.

DESIGN AND APPLICATIONS

There are very specific and detailed guidelines for each area of ergonomic design. A brief discussion of some of the most common areas where ergonomic design is used is given in the following sections: material handling, hand tools, chairs, video terminal displays, lighting, noise, temperature, and workstations. Two design elements common to many of these areas, designing for percentiles and population stereotypes, are discussed first.

Designing for Percentiles

Designing for percentiles means designing for a certain portion of the population instead of just the average person. A typical practice is to design for 90% of users, ranging from the 5th percentile, representing the small body (usually a woman), to the 95th percentile, representing the large body (usually a man). To accommodate such a broad range in sizes, the design may incorporate adjustable controls for height, position, or other features. Table 20.1 gives a sample of population measurements for the 5th and 95th percentiles. Chapter 8, "Statistical Applications," describes the meaning of percentiles in more detail and shows how to calculate percentiles for a given set of data.

TABLE 20.1 Sample Population Measurements

	Population Percentiles (inches) 50/50 Males/Females		
	5th	50th	95th
Measurement			
STANDING			
• Forward functional reach, including body depth at shoulder	27.2	30.7	35.0
• Shoulder height	48.4	54.4	59.7
• Stature	60.8	66.2	72.0
• Functional overhead reach	74.0	80.5	86.9
SEATED			
• Eye height	27.4	29.9	32.8
• Sitting height, normal	32.0	34.6	37.4
• Functional overhead reach	43.6	48.7	54.8
• Hip breadth	12.8	14.5	16.3
HAND			
• Hand length	6.7	7.4	8.0
• Hand breadth	2.8	3.2	3.6
• Grip breadth, inside diameter	1.5	1.8	2.2
• Hand spread, digit one to two, 1st phalangeal joint	3.0	4.3	6.1
OTHER			
• Head breadth	5.4	5.9	6.3
• Weight, in pounds	105.3	164.1	226.8

Source: Information from NSC (1992).

Population Stereotypes

Controls should be designed to operate the way people expect them to. For example, people in the United States naturally assume that turning a knob clockwise or flipping a switch up will turn on a machine (in Europe, flipping a light switch *down* turns the light on). These typical reactions of people are called population stereotypes. The following are examples of up/down control stereotypes in the United States (NSC 1992):

- Up—Expected control responses are on, start, high, in, fast, raise, increase, open, engage, automatic, forward, alternating, and positive.
- Down—Expected control responses are off, stop, low, out, slow, lower, decrease, close, disengage, manual, reverse, direct, and negative.

Material Handling

Material handling, whereby a person lifts, pushes, pulls, or otherwise manually moves a load, is an area susceptible to producing back injuries. In this subsection, the NIOSH guidelines for lifting and recommendations by Snook (NSC 1988) for material handling are described and then illustrated by examples. In addition to these specific guidelines, some general DO's and DON'Ts are presented in Table 20.2.

NIOSH Lifting Equation. NIOSH has established lifting guidelines whereby the lifting is limited to two-hand symmetrical lifting in front of the body that does not involve twisting. These guidelines apply only to lifting; they do not apply to lowering, pushing, pulling, holding, or carrying loads. There are other conditions that must be satisfied before the guidelines are used:

- Lifting must be slow and smooth, without jerking or sudden movement.
- When grasping the load, the hand spread must be 30 inches (76 cm) or less.
- Lifting posture must be unrestricted.
- Couplings must be good between hands and material handles and between shoes and flooring.
- The work environment must be favorable, with good lighting, comfortable temperature, etc.
- Other physical activities associated with the lifting (holding, carrying, pushing, pulling) should be so minimal that the worker may be assumed to be at rest when not lifting.
- The worker must be physically fit and accustomed to physical labor.

TABLE 20.2   General Guidelines for Material Handling

For the Designer

DO . . .   design the task to eliminate manual lifting and lowering. If, instead, the material is handled manually, have the work performed between knuckle and shoulder height.

DO . . .   position the load to reduce the distance a worker would have to reach or bend. Engineering controls could include raising the materials off the floor, setting them on tilt dollies, using containers with removable or drop fronts, or using scissor or platform lifts or hoists.

DON'T . . .   rely solely on training for safe material handling practices. People tend to revert to their own way of doing things if training is not reinforced from time to time. This is a particular problem in an emergency, when people may act quickly without thinking.

For the Worker

DO . . .   be in good physical shape.

DO . . .   think before you act. Place material conveniently within your reach. Have handling aids available. Make sure you have sufficient space and clearance.

DO . . .   get a good grip on the load. Test the weight before trying to move it. If it is too bulky or heavy, get a mechanical lifting aid or another person to help you, or both.

DO . . .   get the load close to your body. Place your feet close to the load. Stand in a stable position with your feet pointing in the direction of movement. Lift mostly by straightening the legs.

DO . . .   keep your work area well organized and free of clutter or materials that could cause tripping, slipping, or falling.

DO . . .   use mechanical lifting aids or help from another person.

DON'T . . .   attempt to lift or lower a difficult load if you are not used to lifting and vigorous exercise.

DON'T . . .   twist your back or bend sideways.

DON'T . . .   lift or lower awkwardly.

DON'T . . .   lift or lower with your arms extended.

DON'T . . .   continue heaving when the load is too heavy.

Source: Information from NSC (1988).

NIOSH has two guidelines for lifting loads, the action limit (AL) and the maximum permissible limit (MPL). The AL is the amount of weight that more than 75% of women and 99% of men could be expected to lift. The MPL is three times the AL and represents the weight that only 1% of women and 25% of men could lift.

Loads below the AL are considered acceptable inasmuch as the risk of injury is nominal for most workers. Loads between the AL and the MPL are considered unacceptable unless administrative or engineering controls are implemented. Such controls may include selecting persons capable of the task; training workers to perform the job safely; reducing the weight, frequency, or distance traveled in lifting; or changing the position of the load when grasped by the worker. Loads above the MPL are considered hazardous and unacceptable and would require engineering controls such as automation; mechanization; reducing the weight, frequency, or distance traveled in lifting; or changing the position of the load when grasped by the worker.

The AL is calculated with Equations 20.1 and 20.2 for load weights in pounds or kilograms, respectively. The AL varies with the horizontal and vertical positions of the load, the distance the load must be lifted, and how often and how long the load is lifted.

For a load in pounds using inches for H, V, and D, the action limit, AL, is:

$$AL\,(\text{lb}) = 90\left(\frac{6}{H}\right)(1 - 0.01|V - 30|)\left(0.7 + \frac{3}{D}\right)\left(1 - \frac{F}{F_{\max}}\right) \qquad (20.1)$$

For a load in kilograms using centimeters for H, V, and D, the action limit, AL, is:

$$AL\,(\text{kg}) = 40\left(\frac{15}{H}\right)(1 - 0.004|V - 75|)\left(0.7 + \frac{7.5}{D}\right)\left(1 - \frac{F}{F_{\max}}\right) \qquad (20.2)$$

The *MPL* is:

$$MPL = 3 \times AL \qquad (20.3)$$

Specific definitions of the equation variables are described as follows (NSC 1992):

- H is the horizontal distance between the center of the handgrasp and the midpoint between ankles at the start of the lift. This distance must be between 6 and 32 in. (15 and 81 cm). If held too close, the load interferes with body movements; if held too far away, the load cannot be grasped by most people.
- V is the vertical distance measured between the handgrasp and the floor or standing surface at the start of the lift. This distance must be between 0 and 70 in. (0 and 178 cm), which is within the vertical reach of most people.
- D is the vertical distance the load travels between the start and the end of the lift. This distance is limited to 80 in. (203 cm). When the travel distance is less than 10 in. (25 cm), set D equal to 10 in. (25 cm) in the AL equation.

- F_{max} is the maximum frequency of lifting that can be sustained, given as lifts per minute. This maximum depends on the vertical reach distance (V) and how long the work is performed. Table 20.3 contains values for F_{max} for a 1-hour and an 8-hour work period.
- F is the frequency of the lift in lifts per minute. For example, one load every 5 minutes is equal to 0.2 (1 divided by 5). For a frequency of less than one load per minute, set F equal to 0. F cannot exceed the maximum frequency, F_{max}.

EXAMPLE 20.1
Action Limit for Load Lifting

You want to design a lifting task to fall below the NIOSH action limit. The load to be lifted is 15 lb. A lifting frequency of three loads per minute throughout an 8-hour shift is required in order to keep up with manufacturing. If you use a horizontal distance of 30 in., a vertical reach distance of 20 in., and a vertical travel distance of 30 in., will the load be below the NIOSH action limit? If not, how much would the horizontal distance have to change to have the load fall below the AL?

Solution. Using Equation 20.1, the AL is 9.72 lb.

$$AL = 90\left(\frac{6}{30}\right)(1 - 0.01|20 - 30|)\left(0.7 + \frac{3}{30}\right)\left(1 - \frac{3}{12}\right)$$

$$= 9.72 \text{ lb}$$

The 15-lb load exceeds the AL, so the task, as designed, is unacceptable under NIOSH guidelines. Because you have the option to change the horizontal distance, you can solve for H as follows:

$$H = \frac{(90)(6)(0.9)(0.8)(0.75)}{15}$$

$$= 19.44 \text{ lb}$$

TABLE 20.3 Maximum Lift Frequency for NIOSH Allowable Load

	Vertical Reach Distance, V, in in. (cm)	
	$V \leqslant 30$ in. ($\leqslant 75$ cm)	$V > 30$ in. (> 75 cm)
Duration of Lifting	Stooped Position	Standing Position
1 hour	15 lifts/minute	18 lifts/minute
8 hours	12 lifts/minute	15 lifts/minute

Source: Information from NSC (1992).

Note in this example how each variable represents a *reducing factor* for the AL. For instance, the vertical reach distance reduces the AL by a factor of 0.9, the vertical travel distance reduces the AL by a factor of 0.8, and the lift frequency combination reduces the AL by a factor of 0.75. Notice how large an impact the original horizontal distance had on the AL (a reducing factor of 0.2).

Snook's Guidelines. Snook (1978) developed a series of tables to determine acceptable loads for lifting, lowering, pushing, pulling, and carrying. The work conditions that must be satisfied before using these guidelines are similar to those for the NIOSH lifting guidelines. Snook's conditions are as follows:

- Material handling is two-handed, symmetric, and directly in front of the body with no twisting.
- When grasping the load, the hand spread must be 30 in. (76 cm) or less.
- Working posture must be unrestricted.
- Couplings must be good between hands and material handles and between shoes and flooring.
- Work environment must be favorable, with good lighting, comfortable temperature, etc.
- Other physical activities associated with the work are minimal.
- The worker is physically fit and accustomed to physical labor.

Instead of using an equation to determine the allowable weight or force, you can look up the acceptable value in Snook's tables. Snook's tables cover a broader range of conditions than the NIOSH lifting equation. With Snook's tables, you can find acceptable values for males or females and for 50%, 75%, or 90% of either gender. Lifting forces are also divided into three different lifting regimes: floor level to knuckle height, knuckle height to shoulder height, and shoulder height to arm reach (NSC 1988). The following example uses Snook's tables to determine an acceptable load for pushing material in a cart.

EXAMPLE 20.2
Snook's Guidelines for Pushing Force

Determine the allowable weight of material that can be pushed in a cart by 90% of the male work force if a loaded cart must be pushed for 100 feet every 5 minutes. The maximum pushing force recommended by Snook under these conditions is 26.4 lb. Assume that the cart weighs 100 lb when empty and that the coefficient of friction is 0.09 while the cart is rolling.

Solution. The sustained force required to push the loaded cart must be greater than the friction force. The friction force is equal to the weight of the loaded cart times the coefficient of friction. Setting up the equation and solving for W, the weight of material in the cart, you find that no more than 193 lb should be pushed in the cart.

$$26.4 \text{ lb} > 0.09(W + 100 \text{ lb})$$

$$W < \frac{26.4}{0.09} - 100$$

$$W < 193 \text{ lb}$$

Hand Tools

Repetitive use of poorly designed hand tools makes a worker susceptible to cumulative trauma disorders, such as carpal tunnel syndrome, Raynaud's syndrome, and others, as discussed earlier in this chapter. Designing a hand tool for a particular job is naturally very specific to the type of tool and the task to be done. There are, however, a few general guidelines that can be followed where they are appropriate for the tool and task at hand (Robinson and Lyon 1994, NSC 1992, NSC 1988):

- Put the bend in the tool instead of forcing the wrist to bend when the tool is used. The wrist should stay at or near a straight position, neither bent nor rotated.
- Fit the tool to the form and functions of the hand; however, in order to accommodate different hand shapes and sizes, do not design the tool to be too formfitting.
- Instead of a finger pinch grip, use a hand grip, which requires less force and reduces the stress on the tendons and other parts of the hand. Design for a power grip when forceful exertions are required.
- Avoid trigger designs that concentrate pressure on the last joint of the finger; instead, design for simultaneous movement of the middle and end finger sections.
- Avoid pressure points where the tool contacts the hand. Avoid sharp, hard, narrow, or fluted edges. Design the handle to spread the stress more evenly across the palm of the hand.
- The cross-sectional shape of the handle should be designed for the type and direction of force exerted in using the tool. Elliptical or rectangular shapes (with well-rounded edges) are useful for twisting, torquing, or turning. Circular shapes may be more useful if the tool will be used differently for a variety of tasks.
- Handles should have sufficient cross section so that the hand nearly circles the handle.
- Handles should accommodate the length of the hand.
- Use flanges at the end of the handle to guide the hand into position and prevent the hand from slipping off the tool. Flared handles, finger stops, and thumb stalls can also help prevent slippage.

- To help prevent slippery conditions, design roughened or textured handles or handles with longitudinal grooves or ridges (avoiding sharp or hard edges).
- Use materials for handles that are imprevious to oil and other liquids.
- Reduce tool vibration.
- Avoid tool design or work conditions that cool the hand too much, such as cold environments or air movement in air-powered tools.
- Help support the weight of the tool during use to reduce stress on the hands, arms, and shoulders.
- Design the tool to be well balanced during use. In general, the center of gravity should be located close to the hand to reduce fatigue and prevent the tool from slipping out of the hand.
- Attach power hoses or cables to top of the tool to avoid tripping hazards.
- Use gloves that do not hinder movement, require excessive force to use the tool, or force the hand, arm, or body into an awkward position.

Chairs

Good chair design can help prevent back stress and injuries associated with cumulative trauma disorders. The type and design of a chair depends on the type of job to be done. A chair should allow the worker a full range of motion—to sit down and stand up easily, move around in the seat, and lean forward. Here are some general guidelines for ergonomic chair design, with particular emphasis on office chairs (Braganza 1994, NSC 1992, NSC 1988):

- The chair should accommodate a normal range of movements and changes in body posture. Both forward and reclined positions should be accommodated.
- The chair should be adjustable to accommodate a wide range in body shapes and sizes. Adjustable controls should be natural and easy to use.
- The seat of the chair should be adjustable in height. The depth of the chair seat should be long enough for support, but short enough so that there is clearance between the seat front and the back of the knees. The front of the seat should be contoured and rounded to avoid excessive pressure on the thighs and turned upward slightly to prevent sliding forward. The seat should be slightly hollow to accommodate the underside of the body. The seat surface should be slip resistant, lightly padded, and covered with a permeable material.
- The backrest should be contoured and large enough to support the back and the neck. The upper part should be slightly concave. The lower part should allow room for the buttocks and contain a lumbar pad to support the lower back. The height of the lumbar pad should be adjustable. The backrest must be capable of being inclined backward.

- For greater stability, use a five-armed chair base.
- Where a worker's feet do not touch the floor after properly adjusting the chair to the workstation, use a footrest to bring the upper leg parallel to the floor.

Video Terminal Displays

Poor design, use, and placement of video terminal displays (VDTs) can result in eye fatigue and irritation, visual discomfort, headaches, and blurred vision. Visual discomfort results from poor contrast between display characters and screen background, high contrast between the screen and other surfaces, screen glare and flicker, eye fixation or continuous looking back and forth between the screen and documents, and refractive errors (OSHA 1991, NSC 1988). The following are general guidelines for VDTs (OSHA 1991, NSC 1988, IBM 1984):

- Both the position and the appearance of the VDT should be adjustable. The screen should tilt up and down and rotate from side to side. The worker should be able to adjust screen brightness and contrast. The worker should be able to adjust his or her position relative to the VDT, keyboard, copy or document holder, and work surface.
- The VDT should be at or below eye level so that the worker does not have to tilt the head at a considerable angle.
- For illumination, use low, diffuse ambient lighting with special area task lighting. Task lighting should be low power and low heat generating. Avoid direct lighting that produces glare and reflection on the screen. Do not place the VDT directly in front of or behind window lighting. Control window lighting with drapes, blinds, or shades. Reduce reflections from surrounding bright surfaces such as walls, windows, furniture, and light fixtures. Use antiglare coatings or filters on screens.
- Use special corrective lenses where required for certain workers, such as those wearing bifocals. Usually the distance between the screen and eye accommodates neither near nor far vision.
- Use screen characters that are neither too small nor too large. People generally recognize words by character and word groupings and word shapes.
- Locate electrical outlets and cables so that cables are uncluttered and unobtrusive.
- Design the task so that the worker can vary position or take short breaks.

Lighting and Light Signals

Improper lighting can cause visual discomfort, eyestrain, eye and muscle fatigue, and headaches (Braganza 1994). Both too little and too much lighting can be problematic. Glare, direct light, and highly reflective surfaces in the field of vision should be avoided.

Lighting should be designed for the task at hand. A combination of ambient and task lighting works well because it matches lighting requirements with specific activities, and because it puts light only where it is needed, it is cost-effective as well. Ambient lighting is used for general illumination and low-visual-demand tasks. Task lighting is used to supplement ambient lighting where a particular task demands it within the work area (Braganza 1994). Recommended light levels for different types of tasks have been published by the Illuminating Engineering Society of North America (Kaufman 1981).

For light signals, the National Safety Council (1988) recommends the following color/signal combinations:

- Red—System is inoperative or action cannot be taken until corrective or override action is taken. Examples are conditions such as "no go," "error," "failure," and "malfunction." A flashing red signal should be used only to signal emergency conditions that require immediate action to prevent personal injury or equipment damage.
- Yellow—Conditions are marginal; extreme caution is required; checking or rechecking, or unexpected delay, is necessary.
- Green—Conditions are satisfactory and it is OK to proceed. Examples are conditions such as "go ahead," "in tolerance," "ready," "function activated," and "power on."
- White—Conditions that are neither right or wrong and do not imply success or failure of an operation; functional or physical position; action in progress. Examples are alternative functions such as "rear steering on" and transitory conditions such as "fan on."
- Blue—Advisory conditions; however, preferential use of blue should be avoided.

Noise and Auditory Signals

Excessive noise can cause both temporary and permanent hearing loss. At a minimum, unwelcome noise can be distracting and aggravating to workers. Noise in the workplace can interfere with communication and response to auditory displays. Noise-induced stresses can result in increased muscle tension, faster pulse rates, and increased blood pressure (OSHA 1991).

Noise generated by power tools should be considered in choosing among available models. Some tools have built-in silencing features (Robinson and Lyon 1994). Direct noise from machinery or an activity can be reduced by enclosing the area or isolating the work from the rest of the work area. Reflected noise can be reduced by absorbing materials such as sound-absorption panels, acoustic ceiling tiles, carpet, and curtains (Braganza 1994).

The National Safety Council (1992) recommends the following guidelines for auditory displays:

- Make the noise intensity higher than the lowest threshold level so that the display can be heard.
- Do not overload the worker with too many auditory signals or other hearing demands.
- Make the auditory display compatible with similar displays, controls, and machine movements.
- Make sure that other environmental factors do not mask the display.

Temperature

Work areas that are too cold, too hot, too humid, or too drafty can cause worker discomfort and affect performance and health. Cold stress can aggravate conditions caused by blood vessel abnormalities such as Raynaud's syndrome or result in hypothermia or frostbite. Heat stress can result in heat fatigue, fainting, heat exhaustion, and heatstroke (NSC 1988). The following lists are samples of temperature control measures recommended by the National Safety Council (1988):

Cold Stress

- If fine work is done with bare hands for 10 to 20 minutes or more, keep hands warm with warm air jets, radiant heaters, or contact warm plates.
- Shield work areas from drafts.
- At temperatures below 1°C (30°F), cover metal tool handles and control bars with thermal insulating material.
- Provide warming shelters for continuous work at −7°C (20°F) and below.
- Schedule work and rest periods to reduce the peak of cold stress.
- Arrange work to minimize sitting still or standing for long periods.

Heat Stress

- Modify the heat load through engineering controls such as ventilation, air-conditioning, screening, insulation, and process modification or operation; by using protective clothing and equipment; and by modifying work practices and using labor-reducing devices.
- Enhance convective heat exchange by modifying air temperature and movement. Spot cooling can be more efficient than cooling large work

areas; however, such cooling should not interfere with ventilation needed to control toxic chemicals.

- If a radiant heat source cannot be reduced, shield the worker from the source. Exposure to radiant heat can be reduced by lowering the process temperature; relocating, insulating, or cooling the heat source; placing line-of-sight reflective shielding between the source and the worker; and coating the hot surface to change its emissivity.

- Enhance evaporative heat control (sweating) by increasing air movement and decreasing ambient water vapor pressure. A simple way to increase air movement is to use fans and blowers. Spot air-conditioning to reduce water vapor pressure can be more efficient than air-conditioning the entire work area. Water vapor pressure can also be reduced by controlling water vapor leaks from manufacturing equipment and evaporation from wet floors.

- Where engineering controls are not practical, modify work practices. Work practices can be modified by limiting exposure time, reducing metabolic heat loads, building up worker heat tolerance through acclimatization and physical conditioning, training workers, and screening workers for heat intolerance and physical fitness.

- Ensure that workers know about and use proper clothing and personal protective equipment. Workers should wear loose-fitting clothing in humid areas. Minimize the amount of skin exposed to medium radiant heat loads, and use reflective garments with high radiant heat loads. Where complete isolation of the worker from the heat source is required, use specialized suits cooled by water, air, or ice.

Workstations

Workstation design requires the integration of many different elements, many of are discussed in previous sections of this chapter. Basic design aspects of workstations involve body clearance and position, material handling, tools and controls, visual displays, and auditory signals.

The National Safety Council (1988) lists six general principles for workstation design:

- Plan the ideal, then the practical.
- Plan the whole, then the detail.
- Plan the work process and equipment around the system requirements.
- Plan the workplace layout around the process and equipment.
- Plan the final enclosure around the workplace layout.
- Use mockups to evaluate alternative solutions and to check the final design.

ERGONOMIC PROGRAMS

OSHA (1991) recommends a four-part ergonomic program for employers, consisting of:

- Worksite analysis;
- Hazard prevention and control;
- Medical management; and
- Training and education.

Worksite Analysis

The purpose of worksite analysis is to identify ergonomic problems and their related risk factors. Records of job-related injuries and illnesses are important sources of information but should be supplemented by other data-gathering activities. In evaluating risk factors, it is important to remember that several different factors can cause the same symptom and that some symptoms result from a combination of factors (NSC 1992, OSHA 1991).

Hazard Prevention and Control

Ergonomic hazard prevention and control can include engineering controls, work practices, administrative controls, and personal protective equipment (OSHA 1991).

Engineering controls are preferred because they can design a job to fit worker capabilities and characteristics. Designs that are adjustable or fit a range of worker characteristics (see "Designing for Percentiles" earlier in this chapter) are best. Engineering controls can improve the design in such areas as material handling systems, hand tools, chairs, workstations, computer terminals, lighting, noise abatement, and temperature adjustment.

Good work practices include proper work techniques, employee training and conditioning, regular monitoring, and a system for maintenance, feedback, adjustments, and modification.

Administrative controls, such as frequent breaks and job rotation, reduce the potential for problems by reducing the duration, frequency, and severity of exposure.

Personal protective equipment should not be the first method of choice for reducing ergonomic hazards, as it is always best to first try eliminating the hazard through engineering controls. Examples of personal protective equipment include gloves, slip-on tool handles, and padded mats to reduce vibration. Personal protective equipment should be adjustable or come in a variety sizes to accommodate differences among workers. It should be chosen judiciously because, in some cases, such equipment can increase a hazard. For example, heavy gloves may cause the worker to apply a more forceful grip and padded tool handles can make a tool too large to grip properly.

Medical Management

Medical management can identify early symptoms of CTDs and develop standard procedures for ergonomic problems and injuries. A medical management program should include (OSHA 1991) the following:

- Injury and illness recordkeeping;
- Early recognition and reporting of CTD symptoms;
- Systematic evaluation and referral;
- Conservative treatment, restricted-duty jobs when necessary;
- Conservative return to work;
- Systematic monitoring, including periodic workplace walk-throughs;
- Adequate staffing and facilities, including health care providers for each work shift; and
- Employee training and education.

Training and Education

Training and education increase awareness of ergonomic hazards. Workers should be trained in the proper use of equipment, tools, and controls. They should also be taught how to use their own bodies correctly, such as for lifting and maintaining comfortable postures, and should be made aware of the potential problems that can develop otherwise.

REFERENCES

Braganza, Barry J., "Ergonomics in the Office," *Professional Safety*, American Society of Safety Engineers, Vol. 39/No. 8, August 1994.

International Business Machines Corporation, "Human Factors of Workstations with Visual Displays," (technical manual), IBM, 1984.

International Business Machines Corporation, *Ergonomics Handbook*, IBM, Armonk, NY, undated.

Kaufman, J.E., ed., "IES Lighting Handbook: Application Volume," Illuminating Engineering Society, Waverly Press, Baltimore, MD, 1981.

National Safety Council, "Accident Prevention Manual for Business & Industry: Engineering & Technology," 10th ed., edited by Patricia M. Laing, NSC, Chicago, 1992.

National Safety Council, "Fundamentals of Industrial Hygiene," 3d ed., edited by Barbara A. Plog, NSC, Chicago, 1988.

Robinson, Fred, Jr., and Bruce K. Lyon, "Ergonomic Guidelines for Hand-Held Tools," *Professional Safety*, American Society of Safety Engineers, Vol. 39/No. 8, August 1994.

Snook, S.H., "The Design of Manual Handling Tasks," *Ergonomics*, Vol. 21, 1978, pp. 963–985.

U.S. Department of Labor, Occupational Safety and Health Administration, "Ergonomics: The Study of Work," U.S. Government Printing Office, Washington, DC, 1991.

APPENDIX A

SELF-INSPECTION CHECKLISTS FOR OSHA COMPLIANCE

The following checklists have been adapted from the *OSHA Handbook for Small Businesses*, OSHA 2209 (1992). They are sample checklists that can serve as a starting point and are not meant to be all inclusive. A safety specialist or other competent person should review and modify each checklist to meet the requirements of a specific facility or operation.

EMPLOYER POSTING AND SIGNS

( ) Required OSHA workplace poster is displayed in a prominent location where all employees are likely to see it.

( ) Emergency telephone numbers are posted so they are readily available in case of emergency.

( ) Information about employee access to medical and exposure records has been posted or otherwise made readily available to those employees who may be exposed to toxic substances or harmful agents.

( ) Exit signs, room capacities, floor loading, biohazards, exposure to X-ray, microwave, or other harmful radiation or substance are posted where appropriate.

( ) Eyewashes and emergency showers are clearly marked.

( ) "No Eating, Drinking, or Smoking" signs are posted in chemical-use areas.

( ) Signs specifying personal protective equipment are posted, as necessary.

( ) "DANGER—Unauthorized Persons, Keep Out," "POWDER-ACTU-ATED TOOL IN USE," and other warning signs are displayed, as appropriate.

( ) Areas under construction are roped off and clearly signed.

RECORDKEEPING

( ) All occupational injury or illnesses, except minor injuries requiring only first aid, are recorded as required on the OSHA 200 log.

( ) Employee medical records and records of employee exposure to hazardous substances or harmful physical agents are up-to-date and in compliance with current OSHA standards.

( ) Employee training records are kept and are accessible for review by employees, when required by OSHA standards.

( ) Arrangements have been made to maintain required records for the legal period of time for each specific type of record.

( ) Operating permits and records are up-to-date for such items as elevators, air pressure tanks, liquefied petroleum gas tanks, and other items, as required.

SAFETY AND HEALTH PROGRAM

( ) There is an active safety and health program that allows for management of general safety and health program elements as well as the management of hazards specific to the worksite.

( ) There is a designated competent person(s) responsible for the safety and health program.

( ) There is a means of communication for employees to voice safety concerns.

( ) There are established safety and health procedures available to managers and employees.

( ) There are documented safety programs, such as a hazardous communication program, medical surveillance program, process safety management program, laser safety program, personal protective equipment programs, lockout/tagout program, and others, as required.

( ) Employees are advised of successful efforts and accomplishments in the safety area.

( ) Emphasis is placed on safety from top management down, throughout the organization.

MEDICAL SERVICES AND FIRST AID

( ) There is a hospital, clinic, or infirmary for medical care at the worksite or in proximity of the workplace.

( ) If medical and first aid facilities are not in proximity of the workplace, there is at least one person on each shift currently qualified to render first aid.

( ) All employees who respond to medical emergencies as part of their work:

( )   have received first aid.

　　( ) have had hepatitis B vaccination made available to them (see *Note*).

　　( ) have had appropriate training on procedures to protect them from bloodborne pathogens, including universal precautions.

　　( ) have available and understand how to use appropriate personal protective equipment to protect against exposure to bloodborne diseases.

( ) Where employees have experienced an exposure incident involving bloodborne pathogens, they were provided immediate postexposure medical evaluation and follow-up.

( ) Medical personnel are readily available for advice and consultation on matters of employees' health.

( ) Emergency phone numbers are posted.

( ) First aid kits are easily accessible to each work area, with necessary supplies available, and they are periodically inspected and replenished as needed.

( ) First aid kit supplies have been approved by a physician, indicating that they are adequate for a particular area or operation.

Note: Pursuant to an OSHA memorandum of July 1, 1992, employees who render first aid only as a collateral duty do not have to be offered a preexposure hepatitis B vaccine if the employer puts the following requirements into his or her exposure control plan and implements them: (1) the employer must record all first aid incidents involving the presence of blood or other potentially infectious materials before the end of the work shift during which the first aid incident occurred; (2) the employer must comply with postexposure evaluation, prophylaxis, and follow-up requirements of the standards with respect to exposure incidents, as defined by the standard; (3) the employer must train designated first aid providers in the reporting procedure; (4) the employer must offer to initiate the hepatitis B vaccination series within 24 hours to all unvaccinated first aid providers who have rendered assistance in any situation involving the presence of blood or other potentially infectious materials.

( ) There are emergency eyewashes and emergency showers for quick drenching or flushing of the eyes and body in areas where corrosive liquids or materials are handled.

GENERAL WORK ENVIRONMENT

( ) The workplace is clean, sanitary, and orderly.
( ) Work surfaces are kept dry, or appropriate means are taken to assure the surfaces are slip resistant.
( ) All spilled hazardous materials or liquids, including blood and other potentially infectious materials, are cleaned up immediately and according to proper procedures.
( ) Combustible scrap, debris, and waste are stored safely and are removed from the worksite promptly.
( ) All regulated waste, as defined in the OSHA bloodborne pathogens standard (29 CFR 1910.1030), is discarded according to federal, state, and local regulations.
( ) Accumulation of combustible dust is routinely removed from elevated surfaces, including the overhead structure of buildings.
( ) Combustible dust is cleaned up with a vacuum system to prevent the dust from going into suspension.
( ) Metallic or conductive dust is prevented from entering or accumulating on or around electrical enclosures or equipment.
( ) Metal waste cans used for oily and paint-soaked waste are covered.
( ) All oil- and gas-fired devices are equipped with flame failure controls that will prevent a flow of fuel if pilots or main burners are not working.
( ) Paint spray booths, dip tanks, etc., are cleaned regularly.
( ) The minimum number of toilets and washing facilities are provided.
( ) Work areas are adequately illuminated.
( ) Pits and floor openings are covered or otherwise guarded.

FIRE PROTECTION

( ) The local fire department is well acquainted with the facility, its location, and specific hazards.
( ) The fire alarm system (if required) is certified.
( ) The fire alarm system (if required) is tested at least annually.
( ) If there are interior standpipes and valves, they are inspected regularly.
( ) If there are outside private fire hydrants, they are flushed at least once a year and are on a routine preventive maintenance schedule.

( ) Fire doors and shutters are in good operating condition.

( ) Fire doors and shutters are unobstructed and protected against obstructions, including their counterweights.

( ) Fire door and shutter fusable links are in place.

( ) Automatic sprinkler system water control valves, air pressure, and water pressure are checked weekly/periodically, as required.

( ) Maintenance of automatic sprinkler systems is assigned to responsible person(s) or to a sprinkler contractor.

( ) Sprinkler heads are protected by metal guards, when exposed to physical damage.

( ) There is proper clearance maintained below sprinkler heads.

( ) Portable fire extinguishers are provided in adequate numbers and type.

( ) Fire extinguishers are mounted in readily accessible locations.

( ) Fire extinguishers are recharged regularly, which is so noted on the inspection tags.

( ) Employees are instructed periodically in the use of extinguishers and fire protection procedures.

PERSONAL PROTECTIVE EQUIPMENT AND CLOTHING

( ) Protective goggles or face shields are provided and worn where there is any danger of flying particles or corrosive materials.

( ) Approved safety glasses are required to be worn at all times in areas where there is a risk of eye injuries such as punctures, abrasions, contusions, or burns.

( ) Employees who need corrective lenses in working environments having harmful exposures are required to wear only approved safety glasses or protective goggles, or to use other medically approved precautionary procedures.

( ) Protective gloves, aprons, shields, or other means are provided and required where employees could be cut or where there is reasonable anticipated exposure to corrosive liquids, chemicals, blood, or other potentially infectious materials. (See 29 CFR 1910.1030 (b) for the definition of "other potentially infectious materials.")

( ) Hard hats are provided and worn where danger of falling objects exists.

( ) Hard hats are inspected periodically for damage to the shell and suspension system.

( ) Appropriate foot protection is required where there is the risk of foot injuries resulting from hot, corrosive, or poisonous substances, falling objects, or crushing or penetrating actions.

( ) Approved respirators are provided for regular or emergency use where needed.

( ) All protective equipment is maintained in a sanitary condition, is stored properly, and is ready for use.

( ) Eyewash and emergency shower facilities are within the work area where employees are exposed to injurious corrosive materials.

( ) Special equipment is available for electrical workers, as required.

( ) Food or beverages are consumed only in areas where there is no exposure to toxic material, blood, or other potentially infectious materials.

( ) Protection against the effects of occupational noise exposure is provided when sound levels exceed those of the OSHA noise standard.

( ) Adequate work procedures, protective clothing, and equipment are provided and used in cleaning up spilled toxic or otherwise hazardous materials or liquids.

( ) There are appropriate procedures in place for disposing of or decontaminating personal protective equipment contaminated with, or reasonable anticipated to be contaminated with, blood or other potentially infectious materials.

WALKWAYS

( ) Aisles and passageways are kept clear.

( ) Aisles and walkways are marked, as appropriate.

( ) Wet surfaces are covered with nonslip material.

( ) Holes in the floor, sidewalk, or other walking surfaces are repaired properly, covered, or otherwise made safe.

( ) There is safe clearance for walking in aisles where motorized or mechanical handling equipment is operating.

( ) Materials and equipment are stored in such as way that sharp projectiles will not interfere with a walkway.

( ) Spilled materials are cleaned up immediately.

( ) Changes in direction or elevation are readily identifiable.

( ) Aisles or walkways that pass near moving or operating machinery, welding operations, or similar operations are arranged so that employees will not be subjected to potential hazards.

( ) There is adequate headroom provided for the entire length of any aisle or walkway.

( ) There are standard guardrails provided wherever aisle or walkway surfaces are elevated more than 30 inches above any adjacent floor or the ground.

( ) Bridges are provided over conveyors and similar hazards.

STAIRS AND STAIRWAYS

( ) There are standard stair rails or handrails on all stairways having four or more risers.

( ) All stairways are at least 22 inches wide.

( ) Stairs have landing platforms no less than 30 inches in the direction of travel and extend 22 inches in width at every 12 feet or less of vertical rise.

( ) Stairs angle between 30 and 50 degrees.

( ) Stairs of hollow-pan type treads and landings are filled to the top edge of the pan with solid material.

( ) Step risers on stairs are uniform from top to bottom.

( ) Steps on stairs and stairways are designed or provided with a surface that renders them slip resistant.

( ) Stairway handrails are located between 30 and 34 inches above the leading edge of the stair treads.

( ) Stairway handrails have at least 3 inches of clearance between the handrails and the wall or surface they are mounted on.

( ) Where doors or gates open directly onto a stairway, there is a platform provided so the swing of the door does not reduce the width of the platform to less than 21 inches.

( ) Stairway handrails are capable of withstanding a load of 200 pounds, applied within 2 inches of the top edge, in any downward or outward direction.

( ) Where stairs or stairways exit directly into any area where vehicles may be operated, there are adequate barriers and warnings provided to prevent employees from stepping into the path of traffic.

( ) Stairway landings have a dimension measured in the direction of travel, at least equal to the width of the stairway.

( ) The vertical distance between stairway landings is limited to 12 feet or less.

FLOOR AND WALL OPENINGS

( ) Floor openings are guarded by a cover, a guardrail, or equivalent on all sides (except at entrances to stairways or ladders).

( ) Toeboards are installed around the edges of permanent floor openings (where persons may pass below the openings)

( ) Skylight screens are of such construction and mounting that they will withstand a load of at least 200 pounds.

( ) Glass in windows, doors, glass walls, etc. that are subject to human impact are of sufficient thickness and type for the condition of use.

( ) Grates or similar type covers over floor openings such as floor drains are of such design that foot traffic or rolling equipment will not be affected by the grate spacing.

( ) Unused portions of service pits and pits not actually in use are either covered or protected by guardrails or equivalent.

( ) Manhole covers, trench covers, and similar covers, plus their supports, are designed to carry a truck rear axle load of at least 20,000 pounds when located in roadways and are subject to vehicle traffic.

( ) Floor or wall openings in fire resistive construction are provided with door or covers compatible with the fire rating of the structure and are provided with a self-closing feature, when appropriate.

ELEVATED SURFACES

( ) Signs are posted, when appropriate, showing the elevated surface load capacity.

( ) Surfaces elevated more than 30 inches above the floor or ground are provided with standard guardrails.

( ) All elevated surfaces (beneath which people or machinery could be exposed to falling objects) are provided with standard 4-inch toeboards.

( ) A permanent means of access and egress is provided to elevated storage and work surfaces.

( ) Required headroom is provided where necessary.

( ) Material on elevated surfaces is piled, stacked, or racked in a manner to prevent it from tipping, falling, collapsing, rolling, or spreading.

( ) Dock boards or bridge plates are used in transferring materials between docks and trucks or rail cars.

EXITING OR EGRESS

( ) All exits are marked with an exit sign and illuminated by a reliable light source.

( ) The directions to exits, when not immediately apparent, are marked with visible signs.

( ) Doors, passageways, or stairways that are neither exits nor access to exits and that could be mistaken for exits are appropriately marked "NOT AN EXIT" or the equivalent.

( ) Exit signs are provided with the word "EXIT" in lettering at least 5 inches high and the stroke of the lettering at least ½ inch wide.

( ) Exit doors are side hinged.

( ) All exits are kept free of obstructions.

( ) At least two means of egress are provided from elevated platforms, pits, or rooms where the absence of a second exit would increase the risk of injury from hot, poisonous, corrosive, suffocating, flammable, explosive, or other harmful substances.

( ) There are sufficient exits to permit prompt escape in case of emergency.

( ) Special precautions are taken to protect employees during construction and repair operations.

( ) The number of exits from each floor of a building and the number of exits from the building itself are appropriate for the building occupancy load.

( ) Exit stairways that are required to be separated from other parts of a building are enclosed by at least two-hour fire-resistive construction in buildings more than four stories in height, and not less than one-hour fire-resistive construction elsewhere.

( ) Where a ramp are is as part of required exiting from a building, the ramp slope is limited to 1 foot vertical and 12 feet horizontal.

( ) Where exiting will be through frameless glass doors, glass exit doors, storm doors, etc., the doors are fully tempered and meet the safety requirements for human impact.

EXIT DOORS

( ) Doors that are required to serve as exits are designed and constructed so that the way of exit travel is obvious and direct.

( ) Windows that could be mistaken for exit doors are made inaccessible by means of barriers or railings.

( ) Exit doors can be opened from the direction of exit travel, without the use of a key or any special knowledge or effort, when the building is occupied.

( ) A revolving, sliding, or overhead door is prohibited from serving as a required exit door.

( ) Where panic hardware is installed on a required exit door, it will allow the door to open with the application of a force of 15 pounds or less in the direction of the exit traffic.

( ) Doors on cold storage rooms are provided with an inside release mechanism that will release the latch and open the door even if it is padlocked or otherwise locked on the outside.

( ) Where exit doors open directly onto any street, alley, or other area where vehicles may be operated, there are adequate barriers and warnings provided to prevent employees from stepping into the path of traffic.

( ) Doors that swing in both directions and are located between rooms where there is frequent traffic are provided with viewing panels.

PORTABLE LADDERS

( ) All ladders are maintained in good conditions joints between steps and side rails are tight, all hardware and fittings are securely attached, and moveable parts operate freely without binding or undue play.

( ) Nonslip safety feet are provided on each metal or rung ladder.

( ) Ladder rungs and steps are free of grease and oil.

( ) It is prohibited to place a ladder in front of a door opening toward the ladder except when the door is blocked open, locked, or guarded.

( ) It is prohibited to place ladders on boxes, barrels, or other unstable bases to obtain addition height.

( ) Employees are instructed to face the ladder when ascending or descending.

( ) Employees are prohibited from using ladders that have missing steps, rungs, or cleats; ladders that have broken side rails; or ladders that have other faulty equipment or are otherwise broken.

( ) Employees are instructed not to use the top step of an ordinary stepladder as a step.

( ) When portable rung ladders are used to gain access to elevated platforms, roofs, etc., the ladder always extends at least 3 feet above the elevated surface.

( ) It is required that when a portable rung or cleat-type ladder is used, the base is so placed that slipping will not occur, or it is lashed or otherwise held in place.

( ) Portable metal ladders are legibly marked with signs reading "CAUTION—Do not Use Around Electrical Equipment," or equivalent wording.

( ) Employees are prohibited from using ladders as guys, braces, skids, gin poles, or for other than their intended purposes.

( ) Employees are instructed to adjust extension ladders only while standing at a base (not while standing on the ladder or from a position above the ladder.)

( ) Metal ladders are inspected for damage.

( ) Rungs of ladders are uniformly spaced at 23 inches, center to center.

HAND TOOLS AND EQUIPMENT

( ) All tools and equipment (both company and employee owned) used by employees at their workplace are in good condition.

( ) Hand tools such as chisels, punches, etc. that develop mushroomed heads during use are reconditioned or replaced, as necessary.

( ) Broken or fractured handles on hammers, axes, and similar equipment are replaced promptly.

( ) Worn or bent wrenches are replaced regularly.

( ) Appropriate handles are used on files and similar tools.

( ) Employees are made aware of the hazards caused by faulty or improperly used hand tools.

( ) Appropriate safety glasses, face shields, etc., are used by employees while using hand tools or equipment that may produce flying materials or be subject to breakage.

( ) Jacks are checked periodically to assure that they are in good operating condition.

( ) Tool handles are wedged tightly in the heads of all tools.

( ) Tool cutting edges are kept sharp so that the tools will move smoothly without binding or skipping.

( ) Tools are stored in a dry, secure location.

( ) Eye and face protection is used when driving hardened or tempered studs or nails.

PORTABLE (POWER-OPERATED) TOOLS AND EQUIPMENT

( ) Grinders, saws, and similar equipment are provided with appropriate safety guards.

( ) Power tools are used with the correct shield, guard, or attachment, as recommended by the manufacturer.

( ) Portable circular saws are equipped with guards above and below the base shoe.

( ) Circular saw guards are checked to ensure that they are not wedged up, thus leaving the lower portion of the blade unguarded.

( ) Rotating or moving parts of equipment are guarded to prevent physical contact.

( ) All cord-connected, electrically operated tools and equipment are effectively grounded or are of the approved double insulated type.

( ) Effective guards are in place over belts, pulleys, chains, and sprockets for equipment such as concrete mixers, air compressors, etc.

( ) Portable fans are provided with full guards or screens having openings of ½ inch or less.

( ) Hoisting equipment is available and used for lifting heavy objects, and hoist ratings and characteristics are appropriate for the task.

( ) Ground-fault circuit interrupters are provided on all temporary electrical 15- and 20-ampere circuits used during a period of construction

( ) Pneumatic and hydraulic hoses on power-operated tools are checked regularly for deterioration and damage.

ABRASIVE WHEEL EQUIPMENT—GRINDERS

( ) The work rest is used and kept adjusted to within ⅛ inch of the wheel.
( ) The adjustable tongue on the top side of the grinder is used and kept adjusted to within ¼ inch of the wheel.
( ) Side guards cover the spindle, nut, flange, and 75% of the wheel diameter.
( ) Bench and pedestal grinders are permanently mounted.
( ) Goggles or face shields are always worn by employees when grinding.
( ) The maximum RPM rating of each abrasive wheel is compatible with the RPM rating of the grinder motor.
( ) Fixed or permanently mounted grinders are connected to their electrical supply system with a metallic conduit or other permanent wiring method.
( ) Each grinder has an individual on-and-off control switch.
( ) Each electrically operated grinder is effectively grounded.
( ) Before new abrasive wheels are mounted, they are visually inspected and ring tested.
( ) Dust collectors and powered exhausts are provided on grinders used in operations that produce large amounts of dust.
( ) Splash guards are mounted on grinders that use coolant, so as to prevent the coolant from reaching employees.
( ) Cleanliness is maintained around grinders.

POWDER-ACTUATED TOOLS

( ) Employees who operate powder-actuated tools are trained in their use and carry valid operators' cards.
( ) Each powder-actuated tool is stored in its own locked container when it is not being used.
( ) A sign at least 7 inches by 10 inches with boldface type reading, "POWDER-ACTUATED TOOL IN USE," is conspicuously posted when such a tool is being used.
( ) Powder-actuated tools are left unloaded until they are actually ready to be used.
( ) Powder-actuated tools are inspected for obstruction or defects each day before use.

( ) Powder-actuated tool operators have and use appropriate personal protective equipment such as hard hats, safety goggles, safety shoes, and ear protectors.

MACHINE GUARDING

( ) There is a training program to instruct employees on safe methods of machine operation.

( ) There is adequate supervision to ensure that employees are following safe machine-operating procedures.

( ) There is a regular program of safety inspection of machinery and equipment.

( ) All machinery and equipment are kept clean and properly maintained.

( ) There is sufficient clearance provided around and between machines to allow for safe operations, setup and servicing, material handling, and waste removal.

( ) Equipment and machinery are securely placed and anchored, when necessary, to prevent tipping or other movement that could result in personal injury.

( ) There is a power shutoff switch within reach of the operator's position at each machine.

( ) Electric power to each machine can be locked out for maintenance, repair, or security.

( ) Noncurrent-carrying metal parts of electrically operated machines are bonded and grounded.

( ) Foot-operated switches are guarded or arranged to prevent accidental actuation by personnel or falling objects.

( ) Manually operated valves and switches controlling the operation of equipment and machines are clearly identified and readily accessible.

( ) All emergency stop buttons are colored red.

( ) All pulleys and belts that are within 7 feet of the floor or working level are properly guarded.

( ) All moving chains and gears are properly guarded.

( ) Splash guards are mounted on machines that use coolant so as to prevent the coolant from reaching employees.

( ) Methods are provided to protect the operator and other employees in the machine area from hazards created at the point of operation, ingoing rip points, rotating parts, flying chips, and sparks.

( ) Machinery guards are secured and so arranged that they do not create a hazard in use.

( ) If special hand tools are used for placing and removing material, they protect the operator's hands.

( ) Revolving drums, barrels, and containers are required to be guarded by an enclosure that is interlocked with the drive mechanism, so that revolution cannot occur unless the guard enclosure is in place.

( ) Arbors and mandrels have firm and secure bearings and are free from play.

( ) Provisions are made to prevent machines from automatically starting when power is restored after a power failure or shutdown.

( ) Machines are constructed so they are free from excessive vibration when the largest-size tool is mounted and run at full speed.

( ) If machinery is cleaned with compressed air, the air pressure is controlled and personal protective equipment or other safeguards are utilized to protect operators and other workers from eye and body injury.

( ) Fan blades, when operating within 7 feet of the floor, are protected with a guard having openings no larger than ½ inch.

( ) Saws used for ripping are equipped with antikickback devices and spreaders.

( ) Radial arm saws are so arranged that the cutting head will gently return to the back of the table when released.

LOCKOUT/TAGOUT PROCEDURES

( ) All machinery or equipment capable of movement is required to be deenergized and blocked or locked out during cleaning, servicing, adjusting, or setting-up operations, whenever required.

( ) Where the means of disconnecting power for equipment does not also disconnect the electrical control circuit:
  ( ) The appropriate electrical enclosures are identified.
  ( ) There is a means provided to ensure that the control circuit can also be disconnected and locked out.

( ) Locking out control circuits in lieu of locking out the main power disconnects is prohibited.

( ) All equipment control valve handles are provided with a means of locking out.

( ) The lockout procedure requires that stored energy (mechanical, hydraulic, air, etc.) be released or blocked before equipment is lockedout for repairs.

( ) Appropriate employees are provided with individually keyed personal safety locks.

( ) Employees are required to keep personal control of their key(s) while they have safety locks in use.

( ) It is required that only an employee exposed to a hazard can place or remove the safety lock.

( ) It is required that employees check the safety of a lockout by attempting a start up after making sure that no one is exposed.

( ) Employees are instructed to always push the control circuit stop button prior to reenergizing the main power switch.

( ) There is a means provided to identify, by their locks or accompanying tags, any employees who are working on locked-out equipment.

( ) There are a sufficient number of accident preventive signs or tags and safety padlocks provided for any reasonable foreseeable repair emergency.

( ) When a machine's operations, configuration, or size requires the operator to leave his or her control station to install tools or perform other operations, and a part of the machine could move if accidentally activated, there is a requirement to separately lock out or block out that element.

( ) In the event that equipment or lines cannot be shut down, locked out, and tagged, there is a safe job procedure established and rigidly followed.

WELDING, CUTTING, AND BRAZING

( ) Only authorized and trained personnel are permitted to use welding, cutting, or brazing equipment.

( ) Each operator has a copy of the appropriate operating instructions and is directed to follow them.

( ) Compressed gas cylinders are regularly examined for obvious signs of defects, deep rusting, or leakage.

( ) Care is used in the handling and storage of cylinders, safety valves, relief valves, etc., to prevent damage.

( ) Only approved apparatus (such as torches, regulators, pressure-reducing valves, acetylene generators, manifolds) are used.

( ) Cylinders are kept away from sources of heat.

( ) Cylinders are kept away from elevators, stairs, and gangways.

( ) It is prohibited to use cylinders as rollers or supports.

( ) Empty cylinders are appropriately marked and their valves are closed.

( ) There are signs reading, "DANGER—No Smoking, Matches, or Open Lights" or the equivalent, posted.

( ) Cylinders, cylinder valves, couplings, regulators, hoses, and other apparatus are kept free of oily or greasy substances.

( ) Care is taken not to drop or strike cylinders.

( ) Unless secured on special trucks, regulators are removed and valve-protection caps are put in place before moving cylinders.

( ) Cylinders without fixed hand wheels have keys, handles, or nonadjustable wrenches on stem valves when in service.

( ) Liquified gases are stored and shipped valve-end up with valve covers in place.

( ) Procedures are in place to ensure that a fuel-gas cylinder valve is never cracked near sources of ignition.

( ) Before a regulator is removed, the valve is closed and gas is released from the regulator.

( ) Red is used to identify an acetylene (or other fuel-gas) hose, green is used for an oxygen hose, and black is used for an inert gas or air hose.

( ) Pressure-reducing regulators are used only for the gases and pressures for which they are intended.

( ) The open circuit (no load) voltage of arc welding and cutting machines is as low as possible and is not in excess of the recommended limits.

( ) Under wet conditions, automatic controls for reducing no load voltage are used.

( ) Grounding of the machine frame and safety ground connections of portable machines are checked periodically.

( ) Electrodes are removed from the holders when not in use.

( ) It is required that electric power to the welder be shut off when no one is in attendance.

( ) Suitable fire-extinguishing equipment is available for immediate use.

( ) A welder is forbidden to coil or loop the electrode cable around his body.

( ) Wet machines are thoroughly dried and tested before being used.

( ) Work and electrode lead cables are frequently inspected for wear and damage and replaced when necessary.

( ) Means for connecting cable lengths have adequate insulation.

( ) When the object to be welded cannot be moved and fire hazards cannot be removed, shields are used to confine heat, sparks, and slag.

( ) Fire watchers are assigned when welding or cutting is performed in locations where a serious fire might develop.

( ) Combustible floors are kept wet, covered by damp sand, or protected by fire-resistant shields.

( ) When floors are wet, personnel are protected from possible electric shock.

( ) When welding is performed on metal walls, precautions are taken to protect combustibles on the other side.

( ) Before hot work is begun, used drums, barrels, tanks, and other containers are thoroughly cleaned so that no substances remain that could explode, ignite, or produce toxic vapors.

( ) It is required that eye-protection helmets, hand shields, and goggles meet appropriate standards.

( ) Employees exposed to the hazards caused by welding, cutting, or brazing operations are protected with personal protective equipment and clothing.

( ) A check is made for adequate ventilation where welding or cutting is performed.

( ) When employees are working in confined spaces, environmental monitoring tests are taken and other requirements for confined spaces are met.

( ) There is a means of quick removal of welders from confined spaces.

COMPRESSORS AND COMPRESSED AIR

( ) Compressors are equipped with pressure relief valves and pressure gauges.

( ) Compressor air intakes are installed and equipped so as to ensure that only clean, uncontaminated air enters the compressor.

( ) Air filters are installed on the compressor intake.

( ) Compressors are operated and lubricated in accordance with the manufacturers recommendations.

( ) Safety devices on compressed air systems are checked frequently.

( ) Before any repair work is done on the pressure system of a compressor, the pressure is bled off and the system is locked-out.

( ) Signs are posted to warn of the automatic starting feature of the compressors.

( ) The belt drive system is totally enclosed to provide protection for the front, back, top, and sides.

( ) It is strictly prohibited to direct compressed air toward a person.

( ) Employees are prohibited from using highly compressed air for cleaning purposes.

( ) If compressed air is used for cleaning off clothing, the pressure is reduced to less than 10 psi.

( ) When using compressed air for cleaning, employees wear protective chip-guarding and personal protective equipment.

( ) Safety chains or other suitable locking devices are used at couplings of high-pressure hose lines where a connection failure would create a hazard.

( ) Before compressed air is used to empty containers of liquid, the safe working pressure of the container is checked.

( ) When compressed air is used with abrasive blast cleaning equipment, the operating valve is a type that must be held open manually.

( ) When compressed air is used to inflate auto tires, a clip-on check and an in-line regulator preset to 40 psi are required.

( ) It is prohibited to use compressed air to clean up or move combustible dust if such action could cause the dust to be suspended in the air and cause a fire or explosion hazard.

COMPRESSORS—AIR RECEIVERS

( ) Every receiver is equipped with a pressure gauge and with one or more automatic, spring-loaded safety valves.

( ) The total relieving capacity of the safety valve is capable of preventing pressure in the receiver from exceeding the maximum allowable working pressure of the receiver by more than 10%.

( ) Every air receiver is provided with a drainpipe and a valve at the lowest point for the removal of accumulated oil and water.

( ) Compressed air receivers are periodically drained of moisture and oil.

( ) All safety valves are tested frequently and at regular intervals to determine whether they are in good operating condition.

( ) There is a current operating permit, as required.

( ) The inlet of air receivers and piping systems are kept free of accumulated oil and carbonaceous materials.

COMPRESSED GAS CYLINDERS

( ) Cylinders with a water weight capacity of more than 30 pounds are equipped with a means for connecting a valve protector device or with a collar or recess to protect the valve.

( ) Cylinders are legibly marked to clearly identify the gas contained.

( ) Compressed gas cylinders are stored in areas that are protected from external heat sources such as flame impingement, intense radiant heat, electric arcs, and high-temperature lines.

( ) Cylinders are located or stored in areas where they will not be damaged by passing or falling objects or subject to tampering by unauthorized persons.

( ) Cylinders are stored or transported in such a manner to prevent them from creating a hazard by tipping, falling, or rolling.

( ) Cylinders containing liquefied fuel gas are stored or transported in such a position that the safety relief device is always in direct contact with the vapor space in the cylinder.

( ) Valve protectors are always placed on cylinders when the cylinders are not in use or connected for use.

( ) All valves are closed off before a cylinder is moved, when the cylinder is empty, and at the completion of each job.

( ) Low-pressure fuel-gas cylinders are checked periodically for corrosion, general distortion, cracks, or any other defect that might indicate a weakness or render them unfit for service.

( ) The periodic check of low pressure fuel-gas cylinders includes a close inspection of each cylinder's bottom.

HOIST AND AUXILIARY EQUIPMENT

( ) Each overhead electric hoist is equipped with a limit device to stop the hook travel at its highest and lowest points of safe travel.

( ) Each hoist automatically stops and holds any load of up to 125% of its rated load if its actuating force is removed.

( ) The rated load of each hoist is legibly marked and visible to the operator.

( ) Stops are provided at the safe limits of travel for trolley hoists.

( ) Controls of the hoist are plainly marked to indicate the direction of travel or motion.

( ) Each cage-controlled hoist is equipped with an effective warning device.

( ) Close-fitting guards or other suitable devices are installed on the hoist to ensure that hoist ropes will be maintained in the sheave grooves.

( ) All hoist chains or ropes are of sufficient length to handle the full range of movement of the application while still maintaining two full wraps on the drum at all times.

( ) Nip points or contact points are guarded when between hoist ropes and sheaves that are permanently located within 7 feet of the floor, ground, or working platform.

( ) It is prohibited to use chains or rope slings that are kinked or twisted.

( ) It is prohibited to use a hoist rope or chain wrapped around the load as a substitute for a sling.

( ) The operator is instructed to avoid carrying loads over people.

( ) Only employees who have been trained in the proper use of hoists are allowed to operate them.

INDUSTRIAL TRUCKS—FORKLIFTS

( ) Only trained personnel are allowed to operate industrial trucks.

( ) Substantial overhead protective equipment is provided on high-lift rider equipment.

( ) Required lift truck operating rules are posted and enforced.

( ) Directional lighting is provided on each industrial truck that operates in an area with less than 2 footcandles per square foot of general lighting.

( ) Each industrial truck has a warning horn, whistle, gong, or other device that can be clearly heard above the normal noise in the areas where operated.

( ) The brakes on each industrial truck are capable of bringing the vehicle to a complete and safe stop when fully loaded.

( ) An industrial truck's parking brake will effectively prevent the vehicle from moving when unattended.

( ) Industrial trucks operating in areas where flammable gases or vapors, or combustible dust or ignitable fibers, may be present in the atmosphere, are approved for such locations.

( ) Motorized hand and hand/rider trucks are designed so that the brakes are applied and the power to the drive motor shuts off when the operator releases his or her grip on the device that controls the travel.

( ) Industrial trucks with an internal combustion engines, operated in buildings or enclosed areas, are carefully checked to ensure such operations do not cause harmful concentrations of dangerous gases or fumes.

SPRAYING OPERATIONS

( ) Adequate ventilation is ensured before spray operations are started.

( ) Mechanical ventilation is provided when spraying operations are done in enclosed areas.

( ) When mechanical ventilation is provided during spraying operations, it is ensured that it will not circulate the contaminated air.

( ) The spray area is free of hot surfaces.

( ) The spray area is at least 20 feet from flames, sparks, operating motors, and other ignition sources.

( ) Portable lamps, suitable for use in a hazardous location, are used to illuminate spray areas.

( ) Approved respiratory equipment is provided and used when appropriate during spraying operations.

( ) Solvents used for cleaning have a flash point of 100°F or greater.

( ) Fire-control sprinkler heads are kept clean.

( ) "NO SMOKING" signs are posted in spray areas, paint rooms, paint booths, and paint storage areas.

( ) The spray area is kept clean of combustible residue.

( ) Spray booths are constructed of metal, masonry, or other substantial noncombustible material.

( ) Spray booth floors and baffles are noncombustible and easily cleaned.

( ) Infrared drying apparatus is kept out of a spray area during spraying operations.

( ) The spray booth is completely ventilated before using the drying apparatus.

( ) The electric drying apparatus is properly grounded.

( ) Lighting fixtures for a spray booth are located outside the booth, and the interior is lighted through sealed clear panels.

( ) The electric motors for exhaust fans are placed outside booths or ducts.

( ) Belts and pulleys inside the booth are fully enclosed.

( ) Ducts have access doors to allow cleaning.

( ) All drying spaces have adequate ventilation.

CONFINED SPACE ENTRY

( ) Confined spaces are thoroughly emptied of any corrosive or hazardous substances, such as acids or caustics, before entry.

( ) All lines to a confined space, containing inert, toxic, flammable, or corrosive materials, are valved off and blanked or disconnected and separated before entry.

( ) It is required that all impellers, agitators, or other moving equipment inside confined spaces be locked out if they present a hazard.

( ) Either natural or mechanical ventilation is provided prior to confined space entry.

( ) Appropriate atmospheric tests are performed to check for oxygen deficiency, toxic substances, and explosive concentrations in a confined space before entry.

( ) Adequate illumination is provided for the work to be performed in a confined space.

( ) The atmosphere inside a confined space is frequently tested or continuously monitored while work is being performed.

( ) There is an assigned safety standby employee outside the confined space, when required, whose sole responsibility is to watch the work in progress, sound an alarm if necessary, and render assistance.

( ) The standby employee is appropriately trained and equipped to handle an emergency.

( ) The standby employee and other employees are prohibited from entering the confined space without lifelines and respiratory equipment if there is any question of an emergency.

( ) Approved respiratory equipment is required if the atmosphere inside the confined space cannot be made acceptable.

( ) All portable electrical equipment used inside confined spaces is either grounded and insulated, or equipped with ground fault protection.

( ) Before gas welding or burning is started in a confined space, hoses are checked for leaks, compressed gas bottles are forbidden inside of the confined space, torches can be lighted only outside the confined area, and the confined area must be tested for an explosive atmosphere each time before a lighted torch is to be taken into the confined space.

( ) If employees will be using oxygen-consuming equipment such as salamanders, torches, furnaces, etc., in a confined space, there is sufficient air provided to assure combustion without reducing the oxygen concentration of the atmosphere below 19.5% by volume.

( ) There is a written confined space program, and employees are trained on the written program and on the specific duties required of them during the confined space entry.

( ) A confined space permit is provided and posted at the entry point to the confined space.

( ) Communications equipment and procedures to be used for maintaining contact during entry are set in place before an entry takes place.

( ) Special equipment, such as personal protective equipment and safety equipment, is specified and available before entry takes place.

( ) Whenever combustion-type equipment is used in a confined space, there are provisions made to ensure that the exhaust gases are vented outside the enclosure.

( ) Each confined space is checked for decaying vegetation or animal matter that may produce methane.

( ) The confined space is checked for possible industrial waste that could have toxic properties.

ENVIRONMENTAL CONTROLS

( ) All work areas are properly illuminated.

( ) Employees are instructed in proper first aid and other emergency procedures.

( ) Hazardous substances, blood, and other potentially infectious materials, which may cause harm by inhalation, ingestion, or skin absorption or contact, are identified.

( ) Employees are aware of the hazards involved with the various chemicals they may be exposed to in their work environment, such as ammonia, chlorine, epoxies, caustics, etc.

( ) Employee exposure to chemicals in the workplace is kept within acceptable levels.

( ) A less harmful chemical or production method is used when possible.

( ) The work areas ventilation system is appropriate for the work being performed.

( ) Spray-painting operations are done in spray rooms or booths equipped with an appropriate exhaust system.

( ) Employee exposure to welding fumes is controlled by ventilation, use of a respirator, limiting exposure time, or other means.

( ) Welders and other workers nearby are provided with a flash shield during welding operations.

( ) If forklifts and other vehicles are used in buildings or other enclosed areas, the carbon monoxide levels are kept below maximum acceptable concentrations.

( ) It has been determined that noise levels at the facility are within acceptable levels.

( ) Steps are being taken to use engineering controls to reduce excessive noise levels.

( ) Proper precautions are being taken for the handling of asbestos and other fibrous materials, per 29 CFR 1910.1001 or 29 CFR 1926.1101.

( ) Asbestos removal is performed by trained personnel and according to all proper procedures.

( ) Caution labels and signs are used to warn of hazardous substances (e.g., asbestos) and biohazards (e.g., bloodborne pathogens).

( ) Engineering controls are examined and maintained or replaced on a scheduled basis.

( ) Vacuuming with appropriate equipment is used whenever possible, rather than blowing or sweeping dust.

( ) Grinders, saws, and other machines that produce respirable dusts are vented to an industrial collector or a central exhaust system.

( ) All local exhaust ventilation systems are designed and operating properly so as to ensure air flow and volume necessary for the application, ducts not plugged, belts not slipping, etc.

( ) Personal protective equipment is provided, used, and maintained wherever required.

( ) There are written standard operating procedures for the selection and use of respirators where needed.

( ) Restrooms and washrooms are kept clean and sanitary.

( ) All water provided for drinking, washing, and cooking is potable.

( ) All outlets for water not suitable for drinking are clearly identified.

( ) Employees' physical capacities are assessed before their being assigned to jobs requiring heavy work.

( ) Employees are instructed in the proper manner of lifting heavy objects.

( ) Where heat is a problem, all fixed work areas have been provided with spot cooling or air-conditioning.

( ) Employees are screened before assignment to areas of high heat to determine whether their health condition might make them more susceptible to having an adverse reaction.

( ) Employees working on streets and roadways where they are exposed to the hazards of traffic are required to wear bright-colored (traffic orange) warning vests.

( ) Exhaust stacks and air intakes are located so that contaminated air will not be recirculated within a building or other enclosed area.

( ) Equipment producing ultraviolet radiation is properly shielded.

( ) Universal precautions are observed where occupational exposure to blood or other potentially infectious materials can occur and in all instances where differentiation of types of body fluids or potentially infectious materials is difficult or impossible.

FLAMMABLE AND COMBUSTIBLE MATERIALS

( ) Combustible scrap, debris, and waste materials (oily rags, etc.) are stored in covered metal receptacles and are removed from the worksite promptly.

( ) Proper storage is practiced to minimize the risk of fire, including spontaneous combustion.

( ) Approved containers and tanks are used for the storage and handling of flammable and combustible liquids.

( ) All connections on drums and piping are tight.

( ) All flammable liquids are kept in closed containers when not in use (e.g., parts cleaning tanks, pens, etc.)

( ) Bulk drums of flammable liquids are grounded and bonded to containers during dispensing.

( ) Storage rooms for flammable and combustible liquids have explosion-proof lights.

( ) Storage rooms for flammable and combustible liquids have mechanical or gravity ventilation.

( ) Liquefied petroleum gas is stored, handled, and used in accordance with safe practices and standards.

( ) "NO SMOKING" signs are posted on liquefied petroleum gas tanks.

( ) Liquefied petroleum storage tanks are guarded to prevent damage from vehicles.

( ) All solvent wastes and flammable liquids are kept in fire-resistant, covered containers until they are removed from the worksite.

( ) Vacuuming is used whenever possible, rather than blowing or sweeping combustible dust.

( ) Firm separators are placed between containers of combustibles or flammables when they are stacked one upon another, to assure their support and stability.

( ) Fuel gas cylinders and oxygen cylinders are separated by distance, fire resistant barriers, etc., while in storage.

( ) Fire extinguishers are selected and provided for the types of materials in areas where they are to be used:

    ( ) Class A—Ordinary combustible material fires

    ( ) Class B—Flammable liquid, gas, or grease fires

    ( ) Class C—Energized-electrical equipment fires

( ) Appropriate fire extinguishers are mounted within 75 feet of outside areas containing flammable liquids and within 10 feet of any inside storage area for such materials.

( ) Extinguishers are free from obstructions or blockage.

( ) All extinguishers are serviced, maintained, and tagged at intervals not to exceed one year.

( ) Where sprinkler systems are permanently installed, the nozzle heads are directed or arranged so that water will not be sprayed into operating electrical switchboards and other equipment.

( ) "NO SMOKING" signs are posted, where appropriate, in areas where flammable or combustible materials are used or stored.

( ) Safety cans are used for dispensing flammable or combustible liquids at a point of use.

( ) All spills of flammable or combustible liquids are cleaned up promptly.

( ) Storage tanks are adequately vented to prevent the development of excessive vacuum or pressure as a result of filling, emptying, or atmosphere temperature changes.

( ) Storage tanks are equipped with emergency venting that will relieve excessive internal pressure caused by fire exposure.

( ) No smoking rules are enforced in areas involving the storage and use of combustible materials.

HAZARDOUS CHEMICAL EXPOSURE

( ) Employees are trained in safe handling practices for hazardous chemicals.

( ) Employees are aware of the potential hazards involving various chemicals stored or used in the workplace.

( ) Employee exposure to chemicals is kept within acceptable levels.

( ) Eyewash fountains and safety showers are provided in areas where corrosive chemicals are handled.

( ) All containers, such as vats, storage tanks, etc., are labeled as to their contents.

( ) All employees are required to use personal protective clothing and equipment (gloves, eye protection, respirators, etc.) when handling chemicals.

( ) Flammable or toxic chemicals are kept in closed containers when not in use.

( ) Chemical piping systems are clearly marked as to their contents.

( ) Where corrosive liquids are frequently handled in open containers or drawn from storage vessels or pipelines, there are adequate means readily available for neutralizing or disposing of spills or overflows properly and safely.

( ) Standard operating procedures have been established and are being followed in cleaning up chemical spills.

( ) Where needed for emergency use, respirators are stored in a convenient, clean, and sanitary location.

( ) Respirators intended for emergency use are adequate for the various uses for which they may be needed.

( ) Employees are prohibited from eating in areas where hazardous chemicals are present.

( ) Personal protective equipment is provided, used, and maintained whenever necessary.

( ) There are written standard operating procedures for the selection and use of respirators, where needed.

( ) If a respirator program is required, employees are instructed on the correct usage and limitations of the respirators, and respirators are NIOSH or NIOSH-MSHA-approved for the particular application.

( ) Respirators are regularly inspected, cleaned, sanitized, and maintained.

( ) If hazardous substances are used at the facility, there is a medical or biological monitoring system in place.

( ) Threshold limit values and permissible exposure limits have been investigated and documented for chemicals used at the facility.

( ) Control procedures have been instituted for hazardous materials, where appropriate, such as the use of respirators, ventilation systems, handling practices, etc.

( ) Whenever possible, hazardous substances are handled in properly designed and exhausted booths or similar locations.

( ) A general dilution or local exhaust ventilation system is used to control dusts, vapors, gases, fumes, smoke, solvents, or mists that may be generated in the workplace.

( ) Ventilation equipment is provided for the removal of contaminants from operations such as production grinding, buffing, spray painting, and/or vapor degreasing, and it is operating properly.

( ) Employees have not complained about dizziness, headaches, nausea, irritation, or other factors of discomfort when they use solvents or other chemicals. If they have complained, a competent person (i.e., an industrial hygienist) is investigating the cause.

( ) Employees have not complained of dryness, irritation, or sensitization of the skin. If they have complained, a competent person (i.e., an industrial hygienist) is investigating the cause.

( ) There is a trained safety professional and/or industrial hygienist at the facility who evaluates operations at the facility.

( ) If internal combustion engines are used, carbon monoxide is kept within acceptable levels.

( ) Vacuuming is used, rather than blowing or sweeping dusts, whenever possible.

( ) Materials that give off toxic, asphyxiant, suffocating, or anesthetic fumes are stored in remote or isolated locations when not in use.

HAZARDOUS SUBSTANCES COMMUNICATION

( ) There is list of hazardous substances used at the facility.

( ) There is a current written exposure control plan for occupational exposure to bloodborne pathogens and for other potentially infectious materials, where applicable.

( ) There is a written hazard communication program dealing with material safety data sheets (MSDS), labeling, and employee training.

( ) Each container for a hazardous substance (i.e., vats, bottles, storage tanks, etc.) is labeled with product identity and a hazard warning that communicates the specific health hazards and physical hazards.

( ) There is a MSDS readily available for each hazardous substance used.

( ) There is an employee training program for hazardous substances, which includes:

    ( ) An explanation of what an MSDS is and how to use and obtain one.

    ( ) MSDS contents for each hazardous substance or class of substances.

( ) Explanation of worker's "Right-to-Know."

( ) Identification of where an employee can see the employer's written hazard communication program and where hazardous substances are present in their work areas.

( ) The physical and health hazards of substances in the work area and specific protective measures to be used.

( ) Details of the hazard communication program, including how to use the labeling system and MSDSs.

( ) The employee training program on bloodborne pathogens includes the following elements:

( ) An accessible copy of the standard and an explanation of the contents.

( ) A general explanation of the epidemiology and symptoms of bloodborne diseases.

( ) An explanation of the modes of transmission of bloodborne pathogens.

( ) An explanation of the employers exposure control plan and the means by which employees can obtain a copy of the written plan.

( ) An explanation of the appropriate methods for recognizing tasks and the other activities that may involve exposure to blood and other potentially infectious materials.

( ) An explanation of the use and limitations of methods that will prevent or reduce exposure, including appropriate engineering controls, work practices, and personal pretective equipment.

( ) Information on the types, proper use, location, removal, handling, decontamination, and disposal of personal protective equipment.

( ) Information on the hepatitis B vaccine.

( ) Information on the appropriate actions to take and persons to contact in an emergency involving blood or other potentially infectious materials.

( ) An explanation of the procedure to follow if an exposure incident occurs, including the methods of reporting the incident and the medical follow-up that will be made available.

( ) Information on postexposure evaluations and follow-up.

( ) An explanation of signs, labels, and color coding.

( ) Employees are trained in the following:

( ) How to recognize tasks that may result in occupational exposure.

( ) How to use work practice and engineering controls and personal protective equipment and to know their limitations.

( ) How to obtain information on the types, selection, proper use, location, removal, handling, decontamination, and disposal of personal protective equipment.

( ) The contact number and what to do in an emergency.

ELECTRICAL

( ) Compliance with OSHA is specified for all contract electrical work.

( ) All employees are required to report as soon as practicable any obvious hazard to life or property observed in connection with electrical equipment or lines.

( ) Employees are instructed to make preliminary inspection and/or appropriate tests to determine existing conditions before starting work on electrical equipment or lines.

( ) When electrical equipment or lines are to be serviced, maintained, or adjusted, the necessary switches are opened, locked out, and tagged whenever possible.

( ) Portable electrical tools and equipment are grounded or are of the double-insulated type.

( ) Electrical appliances such as vacuum cleaners, polishers, vending machines, etc., are grounded.

( ) Extension cords used have a grounding conductor.

( ) Multiple-plug adapters are prohibited.

( ) Ground-fault circuit interrupters are installed on each temporary 15 or 20 ampere, 120 volt AC circuit at locations where construction, demolition, modifications, alterations, or excavations are being performed.

( ) Temporary circuits are protected by suitable disconnecting switches or plug connectors at the junction with permanent wiring.

( ) Electrical installations in hazardous dust or vapor areas meet the National Electrical Code (NEC) for hazardous locations.

( ) Exposed wiring and cords with frayed or deteriorated insulation are repaired or replaced promptly.

( ) Flexible cords and cables are free of splices or taps.

( ) Clamps or other securing means are provided on flexible cords or cables at plugs, receptacles, tools, equipment, etc., and the cord jacket is securely held in place.

( ) All cord, cable, and raceway connections are intact and secure.

( ) In wet or damp locations, electrical tools and equipment appropriate for the use or location are used.

( ) The location of electrical power lines and cables (overhead, underground, under floors, other side of walls, etc.) is determined before digging, drilling, or similar work is begun.

( ) Metal measuring tapes, ropes, hand lines, or similar devices with metallic thread woven into the fabric are prohibited where they could come in contact with energized parts of equipment or circuit conductors.

( ) The use of metal ladders is prohibited in areas where a ladder or a person using the ladder could come in contact with energized parts of equipment, fixtures, or circuit conductors.

( ) All disconnecting switches and circuit breakers are labeled to indicate their use or the equipment served.

( ) Disconnecting means are always opened before fuses are replaced.

( ) All interior wiring systems include provision for grounding metal parts of electrical raceways, equipment, and enclosures.

( ) All electrical raceways and enclosures are securely fastened in place.

( ) All energized parts of electrical circuits and equipment are guarded against accidental contact by approved cabinets or enclosures.

( ) Sufficient access and working space is provided and maintained about all electrical equipment to permit ready and safe operations and maintenance.

( ) All unused openings (including conduit knockouts) in electrical enclosures and fittings are closed with appropriate covers, plugs, or plates.

( ) Electrical enclosures such as switches, receptacles, junction boxes, etc., are provided with light-fitting covers or plates.

( ) Disconnecting switches for electrical motors in excess of two horsepower are capable of opening the circuit, without exploding, when the motor is in a stalled condition. (Switches must be horsepower rated equal to or in excess of the motor hp rating.)

( ) Low-voltage protection is provided in the control device of motors driving machines or equipment that could cause injury, as a result of inadvertent starting.

( ) Each motor disconnecting switch or circuit breaker is located within sight of the motor control device.

( ) Each motor is located within sight of its controller, or the controller disconnecting means is capable of being locked in the open position or is a separate disconnecting means installed in the circuit within sight of the motor.

( ) The controller for each motor in excess of two horsepower, is rated in horsepower equal to or in excess of the rating of the motor it serves.

( ) Employees who regularly work on or around energized electrical equipment or lines are instructed in cardiopulmonary resuscitation (CPR) methods.

( ) Employees are prohibited from working alone on energized lines or equipment of more than 600 volts.

NOISE

( ) If there are areas in the workplace where continuous noise levels exceed 85 dBA, these are identified.

( ) There is an ongoing preventive health program to educate employees in safe levels of noise, exposures, effects of noise on their health, and the use of personal protection.

( ) Work areas where noise levels make voice communication between employees difficult have been identified, tested for noise exposure levels, and signed for hearing protection, as appropriate.

( ) Noise levels are being measured, using a sound-level meter or an octave band analyzer, and records of noise levels in specific processes or operations are being recorded.

( ) Engineering controls are used to reduce excessive noise levels. Where engineering controls are determined not to be feasible, there are administrative controls (i.e., worker rotation) being used to minimize individual employee exposure to noise.

( ) Approved hearing protective equipment (noise attenuating devices) are available to every employee working in noisy areas.

( ) Noisy machinery is isolated from other operations when possible.

( ) When hearing protection is used, employees are properly fitted and instructed in the use of protective devices.

( ) Employees in high noise areas are given periodic audiometric testing to ensure that the hearing protection program is effective.

FUELING

( ) It is prohibited to fuel an internal combustion engine with a flammable liquid while the engine is running.

( ) Fueling operations are done in such a manner that likelihood of spillage is minimal.

( ) If spillage occurs during fueling operations, the spilled fuel is washed away completely or evaporated, or other measures are taken to control vapors before restarting the engine.

( ) Fuel tank caps are replaced and secured before starting the engine.

( ) In fueling operations, there is always metal contact between the container and the fuel tank.

( ) Fueling hoses are of a type designed to handle the specific type of fuel.

( ) It is prohibited to handle or transfer gasoline in open containers.

( ) Open lights, open flames, and sparking or arcing equipment are prohibited near fueling or transfer of fuel operations.

( ) Smoking is prohibited in the vicinity of fueling operations.
( ) Fueling operations are prohibited in buildings or other enclosed areas that are not specifically ventilated for this purpose.
( ) Where fueling or transfer of fuel is done through a gravity-flow system, the nozzles are of the self-closing type.

IDENTIFICATION OF PIPING SYSTEMS

( ) When nonpotable water is piped through a facility, the outlets or taps are posted to alert employees that the water is unsafe and not to be used for drinking, washing, or other personal uses.
( ) When hazardous substances are transported through aboveground piping, each pipeline is identified at points where confusion could result in hazards to employees.
( ) When pipelines are identified by color painting, all visible parts of the line are so identified.
( ) When pipelines are identified by color, the color code is posted at all locations where confusion could result in hazards to employees.
( ) When the contents of pipelines are identified by name or name abbreviation, the information is readily visible on the pipe near each valve or outlet.
( ) When pipelines carrying hazardous substances are identified by tags, the tags are constructed of durable materials, the message carried is clearly and permanently distinguishable, and tags are installed at each valve or outlet.
( ) When pipelines are heated by electricity, steam, or another external source, suitable warning signs or tags are placed at unions, valves, or other serviceable parts of the system.

MATERIAL HANDLING

( ) There is safe clearance for equipment through aisles and doorways.
( ) Aisleways are designed, permanently marked, and kept clear to allow unhindered passage.
( ) Motorized vehicles and mechanized equipment are inspected daily or prior to use.
( ) Vehicles are shut off and brakes are set prior to loading and unloading.
( ) Containers of combustibles or flammables, when stacked while being moved, are always separated by pallets to provide sufficient stability.
( ) Dock boards (bridge plates) are used when loading or unloading operations are taking place between vehicles and docks.

( ) Trucks and trailers are secured against movement during loading and unloading.

( ) Dock plates and loading ramps are constructed and maintained with sufficient strength to support imposed loading.

( ) Hand trucks are maintained in safe operating condition.

( ) Chutes are equipped with sideboards of sufficient height to prevent the materials being handled from falling off.

( ) Chutes and gravity roller sections are firmly placed or secured to prevent displacement.

( ) At the delivery end of the rollers or chutes, provisions are made to brake the movement of the handled materials.

( ) Pallets are inspected before being loaded or moved.

( ) Hooks with safety latches or other arrangements are used when hoisting materials so that slings or load attachments will not accidentally slip off the hoist hooks.

( ) Securing chains, ropes, chockers, or slings are adequate for the job to be performed.

( ) When material or equipment is being hoisted, provisions are made to ensure that no one will be passing under the suspended loads.

( ) Material safety data sheets are available to employees handling hazardous substances.

TRANSPORTING EMPLOYEES AND MATERIALS

( ) Employees who operate vehicles on public thoroughfares have valid operators' licenses.

( ) When seven or more employees are regularly transported in a van, bus, or truck, the operators license is appropriate for the class of vehicle being driven.

( ) Each van, bus, or truck used regularly to transport employees is equipped with an adequate number of seats.

( ) When employees are transported by truck, provisions are made to prevent their falling from the vehicle.

( ) Vehicles used to transport employees are equipped with lamps, brakes, horns, mirrors, windshields, and turn signals in good repair.

( ) Transport vehicles provided with handrails, steps, stirrups, or similar devices are placed and arranged so that employees can safely mount and dismount.

( ) Employee transport vehicles are equipped at all times with at least two reflective type flares.

( ) A fully charged fire extinguisher, in good condition, with at least a 4 B:C rating is maintained in each employee transport vehicle.

( ) When cutting tools with sharp edges are carried in passenger compartments of employee transport vehicles, they are placed in closed boxes or containers that are secured in place.

( ) Employees are prohibited from riding on top of any load that can shift, topple, or otherwise become unstable.

CONTROL OF HAZARDOUS SUBSTANCES BY VENTILATION

( ) The volume and velocity of air in each exhaust system is sufficient to gather the dusts, fumes, mists, vapors, or gases to be controlled and to convey them to a suitable point of disposal.

( ) Exhaust inlets, ducts, and plenums are designed, constructed, and supported to prevent collapse or failure of any part of the system.

( ) Cleanout ports or doors are provided at intervals not to exceed 12 feet in all horizontal runs of exhaust ducts.

( ) Where two or more different types of operations are being controlled through the same exhaust system, the combination of substances being controlled will not cause a fire, explosion, or chemical reaction hazard in the duct.

( ) Adequate makeup air is provided to areas where exhaust systems are operating.

( ) The source point for makeup air is located so that only clean, fresh air, which is free of contaminants, will enter the work environment.

( ) Where two or more ventilation systems are serving a work area, their operation is such that one will not offset the functions of the other.

( ) Air flow in a local exhaust hood is monitored to ensure adequate air flow, as necessary, and, if necessary for particularly harmful substances, an alarm or interlock is in place to indicate when air flow is below the required velocity.

SANITIZING EQUIPMENT AND CLOTHING

( ) Personal protective clothing and equipment that employees are required to wear or use is of a type capable of being cleaned and disinfected easily.

( ) Employees are prohibited from interchanging personal protective clothing or equipment unless it has been properly cleaned.

( ) Machines and equipment that process, handle, or apply materials that could be injurious to employees are cleaned and/or decontaminated before being overhauled or placed in storage.

( ) Employees are prohibited from smoking or eating in any area where there are contaminants that could be injurious if ingested.

( ) When employees are required to change from street clothing into protective clothing, a clean change room with separate storage facilities for street and protective clothing is provided.

( ) Employees are required to shower and wash their hair as soon as possible after a known contact has occurred with a carcinogen or other hazardous substance that could be injurious to health upon contact.

( ) When equipment, materials, or other items are taken into or removed from a carcinogen regulated area, it is done in a manner that will not contaminate nonregulated areas or the external environment.

TIRE INFLATION

( ) Where tires are mounted and/or inflated on drop center wheels, a safe practice procedure is posted and enforced.

( ) Where tires are mounted and/or inflated on wheels with split rims and/ or retainer rings, a safe practice procedure is posted and enforced.

( ) Each tire inflation hose has a clip-on chuck with at least 24 inches of hose between the chuck and an in-line hand valve and gauge.

( ) The tire inflation control valve automatically shuts off the air flow when the valve is released.

( ) A tire-restraining device such as a cage, rack, or other effective means is used while inflating tires mounted on split rims or on rims using retainer rings.

( ) Employees are prohibited from positioning themselves directly over or in front of a tire while it is being inflated.

( ) Employees receive proper instruction before mounting or inflating tires.

APPENDIX B

PUBLICATIONS OF THE DEPARTMENT OF ENERGY—SYSTEM SAFETY DEVELOPMENT CENTER

CONSTRUCTION SAFETY MONOGRAPHS

SSDC-26.1—*Excavation*
SSDC-26.2—*Scaffolding*
SSDC-26.3—*Steel Erection*
SSDC-26.4—*Electrical*
SSDC-26.5—*Housekeeping*
SSDC-26.6—*Welding/Cutting*
SSDC-26.7—*Confined Spaces*
SSDC-26.8—*Heating of Work Spaces*
SSDC-26.9—*Use of Explosives*
SSDC-26.10—*Medical Services*
SSDC-26.11—*Sanitation*
SSDC-26.12—*Ladders*
SSDC-26.13—*Painting/Special Coatings*
SSDC-26.14—*Fire Protection*
SSDC-26.15—*Project Layout*
SSDC-26.16—*Emergency Action Plans*
SSDC-26.17—*Heavy Equipment*
SSDC-26.18—*Air Quality*

PUBLICATIONS RELATED TO HAZARD ASSESSMENT

SSDC-4—*MORT User's Manual*

SSDC-8—*Standardization Guide for Construction and Use of MORT-Type Analytic Trees*

SSDC-14—*Events and Causal Factors Charting*

SSDC-15—*Work Process Control Guide*

SSDC-17—*Applications of MORT to Review of Safety Analyses*

SSDC-19—*Job Safety Analysis*

SSDC-20—*Management Evaluation and Control of Release of Hazardous Materials*

SSDC-22—*Reliability and Fault Tree Analysis Guide*

SSDC-24—*Safety Assurance System Summary (SASS) Manual for Appraisal*

SSDC-29—*Barrier Analysis*

SSDC-31—*The Process of Task Analysis*

SSDC-33—*The MORT Program and the Safety Performance Measurement System*

PUBLICATIONS RELATED TO INCIDENT INVESTIGATION AND REPORTING

SSDC-3—*A Contractor Guide to Advance Preparation for Accident Investigation*

SSDC-5—*Reported Significant Observation (RSO) Studies*

SSDC-7B—*DOE Guide to the Classification of Recordable Accidents*

SSDC-27—*Accident/Incident Investigation Manual (2d Edition)*

SSDC-41—*Investigating and Reporting Accidents Effectively*

PUBLICATIONS RELATED TO SAFETY MANAGEMENT

SSDC-1—*Occupancy-Use Readiness Manual*

SSDC-6—*Training as Related to Behavioral Change*

SSDC-9—*Safety Information System Guide*

SSDC-10—*Safety Information System Cataloging*

SSDC-11—*Risk Management Guide*

SSDC-12—*Safety Considerations in Evaluation of Maintenance Programs*

SSDC-13—*Management Factors in Accident/Incidents (Including Management Self-Evaluation Checksheets)*

SSDC-18—*Safety Performance Measurement System*

SSDC-23—*Safety Appraisal Guide*

SSDC-25—*Effective Safety Review*

SSDC-36—*MORT-Based Safety Professional Program Development and Improvement*

SSDC-38—*Safety Considerations for Security Programs*

SSDC-39—*Process Operational Readiness and Operational Readiness Follow-On*

PUBLICATIONS RELATED TO HUMAN FACTORS

SSDC-2—*Human Factors in Design*

SSDC-30—*Human Factors Management*

SSDC-32—*The Impact of the Human on System Safety Analysis*

SSDC-34—*Basic Human Factors Considerations*

SSDC-40—*The Assessment of Behavioral Climate*

OTHER PUBLICATIONS

SSDC-16—*SPR0 Drilling and Completion Operations*

SSDC-21—*Change Control and Analysis*

SSDC-28—*Glossary of SSDC Terms and Acronyms*

SSDC-35—*A Guide for the Evaluation of Displays*

SSDC-37—*Time Loss Analysis*

APPENDIX C

SELECTED PUBLICATIONS OF THE NATIONAL COUNCIL ON RADIATION PROTECTION AND MEASUREMENTS

PUBLICATIONS RELATED TO RADIATION EXPOSURE

NCRP Report No. 22 (1959); Addendum 1 (1963)—*Maximum Permissible Body Burdens and Maximum Permissible Concentrations of Radionuclides in Air and in Water for Occupational Exposure*

NCRP Report No. 46 (1975)—*Alpha-Emitting Particles in the Lungs*

NCRP Report No. 52 (1977)—*Cesium-137 from the Environment to Man: Metabolism and Dose*

NCRP Report No. 64 (1980)—*Influence of Dose and Its Distribution in Time on Dose-Response Relationships for Low-LET Radiations*

NCRP Report No. 77 (1984)—*Exposures from the Uranium Series with Emphasis on Radon and Its Daughters*

NCRP Report No. 78 (1984)—*Evaluation of Occupational and Environmental Exposures to Radon and Radon Daughters in the United States*

NCRP Report No. 86 (1986)—*Biological Effects and Exposure Criteria for Radiofrequency Electromagnetic Fields*

NCRP Report No. 89 (1987)—*Genetic Effects from Internally Deposited Radionuclides*

NCRP Report No. 92 (1987)—*Public Radiation Exposure from Nuclear Power Generation in the United States*

NCRP Report No. 93 (1987)—*Ionizing Radiation Exposure of the Population of the United States*

NCRP Report No. 94 (1987)—*Exposure of the Population in the United States and Canada from Natural Background Radiation*

NCRP Report No. 95 (1987)—*Radiation Exposure of the U.S. Population from Consumer Products and Miscellaneous Sources*

NCRP Report No. 100 (1989)—*Exposure of the U.S. Population from Diagnostic Medical Radiation*

NCRP Report No. 101 (1989)—*Exposure of the U.S. Population from Occupational Radiation*

NCRP Report No. 106 (1989)—*Limit for Exposure to "Hot Particles" on the Skin*

NCRP Report No. 116 (1993)—*Limitation of Exposure to Ionizing Radiation*

PUBLICATIONS RELATED TO RADIATION MEASUREMENT AND ASSESSMENT

NCRP Report No. 23 (1960)—*Measurement of Neutron Flux and Spectra for Physical and Biological Applications*

NCRP Report No. 25 (1961)—*Measurement of Absorbed Dose of Neutrons, and of Mixtures of Neutrons and Gamma Rays*

NCRP Report No. 47 (1976)—*Tritium Measurement Techniques*

NCRP Report No. 50 (1976)—*Environmental Radiation Measurements*

NCRP Report No. 57 (1978)—*Instrumentation and Monitoring Methods for Radiation Protection*

NCRP Annual Meeting Proceedings No. 1 (1980)—*Perceptions of Risk*

NCRP Annual Meeting Proceedings No. 3 (1982)—*Critical Issues in Setting Radiation Dose Limits*

NCRP Report No. 58 (1985)—*A Handbook of Radioactivity Measurements Procedures, 2d ed.*

NCRP Report No. 67 (1981)—*Radiofrequency Electromagnetic Fields—Properties, Quantities and Units, Biophysical Interaction, and Measurements*

NCRP Report No. 75 (1983)—*Iodine-129: Evaluation of Releases from Nuclear Power Generation*

NCRP Report No. 76 (1984)—*Radiological Assessment: Predicting the Transport, Bioaccumulation, and Uptake by Man of Radionuclides Released to the Environment*

NCRP Report No. 84 (1985)—*General Concepts for the Dosimetry of Internally Deposited Radionuclides*

NCRP Report No. 87 (1987)—*Use of Bioassay Procedures for Assessment of Internal Radionuclide Deposition*

NCRP Report No. 97 (1988)—*Measurement of Radon and Radon Daughters in Air*

NCRP Report No. 96 (1989)—*Comparative Carcinogenicity of Ionizing Radiation and Chemicals*

NCRP Commentary No. 3 (1986); Revised (1989)—*Screening Techniques for Determining Compliance with Environmental Standards—Releases of Radionuclides to the Atmosphere*

NCRP Annual Meeting Proceedings No. 9 (1988)—*New Dosimetry at Hiroshima and Nagasaki and Its Implications for Risk Estimates*

NCRP Report No. 108 (1991)—*Conceptual Basis for Calculations of Absorbed-Dose Distributions*

NCRP Report No. 112 (1991)—*Calibration of Survey Instruments Used in Radiation Protection for the Assessment of Ionizing Radiation Fields and Radioactive Surface Contamination*

NCRP Commentary No. 6 (1991)—*Radon Exposure of the U.S. Population—Status of the Problem*

NCRP Report No. 115 (1993)—*Risk Estimates for Radiation Protection*

NCRP Commentary No. 8 (1993)—*Uncertainty in NCRP Screening Models Relating to Atmospheric Transport, Deposition and Uptake by Humans*

PUBLICATIONS RELATED TO RADIATION PROTECTION

NCRP Report No. 32 (1966)—*Radiation Protection in Educational Institutions*

NCRP Report No. 38 (1971)—*Protection Against Neutron Radiation*

NCRP Report No. 51 (1977)—*Radiation Protection Guidelines for 0.1-100 MeV Particle Accelerator Facilities*

NCRP Report No. 55 (1977)—*Protection of the Thyroid Gland in the Event of Releases of Radioiodine*

NCRP Report No. 60 (1978)—*Physical, Chemical, and Biological Properties of Radiocerium Relevant to Radiation Protection Guidelines*

NCRP Report No. 72 (1983)—*Radiation Protection and Measurement for Low-Voltage Neutron Generators*

NCRP Report No. 90 (1988)—*Neptunium: Radiation Protection Guidelines*

NCRP Annual Meeting Proceedings No. 6 (1985)—*Some Issues Important in Developing Basic Radiation Protection Recommendations*

NCRP Annual Meeting Proceedings No. 11 (1990)—*Radiation Protection Today*

NCRP Annual Meeting Proceedings No. 13 (1992)—*Genes, Cancer and Radiation Protection*

NCRP Report No. 117 (1993)—*Research Needs for Radiation Protection*

NCRP Report No. 118 (1993)—*Radiation Protection in the Mineral Extraction Industry*

PUBLICATIONS RELATED TO ENGINEERING AND ADMINISTRATIVE CONTROLS

NCRP Report No. 8 (1951)—*Control and Removal of Radioactive Contamination in Laboratories*

NCRP Report No. 30 (1964)—*Safe Handling of Radioactive Materials*

NCRP Report No. 42 (1974)—*Radiological Factors Affecting Decision Making in a Nuclear Attack*

NCRP Report No. 58 (1978)—Operational Radiation Safety Program

NCRP Report No. 61 (1978)—*Radiation Safety Training Criteria for Industrial Radiography*

NCRP Report No. 65 (1980)—*Management of Persons Accidentally Contaminated with Radionuclides*

NCRP Commentary No. 2 (1982)—*Preliminary Evaluation of Criteria for the Disposal of Transuranic Waste*

NCRP Report No. 71 (1983)—*Operational Radiation Safety—Training*

NCRP Report No. 82 (1985)—*SI Units in Radiation Protection and Measurements*

NCRP Report No. 88 (1986)—*Radiation Alarms and Access Control Systems*

NCRP Report No. 103 (1989)—*Control of Radon in Houses*

NCRP Report No. 111 (1991)—*Developing Radiation Emergency Plans for Academic, Medical, or Industrial Facilities*

NCRP Report No. 114 (1992)—*Maintaining Radiation Protection Records*

PUBLICATIONS RELATED TO MEDICINE/HEALTH FIELDS

NCRP Report No. 35 (1970)—*Dental X-ray Protection*

NCRP Report No. 37 (1970)—*Precautions in the Management of Patients Who Have Received Therapeutic Amounts of Radionuclides*

NCRP Report No. 40 (1972)—*Protection Against Radiation from Brachytherapy Sources*

NCRP Report No. 41 (1974)—*Specification of Gamma-Ray Brachytherapy Sources*

NCRP Report No. 49 (1976)—*Structural Shielding Design and Evaluation for Medical Use of X Rays and Gamma Rays of Energies up to 10 MeV*

NCRP Report No. 54 (1977)—*Medical Radiation Exposure of Pregnant and Potentially Pregnant Women*

NCRP Report No. 68 (1981)—*Radiation Protection in Pediatric Radiology*

NCRP Report No. 69 (1981)—*Dosimetry of X-Ray and Gamma-Ray Beams for Radiation Therapy in the Energy Range 10 keV to 50 MeV*

NCRP Report No. 70 (1982)—*Nuclear Medicine—Factors Influencing the Choice and Use of Radionuclides in Diagnosis and Therapy*

NCRP Report No. 73 (1983)—*Protection in Nuclear Medicine and Ultrasound Diagnostic Procedures in Children*

NCRP Report No. 74 (1983)—*Biological Effects of Ultrasound: Mechanisms and Clinical Implications*

NCRP Annual Meeting Proceedings No. 4 (1983)—*Radiation Protection and New Medical Diagnostic Approaches*

NCRP Report No. 79 (1984)—*Neutron Contamination from Medical Electron Accelerators*

NCRP Report No. 80 (1985)—*Induction of Thyroid Cancer by Ionizing Radiation*

NCRP Report No. 83 (1985)—*The Experimental Basis for Absorbed-Dose Calculations in Medical Uses of Radionuclides*

NCRP Report No. 85 (1986)—*Mammography—A User's Guide*

NCRP Report No. 99 (1988)—*Quality Assurance for Diagnostic Imaging*

NCRP Report No. 102 (1989)—*Medical X-Ray, Electron Beam and Gamma-Ray Protection for Energies up to 50 MeV (Equipment Design, Performance and Use)*

NCRP Report No. 104 (1990)—*The Relative Biological Effectiveness of Radiations of Different Quality*

NCRP Report No. 105 (1989)—*Radiation Protection for Medical and Allied Health Personnel*

NCRP Report No. 107 (1990)—*Implementation of the Principle of As Low As Reasonably Achievable (ALARA) for Medical and Dental Personnel*

NCRP Commentary No. 7 (1991)—*Misadministration of Radioactive Material in Medicine—Scientific Background*

NCRP Report No. 113 (1992)—*Exposure Criteria for Medical Diagnostic Ultrasound: I. Criteria Based on Thermal Mechanisms*

NCRP Annual Meeting Proceedings No. 14 (1993)—*Radiation Protection in Medicine*

OTHER PUBLICATIONS RELATED TO RADIATION

NCRP Report No. 44 (1975)—*Krypton 84 in the Atmosphere—Accumulation, Biological Significance, and Control Technology*

NCRP Report No. 62 (1979)—*Tritium in the Environment*

NCRP Report No. 63 (1979)—*Tritium and Other Radionuclide Labeled Organic Compounds Incorporated in Genetic Material*

NCRP Commentary No.1 (1980)—*Krypton-85 in the Atmosphere—With Specific Reference to the Public Health Significance of the Proposed Controlled Release at Three Mile Island*

NCRP Annual Meeting Proceedings No. 5 (1983)—*Environmental Radioactivity*

NCRP Report No. 81 (1985)—*Carbon-14 in the Environment*

NCRP Annual Meeting Proceedings No. 7 (1986)—*Radioactive Waste*

NCRP Commentary No. 4 (1987)—*Guidelines for the Release of Waste Water from Nuclear Facilities with Special Reference to the Public Health Significance of the Proposed Release of Treated Waste Waters at Three Mile Island*

NCRP Annual Meeting Proceedings No. 8 (1988)—*Nonionizing Electromagnetic Radiations and Ultrasound*

NCRP Report No. 98 (1989)—*Guidance on Radiation Received in Space Activities*

NCRP Commentary No. 5 (1989)—*Review of the Publication, Living Without Landfills*

NCRP Annual Meeting Proceedings No. 10 (1989)—*Radon*

NCRP Report No. 109 (1991)—*Effects of Ionizing Radiation on Aquatic Organisms*

NCRP Report No. 110 (1991)—*Some Aspects of Strontium Radiobiology*

NCRP Annual Meeting Proceedings No. 12 (1991)—*Health and Ecological Implications of Radioactively Contaminated Environments*

INDEX